Analysis of Complex Networks

From Biology to Linguistics

Edited by

Matthias Dehmer and

Frank Emmert-Streib

Related Titles

B.H. Junker, F. Schreiber

Analysis of Biological Networks

2008

ISBN 978-0-470-04144-4

F. Emmert-Streib, M. Dehmer (Eds.)

Analysis of Microarray Data
A Network-Based Approach

2008

ISBN 978-3-527-31822-3

E. Keedwell, A. Narayanan

Intelligent Bioinformatics
The Application of Artificial Intelligence Techniques to Bioinformatics Problems

2005

ISBN 978-0-470-02175-0

F. Azuaje, J. Dopazo (Eds.)

Data Analysis and Visualization in Genomics and Proteomics

2005

ISBN 978-0-470-09439-6

Analysis of Complex Networks

From Biology to Linguistics

Edited by
Matthias Dehmer and Frank Emmert-Streib

WILEY-VCH
WILEY-VCH Verlag GmbH & Co. KGaA

The Editors

PD Dr. habil. Matthias Dehmer
Vienna University of Technology
Discrete Mathematics and Geometry
Wiedner Hauptstraße 8–10
1040 Vienna
Austria
and
University of Coimbra
Center for Mathematics
Apartado 3008
3001-454 Coimbra
Portugal

Prof. Dr. Frank Emmert-Streib
Computational Biology and Machine Learning
Center for Cancer Research and Cell Biology
School of Medicine, Dentistry and Biomedical
Sciences Queen's University Belfast
97 Lisburn Road
Belfast, BT9 7BL
UK

Library of Congress Card No.: applied for

British Library Cataloguing-in-Publication Data: A catalogue record for this book is available from the British Library.

Bibliographic information published by the Deutsche Nationalbibliothek
The Deutsche Nationalbibliothek lists this publication in the Deutsche Nationalbibliografie; detailed bibliographic data are available on the Internet at http://dnb.d-nb.de.

Printed in the Federal Republic of Germany
Printed on acid-free paper

Cover design Adam Design, Weinheim
Typesetting le-tex publishing services oHG, Leipzig
Printing Strauss GmbH, Mörlenbach
Bookbinding Litges & Dopf Buchbinderei GmbH, Heppenheim

ISBN 978-3-527-32345-6

Contents

Preface *XIII*

List of Contributors *XV*

1 **Entropy, Orbits, and Spectra of Graphs** *1*
Abbe Mowshowitz and Valia Mitsou
1.1 Introduction *1*
1.2 Entropy or the Information Content of Graphs *2*
1.3 Groups and Graph Spectra *4*
1.4 Approximating Orbits *11*
1.4.1 The Degree of the Vertices *13*
1.4.2 The Point-Deleted Neighborhood Degree Vector *13*
1.4.3 Betweenness Centrality *15*
1.5 Alternative Bases for Structural Complexity *19*
References *21*

2 **Statistical Mechanics of Complex Networks** *23*
Stefan Thurner
2.1 Introduction *23*
2.1.1 Network Entropies *25*
2.1.2 Network Hamiltonians *27*
2.1.3 Network Ensembles *28*
2.1.4 Some Definitions of Network Measures *30*
2.2 Macroscopics: Entropies for Networks *31*
2.2.1 A General Set of Network Models Maximizing Generalized Entropies *32*
2.2.1.1 A Unified Network Model *32*
2.2.1.2 Famous Limits of the Unified Model *35*
2.2.1.3 Unified Model: Additional Features *35*
2.3 Microscopics: Hamiltonians of Networks – Network Thermodynamics *35*

Analysis of Complex Networks: From Biology to Linguistics. Edited by Matthias Dehmer and Frank Emmert-Streib

ISBN: 978-3-527-32345-6

2.3.1 Topological Phase Transitions *36*
2.3.2 A Note on Entropy *37*
2.4 Ensembles of Random Networks – Superstatistics *39*
2.5 Conclusion *42*
References *43*

3 A Simple Integrated Approach to Network Complexity and Node Centrality *47*
Danail Bonchev
3.1 Introduction *47*
3.2 The Small-World Connectivity Descriptors *49*
3.3 The Integrated Centrality Measure *52*
References *53*

4 Spectral Theory of Networks: From Biomolecular to Ecological Systems *55*
Ernesto Estrada
4.1 Introduction *55*
4.2 Background on Graph Spectra *56*
4.3 Spectral Measures of Node Centrality *58*
4.3.1 Subgraph Centrality as a Partition Function *60*
4.3.2 Application *61*
4.4 Global Topological Organization of Complex Networks *62*
4.4.1 Spectral Scaling Method *63*
4.4.2 Universal Topological Classes of Networks *65*
4.4.3 Applications *68*
4.5 Communicability in Complex Networks *69*
4.5.1 Communicability and Network Communities *71*
4.5.2 Detection of Communities: The Communicability Graph *73*
4.5.3 Application *74*
4.6 Network Bipartivity *76*
4.6.1 Detecting Bipartite Substructures in Complex Networks *77*
4.6.2 Application *80*
4.7 Conclusion *80*
References *81*

5 On the Structure of Neutral Networks of RNA Pseudoknot Structures *85*
Christian M. Reidys
5.1 Motivation and Background *85*
5.1.1 Notation and Terminology *87*
5.2 Preliminaries *88*
5.3 Connectivity *90*

5.4 The Largest Component *93*
5.5 Distances in n-Cubes *105*
5.6 Conclusion *110*
References *111*

6 Graph Edit Distance – Optimal and Suboptimal Algorithms with Applications *113*
Horst Bunke and Kaspar Riesen
6.1 Introduction *113*
6.2 Graph Edit Distance *115*
6.3 Computation of GED *118*
6.3.1 Optimal Algorithms *118*
6.3.2 Suboptimal Algorithms *121*
6.3.2.1 Bipartite Graph Matching *121*
6.4 Applications *125*
6.4.1 Graph Data Sets *125*
6.4.2 GED-Based Nearest-Neighbor Classification *129*
6.4.3 Dissimilarity-Based Embedding Graph Kernels *129*
6.5 Experimental Evaluation *132*
6.5.1 Optimal vs. Suboptimal Graph Edit Distance *133*
6.5.2 Dissimilarity Embedding Graph Kernels Based on Suboptimal Graph Edit Distance *136*
6.6 Summary and Conclusions *139*
References *140*

7 Graph Energy *145*
Ivan Gutman, Xueliang Li, and Jianbin Zhang
7.1 Introduction *145*
7.2 Bounds for the Energy of Graphs *147*
7.2.1 Some Upper Bounds *147*
7.2.2 Some Lower Bounds *154*
7.3 Hyperenergetic, Hypoenergetic, and Equienergetic Graphs *156*
7.3.1 Hyperenergetic Graphs *156*
7.3.2 Hypoenergetic Graphs *157*
7.3.3 Equienergetic Graphs *157*
7.4 Graphs Extremal with Regard to Energy *162*
7.5 Miscellaneous *168*
7.6 Concluding Remarks *169*
References *170*

8 Generalized Shortest Path Trees: A Novel Graph Class by Example of Semiotic Networks *175*
Alexander Mehler
8.1 Introduction *175*
8.2 A Class of Tree-Like Graphs and Some of Its Derivatives *178*
8.2.1 Preliminary Notions *178*
8.2.2 Generalized Trees *180*
8.2.3 Minimum Spanning Generalized Trees *186*
8.2.4 Generalized Shortest Path Trees *190*
8.2.5 Shortest Paths Generalized Trees *193*
8.2.6 Generalized Shortest Paths Trees *195*
8.2.7 Accounting for Orientation: Directed Generalized Trees *198*
8.2.8 Generalized Trees, Quality Dimensions, and Conceptual Domains *204*
8.2.9 Generalized Forests as Multidomain Conceptual Spaces *208*
8.3 Semiotic Systems as Conceptual Graphs *212*
References *218*

9 Applications of Graph Theory in Chemo- and Bioinformatics *221*
Dimitris Dimitropoulos, Adel Golovin, M. John, and Eugene Krissinel
9.1 Introduction *221*
9.2 Molecular Graphs *222*
9.3 Common Problems with Molecular Graphs *223*
9.4 Comparisons and 3D Alignment of Protein Structures *225*
9.5 Identification of Macromolecular Assemblies in Crystal Packing *229*
9.6 Chemical Graph Formats *231*
9.7 Chemical Software Packages *232*
9.8 Chemical Databases and Resources *232*
9.9 Subgraph Isomorphism Solution in SQL *232*
9.10 Cycles in Graphs *235*
9.11 Aromatic Properties *236*
9.12 Planar Subgraphs *237*
9.13 Conclusion *238*
References *239*

10 Structural and Functional Dynamics in Cortical and Neuronal Networks *245*
Marcus Kaiser and Jennifer Simonotto
10.1 Introduction *245*
10.1.1 Properties of Cortical and Neuronal Networks *246*
10.1.1.1 Modularity *247*
10.1.1.2 Small-World Features *247*

10.1.1.3 Scale-Free Features *248*
10.1.1.4 Spatial Layout *250*
10.1.2 Prediction of Neural Connectivity *252*
10.1.3 Activity Spreading *254*
10.2 Structural Dynamics *255*
10.2.1 Robustness Toward Structural Damage *255*
10.2.1.1 Removal of Edges *256*
10.2.1.2 Removal of Nodes *257*
10.2.2 Network Changes During Development *258*
10.2.2.1 Spatial Growth Can Generate Small-World Networks *258*
10.2.2.2 Time Windows Generate Multiple Clusters *259*
10.3 Functional Dynamics *260*
10.3.1 Spreading in Excitable Media *260*
10.3.1.1 Cardiac Defibrillation as a Case Study *261*
10.3.1.2 Critical Timing for Changing the State of the Cardiac System *261*
10.3.2 Topological Inhibition Limits Spreading *262*
10.4 Summary *264*
References *266*

11 Network Mapping of Metabolic Pathways *271*
Qiong Cheng and Alexander Zelikovsky
11.1 Introduction *271*
11.2 Brief Overview of Network Mapping Methods *273*
11.3 Modeling Metabolic Pathway Mappings *275*
11.3.1 Problem Formulation *277*
11.4 Computing Minimum Cost Homomorphisms *277*
11.4.1 The Dynamic Programming Algorithm for Multi-Source Tree Patterns *278*
11.4.2 Handling Cycles in Patterns *280*
11.4.3 Allowing Pattern Vertex Deletion *281*
11.5 Mapping Metabolic Pathways *282*
11.6 Implications of Pathway Mappings *285*
11.7 Conclusion *291*
References *291*

12 Graph Structure Analysis and Computational Tractability of Scheduling Problems *295*
Sergey Sevastyanov and Alexander Kononov
12.1 Introduction *295*
12.2 The Connected List Coloring Problem *296*
12.3 Some Practical Problems Reducible to the CLC Problem *298*
12.3.1 The Problem of Connected Service Areas *298*
12.3.2 No-Idle Scheduling on Parallel Machines *300*

12.3.3 Scheduling of Unit Jobs on a p-Batch Machine 301
12.4 A Parameterized Class of Subproblems of the CLC Problem 302
12.5 Complexities of Eight Representatives of Class CLC($\mathcal{X}$) 304
12.5.1 Three NP-Complete Subproblems 304
12.5.2 Five Polynomial-Time Solvable Subproblems 305
12.6 A Basis System of Problems 317
12.7 Conclusion 320
References 322

13 Complexity of Phylogenetic Networks: Counting Cubes in Median Graphs and Related Problems 323
Matjaž Kovše
13.1 Introduction 323
13.2 Preliminaries 324
13.2.1 Median Graphs 325
13.2.1.1 Expansion Procedure 328
13.2.1.2 The Canonical Metric Representation and Isometric Dimension 328
13.3 Treelike Equalities and Euler-Type Inequalities 330
13.3.1 Treelike Eequalities and Euler-Type Inequalities for Median Graphs 330
13.3.1.1 Cube-Free Median Graphs 332
13.3.1.2 Q_4-Free Median Graphs 333
13.3.1.3 Median Grid Graphs 333
13.3.2 Euler-Type Inequalities for Quasi-Median Graphs 334
13.3.3 Euler-Type Inequalities for Partial Cubes 335
13.3.4 Treelike Equality for Cage-Amalgamation Graphs 336
13.4 Cube Polynomials 337
13.4.1 Cube Polynomials of Cube-Free Median Graphs 339
13.4.2 Roots of Cube Polynomials 340
13.4.2.1 Rational Roots of Cube Polynomials 340
13.4.2.2 Real Roots of Cube Polynomials 341
13.4.2.3 Graphs of Acyclic Cubical Complexes 341
13.4.2.4 Product Median Graphs 342
13.4.3 Higher Derivatives of Cube Polynomials 342
13.5 Hamming Polynomials 343
13.5.1 A Different Type of Hamming Polynomial for Cage-Amalgamation Graphs 344
13.6 Maximal Cubes in Median Graphs of Circular Split Systems 345
13.7 Applications in Phylogenetics 346
13.8 Summary and Conclusion 347
References 348

14 Elementary Elliptic (R, q)-Polycycles *351*
Michel Deza, Mathieu Dutour Sikirić, and Mikhail Shtogrin
14.1 Introduction *351*
14.2 Kernel Elementary Polycycles *355*
14.3 Classification of Elementary $(\{2, 3, 4, 5\}, 3)$-Polycycles *356*
14.4 Classification of Elementary $(\{2, 3\}, 4)$-Polycycles *359*
14.5 Classification of Elementary $(\{2, 3\}, 5)$-Polycycles *359*
14.6 Conclusion *361*
Appendix 1: 204 Sporadic Elementary $(\{2,3,4,5\},3)$-Polycycles *364*
Appendix 2: 57 Sporadic eLementary $(\{2, 3\}, 5)$-polycycles *371*
References *375*

15 Optimal Dynamic Flows in Networks and Algorithms for Finding Them *377*
Dmitrii Lozovanu and Maria Fonoberova
15.1 Introduction *377*
15.2 Optimal Dynamic Single-Commodity Flow Problems and Algorithms for Solving Them *378*
15.2.1 The Minimum Cost Dynamic Flow Problem *378*
15.2.2 The Maximum Dynamic Flow Problem *380*
15.2.3 Algorithms for Solving the Optimal Dynamic Flow Problems *380*
15.2.4 The Dynamic Model with Flow Storage at Nodes *384*
15.2.5 The Dynamic Model with Flow Storage at Nodes and Integral Constant Demand–Supply Functions *384*
15.2.6 Approaches to Solving Dynamic Flow Problems with Different Types of Cost Functions *386*
15.2.7 Determining the Optimal Dynamic Flows in Networks with Transit Functions That Depend on Flow and Time *390*
15.3 Optimal Dynamic Multicommodity Flow Problems and Algorithms for Solving Them *392*
15.3.1 The Minimum Cost Dynamic Multicommodity Flow Problem *392*
15.3.2 Algorithm for Solving the Minimum Cost Dynamic Multicommodity Flow Problem *394*
15.4 Conclusion *398*
References *398*

16 Analyzing and Modeling European R&D Collaborations: Challenges and Opportunities from a Large Social Network *401*
Michael J. Barber, Manfred Paier, and Thomas Scherngell
16.1 Introduction *401*
16.2 Data Preparation *402*
16.3 Network Definition *404*

16.4 Network Structure *405*
16.5 Community Structure *407*
16.5.1 Modularity *408*
16.5.2 Finding Communities in Bipartite Networks *409*
16.6 Communities in the Framework Program Networks *409*
16.6.1 Topical Profiles of Communities *411*
16.7 Binary Choice Model *413*
16.7.1 The Empirical Model *413*
16.7.2 Variable Construction *415*
16.7.2.1 The Dependent Variable *415*
16.7.2.2 Variables Accounting for Geographical Effects *415*
16.7.2.3 Variables Accounting for FP Experience of Organizations *415*
16.7.2.4 Variables Accounting for Relational Effects *416*
16.7.3 Estimation Results *417*
16.8 Summary *420*
References *421*

17 Analytic Combinatorics on Random Graphs *425*
Michael Drmota and Bernhard Gittenberger
17.1 Introduction *425*
17.2 Trees *426*
17.2.1 The Degree Distribution *429*
17.2.2 The Height *430*
17.2.3 The Profile *431*
17.2.4 The Width *434*
17.3 Random Mappings *436*
17.4 The Random Graph Model of Erdős and Rényi *438*
17.4.1 Counting Connected Graphs with Wright's Method *438*
17.4.2 Emergence of the Giant Component *440*
17.5 Planar Graphs *445*
References *449*

Index *451*

Preface

This book, *Analysis of Complex Networks: From Biology to Linguistics*, presents theoretical and practical results on graph-theoretic methods that are used for modeling as well as structurally investigating complex networks. Instead of focusing exclusively on classical graph-theoretic approaches, its major goal is to demonstrate the importance and usefulness of network-based concepts for scientists in various disciplines. Further, the book advocates the idea that theoretical as well as applied results are needed to enhance our knowledge and understanding of networks in general and as representations for various problems. We emphasize methods for analyzing graphs structurally because it has been shown that especially data-driven areas such as web mining, computational and systems biology, chemical informatics, and cognitive sciences profit tremendously from this field.

The main topics treated in this book can be summarized as follows:

- Information-theoretic methods for analyzing graphs
- Problems in quantitative graph theory
- Structural graph measures
- Investigating novel network classes
- Metrical properties of graphs
- Aspects in algorithmic graph theory
- Analytic methods in graph theory
- Network-based applications

Analysis of Complex Networks: From Biology to Linguistics is intended for an interdisciplinary audience ranging from applied discrete mathematics, artificial intelligence, and applied statistics to computer science, computational and systems biology, cognitive science, computational linguistics, machine learning, mathematical chemistry, and physics. Many colleagues, whether consciously or unconsciously, provided us with input, help, and support before and during the development of the present book. In particular we would like to thank Andreas Albrecht, Rute Andrade, Gökhan Bakır, Alexandru T. Balaban, Subhash Basak, Igor Bass, Natália Bebiano,

Analysis of Complex Networks: From Biology to Linguistics. Edited by Matthias Dehmer and Frank Emmert-Streib

ISBN: 978-3-527-32345-6

Danail Bonchev, Stefan Borgert, Mieczyslaw Borowiecki, Michael Drmota, Abdol-Hossein Esfahanian, Bernhard Gittenberger, Earl Glynn, Elena Konstantinova, Dmitrii Lozovanu, Alexander Mehler, Abbe Mowshowitz, Max Mühlhäuser, Arcady Mushegian, Paolo Oliveira, João da Providência, Horst Sachs, Heinz Georg Schuster, Helmut Schwegler, Chris Seidel, Fred Sobik, Doru Stefanescu, Thomas Stoll, John Storey, Kurt Varmuza, Bohdan Zelinka, and all the coauthors of this book and apologize to all those whose names have been mistakenly omitted. We would also like to thank our editors Andreas Sendtko and Gregor Cicchetti from Wiley-VCH; they were always available and extremely helpful. Last but not least, we would like to thank our families for their support and encouragement throughout the writing of the book.

Finally, we hope that this book helps the reader to understand that the presented field is multifaceted in depth and breadth and as such is inherently interdisciplinary. This is important to realize because it allows one to pursue a problem-oriented rather than field-oriented approach to efficiently tackling state-of-the-art problems in modern sciences.

Vienna and Belfast,
March 2009

Matthias Dehmer
Frank Emmert-Streib

List of Contributors

Michael J. Barber
Austrian Research Centers – ARC
Division Systems Research
Donau-City-Straße 1
1220 Vienna
Austria

Danail Bonchev
Virginia Commonwealth University
Center for the Study of
Biological Complexity
P.O. Box 842030
Richmond, VA 23284-2030
USA

Horst Bunke
University of Bern
Institute of Computer Science
and Applied Mathematics
Neubrückstraße 10
3012 Bern
Switzerland

Qiong Cheng
Georgia State University
Department of Computer Science
Atlanta, GA 30303
USA

Michel Deza
École Normale Supérieure
45 rue d'Ulm
75005 Paris
France
and
Japan Advanced Institute of
Science and Technology
1-1 Asahidai
Nomi
Ishikawa
Japan

Dimitris Dimitropoulos
European Bioinformatics Institute
Genome Campus, Hinxton
Cambridge CB10 1SD
UK

Michael Drmota
Vienna University of Technology
Institute of Discrete Mathematics
and Geometry
Wiedner Hauptstraße 8/104
1040 Vienna
Austria

Analysis of Complex Networks: From Biology to Linguistics. Edited by Matthias Dehmer and Frank Emmert-Streib

ISBN: 978-3-527-32345-6

Ernesto Estrada
University of Strathclyde
Institute of Complex Systems
at Strathclyde
Department of Physics and
Department of Mathematics
Glasgow G1 1XH
UK

Maria Fonoberova
AIMdyn, Inc.
Santa Barbara, CA 93101
USA

Bernhard Gittenberger
Vienna University of Technology
Institute of Discrete Mathematics
and Geometry
Wiedner Hauptstraße 8/104
1040 Vienna
Austria

Adel Golovin
European Bioinformatics Institute
Genome Campus, Hinxton
Cambridge CB10 1SD
UK

Ivan Gutman
University of Kragujevac
Faculty of Science
P.O. Box 60
34000 Kragujevac
Serbia

M. John
European Bioinformatics Institute
Genome Campus, Hinxton
Cambridge CB10 1SD
UK

Marcus Kaiser
Newcastle University
School of Computing Science
Newcastle-upon-Tyne NE1 7RU
UK
and
Newcastle University
Institute of Neuroscience
Newcastle-upon-Tyne NE2 4HH
UK

Alexander Kononov
Russian Academy of Sciences
Sobolev Institute of Mathematics
Novosibirsk
Russia

Matjaž Kovše
University of Maribor
Faculty of Natural Sciences
and Mathematics
Koroška cesta 160
2000 Maribor
Slovenia
and
Institute of Mathematics,
Physics and Mechanics
Jadranska 19
1000 Ljubljana
Slovenia

Eugene Krissinel
European Bioinformatics Institute
Genome Campus, Hinxton
Cambridge CB10 1SD
UK

Xueliang Li
Nankai University
Center for Combinatorics
LPMC-TJKLC
Tianjin 300071
P.R. China

Dmitrii Lozovanu
Moldovan Academy of Sciences
Institute of Mathematics
and Computer Science
Academiei str., 5
Chisinau, MD-2005
Moldova

Alexander Mehler
Goethe-Universität
Frankfurt am Main
Abteilung für geisteswis-
senschaftliche Fachinformatik/
Department for Computing
in the Humanities
Georg-Voigt-Straße 4
60325 Frankfurt am Main
Germany

Valia Mitsou
The City University of New York
Doctoral Program in Computer
Science
365 Fifth Avenue
New York, NY 10016
USA

Abbe Mowshowitz
The City College of New York
Department of Computer Science
Convent Avenue at 138th Street
New York, NY 10031
USA
and
The City University of New York
Doctoral Program in Computer
Science
365 Fifth Avenue
New York, NY 10016
USA

Manfred Paier
Austrian Research Centers – ARC
Division Systems Research
Donau-City-Straße 1
1220 Vienna
Austria

Christian M. Reidys
Nankai University
Center for Combinatorics
LPMC-TJKLC
Tianjin 300071
P.R. China

Kaspar Riesen
Universtity of Bern
Institute of Computer Science and
Applied Mathematics
Neubrückstraße 10
3012 Bern
Switzerland

Thomas Scherngell
Austrian Research Centers – ARC
Division Systems Research
Donau-City-Straße 1
1220 Vienna
Austria

Sergey Sevastyanov
Russian Academy of Sciences
Sobolev Institute of Mathematics
Novosibirsk
Russia

Mikhail Shtogrin
Steklov Mathematical Institute
Gubkina str. 8
117966 Moscow
Russia

Mathieu Dutour Sikirić
Institute Rudjer Bošković
Group for Satellite Oceanography
10000 Zagreb
Croatia

Jennifer Simonotto
Newcastle University
School of Computing Science
Newcastle-upon-Tyne NE1 7RU
UK
and
Newcastle University
Institute of Neuroscience
Newcastle-upon-Tyne NE2 4HH
UK

Stefan Thurner
Medical University of Vienna
Complex Systems Research Group
Währinger Gürtel 18–20
1090 Vienna
Austria
and
Santa Fe Institute
1399 Hyde Park Road
Santa Fe, NM 87501
USA

Alexander Zelikovsky
Georgia State University
Department of Computer Science
Atlanta, GA 30303
USA

Jianbin Zhang
Nankai University
Center for Combinatorics
LPMC-TJKLC
Tianjin 300071
P.R. China

1
Entropy, Orbits, and Spectra of Graphs

Abbe Mowshowitz and Valia Mitsou

1.1
Introduction

This chapter is concerned with the notion of entropy as applied to graphs for the purpose of measuring complexity.

Most studies of complexity focus on the execution time or space utilization of algorithms. The execution time of an algorithm is proportional to the number of operations required to produce the output as a function of the input size. Space utilization measures the amount of storage required for computation. Both time and space complexity measure the resources required to perform a computation for a specified input. Measuring the complexity of a mathematical object such as a graph is an exercise in structural complexity. This type of complexity does not deal directly with the costs of computation; rather, it offers insight into the internal organization of an object. The structural complexity of a computer program, for example, may indicate the difficulty of modifying or maintaining the program.

One approach to structural complexity involves the length of a code needed to specify an object uniquely (Kolmogorov complexity). The complexity of a string, for example, is the length of the string's shortest description in a given description language [27]. The approach taken in this chapter centers on finding indices of structure, based on Shannon's entropy measure. Unlike Kolmogorov complexity, such an index captures a particular feature of the structure of an object. The symmetry structure of a graph provides the basis for the index explored here.

The choice of symmetry is dictated by its utility in many scientific disciplines. D'Arcy Thompson's classic work [25] showed the relevance of symmetry in the natural world. Structure-preserving transformations based on symmetry play a role in physics, chemistry, and sociology as well as in biology. A symmetry transformation of a graph is typically an edge-preserving bijection of the vertices, i.e., an isomorphism of the graph onto itself. Such a transformation is called an *automorphism*. If the vertices of the graph are labeled, an automorphism can be viewed as a permutation of the vertices that preserves adjacencies. The set of all automorphisms forms a group

Analysis of Complex Networks: From Biology to Linguistics. Edited by Matthias Dehmer and Frank Emmert-Streib

ISBN: 978-3-527-32345-6

whose orbits provide the foundation for applying Shannon's entropy measure.

The collection of orbits of the automorphism group constitutes a partition and thus defines an equivalence relation on the vertices of a graph. Two vertices in the same orbit are similar in some sense. In a social network, collections of similar vertices can be used to define communities with shared attributes. The identification of such communities is of interest in applications such as advertising, intelligence, and sensor networks.

Measures of structural complexity are useful for classifying graphs and networks represented by graphs. One is led to conjecture, for example, that the more symmetric a network is (or the lower its automorphism-based complexity is), the more vulnerable to attack it will be. These related issues are explored in [19] in relation to sensor networks modeled as dynamic distributed federated databases [2].

In what follows we define the measure of graph complexity, discuss algorithms and heuristics for computing it, and examine its relationship to another prominent entropy measure [11] defined on graphs.

1.2
Entropy or the Information Content of Graphs

Given a decomposition of the vertices or edges of a graph, one can construct a *finite probability scheme* [10] and compute its entropy. A finite probability scheme assigns a probability to each subset in the decomposition. Such a numerical measure can be seen to capture the information contained in some particular aspect of the graph structure.

The orbits of the automorphism group of a graph constitute a decomposition of the vertices of the graph. As noted above, this decomposition captures the symmetry structure of the graph, and the entropy of the finite probability scheme obtained from the automorphism group provides an index of the complexity of the graph relative to the symmetry structure.

Let $G = (V, E)$ be a graph with vertex set V (with $|V| = n$) and edge set E. The *automorphism group* of G, denoted by $Aut(G)$, is the set of all adjacency-preserving bijections of V. Let $\{V_i | 1 \le i \le k\}$ be the collection of orbits of $Aut(G)$, and suppose $|V_i| = n_i$ for $1 \le i \le k$. The *entropy* or *information content* of G is given by the following formula ([13]):

$$I_a(G) = -\sum_{i=1}^{k} \frac{n_i}{n} \log \frac{n_i}{n} .$$

For example, the orbits of the graph of Figure 1.1 are $\{1\}$, $\{2,5\}$, and $\{3,4\}$, so the information content of the graph is $I_a(G) = -\frac{1}{5}\log\frac{1}{5} - 2(\frac{2}{5}\log\frac{2}{5}) = \log 5 - \frac{4}{5}\log 2$.

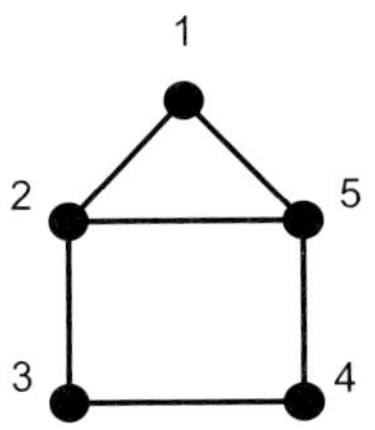

Figure 1.1 Information content of a graph.

Clearly, $I_a(G)$ satisfies $0 \leq I_a(G) \leq \log n$, where the minimum value occurs for graphs with the transitive automorphism group, such as the cycle C_n and the complete graph K_n on n vertices; the maximum is achieved for graphs with the identity group. The smallest nontrivial, undirected graph with an identity group is shown in Figure 1.2.

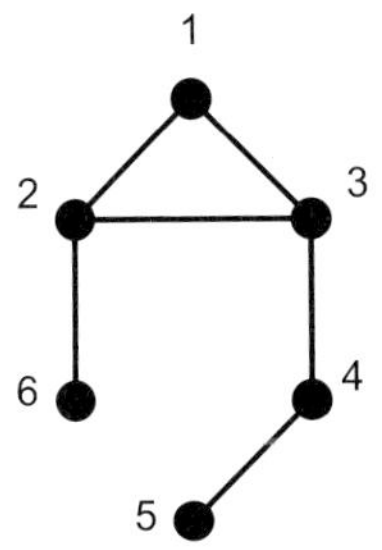

Figure 1.2 Smallest nontrivial graph with identity group.

The idea of measuring the information content of a graph was first presented in [21]; it was formalized in [26] and further developed in [13–16]. $I_a(G)$ is a function of the partition of the vertices of G determined by the orbits of $Aut(G)$. As such the measure captures the structure of vertex similarity. In the case of organic molecules, the lower the information content (or the greater the symmetry), the fewer the possibilities for different interactions with other molecules. If all the atoms are in the same equivalence class, then it makes no difference which one interacts with an atom of another molecule. The same can be said for social networks. Any member of an equivalence class of similar individuals can serve as a representative of the class.

The utility of the measure $I_a(G)$ can be seen from the following special case. The cartesian product $G \times H$ of graphs G and H is defined by $V(G \times H) = V(G) \times V(H)$ and for $(a, b), (c, d) \in V(G \times H)$, $[(a, b), (c, d)] \in E(G \times H)$ if $a = c$ and $[b, d] \in E(H)$ or if $b = d$ and $[a, c] \in E(G)$.

The hypercube Q_n with 2^n vertices is defined recursively by $Q_1 = K_2$ and for $n \geq 2$, $Q_n = K_2 \times Q_{n-1}$. Since Q_n has a transitive automorphism group, $I(Q_n) = 0$. The hypercube Q_n offers a desirable configuration for parallel computation because processors must exchange messages in executing an algorithm, and the distance between any two vertices (representing processors) in the hypercube is at most n.

By contrast, an $m \times m$ mesh configuration (formed by taking the cartesian product of two isomorphic line graphs, each with m vertices) consists of m^2 vertices and has a maximum distance of $2m$. A $2^{\frac{n}{2}} \times 2^{\frac{n}{2}}$ mesh for even n having the same number of vertices as Q_n has a maximum distance between vertices of $2(2^{\frac{n}{2}}-1)$. At the same time the information content of such a mesh is $\frac{n}{2}-1$ [13].

This example suggests that good graph configurations for parallel computation score low on information complexity or, alternatively, are highly symmetric. Information complexity is a coarse filter, but it is useful nonetheless.

Computing the group-based entropy or information content of a graph requires knowledge of the orbits of the automorphism group. An obvious approach to computing the orbits is to determine the automorphism group and then to observe the action of automorphisms on the vertices of the graph. This is not an efficient method in general, but the algebraic structure of a graph can be exploited to find the automorphism group efficiently in some cases. The general question of determining the automorphism group is taken up in Section 1.3; heuristics for finding the orbits of $Aut(G)$ are surveyed in Section 1.4.

1.3 Groups and Graph Spectra

Let $G = (V, E)$ be a graph with vertex set V of size n, edge set E of size m, and automorphism group $Aut(G)$. (See [3] for general coverage of algebraic aspects of graph theory and [12] for specific treatment of the automorphism group of a graph.) Since automorphisms are in effect relabelings of the vertices, they can be represented as permutation matrices. Let $A = A(G)$ be the adjacency matrix of G. Then a permutation matrix P is an automorphism of G if and only if $P^TAP = A$ or $PA = AP$.

Thus, one way to construct the automorphism group of a graph G is to solve the matrix equation $AX = XA$ for permutation matrices X. The Jordan canonical form of A as a matrix over the reals can be used to obtain the general solution X. Taking G to be undirected and thus A symmetric and letting $\tilde{A} = U^TAU$ be the Jordan form of A, we have $(U\tilde{A}U^T)X = X(U\tilde{A}U^T)$ or $\tilde{A}\tilde{X} = \tilde{X}\tilde{A}$, where $\tilde{X} = U^TXU$.

Thus the construction of $Aut(G)$ requires computing the orthogonal matrix U and finding all $\tilde{X}$ that commute with $\tilde{A}$. The matrix $\tilde{X}$ depends on the elementary divisors of A. With no additional information, this method of constructing the group is not too promising since it is necessary to find all those solutions that are permutation matrices.

In the special case where A has all distinct eigenvalues, $\tilde{X}$ has the form of a diagonal matrix with arbitrary parameters on the main diagonal. In this case, $X = U\tilde{X}U^T$. Clearly $U\tilde{X}U^T$ is symmetric, so if it is a permutation matrix, it must correspond to a product of disjoint transpositions. This means that every element of $Aut(G)$ has order 2 and the group is therefore abelian [12, 17]. The converse is not true since, for example, the graph G of Figure 1.3 has the characteristic polynomial $(x+1)^2(x^3-2x^2-5x+2)$.

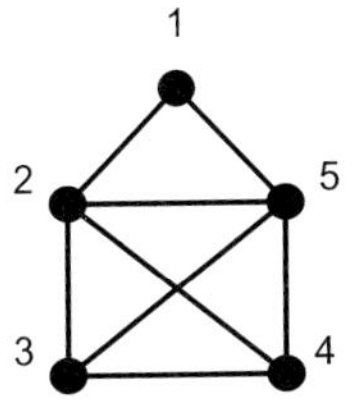

Figure 1.3 $Aut(G)$ is abelian, every element is of order 2, but the characteristic polynomial has repeated roots.

An analogous result holds for digraphs. Using the same analysis, Chao [5] showed that if the adjacency matrix of a digraph has all distinct eigenvalues, then its automorphism group is abelian. However, the automorphisms need not be of order 2. For example, the adjacency matrix of digraph D in Figure 1.4 has the characteristic polynomial $(x^3-1) = (x-1)(x^2+x+1)$ but the permutation (123) is an automorphism of D.

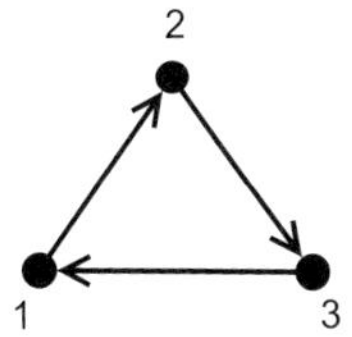

Figure 1.4 $Aut(D) = \langle(123)\rangle$, abelian but not every element has order 2.

Both of these results are special cases of the following:

Theorem 1.1 *Suppose the adjacency matrix $A = A(D)$ of a digraph D is non-derogatory with respect to a field F, i.e., its characteristic polynomial coincides with its minimal polynomial over F. Then $Aut(D)$ is abelian.*

Proof. The result is an immediate consequence of the fact that under the hypothesis of the theorem, every matrix over F commuting with A can be expressed as a polynomial in A.

In particular, if A has all distinct eigenvalues, it is non-derogatory over the complex number field. To see that every automorphism of an (undirected) graph has order 2 under this condition, it suffices to observe that any polynomial in a symmetric matrix is again symmetric.

If the adjacency matrix fails to be nonderogatory, then some leverage in constructing the automorphism group can be obtained by taking advantage of the fact that the matrix consists of zeroes and ones. In particular, the adjacency matrix can be interpreted as a matrix over $GF(2)$, thus reducing the solution space of the matrix equation $AX = XA$ to zero-one matrices at the outset.

Thus suppose $A = A(G)$ (for a graph G) is a matrix over $GF(2)$. To demonstrate a method for constructing automorphisms, we revisit the special case of A being nonderogatory over $GF(2)$.

In this case we know that:

1. $M \in Aut(G)$ implies $M^2 = I$ (the identity matrix) and
2. $M \in Aut(G)$ implies $M = \sum_{i=0}^{n-1} a_i A^i$.

So if $M \in Aut(G)$, then we can write

$$M = \sum_{i=0}^{n-1} a_i A^i$$

and

$$I = M^2 = \left(\sum_{i=0}^{n-1} a_i A^i\right)^2 = \sum_{i=0}^{n-1} a_i (A^i)^2.$$

Thus $\{M | M = \sum_{i=0}^{n-1} a_i A^i \text{ and } M^2 = I\} \supseteq Aut(G)$.

Constructing the group in this case reduces to finding all polynomials in A^2 that are equal to the identity matrix. These have the form

$$p(A)\mu_{A^2}(A^2) + I,$$

where $\mu_{A^2}(x)$ is the minimal polynomial of A^2.

Thus, if $M^2 = I$, then $M = p(A)\mu_{A^2}(A) + I$ for some polynomial $p(x)$, since $(p(A)\mu_{A^2}(A) + I)^2 = (p(A^2)\mu_{A^2}(A^2) + I) = 0 + I = I$.

The characteristic and minimal polynomials of graph G in Figure 1.5 coincide over the real numbers, i.e., $\phi(x) = \mu(x) = (x^3 - x^2 - 6x + 2)x(x + 1)$ and over $GF(2)$ with $\phi(x) = \mu(x) = x^3(x + 1)^2$. Hence, the adjacency matrix of G is nonderogatory over both fields. The minimal polynomial of A^2 is $\mu_{A^2}(x) = x^2(x + 1)$, which is of degree 3.

Therefore, $M \in Aut(G)$ implies $M = \mu_{A^2}(A)(b_0 I + b_1 A) + I$. There are four possible solutions for M corresponding to the four possible values for b_0 and b_1. All of these solutions, namely,

$$I, A^3 + A^2 + I, A^4 + A^3 + I, A^4 + A^2 + I,$$

turn out to be permutation matrices so that the automorphism group of G contains precisely these four elements.

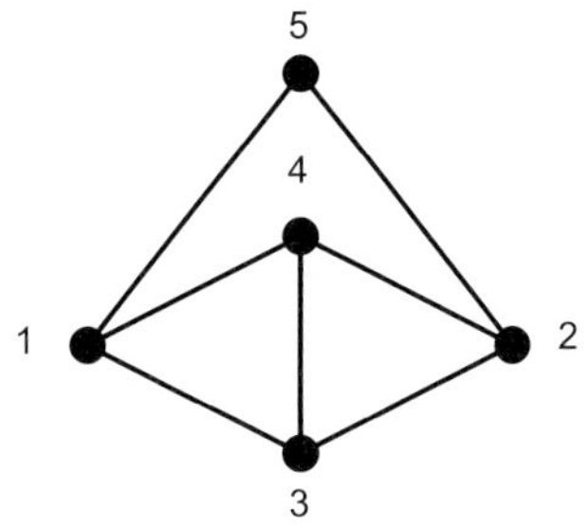

Figure 1.5 Computation of automorphisms over $GF(2)$.

Note that $\mu_{A^2}^2(x) = x\phi_A(x)$ if n is odd, or $\mu_{A^2}^2(x) = \phi_A(x)$ if n is even. Hence, if $m = \deg \mu_{A^2}(x)$ and M satisfies $AM = MA$ and $M^2 = I$, then $M = \mu_{A^2}(x) \sum_{i=0}^{n-m-1} b_i A^i + I$, where $b_i \in GF(2)$.

To determine $Aut(G)$, it suffices to examine $2^{n-m-1} \approx 2^{n/2}$ values of the parameters b_i, to pick out the permutation matrices (i.e., elements of $Aut(G)$).

However, some further simplification is possible. Let $Q = \mu_{A^2}(A)$ and $Z(b) = \sum_{i=0}^{n-m-1} b_i A^i$. Then $M = QZ(b) + I$. Multiplying by M on the right gives $MQ = Q^2 Z(b) + Q = Q$. Thus, if M is an automorphism of G, then $MQ = Q$, which means that similar vertices of G correspond to identical rows of Q. In addition, the identical rows must occur in *minimal pairs*, which gives a sufficient condition for $Aut(G)$ to be trivial.

If $\mu_{A^2}(A)$ has all distinct rows or no minimal pairs of identical rows, then $Aut(G)$ is trivial. The converse is not true. Both graphs in Figure 1.6 have trivial groups, but $\mu_{A^2}(A(G_1))$ has all distinct rows while $\mu_{A^2}(A(G_2))$ has three pairs of identical rows.

Theorem 1.2 *[18]; see also [6]. Let D be a digraph and $A = A(D)$ be its adjacency matrix. If $\phi_A(x)$ is irreducible over the integers, then $Aut(D)$ is trivial.*

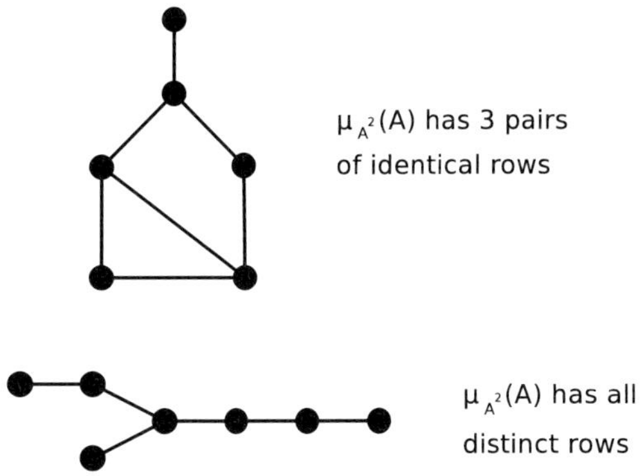

Figure 1.6 Identity graphs.

Proof. Suppose there is an $M(\neq I) \in Aut(D)$, and that the permutation corresponding to M consists of r disjoint cycles of lengths $k_1, \ldots, k_r$. Let z be a nonzero vector consisting of k_1 components equal to c_1, followed by k_2 components equal to c_2, followed by $\ldots k_r$ components equal to c_r. Consider $Az = xz$. This gives a system of r equations in the r unknowns $c_1, c_2, \ldots, c_r$. Thus $Az = xz$ reduces to $Bc = xc$, where $c = (c_1, c_2, \ldots, c_r)^T$. Now z and c are eigenvectors of A and B, respectively, and $\det(B - xc) | \det(A - xz)$, where $\deg(\det(B - xc)) < \deg(\det(A - xz))$. Hence, $\phi_A(x)$ has a nontrivial factorization, which completes the proof.

Figure 1.7 shows a digraph (D) and graph (G) (with the smallest number of vertices) satisfying the condition of the theorem. $\phi_{A(D)}(x) = x^3 - x - 1$ and $\phi_{A(G)}(x) = x^6 - 6x^4 - 2x^3 + 7x^2 + 2x - 1$.

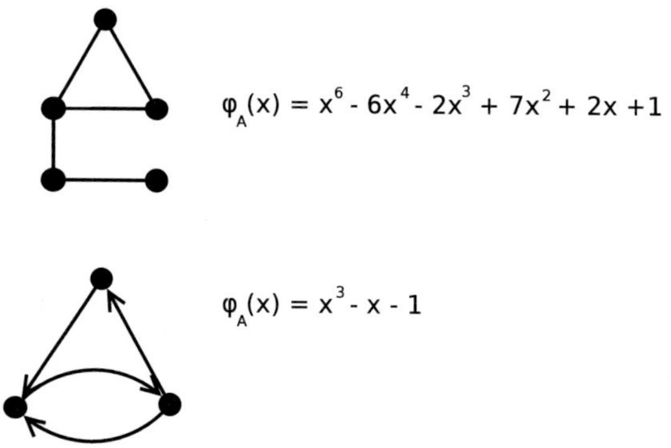

Figure 1.7 Smallest graph and digraph whose characteristic polynomials are irreducible over the integers.

Note that the theorem also holds if $\phi_A(x)$ is taken as a polynomial over a finite field. For example, over $GF(2)$, $x^3 - x - 1$ is irreducible, but $x^6 - 6x^4 - 2x^3 + 7x^2 + 2x - 1 = x^6 + x^2 + 1 = (x^3 + x + 1)^2$.

For graphs this criterion is not very useful since the characteristic polynomial of any graph is reducible over $GF(2)$. There are regular graphs and trees that have the trivial group, but the characteristic polynomial of any regular graph has a linear factor, as does the characteristic polynomial of a tree with an odd number of vertices.

The foregoing discussion suggests the utility of trying to relate the factorization of the characteristic polynomial to the structure of the automorphism group. For example, if G is a graph with an even number n of vertices and adjacency matrix $A = A(G)$, and if $\phi_A(x) = \alpha(x)\beta(x)$ with $\deg \alpha = \deg \beta$ and both α and β are irreducible, then either $Aut(G)$ is trivial or it is of order 2 and consists of the identity and (with a suitable labeling) the permutation $(1, 2)(3, 4) \cdots (n/2, n/2)$.

Table 1.1 contains a list of all 156 graphs on six vertices, showing factored characteristic polynomials and the sizes of their respective automorphism group orbits. Each graph is defined by its list of edges, shown as a sequence of pairs of numbers referring to a standard template with the vertices numbered from 1 to 6 in clockwise order. The last column shows the sizes of the orbits. Complements are not given explicitly, but their polynomials are listed. The orbits of G and $\bar{G}$ are the same.

Table 1.1 Characteristic polynomials and orbit sizes of all graphs on six vertices.

G: # edges: list	Polynomial of G	Polynomial of $\bar{G}$	Orbit sizes
0:	x^6	$(x+1)^5(x-5)$	6
3: 16 23 45	$(x+1)^3(x-1)^3$	$x^3(x+2)^2(x-4)$	6
6: 12 16 23 34 45 56	$(x-1)^2(x+1)^2(x+2)(x-2)$	$x^2(x-1)(x+2)^2(x-3)$	6
6: 15 16 23 24 34 56	$(x-2)^2(x+1)^4$	$x^4(x-3)(x+3)$	6
3: 15 16 56	$x^3(x+1)^2(x-2)$	$x^2(x+1)^2(x^2-2x-9)$	33
1: 12	$x^4(x-1)(x+1)$	$x(x+1)^3(x^2-3x-8)$	24
4: 12 15 24 45	$x^4(x-2)(x+2)$	$(x-1)(x+1)^3(x^2-2x-7)$	24
4: 12 16 34 45	$x^2(x^2-2)^2$	$(x+1)^2(x^4-2x^3-8x^2+6x-1)$	24
5: 12 15 16 23 24	$x^2(x-1)(x+1)(x-2)(x+2)$	$(x-1)(x+2)(x+1)^2(x^2-3x-2)$	24
5: 14 16 23 45 56	$x^2(x-1)(x+1)(x-2)(x+2)$	$x(x-1)(x+1)^2(x^2-x-8)$	24
6: 12 14 15 24 25 45	$x^2(x-3)(x+1)^3$	$x^3(x+1)(x^2-x-8)$	24
7: 12 16 23 25 34 45 56	$(x+1)(x-1)(x^2-2x-1)(x^2+2x-1)$	$x(x+2)(x^2-2)(x^2-2x-2)$	24
7: 14 15 16 23 45 46 56	$(x+1)^4(x-3)(x-1)$	$x^4(x^2-8)$	24
7: 15 16 23 24 34 45 56	$(x+1)^2(x^2-3)(x^2-2x-1)$	$x^2(x^2-2x-2)(x^2+2x-2)$	24
2: 12 56	$x^2(x-1)^2(x+1)^2$	$x^2(x+1)(x+2)(x^2-3x-6)$	222
3: 12 16 23	$x^2(x^2-x-1)(x^2+x-1)$	$(x+1)(x^2+x-1)(x^3-2x^2-8x-3)$	222
6: 14 15 16 23 45 56	$x(x-1)(x+1)^2(x^2-x-4)$	$x^2(x+1)(x^3-x^2-8x+4)$	222
7: 13 16 23 26 34 45 56	$x(x^2+x-1)(x^3-x^2-5x+4)$	$(x-1)(x+1)(x^2+x-1)(x^2-x-5)$	222
7: 12 15 23 24 25 45 56	$x^2(x^2+x-1)(x^2-x-5)$	$(x+1)(x^2+x-1)(x^3-2x^2-4x+1)$	222
5: 12 16 24 45 56	$x(x-2)(x^2+x-1)^2$	$(x^2+x-1)^2(x^2-2x-5)$	15
5: 12 23 24 25 26	$x^4(x^2-5)$	$x(x-4)(x+1)^4$	15
2: 12 16	$x^4(x^2-1)$	$(x+1)^3(x^3-3x^2-7x+3)$	123
3: 12 15 16	$x^4(x^2-3)$	$(x+1)^3(x^3-3x^2-6x+4)$	123
4: 15 16 23 56	$x(x+1)(x-2)(x+1)^3$	$x^3(x^3-11x-12)$	123
4: 12 15 16 34	$x^2(x+1)(x-1)(x^2-3)$	$x(x+1)^2(x^3-2x^2-8x+4)$	123

Table 1.1 (continued).

G: # edges: list	Polynomial of G	Polynomial of $\bar{G}$	Orbit sizes
5: 15 16 23 34 56	$x(x-2)(x+1)^2(x^2-2)$	$x^2(x+1)(x^3-x^2-9x+3)$	123
6: 12 15 23 24 35 45	$x^4(x^2-6)$	$(x+1)^3(x^3-3x^2-3x+7)$	123
6: 12 13 14 15 16 56	$x^2(x+1)(x^3-x^2-5x+3)$	$x^2(x+1)^2(x^2-2x-6)$	123
7: 12 14 15 16 25 45 56	$x^3(x-3)(x+1)(x+2)$	$x(x-1)^2(x^3-2x^2-5x+4)$	123
4: 12 14 15 16	$x^4(x-2)(x+2)$	$(x+1)^3(x^3-3x^2-5x+3)$	114
7: 15 16 26 34 35 45 56	$(x-1)(x+1)^2(x^3-x^2-5x+1)$	$x^3(x+2)(x^2-2x-4)$	114
3: 16 23 56	$x^2(x+1)(x-1)(x^2-2)$	$x(x+1)(x^4-x^3-11x^2-7x+4)$	1122
4: 12 15 16 56	$x^2(x+1)(x^3-x^2-3x+1)$	$x(x+1)(x+2)(x^3-3x^2-4x+2)$	1122
4: 12 16 23 56	$x^2(x-1)(x+1)(x^2-3)$	$x(x+1)(x+2)(x^3-3x^2-4x+4)$	1122
5: 12 15 16 45 56	$x^2(x^2-x-3)(x^2+x-1)$	$(x+1)(x^2+x-1)(x^3-2x^2-6x+1)$	1122
5: 12 14 15 16 56	$x^2(x+1)(x^3-x^2-4x+2)$	$x(x+1)(x^4-x^3-9x^2-5x+4)$	1122
5: 12 15 16 23 45	$(x-1)(x+1)(x^4-4x^2+1)$	$x(x+2)(x^4-2x^3-6x^2+2x+4)$	1122
5: 15 16 23 45 56	$(x-1)(x+1)^2(x^3-x^2-3x+1)$	$x^2(x^4-1x^2-8x+4)$	1122
6: 12 15 16 24 45 56	$x^2(x+2)(x^3-2x^2-2x+2)$	$(x-1)(x+1)(x^4-8x^2-8x+1)$	1122
6: 13 16 23 34 45 56	$(x-1)(x^2+x-1)(x^3-4x-1)$	$(x+2)(x^2+x-1)(x^3-3x^2-x+2)$	1122
6: 12 13 14 16 45 56	$x^2(x^4-6x^2+4)$	$(x-1)(x+1)^2(x^3-x^2-7x-3)$	1122
4: 12 16 23 45	$(x-1)(x+1)(x^2-x-1)(x^2+x-1)$	$x(x^2+x-1)(x^3-x^2-9x-4)$	222
5: 12 14 15 24 45	$x^3(x+1)(x^2-x-4)$	$x(x+1)^2(x^3-2x^2-7x+4)$	222
5: 12 16 34 45 56	$(x^3-x^2-2x+1)(x^3+x^2-2x-1)$	$(x^3-2x^2-5x+1)(x^3+2x^2-x-1)$	222
6: 12 14 16 34 45 56	$(x^3-2x^2-x+1)(x^3+2x^2-x-1)$	$(x^3-x^2-6x-3)(x^3+x^2-2x-1)$	222
6: 12 15 16 23 25 45	$(x^2-2x-1)(x^2+x-1)^2$	$(x^2-2x-4)(x^2+x-1)^2$	1122
6: 12 15 16 23 24 56	$x(x+1)(x^4-x^3-5x^2+3x+4)$	$x(x+1)(x^4-x^3-8x^2-2x+6)$	1122
7: 12 15 16 24 26 45 56	$x^2(x+1)(x^3-x^2-6x+2)$	$x(x+1)(x^4-x^3-7x^2+x+8)$	1122
7: 12 14 15 16 24 45 56	$x(x^2+x-1)(x^3-x^2-5x-2)$	$(x^2+x-1)(x^4-x^3-6x^2-x+1)$	1122
7: 12 16 23 24 34 45 56	$(x^2+x-1)(x^4-x^3-5x^2+2x+4)$	$(x^2+x-1)(x^4-x^3-6x^2+3x+1)$	1122
7: 14 16 23 24 34 45 56	$x(x+1)(x^4-x^3-6x^2+4x+4)$	$x(x-1)(x+1)(x^3-7x-4)$	1122
7: 12 13 15 24 34 45 56	$x^2(x^4-7x^2+4)$	$(x+1)^2(x^4-2x^3-5x^2+6x+4)$	1122
7: 12 14 16 23 24 45 56	$x^2(x-1)(x+2)(x^2-x-4)$	$(x+1)(x-1)(x+2)(x^3-2x^2-3x+2)$	1122

Table 1.1 (continued).

G: # edges: list	Polynomial of G	Polynomial of $\bar{G}$	Orbit sizes
7: 14 16 24 34 45 46 56	$x^2(x+2)(x^3-2x^2-3x+2)$	$x(x+1)^2(x^3-2x^2-5x+2)$	1122
7: 12 15 16 24 25 34 45	$(x-1)(x+1)^2(x^3-x^2-5x+1)$	$x^2(x+2)(x^3-2x^2-4x+4)$	1122
7: 12 15 16 23 24 25 56	$x(x-1)(x+1)(x+2)(x^2-2x-2)$	$x(x+1)(x^4-x^3-7x^2+x+4)$	1122
5: 16 25 35 45 56	$x^2(x^4-5x^2+3)$	$(x+1)^2(x^4-2x^3-7x^2+2x+3)$	1113
7: 12 14 15 24 25 34 45	$x(x+1)^2(x^3-2x^2-4x+2)$	$x^2(x^4-8x^2-6x+3)$	1113
7: 12 13 15 16 24 34 45	$x^2(x^4-7x^2+3)$	$(x+1)^2(x^4-2x^3-5x^2+4x+3)$	1113
4: 12 15 16 23	$x^2(x^4-4x^2+2)$	$(x+1)(x^5-x^4-1x^3-6x^2+7x+3)$	11112
5: 12 15 16 23 56	$x(x-1)(x+1)(x^3-4x-2)$	$x(x^5-1x^3-1x^2+5x+4)$	11112
5: 12 15 16 34 45	$x^2(x^4-5x^2+2)$	$(x+1)(x^5-x^4-9x^3-3x^2+1x+4)$	11112
5: 16 24 34 45 56	$x^2(x^4-5x^2+5)$	$(x+1)(x^5-x^4-9x^3-x^2+7x-1)$	11112
6: 12 15 23 24 25 45	$x^2(x^4-6x^2-4x+2)$	$(x+1)(x^5-x^4-8x^3-2x^2+5x-1)$	11112
6: 12 15 24 25 34 45	$x(x+1)(x^4-x^3-5x^2+x+2)$	$x(x+2)(x^4-2x^3-5x^2+2x+2)$	11112
6: 14 16 23 34 45 56	$x^2(x^4-6x^2+6)$	$(x+1)(x^5-x^4-8x^3+2x^2+9x-1)$	11112
6: 12 13 15 16 45 56	$x^2(x^4-6x^2-2x+5)$	$x(x+1)(x^4-x^3-8x^2-2x+6)$	11112
7: 12 14 15 16 23 45 56	$x(x-1)(x+1)(x+2)(x^2-2x-2)$	$(x+1)(x^5-x^4-7x^3+3x^2+3x-1)$	11112
7: 14 15 16 23 24 45 56	$x(x+1)^2(x^3-2x^2-4x+6)$	$x^2(x^4-8x^2-2x+7)$	11112
7: 15 16 24 25 34 45 56	$(x+1)(x^5-x^4-6x^3+2x^2+7x-1)$	$x^2(x^4-8x^2-4x+6)$	11112
7: 12 15 16 23 24 25 45	$x^6-7x^4-4x^3+6x^2+2x-1$	$x^6-8x^4-6x^3+7x^2+4x-1$	111111
7: 12 15 16 24 34 45 56	$x^6-7x^4-2x^3+8x^2+2x-1$	$x^6-8x^4-4x^3+9x^2+4x-1$	111111
7: 12 14 16 24 34 45 56	$x^6-7x^4-2x^3+7x^2-1$	$x(x^5-8x^3-6x^2+8x+6)$	111111

1.4 Approximating Orbits

The automorphism group $Aut(G)$ of a graph G is a subgroup of S_n, the symmetric group on n objects, so $|Aut(G)| \leq n!$. Constructing all the elements of the automorphism group could take exponential time, e.g., K_n has S_n as its automorphism group. However, it may be sufficient to find a relatively small generating set that represents $Aut(G)$. Indeed, it is always possible to find a generating set of size $\log n$ for a group H of size n [1].

Unfortunately it is not known whether or not such a small set representing $Aut(G)$ can be computed in polynomial time, because the problem of determining the automorphism group can be shown to be equivalent to graph isomorphism (i.e., determining whether two graphs are isomorphic). The relationship between the two problems is shown more explicitly in [1].

Since the problem of determining when two graphs are isomorphic has been studied extensively and is not known to be solvable by a polynomial bounded algorithm, heuristics are needed to find the orbits of the automorphism group. If such heuristics are easy to compute and provide a high degree of accuracy, the complexity of a graph can be computed efficiently with a high degree of confidence.

The orbits of a graph consist of vertices with similar properties such as having the same degree. So if it were possible to create a small list of all these properties and if, in addition, there were polynomial time tests for each one, then there would be a polynomial time algorithm for the graph automorphism problem. Of course such a complete list of properties is not known. However, if there exists one such property that does not hold for two vertices, then these vertices are not in the same orbit. So, creating a partial list of polynomial time tests would help to distinguish vertices having different properties and thus to separate them into different orbits. In surveying the literature on heuristic approaches to computing the orbits of the automorphism group of a graph we have made use of [20], which in turn draws on [7].

The procedure adopted here for finding the orbits of a graph is as follows:

1. Identify several polynomially checkable properties designed to distinguish between vertices. At the start of the procedure all the vertices are taken to be in the same orbit.
2. For each property and each pair of vertices u, v in an orbit thus far determined, find whether or not u and v can be distinguished by the property. If yes, then draw the inference "u, v are in different orbits"; otherwise, apply the next test.

If two vertices pass all the tests, then they will be considered to be in the same orbit. This procedure gives rise to a deterministic process with one-sided error, i.e., two vertices in the same putative orbit may in fact be distinguishable.

Critical to developing an efficient procedure is making judicious choices of vertex properties that can serve as tests. The selection of properties used in our procedure has been guided by results in the theory of networks and in sociological theory.

1.4.1
The Degree of the Vertices

The first test is quite simple: If two vertices have different degrees, then they cannot be in the same orbit. The degree of a vertex can be computed in time n, with n being the size of the graph. So this test will take time $O(n)$. The degree of a vertex is an important property in the theory of networks since finding the high-degree vertices in an underlying graph is considered equivalent to determining the so-called "authorities" of the network.

1.4.2
The Point-Deleted Neighborhood Degree Vector

Examining only the degree of the vertices is insufficient. Consider, for example, the path of five vertices, labeled 1, 2, 3, 4, 5. The degree test does not distinguish 2 and 4 from 3, but 3 is in not in the same orbit as 2 and 4; this is obvious since both 2 and 4 have one neighbor (1 and 5, respectively) with degree 1, while 3 does not have such a neighbor. This observation leads naturally to the idea of the second test, namely, to examine the neighborhood of the vertices.

Definition 1.1 The *neighborhood* of a subset $S \subseteq V$ denoted by $N(S)$ is the set of all the neighbors of S, i.e., $N(S) = S \bigcup \{v \in V \setminus S | (v, s) \in E \text{ and } s \in S\}$. The $\{i+1\}^{th}$ *degree neighborhood* of S is defined inductively, $N^{i+1}(S) = N(N^i(S))$. The *point-deleted neighborhood* of v, with v being a vertex of V, is $\tilde{N}^i(x) = N^i(\{x\}) \setminus \{x\}$.

Thus, for the path mentioned above we have that

- $N(\{3\}) = \{2, 3, 4\}$,
- $N^2(\{3\}) = \{1, 2, 3, 4, 5\}$,
- $\tilde{N}(3) = \{2, 4\}$.

The notion of the degree vector is also needed.

Definition 1.2 The *degree vector* $d(S)$ of a subset S of the vertices of a graph can be defined as the ordered sequence of the degrees of these vertices in the induced subgraph.

Again for the above example of the path of five vertices, the degree vector of the whole path is (1,1,2,2,2). The degree vector of $\{2,3\}$ is (1,1).

The test to be considered is comparing the point-deleted neighborhood degree vectors of the vertices to be examined. This technique is presented in [7]. The degree test yields several groups of vertices of equal degree. To apply this test we first compute the point-deleted neighborhood of each of the vertices to be compared and then determine the degree vector. If the degree vector is different, draw the inference "not in the same orbit." The test can be made more subtle if higher-order neighborhoods are taken into account.

The execution time required for this test can be computed as follows. If v and u are the vertices to be compared and k is their degree, we can compute the degree vectors of the point-deleted neighborhoods of u and v in time k^2. Sorting the two vectors, each of size k, and comparing them is at most of this order of complexity. Thus the total execution time required is $\Theta(k^2)$. Since k is bounded by n, the worst-case complexity of the test is $O(n^2)$.

As an example, consider the graph of Figure 1.8. Table 1.2 shows the point-deleted neighborhood of the vertices.

The test defines two groups of vertices that for this example coincide with the two orbits. The sets are $\{1,3,5,7\}$ and $\{2,4,6,8\}$.

Examining the point-deleted neighborhood degree vector has several advantages and disadvantages over the degree sequence of the neighbors of a vertex. The latter would probably give faster negative results when testing whether two vertices belong in the same orbit, since it takes account of the whole graph and not simply the induced subgraph of the neighbors; on the other hand, the former also works in regular graphs. However, the main reason for using the latter test is that it is associated with the concept of clustering coefficients that are widely used in networking theory. Informally

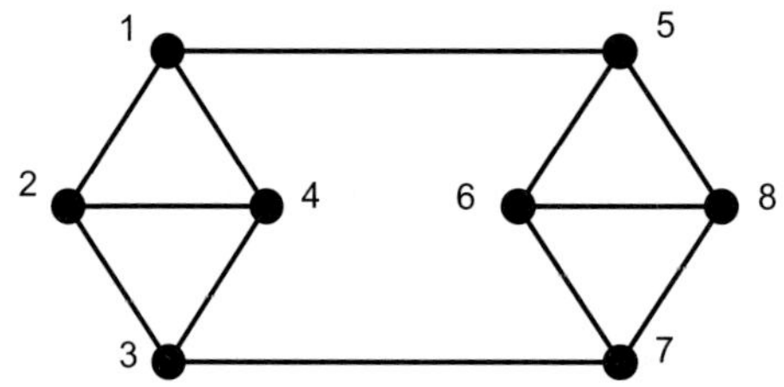

Figure 1.8 Example: graph for point-deleted neighborhood degree vectors test.

Table 1.2 Information about the point-deleted neighborhood of every vertex and its degree vector.

v	$\tilde{N}(v)$	$d(\tilde{N}(v))$
1	{2,4,5}	(0,1,1)
2	{1,3,4}	(1,1,2)
3	{2,4,7}	(0,1,1)
4	{1,2,3}	(1,1,2)
5	{1,6,8}	(0,1,1)
6	{5,7,8}	(1,1,2)
7	{3,6,8}	(0,1,1)
8	{5,6,7}	(1,1,2)

speaking, the clustering coefficient indicates the degree to which the induced subgraph of the neighbors of a vertex resembles a clique. Determining the point-deleted neighborhood degree vector of a vertex would help in computing the clustering coefficient. So vertices that pass this test will also have the same clustering coefficient.

1.4.3
Betweenness Centrality

The above techniques correctly determine the orbits of a large variety of graphs. However, there are cases where they fail. Consider, for example, the graph in Figure 1.9. The point-deleted neighborhood degree vector would place vertices 13 and 14 in the same orbit since $d(\tilde{N}(13)) = d(\tilde{N}(14)) = (2, 2, 2, 2, 3, 3)$. However, it is obvious from the figure that vertices 13 and 14 do not belong in the same orbit.

In this section we will describe one more method of estimating the orbits of a graph. This method is based on the concept of betweenness centrality, which was first introduced in [8] and can be described as follows.

Definition 1.3 The betweenness centrality of a vertex x of a graph $G(V, E)$ is the sum over all pairs of vertices y, z in the graph of the number of shortest paths ($p^x_{y,z}$) from y to z that pass through x divided by the number of all

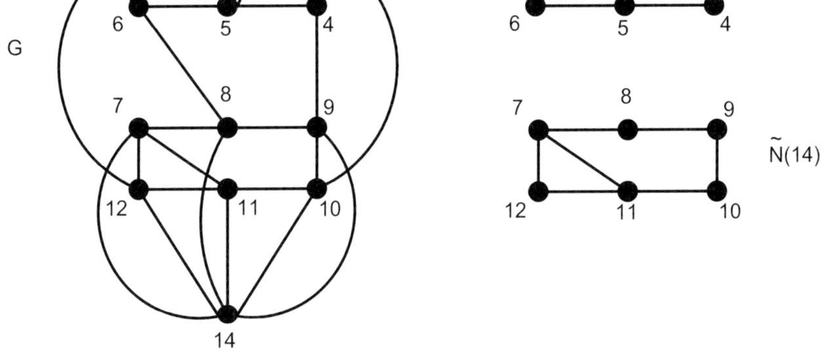

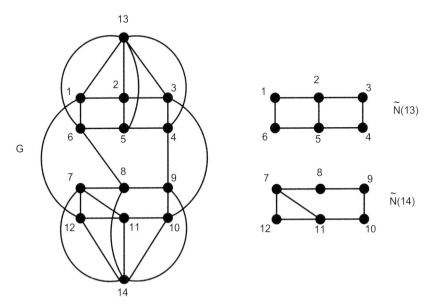

Figure 1.9 Graph where the technique of the point-deleted neighborhood degree vector fails.

the shortest paths from y to z. More precisely, the betweenness centrality of a vertex x is:

$$C_B(x) = \frac{1}{2}\sum_{y\in V} d_{y,x}\,,$$

where

$$d_{y,x} = \sum_{z\in V} b_{y,z}$$

and

$$b_{y,z}(x) = \frac{p^x_{y,z}}{p_{y,z}}\,.$$

This measure is suggested by the sociology of networks of individuals, so-called "social networks." In addition, it captures important structural features of a graph, making it quite useful in approximating orbits.

The method considered in [7] is an extension of the point-deleted neighborhood degree vector. Once the induced subgraph on $\tilde{N}(x)$ is computed for every vertex x, we compute the betweenness vector $C_B(\tilde{N}(x)) = C_B(v_1)$, $C_B(v_2), \ldots, C_B(v_{d(x)})$, where $C_B(v_i) \le C_B(v_{i+1})$, with $v_i \in \tilde{N}(x)$ and $d(x)$ being the degree of vertex x. If the vectors $C_B(\tilde{N}(x))$ and $C_B(\tilde{N}(y))$ are not identical, then x and y belong to different orbits.

Consider again the graph of Figure 1.9. We want to compute the betweenness centrality vectors of vertices 13 and 14, which will help in deciding whether they belong in the same orbit or not. We first find the induced subgraphs on $\tilde{N}(x)$ for every vertex x (also shown in Figure 1.9). Then, for each subgraph, we first compute a table that contains all the intermediate vertices in every shortest path between each pair of vertices. We assume that paths of length 1 have no intermediate vertices and we omit paths that start and end on the same vertex. The tables are shown below (Table 1.3). Then we compute the dependency matrix, the matrix $\boldsymbol{D} = d_{y,z}$ for every y, z. Finally we sum and then halve every column of $\boldsymbol{D}$ to compute the betweenness centrality of every vertex of the induced subgraph. The dependency matrices for both vertices 13 and 14 are computed below (Table 1.4). The computed vector will be the betweenness centrality vector of vertex x. Finally, when all the computations are done, we compute the vectors of each pair of vertices. If they are different, then the vertices belong in different orbits; otherwise the algorithm concludes that the vertices belong to the same orbit.

It is clear that

$$C_B(\tilde{N}(13)) = \left(\frac{5}{6}, \frac{5}{6}, \frac{5}{6}, \frac{5}{6}, 3, 3\right) \neq C_B(\tilde{N}(14)) = \left(0, 1, 1\frac{1}{2}, 1\frac{1}{2}, 2\frac{1}{2}, 2\frac{1}{2}\right).$$

Thus vertices 13 and 14 belong to different orbits.

Table 1.3 Tables containing the intermediate vertices of the shortest paths between each pair of vertices.

	1	2	3	4	5	6
1	–	–	2	2 3 2 5 6 5	2 6	–
2	–	–	–	3 5	–	1 5
3	2	–	–	–	2 4	2 1 2 5 4 5
4	3 2 5 2 5 6	3 5	–	–	–	5
5	2 6	–	2 4	–	–	–
6	–	1 5	1 2 5 2 5 4	5	–	–

	7	8	9	10	11	12
7	–	–	8	11	–	–
8	–	–	–	9	7	7
9	8	–	–	–	10	8 7 10 11
10	11	9	–	–	–	11
11	–	7	10	–	–	–
12	–	7	8 7 10 11	11	–	–

Table 1.4 Dependency matrices for vertices 13 and 14 and the betweenness centrality vectors.

	1	2	3	4	5	6
1	0	$1\frac{5}{6}$	$\frac{1}{3}$	0	$\frac{2}{3}$	$\frac{5}{6}$
2	$\frac{1}{2}$	0	$\frac{1}{2}$	0	1	0
3	$\frac{1}{3}$	$1\frac{5}{6}$	0	$\frac{5}{6}$	$\frac{2}{3}$	0
4	0	$\frac{2}{3}$	$\frac{5}{6}$	0	$1\frac{5}{6}$	$\frac{1}{3}$
5	0	1	0	$\frac{1}{2}$	0	$\frac{1}{2}$
6	$\frac{5}{6}$	$\frac{2}{3}$	0	$\frac{1}{3}$	$1\frac{5}{6}$	0
C_B	$\frac{5}{6}$	3	$\frac{5}{6}$	$\frac{5}{6}$	3	$\frac{5}{6}$

	7	8	9	10	11	12
7	0	1	0	0	1	0
8	2	0	1	0	0	0
9	$\frac{1}{2}$	$1\frac{1}{2}$	0	$1\frac{1}{2}$	$\frac{1}{2}$	0
10	0	0	1	0	2	0
11	1	0	0	1	0	0
12	$1\frac{1}{2}$	$\frac{1}{2}$	0	$\frac{1}{2}$	$\frac{1}{2}$	0
C_B	$2\frac{1}{2}$	$1\frac{1}{2}$	1	$1\frac{1}{2}$	$2\frac{1}{2}$	0

The betweenness centrality of all the vertices of a graph can be computed in $O(n^3)$ time (where n is the number of vertices in the graph) by a modified version of Floyd's algorithm for determining all shortest paths between pairs of vertices. The fastest known exact algorithm for determining the betweenness centrality of all the vertices of a graph is due to Brandes [4], and its complexity is $\Theta(n \cdot m)$, where m is the number of edges of the graph. Thus, if u and v are vertices to be compared and k is their degree, then determining the betweenness centrality vectors of u and v requires $O(k^3)$ time. Thus the worst-case complexity of the test is $O(n^3)$.

It is still possible to compute the orbits of the previous example exactly by examining the degree vector of higher-order point-deleted neighborhoods.

However, there are (rare) cases where the idea of the degree vector does not work at all. The graph (whose adjacency matrix is presented below) is such an example.

0	1	0	1	1	0	0	0	0	1	1	0	1	1	0	0	0	1	0	0	0	0	0	1	0	1
1	0	1	0	1	1	0	0	0	0	1	1	0	1	1	0	0	0	1	0	0	0	0	0	1	0
0	1	0	1	0	1	1	0	0	0	0	1	1	0	1	1	0	0	0	1	0	0	0	0	0	1
1	0	1	0	1	0	1	1	0	0	0	0	1	1	0	1	1	0	0	0	1	0	0	0	0	0
1	1	0	1	0	1	0	1	1	0	0	0	0	0	1	0	1	1	0	0	0	1	0	0	0	0
0	1	1	0	1	0	1	0	1	1	0	0	0	0	0	1	0	1	1	0	0	0	1	0	0	0
0	0	1	1	0	1	0	1	0	1	1	0	0	0	0	0	1	0	1	1	0	0	0	1	0	0
0	0	0	1	1	0	1	0	1	0	1	1	0	0	0	0	0	1	0	1	1	0	0	0	1	0
0	0	0	0	1	1	0	1	0	1	0	1	1	0	0	0	0	0	1	0	1	1	0	0	0	1
1	0	0	0	0	1	1	0	1	0	1	0	1	1	0	0	0	0	0	1	0	1	1	0	0	0
1	1	0	0	0	0	1	1	0	1	0	1	0	0	1	0	0	0	0	0	1	0	1	1	0	0
0	1	1	0	0	0	0	1	1	0	1	0	1	0	0	1	0	0	0	0	0	1	0	1	1	0
1	0	1	1	0	0	0	0	1	1	0	1	0	0	0	0	1	0	0	0	0	0	1	0	1	1
1	1	0	1	0	0	0	0	0	1	0	0	0	0	0	0	0	0	1	1	1	1	0	0	1	0
0	1	1	0	1	0	0	0	0	0	1	0	0	0	0	0	0	0	0	1	1	1	1	0	0	1
0	0	1	1	0	1	0	0	0	0	0	1	0	0	0	0	0	1	0	0	1	1	1	1	0	0
0	0	0	1	1	0	1	0	0	0	0	0	1	0	0	0	0	0	1	0	0	1	1	1	1	0
1	0	0	0	1	1	0	1	0	0	0	0	0	0	0	1	0	0	0	1	0	0	1	1	1	1
0	1	0	0	0	1	1	0	1	0	0	0	0	1	0	0	1	0	0	0	1	0	0	1	1	1
0	0	1	0	0	0	1	1	0	1	0	0	0	1	1	0	0	1	0	0	0	1	0	0	1	1
0	0	0	1	0	0	0	1	1	0	1	0	0	1	1	1	0	0	1	0	0	0	1	0	0	1
0	0	0	0	1	0	0	0	1	1	0	1	0	1	1	1	1	0	0	1	0	0	0	1	0	0
0	0	0	0	0	1	0	0	0	1	1	0	1	0	1	1	1	1	0	0	1	0	0	0	1	0
1	0	0	0	0	0	1	0	0	0	1	1	0	0	0	1	1	1	1	0	0	1	0	0	0	1
0	1	0	0	0	0	0	1	0	0	0	1	1	1	0	0	1	1	1	1	0	0	1	0	0	0
1	0	1	0	0	0	0	0	1	0	0	0	1	0	1	0	0	1	1	1	1	0	0	1	0	0

This graph is 10-regular, so that for every vertex x, $\tilde{N}(x)$ contains ten vertices. Furthermore, its diameter is two, so $\tilde{N}^2$ contains every vertex of the graph except for x. It follows that $d(\tilde{N}(x)) = (3,3,3,3,3,3,3,3,3,3)$ for every vertex x. Thus the above tests will yield one orbit $\{1,\ldots,26\}$. However, it can be proven that the graph has two orbits, $\{1,\ldots,13\}$ and $\{14,\ldots,26\}$ (this example is taken from [7]). The problem is solved by examining the point-deleted neighborhood betweenness centrality vector. We can show that $C_B(\tilde{N}(1)) = \ldots = C_B(\tilde{N}(13)) = (3,3,3,3,4,4,4,5,5,5)$, whereas $C_B(\tilde{N}(14)) = \ldots = C_B(\tilde{N}(26)) = (3,3,3,4,4,4,4,4,4,6)$.

In this section we have only examined the betweenness centrality measure in approximating the orbits of a graph. However, there are several variants of centrality measures that could be taken into account. These variants include

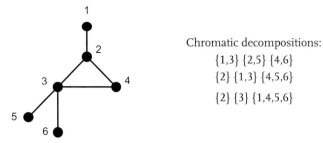

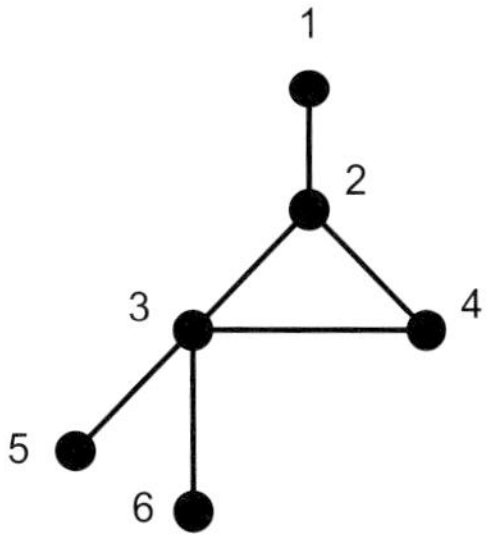

Chromatic decompositions:
{1,3} {2,5} {4,6}
{2} {1,3} {4,5,6}
{2} {3} {1,4,5,6}

Figure 1.10 Graph with multiple chromatic decompositions.

closeness centrality [22], graph centrality [9], and stress centrality [23]. All these concepts are attempts to capture the notion of the relative importance of a vertex in the overall structure of a graph, and thus each of them could play an important role in estimating the orbits of a graph.

1.5
Alternative Bases for Structural Complexity

Colorings of a graph can be used to obtain a decomposition of the vertices. Sets of vertices of the same color (or independent sets) constitute equivalence classes. Unlike the orbits of the automorphism group, a partition of the vertices obtained in this way is not unique. However, an information measure may be defined by taking the minimum value over some set of decompositions linked to colorings [16]. This section explores such a measure, compares it with the symmetry-based measure, and shows its relationship to the graph entropy as defined in [11].

A *coloring* of a graph is an assignment of colors to the vertices so that no two adjacent vertices have the same color.

An *n-coloring* of a graph $G = (V, E)$ is a coloring with n colors or, more precisely, a mapping f of V onto the set $\{1, 2, \ldots, n\}$ such that whenever $[u, v] \in E, f(u) \neq f(v)$.

The *chromatic number* $\kappa(G)$ of a graph G is the smallest value of n for which there is an n-coloring. A graph may have more than one n-coloring.

An n-coloring is *complete* if, for every i, j with $i \neq j$, there exist adjacent vertices u and v such that $f(u) = i$ and $f(v) = j$.

A decomposition $\{V_i\}_{i=1}^{n}$ of the set of vertices V is called a *chromatic decomposition* of G if $u, v \in V_i$ imply that $[u, v] \notin E$. Note that V_i in a chromatic decomposition is a set of *independent* vertices. If f is an n-coloring, the collection of sets $\{v \in V | f(v) = i\}_{i=1}^{n}$ forms a chromatic decomposition; conversely, a chromatic decomposition $\{V_i\}_{i=1}^{n}$ determines an n-coloring f. The sets V_i are thus called *color classes*.

Given a graph $G = (V, E)$ with $|V| = n$ and $h = \kappa(G)$, let $\hat{V} = \{V_i\}_{i=1}^{h}$ be an arbitrary chromatic decomposition of G with $n_i(\hat{V}) = |V_i|$ for $1 \le i \le h$. The *chromatic information content* $I_c(G)$ of G is defined by the following formula [16]:

$$I_c(G) = \min_{\hat{V}} \left\{ -\sum_{i=1}^{h} \frac{n_i(\hat{V})}{n} \log \frac{n_i(\hat{V})}{n} \right\}.$$

Figure 1.10 shows a graph with three different chromatic decompositions whose finite probability schemes are $(1/3, 1/3, 1/3)$, $(1/2, 1/3, 1/6)$, and $(2/3, 1/6, 1/6)$. The minimum entropy is given by $(2/3, 1/6, 1/6)$, so that $I_c(G) = 2/3 \log 3/2 + 1/3 \log 6$.

$I_c(G)$ is defined as the minimum value over chromatic decompositions with $\kappa(G)$ color classes and thus does not necessarily give the minimum over all chromatic decompositions. When the graph does not have a complete k-coloring for $k > \kappa(G)$, $I_c(G)$ does give the minimum over all chromatic decompositions [16]. The restricted minimization in the definition allows for interpreting $I_c(G)$ as the amount of information needed to construct a $\kappa(G)$-coloring.

A related measure called *graph entropy* was introduced in [11] and subsequently applied to a variety of problems in graph theory and combinatorics [24]. This measure is a generalization of $I_c(G)$ formulated as average mutual information between two random variables representing the vertices and independent sets of G, respectively. Let S be the collection of independent sets of $G = (V, E)$ with $|V| = n$, and let $\boldsymbol{P}$ be a probability distribution on V. The graph entropy $H(G, \boldsymbol{P})$ is given by $I(V; S)$, the average mutual information between V and S (treated as random variables). Now, $I(V; S) = H(V) - H(V|S)$, so if $\boldsymbol{P}$ is a uniform probability distribution over V, then $H(G, \boldsymbol{P}) = I(V; S) = \log n - H(S)$. So, $I_c(G) = \log n - H(G, \boldsymbol{P})$. In summary, the essential difference between the two measures is that $I_c(G)$ as-

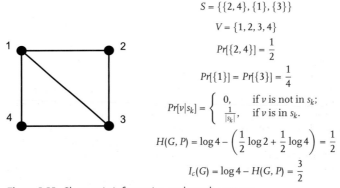

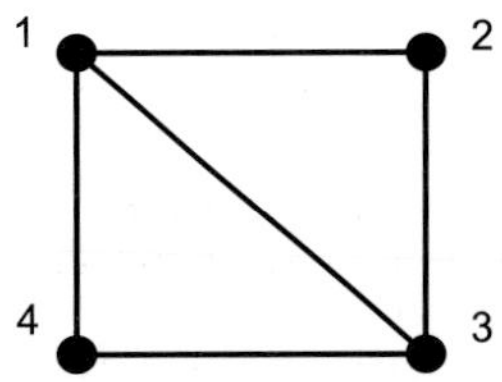

$$S = \{\{2, 4\}, \{1\}, \{3\}\}$$

$$V = \{1, 2, 3, 4\}$$

$$Pr[\{2, 4\}] = \frac{1}{2}$$

$$Pr[\{1\}] = Pr[\{3\}] = \frac{1}{4}$$

$$Pr[v|s_k] = \begin{cases} 0, & \text{if } v \text{ is not in } s_k; \\ \frac{1}{|s_k|}, & \text{if } v \text{ is in } s_k. \end{cases}$$

$$H(G, P) = \log 4 - \left(\frac{1}{2}\log 2 + \frac{1}{2}\log 4\right) = \frac{1}{2}$$

$$I_c(G) = \log 4 - H(G, P) = \frac{3}{2}$$

Figure 1.11 Chromatic information and graph entropy.

sumes a fixed (uniform) probability distribution over V, whereas $H(G, \boldsymbol{P})$ allows the probability distribution over V to vary. Figure 1.11 illustrates the relationship between $I_c(G)$ and $H(G, \boldsymbol{P})$.

The two entropy-based measures of graph complexity, $I_a(G)$ and $I_c(G)$, discussed in this chapter capture different aspects of graph structure. Colorings and symmetries of a graph do not necessarily say much about each other. The difference can be seen from examples such as the cycle C_n on n vertices. This graph has a transitive automorphism group so $I_a(C_n) = 0$ for all n, whereas the cycle has chromatic number two or three, depending on whether n is even or odd, and $I_c(C_n) \approx \log 2$. The divergence between the two measures is unbounded in the case of trees that have chromatic number two but (for $n \geq 7$) can have a trivial automorphism group. In these cases, $I_c(C_n) \leq 1$ but $I_a(C_n) = \log n$.

The foregoing observations support the view that structural complexity is in the eye of the beholder. No single measure can capture all aspects of a graph.

Acknowledgment

Research was sponsored by the U.S. Army Research Laboratory and the U.K. Ministry of Defence and was accomplished under Agreement No. W911NF-06-3-0001. The views and conclusions contained in this document are those of the author(s) and should not be interpreted as representing the official policies, either express or implied, of the U.S. Army Research Laboratory, the U.S. government, the U.K. Ministry of Defence or the U.K. government. The U.S. and U.K. governments are authorized to reproduce and distribute reprints for government purposes notwithstanding any copyright notation hereon.

References

1 Arvind, V. Algebra and Computation. Lecture notes (transcribed by Ramprasad Saptharishi), 2007. http://www.cmi.ac.in/~ramprasad/lecturenotes/algcomp/tillnow.pdf, last viewed 5-30-2008.

2 Bent, G., Dantressangle, P., Vyvyan, D., Mowshowitz, A., and Mitsou, V. A dynamic distributed federated database. Proceedings of the Second Annual Conference of the International Technology Alliance, Imperial College, London, September 2008.

3 Biggs, N.L. *Algebraic Graph Theory*. Cambridge University Press, Cambridge, 1993.

4 Brandes, U. A faster algorithm for betweenness centrality. *J. Math. Sociol.* 25 (**2001**), pp. 163–177.

5 Chao, C.C. A note on the eigenvalues of a graph. *J. Combinator. Theory* (Series B) 10 (**1971**), pp. 301–302.

6 Collatz, L., and Sinogowitz, U. Spektrum endlicher Graphen. *Abh. Math. Sem. Univ. Hamburg* 21 (**1957**), pp. 63–77.

7 Everett, M. G., and Borgatti, S. Calculating Role Similarities: An Algorithm that helps

determine the Orbits of Graph. *Social Networks* 10 (**1988**), pp. 77–91.

8 Freeman, L. C. A set of measures of centrality based on betweenness. *Sociometry* 40 (**1977**), pp. 35–41.

9 Hage, P. and Harary, F. Eccentricity and centrality in networks. *Social Networks* 17 (**1995**), pp. 57–63.

10 Khinchin, A.I. *Mathematical Foundations of Information Theory*. Dover Publications, New York, 1957.

11 Korner, J. Coding of an information source having ambiguous alphabet and the entropy of graphs. *Transactions of Prague Conference on Information Theory, Statistical Decision Functions, Random Processes*. (**1971**), pp. 411–425.

12 Lauri, J. and Scapellato, R. *Topics in Graph Automorphisms and Reconstruction*. Cambridge University Press, Cambridge, 2003.

13 Mowshowitz, A. Entropy and the complexity of graphs: I. An index of the relative complexity of a graph. *Bull. Math. Biophys.* 30 (**1968**), pp. 175–204.

14 Mowshowitz, A. Entropy and the complexity of graphs: II. The information content of digraphs and infinite graphs. *Bulletin of Mathematical Biophysics* 30 (**1968**), pp. 225–240.

15 Mowshowitz, A. Entropy and the complexity of graphs: III. Graphs with prescribed information content. *Bull. Math. Biophys.* 30 (**1968**), pp. 387–414.

16 Mowshowitz, A. Entropy and the complexity of graphs: IV. Entropy measures and graphical structure. *Bull. Math. Biophys.* 30 (**1968**), pp. 533–546.

17 Mowshowitz, A. The group of a graph whose adjacency matrix has all distinct eigenvalues. In F. Harary, editor, *Proof Techniques in Graph Theory*, pp. 109–110. Academic Press, New York, 1969.

18 Mowshowitz, A. Graphs, groups and matrices. In *Proceedings of the Canadian Mathematical Congress* (**1971**), pp. 509–522.

19 Mowshowitz, A., Mitsou, V., and Bent, G. Models of network growth by combination. Proceedings of the Second Annual Conference of the International Technology Alliance, Imperial College, London, September 2008.

20 Papireddy, Y.S. A cost efficient approach for finding the orbits to calculate the Entropy of a graph. Project Presentation, Department of Computer Science, The City College of New York, Spring 2007.

21 Rashevsky, N. Life, information theory, and topology. *Bull. Math. Biophys.* 17 (**1955**), pp. 229–235.

22 Sabidussi, G. The centrality index of a graph. *Psychometrika* 31 (**1966**), pp. 581–603.

23 Shimbel, A. Structural parameters of communication networks. *Bull. Math. Biophys.* 15 (**1953**), pp. 501–507.

24 Simonyi, G. Graph entropy: a survey. In W. Cook and L. Lovasz, editors, *Combinatorial Optimization*. DIMACS: Series in Discrete Mathematics and Theoretical Computer Science, 1995.

25 Thompson, D.W. *On Growth and Form* (abridged version edited by J.T. Bonner). Cambridge University Press, Cambridge, 1961.

26 Trucco, E. A note on the information content of graphs. *Bull. Math. Biophys.* 18 (**1956**), pp. 129–135.

27 Wikepedia. Kolmogorov complexity. http://en.wikipedia.org/wiki/Kolmogorov_complexity, last viewed 5-28-08.

2
Statistical Mechanics of Complex Networks

Stefan Thurner

2.1
Introduction

An explanation for the impressive recent quantitative efforts in network theory might be that it provides a promising tool for understanding complex systems. Network theory is mainly focused on statistical descriptions of discrete large-scale topological structures rather than on microscopic details of interactions of its elements. This viewpoint allows one to naturally treat collective phenomena that are often an integral part of complex systems, such as biological or socioeconomic phenomena. Much of the attraction of network theory arises from the discovery that many networks, natural or manmade, exhibit some sort of universality, meaning that most of them belong to one of three classes: *random*, *scale-free*, and *small-world* networks. Maybe most important, however, is that, due to its conceptually intuitive nature, network theory seems to be within realistic reach of a full and coherent understanding from first principles.

It has become standard practice to describe networks by a set of *macroscopic* parameters. These parameters usually provide a practical understanding about the statistics of linking within the network, the degrees of clustering, or the statistics of occurrence of certain motives. With this knowledge it is in many cases sufficient to reliably characterize a particular network in terms of its structure, robustness, and performance or function. Often networks are not structures that are purposefully designed but that emerge as a consequence of *microscopic* rules that govern the linking and relinking dynamics of individual nodes. These rules can be very general and cover a huge variety, ranging from purely deterministic to fully statistical ones.

One of the milestones in the history of science was the discovery that the laws of thermodynamics could be related to – and based on – a microscopic theory, so-called statistical mechanics. The statistical mechanics of Boltzmann is a framework that relates the properties of microscopic particles to the macroscopic bulk properties of matter, thereby explaining thermodynamics. The formal link between the macro- and the microworld is the concept

Analysis of Complex Networks: From Biology to Linguistics. Edited by Matthias Dehmer and Frank Emmert-Streib

ISBN: 978-3-527-32345-6

of entropy, S. The aim of the following text is to motivate the idea that it is straightforwardly possible to formulate a statistical mechanics of networks, which in principle should allow one to relate macroscopic – emergent structures and phenomena – of networks to a set of microscopic rules. In this view the connectivity of nodes plays the role of particles in classical statistical mechanics.

Many real-world networks differ considerably from pure random graphs [1, 2], leading to the notion of complex networks, which is a well-established concept by now [3, 4]. Networks are discrete objects made up of a set of nodes that are joined by a set of links. If a given set of N nodes is linked by a fixed number of links in a completely random manner, the result is a so-called *random network*, whose characteristics can be easily understood. One of the simplest measures describing a network in statistical terms is its degree distribution, $p(k)$. The degree k_i of a node i is defined by the discrete number of links leading or originating from it, the degree distribution, $p(k)$, is the distribution of degrees over the whole network, i.e., the probability that a randomly chosen node has a specific degree, k. In the case of random networks, the degree distribution is a Poissonian, i.e., the probability (density) that a randomly chosen node has degree k is given by $p(k) = (\lambda^k e^{-\lambda})/k!$, where $\lambda = \bar{k}$ is the average degree of all nodes in the network. However, as soon as more complicated rules for wiring or growing of a network are considered, the seemingly simple concept of a network can become more involved. In particular, in many cases the degree distribution becomes a power law, without any characteristic scale, which raises associations to critical phenomena and scaling phenomena in complex systems. This is the reason why these *scale-free* types of networks are also called complex networks. A further intriguing aspect of dynamical complex networks is that they can naturally provide some sort of toy model for nonergodic systems, in the sense that not all possible states (configurations) are equally probable or homogeneously populated, and thus can violate a key assumption for systems described by classical statistical mechanics. In what follows we will not deal with the third "universality class" of networks, the *small-world* networks, since in a dynamical context they often appear as transients, resulting in either random network or some highly ordered states.

Networks in their purest form are sets of L links connecting a set of N nodes. Every possible connection pattern is called a network state or a microstate. If states are not considered to be constant over time and are allowed to change through the rewirement of nodes, it is clear that the theory of networks reduces to a statistical mechanics problem of counting states under certain constraints. If we assume – as we will do for reasons explained below – that N and L are constants, the only dynamics possible are rewirements that can be governed by deterministic dynamical rules or pure chance.

If it is assumed that links in a network carry no structure or information (i.e., they do not have link weights/strengths or direction), the full information about a network is captured in the adjacency matrix, **c**, whose elements are $c_{ij} = 1$, if a pair of nodes $\{i, j\}$ is linked and $c_{ij} = 0$ otherwise. A network is then called unweighted and undirected. For weighted networks links c_{ij} are associated with a link weight; if the links indicate a direction, matrix **c** will not be symmetric. Each microstate can also be characterized by a degree distribution that is obviously a reduction of information. In general there can be many states leading to the same degree distribution. The prime goal of a statistical mechanics of network theory is to provide an understanding of measured degree distributions, correlation functions between degrees, clustering, motive distributions, etc. from first principles. These principles are a combination of the possible number of states with the physical restrictions on microstates, exclusion of states, restrictions on relinking rules, and overall constraints on the system.

As in classical statistical mechanics, physics enters through the physical characterizations of states, such as the energy dependence of the probabilities of finding certain states, or simply restrictions and the exclusion of states. The same is true for networks. To understand a network in a real-world situation (biological, social, or economic) besides counting possible states, it is essential to identify the "physical" restrictions on the network states. Given that network dynamics is sufficiently fast and that reasonably large phase-space volumes are covered by network dynamics, a statistical mechanics approach to networks certainly seems reasonable. It is possible to think of the "phase space" of networks as the set of all possible adjacency matrices.

2.1.1
Network Entropies

Following a microstate view of networks, the measured degree distribution in a given system should correspond to the largest number of states that have this degree distribution, subject to certain constraints. Given this view, it is natural to think about the existence of some sort of an H-theorem or entropy principle, relating networks in equilibrium (or reasonably steady states) to a maximum of some entropy definition. The existence of such an entropy makes it possible to talk about a "thermodynamics of networks." This establishes the link from microstates to a macroscopic description of networks. As we shall see below, networks even in their simplest form lead to entropies that are nonadditive or nonextensive. This is intuitively clear: if one imagines that one separates a given network into two subnetworks A and B, all links between A and B will be lost. This can be a substantial part of the network and clearly induces nonadditivity.

As mentioned, the degree distributions of complex networks are not of a trivial Poissonian type but very often follow power laws [5] or more complicated forms. It can be shown that there exist entropies that are associated to these degree distributions. In particular, the degree distributions can be derived under the extremization of these entropies – given the constraints that the degree distribution is normalizable and the average degree exists. In the special and important case for q-exponential degree distributions, the entropy coincides with the Tsallis entropy (however, care must be taken for the constraints under maximization). A q-exponential degree distribution is given by

$$P(k) = \mathrm{e}_q(-(k-1)/\kappa) \quad (k = 1, 2, 3, 4, \ldots)\,, \tag{2.1}$$

where the q-exponential function is defined by $\mathrm{e}_q(x) \equiv [1+(1-q)x]^{1/(1-q)}$ and κ is a characteristic degree of the distribution. It has long been noticed that degree distributions of some network models are exactly q-exponentials [6]. The model in [5] describes growing networks with a so-called preferential attachment rule, meaning that any new node being added to the system links itself to an already existing node i in the network with a probability that is proportional to the degree of node i, i.e., $p_{\text{link}} \propto k_i$. In [6] this model was extended to also allow for preferential rewiring. The analytical solution to the model has a q-exponential as a result, with parameter q being fixed uniquely by the model parameters. Recently it has been found that networks exhibiting degree distributions compatible with q-exponentials are not at all limited to growing and preferentially organizing networks. Degree distributions of real-world networks as well as of nongrowing models of various kinds seem to exhibit a universality in this respect [7–9]. A model for non-growing networks which was recently put forward in [9] also unambiguously exhibit q-exponential degree distributions. This model was motivated by interpreting networks as a certain type of "gas" where upon an (inelastic) collision of two nodes, links get transfered in analogy to the energy-momentum transfer in real gases. In this model a fixed number of nodes in an (undirected) network can merge, i.e., two nodes fuse into one single node, which retains the union of links of the two original nodes; the link connecting the two nodes before the merger is removed. At the same time a new node is introduced to the system and is linked randomly to any of the existing nodes in the network [8]. In Figure 2.1 we show a snapshot of this type of network representative for the many models exhibiting q-exponential degree distributions. The corresponding (cumulative) degree distribution is shown in log-log scale, clearly exhibiting an assymptotic power law. The same figure (right) shows q-logarithms of the degree distribution for several values of q. It is clear from the correlation coefficient of the q-logarithm with straight lines (inset) that there exists an optimal value of q, which makes the q-logarithm a linear function in k, showing that the degree distribution is in fact a q-exponential.

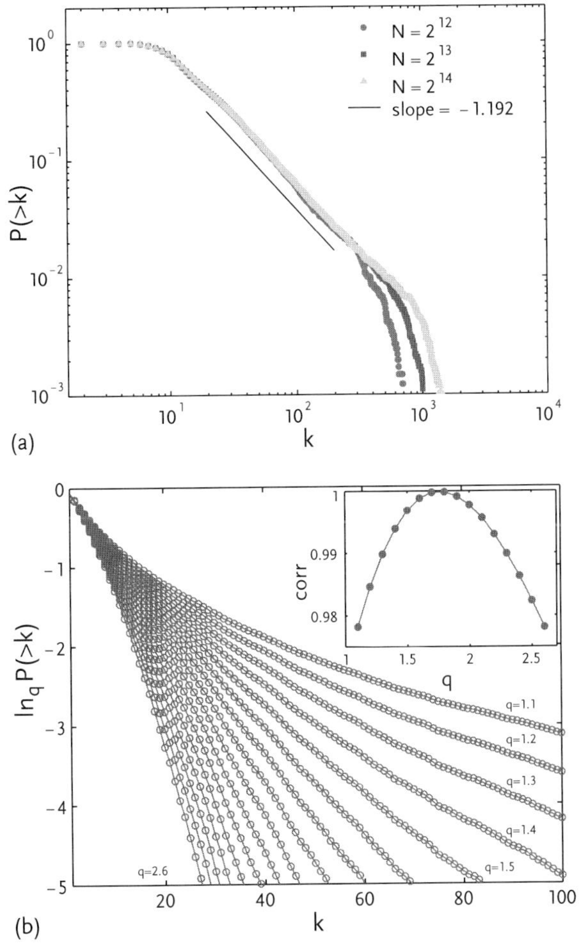

Figure 2.1 (a) Log-log representation of (cumulative) degree distributions for the model discussed in Section 2.1.1 for various system sizes; (b) q-logarithm of the (cumulative) distribution function from the same networks as a function of degree k. Clearly, there exists an optimum q that allows for an optimal linear fit. *Inset*: Linear correlation coefficient of $\ln_q P(\geq k)$ and straight lines for various values of q. The optimum value of q is obtained when $\ln_q P(\geq k)$ is optimally linear, i.e., where the correlation coefficient has a maximum. A linear $\ln_q$ means that the distribution function is a q-exponential; the slope of the linear function determines κ. In this example we get for the optimum $q = 1.84$, which corresponds to the slope $\gamma = 1.19$. Plot after [39].

2.1.2 Network Hamiltonians

A Hamiltonian provides an expression for the energy of a system based on its state variables. The introduction of a network Hamiltonian that determines the energy of a given network, under suitable conditions, formally acts as a temperature-dependent constraint. A Hamiltonian approach can thus be mapped into the category of constraint optimization problems, like, e.g., the traveling salesman problem, the spin-glass problem, etc. See, e.g., [10] for a recent treatment of these analogies; some analysis about optimization and networks can additionally be found in [11, 12]. Almost all of the microscopic models proposed to describe complex – growing or static – networks involve nonequilibrium and evolutionary elements, manifesting themselves in different procedures of preferential attachment [5, 13–15] or other structured rewirement schemes [8, 9, 16]. Further, these procedures often involve the need for nonlocal information, making an statistical mechanics approach impracticable if not impossible. So far, comparatively little has been done to understand complex networks from a purely classical statistical mechanics point of view where the phase space is not constrained. A few serious equilibrium approaches have been proposed [17–20] where topological properties of networks associated with specific Hamiltonians were studied. In [17] an equilibrium partition function of the form (2.10), see below, was established, giving an arbitrary degree distribution. In [19, 20] it was shown that along topological transitions scale-free networks can be recovered at a certain point in time during a relaxation process to equilibrium, implying that scale-free graphs are temporary configurations not typical for equilibrium.

Below we shall review a form of a network Hamiltonian leading to *ensemble averages* of networks that correspond to distinct topological "phases" of networks, depending on the temperature of the system. By increasing the temperature we observe a transition from starlike to scale-free to eventually Poissonian networks. Numerical evidence is presented showing that scale-free networks may indeed be obtained within a pure equilibrium approach, as suggested in [17]. Moreover, we demonstrate that the introduced Hamiltonian leads to nontrivial hierarchic features. The form of the Hamiltonian is derived from simple and general assumptions about individual linking energies of nodes, in a way that is standard in, e.g., economics. Nodes act as utility maximizers, in analogy to physical systems minimizing energy.

2.1.3
Network Ensembles

Many real-world networks are random networks. The simplest random network is the Erdös–Rényi classical random graph [1, 2], where a fixed number of links connect a fixed number of nodes. The linking probability for each node is the same. The resulting networks show that trivial clustering and degree distributions are of the mentioned Poissonian type. The simplest generalization of the classical random graph is to relax the condition that all linking probabilities are the same, but different nodes may be characterized by different probabilities of linking to others. Suppose that the linking probability of nodes is dictated by a distribution function. Then it becomes possible to view any network as a superposition of Poissonian random graphs with different average degrees, $\bar{k}$. Further, this superposition can be seen as a network realization of superstatistics, where a statistical system is not characterized by a single temperature, but by a "superposition of temperatures" [21]. We will explicitly review how the linking probability distribution of the individual nodes will translate to the macroscopic degree distribution of the network ensemble. It is possible to show that instead of the ensemble picture in [22], one could introduce an algorithm, which enables one to produce single realizations of such networks [23].

This ensemble approach is motivated by the observation of the vast abundance of similar degree distributions in a wide variety of static real-world networks, in which nodes may be individuals, animals, chemicals, companies, and so on. Often, nodes are characterized by a linking probability defining how they are connected by links. A linking probability can be nonuniform for all pairs of vertices, in general, unlike in the classical theory of random graphs. It conditions the states of nodes and introduces nontrivial correlation properties. This approach may offer the possibility of finding a common "driving force" leading to the fact that only a few "universality classes" of networks are observed in so many cases. This driving force would be featured by a linking probability distribution common in various systems. For example, many social and economic networks explicitly depend on a wealth of individuals, firms, banks, and so on. Suppose that the wealth distribution, which obeys a power law common in industrialized countries [25, 26], is associated with such a linking probability distribution, and assume that, under this probability, linking is entirely random. Then, the resulting network will have a connectivity distribution, which is uniquely determined by the linking probability distribution [22]. A common driving force could thus provide an understanding of why real static networks tend to belong to only a few "universality classes."

Statistical mechanics is an equilibrium concept. In a few cases it may be relaxed to sufficiently stationary situations; however, it is never wise to apply it to transient or growing systems. Thus in what follows we shall limit ourselves to nongrowing, undirected, and unweighted networks, where no global information exists about the system. The dynamics of the networks is rewirement only, which is considered to take place *rapidly*, such that an ensemble picture can be justified. Note that dozens of models have been introduced for *growing* networks, characterized by growth rules that can often be mapped to simple differential equations whose solutions characterize the degree distribution of such networks. For systems open to such a description, clearly, not much of statistical physics reasoning is necessary.

2.1.4 Some Definitions of Network Measures

In what follows we will use the measures of the clustering coefficient and the neighbor connectivity, which we define here. The clustering coefficient of node i, c_i is defined by

$$c_i = \frac{2e_i}{k_i(k_i - 1)} , \tag{2.2}$$

with e_i being the number of triangles node i is part of. $c(k)$ is obtained by averaging over all c_i with a fixed k. It has been noted that $c(k)$ contains information about hierarchies present in networks [27]. For Erdös–Rényi (ER) networks [1, 2], as well as for pure preferential attachment algorithms without the possibility of rewiring, the clustering coefficient $c(k)$ is constant. The global clustering coefficient is the average over all nodes, $C = \langle c_i \rangle_i$. A large global clustering coefficient is often indicative of a small-world structure [28]. The average nearest-neighbor connectivity (of the neighbors) of node i is

$$k_i^{nn} = \frac{1}{k_i} \sum_{j \text{ neighbor of } i} k_j . \tag{2.3}$$

Again, when plotted as a function of k, $k^{nn}(k)$ is a measure to assess the assortativity of networks. A rising function means assortativity, which is the tendency for well-connected nodes to link to other well-connected ones, while a declining function signals a disassortative structure.

This work is structured as follows. In Section 2.2 we revisit the matter of entropies of networks and review some work on a large class of models that result in q-exponential degree distributions. Section 2.3 introduces network Hamiltonians, ways of simulating their dynamics, and the topic of topological phase transitions in networks. In particular it focuses on restrictions of states imposed by rewirement dynamics dictated by an explicit form of a network Hamiltonian [16, 19, 20, 29, 30]. Introducing a Hamiltonian allows one to relate states to energy (or utility), which in turn enables one to study the thermodynamics of network systems. Further, it is possible to use thermodynamical relations to estimate the entropy of the system and compare it with the microscopic entropy. In Section 2.4 we review an ensemble picture of networks closely related to the recently introduced approach of superstatistics [21]. Following [22], we shall see how degree distributions change by relaxing the restriction of equal linking probabilities of nodes (pure random matrix case) to a *distribution* of linking probabilities.

2.2
Macroscopics: Entropies for Networks

Consider the microcanonical entropy of an unconstrained network with equal a priori probabilities for the microstates. The entropy can be written – in the spirit of Boltzmann – as the logarithm of the number of possible states

$$S_c = \log \binom{\binom{N}{2}}{L} , \tag{2.4}$$

representing the number of possibilities to distribute L indistinguishable links (particles) on $\frac{N(N-1)}{2}$ possible distinguishable positions in the symmetric adjacency matrix. If the N involved nodes are indistinguishable, then this has to be taken care of by an additional factor, $\frac{1}{N!}$, in the argument of the logarithm. As soon as networks become subjected to constraints such as specific linking rules or probabilities, or through the definition of a Hamiltonian (as will be discussed later), the evaluation of the number of possible states (adjacency matrices) becomes more difficult. Clearly, S_c is a nonextensive quantity, regardless of whether or not nodes are distinguishable. Note that this is in contrast to classical statistical mechanics, where only the case of distinguishable particles leaves room for nonextensivity. If S_c is plotted against the number of nodes N for a fixed density of links (average degree) $\lambda = \bar{k} = L/N$, it is obvious that it is not linear in N (Figure 2.2). Nonextensivity becomes more pronounced for more densely populated networks, which is intuitively clear to understand (Section 2.1.1). Networks, even in their simplest realizations, are prototypes of nonextensive systems. For more details and other aspects of nonextensivity in the context of networks, see, e.g., [31].

As indicated previously, most networks are not characterized by exponential degree distributions, but in many cases by q-exponentials, which are naturally related to so-called Tsallis entropies. A Tsallis entropy is an example of a nonextensive entropy. The concept of nonextensive statistical mechanics

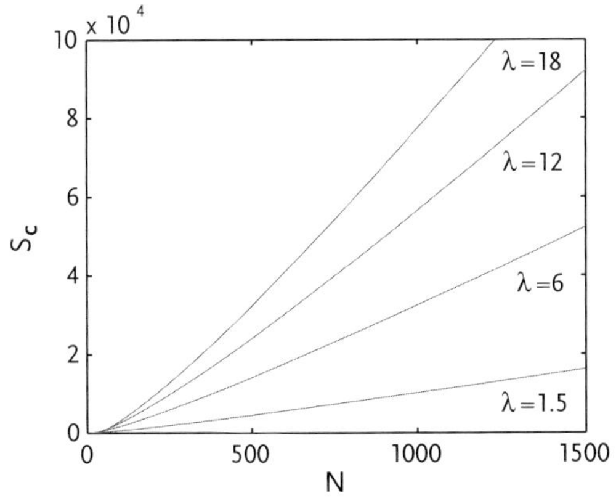

Figure 2.2 Dependence of entropy S_c on the number of (distinguishable) nodes, N, for various values of average degrees $\bar{k} \equiv L/N$. Plot after [30].

has been extremely successful in addressing critical phenomena, complex systems, and nonergodic systems [32–34]. Nonextensive statistical mechanics is a generalization of Boltzmann–Gibbs statistical mechanics, where entropy is defined as

$$S_q \equiv \frac{1-\int_0^\infty \mathrm{d}k\,[p(k)]^q}{q-1} \quad \text{with } q \to 1 \text{ limit } S_1 = S_{\mathrm{BG}} \equiv -\int_0^\infty \mathrm{d}k\,p(k)\ln p(k)\,, \tag{2.5}$$

where BG stands for Boltzmann–Gibbs. If – in the philosophy of the maximum entropy principle – one extremizes S_q under certain constraints, the corresponding distribution is the q-exponential. The issue of these constraints is an important one under current debate. Recently there has been concern that the so-called escort constraints, which have been advocated, e.g., in [35], are not physically relevant [24]. In order to use the usual constraints, it might be necessary to switch to a slightly modified Tsallis entropy, as given in Example 2 in [36, 37]. Another sign of the importance and ubiquity of q-exponentials in nature might be the fact that the most general Boltzmann factor for canonical ensembles (*extensive*) is the q-exponential, as was proved in [38]. Given the above characteristics of networks and the fact that a vast number of real-world and model networks show asymptotic power-law degree distributions, it seems almost obvious to expect a deep connection between networks and nonextensive statistical physics, governed by generalized entropies.

2.2.1
A General Set of Network Models Maximizing Generalized Entropies

Recently there has been brought forward a general model [39] that is a unification and generalization of network generation models presented in [7, 9]. The model in [7] captures preferential growing aspects of networks embedded in a *metric space*, while [9] introduces a static, self-organizing model with a sensitivity to an *internal metric* (chemical distance, Dijkstra distance). The rewiring scheme there can be thought of as having preferential attachment aspects in one of its limits [8] (see below), but none in the other limit. The model contains some of the most important network generation models as special cases. This will be made clear at the end of this section.

2.2.1.1 A Unified Network Model

In the model in [39] the network evolves in time as follows. At $t = 1$, the first node ($i = 1$) is placed at some arbitrary position in a metric space. The next node is placed isotropically on a sphere (in that space) of radius r, which is drawn from a distribution $P_G(r) \propto 1/r^{\alpha_G}$ ($\alpha_G > 0$, G stands for *growth*).

To avoid problems with the singularity, we impose a cutoff at $r_{\min} = 1$. The second node is linked to the first. The third node is placed again isotropically on a sphere with random radius $r \in P_G$; however, the center of the sphere is now the barycenter of the preexisting nodes. From the third added node on there is an ambiguity where the newly positioned node should link to. We choose a generalized preferential attachment process, meaning that the probability that the newly created node i attaches to a previously existing node j is proportional to the degree k_j of the existing node j, and inversely proportional to the metric (Euclidean) distance between i and j, denoted by r_{ij}. In particular, the linking probability is

$$p_{ij}^{A} = \frac{k_j / r_{ij}^{\alpha_A}}{\sum_{j=1}^{N(t)-1} k_j / r_{ij}^{\alpha_A}} , \tag{2.6}$$

where $N(t)$ is the number of nodes at time t. It is not necessary that at each timestep only one node enter the system, so we immediately generalize that a number of $\bar{n}$ nodes are produced and linked to the existing network with $\bar{l}$ links per timestep. Note that $\bar{n}$ and $\bar{l}$ can also be random numbers from an arbitrary distribution. For simplicity and clarity fix $\bar{n} = 1$ and $\bar{l} = 1$.

After every $\bar{\lambda}$ timestep, a different action takes place on the network. At this timestep the network does not grow but a pair of nodes, say i and j, merge to form one single node [8]. This node keeps the name of one of the original nodes, say, for example, i. This node now gains all the links of the other node j, resulting in a change of degree for node i according to

$$\begin{aligned} k_i &\to k_i + k_j - N_{\text{common}} , && \text{if } (i,j) \text{ are not first neighbors} , \\ k_i &\to k_i + k_j - N_{\text{common}} - 2 , && \text{if } (i,j) \text{ are first neighbors} , \end{aligned} \tag{2.7}$$

where N_{common} is the number of nodes that shared links to both i and j before the merger. In the case where i and j were first neighbors before the merger, i.e., they had been previously linked, the removal of this link will be taken care of by the term –2 in Equation (2.7). The probability that two nodes i and j will merge can be made distance dependent, as before. To stay close to the model presented in [9], we randomly choose node i with probability $\propto 1/N(t)$ and then choose the merging partner j with probability

$$p_{ij}^{M} = \frac{d_{ij}^{-\alpha_M}}{\sum_j d_{ij}^{-\alpha_M}} \quad (\alpha_M \geq 0) , \tag{2.8}$$

where d_{ij} is the shortest distance (path) on the network connecting nodes i and j. Obviously, changing α_M from 0 toward large values switches the model from the case where j is picked fully at random ($\propto 1/N(t)$) to a case where only the nearest neighbors of i will have a nonnegligible chance to get chosen for the merger. Note that the number of nodes is reduced by one at that

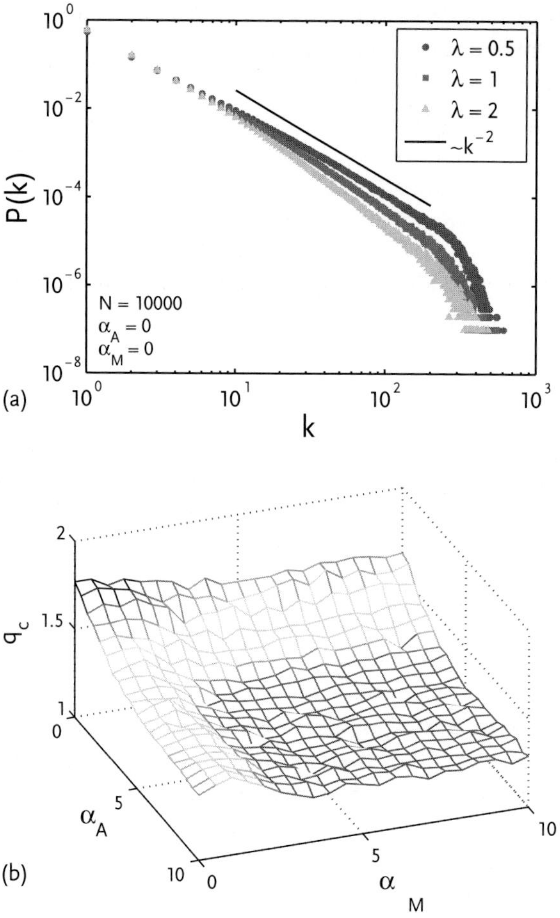

Figure 2.3 (a) Degree distribution $p(k)$ for $N = 10000$, $\alpha_A = 0$, $\alpha_M = 0$, and various values of $\bar{\lambda}$. This case corresponds to a growing network with preferential linking and random merging of nodes. (b) q_c values from q-exponential fits to the cumulative degree distributions $p(>k)$ for $\alpha_G = 1$, $N = 1000$, and $\bar{\lambda} = 1$. Plot after [39].

point. To keep the number of nodes constant at this timestep, a new node is introduced and linked with $\bar{l}$ of the existing nodes with probability given in Equation (2.6).

The relevant model parameters are the merging exponent α_M, the attachment exponent α_A, controlling the sensitivity of "distance" in the network, and the relative rate of merging and growing, $\bar{\lambda}$. The remaining parameters, α_G, $\bar{n}$, $\bar{l}$, and r_{min}, play marginal roles in the dynamics of the model. In Figure 2.3 typical degree distributions are shown for three typical values of $\bar{\lambda}$. Obviously, the distribution is dominated by a power-law decay (see details of the functional form below) ending in an exponential finite size cutoff for large k.

2.2.1.2 Famous Limits of the Unified Model

Depending on the choice of model parameters, some well-known networks result as natural limits of the model:

- *Soares et al. limit.* For the $\lim \bar{\lambda} \to \infty$ we have no merging, and α_M is an irrelevant parameter. The model corresponding to this limit has been proposed and studied in [7].
- *Albert–Barabasi limit.* The $\lim \bar{\lambda} \to \infty$ and $\lim \alpha_A \to 0$ gets rid of the metric in the Soares et al. model and recovers the original Albert–Barabasi preferential attachment model [5].
- *Kim et al. limits.* The limit $\lim \bar{\lambda} \to 0$ allows no preferential growing of the network. If at each timestep after every merger a new node is linked randomly with $\bar{l}$ links to the network, the model reported in [9] is recovered. The $\lim \bar{\lambda} \to 0$ model with $\lim \alpha_M \to 0$ ($\lim \alpha_M \to \infty$) recovers the *random* case ("neighbor" case) in [8].

2.2.1.3 Unified Model: Additional Features

Not only does the model produce q-exponential degree distributions over large regions of parameter space (Figure 2.3a), for many parameter settings the resulting networks exhibit nontrivial clustering (both global and as a function of the degree) and nontrivial neighbor connectivity. For details, see [39]. The fitted value from simulations of parameter q is shown in Figure 2.3b as a function of model parameters α_M and α_A.

2.3 Microscopics: Hamiltonians of Networks – Network Thermodynamics

We now turn to dynamical restrictions on rewirement imposed by a network Hamiltonian. Network Hamiltonians were recently introduced and studied, e.g., in [16, 18–20]. Here we study the Hamiltonian introduced in [29], which was originally motivated by a standard choice of utility functions in the socioeconomic literature [40, 41]. The basic idea is that a node has more utility (less energy) if it has a link to a more "important" node, where the importance of a node is proportional to its degree. The Hamiltonian that we take as a starting point – for a derivation see [29] – reads

$$\mathcal{H}(c) = -\sum_{\ell=1}^{L} \log(b + \Delta k) , \tag{2.9}$$

where c stands for the adjacency matrix, ℓ is the summation over all links, and Δk is the absolute value of the difference in degree between a pair of nodes, one of them being ℓ. b is a free parameter related to the detailed structure of the log-utility [40, 41] needed to derive this Hamiltonian. Given a Hamiltonian, the canonical ensemble, given by the partition function

$$Z(T, N, L) = \sum_{P(c)} \delta\left(L - \frac{\mathrm{Tr}(c^2)}{2}\right) \mathrm{e}^{-\beta\mathcal{H}(c)}, \tag{2.10}$$

using the usual definition of temperature $T \equiv 1/k\beta$, can be simulated, e.g., by the Metropolis algorithm: starting from an adjacency matrix c at time t, a graph $\hat{c}$ is generated by replacing a randomly chosen edge between nodes i and j with a new edge between randomly chosen, previously unconnected, nonidentical nodes k and m. In the next timestep c is replaced by $\hat{c}$ with probability $p_{\mathrm{replace}} = \min(1, \exp[-\beta(\mathcal{H}(\hat{c}) - \mathcal{H}(c))])$. This procedure guarantees that *every possible* configuration of the adjacency matrix is realized with the same *a priori* probability. Note that here nodes are distinguishable whereas links are not. Figure 2.4a shows the ensemble average of the total utility (energy) of the system, $\mathcal{U} \equiv -\sum_{i=1}^{N} U_i$, as a function of T, where U_i is the utility (energy) contribution of node i. One clearly finds a radical change in energy and a characteristic maximum of the specific heat (inset) at about $T_c = 0.8 - 0.9$, indicating the vicinity of a critical point.

2.3.1
Topological Phase Transitions

Changes in energy are associated with considerable restructuring of the underlying networks across the critical temperature. To study this in more detail, ensemble averages of degree distributions along the transition were calculated. Results are presented in Figure 2.4b. At low temperatures, networks are dominated by a so-called star structure, indicating that there are a few highly linked nodes that are practically linked to all the other nodes. For high temperatures the networks are classical random graphs [1,2], with their Poissonian degree distributions. These two phases are separated by a transition region from $T \sim 0.8$ to ~ 1.1, where the network structures are dominated by scale-free degree distributions, with asymptotic power laws, $P(k) \sim k^{-\gamma}$. For $b = 5$ and $T = 0.95$ the exponent is $\gamma \sim 3$. It was found that for values of $b > 5$ the transition gets sharper, for $b < 5$ the transition softens slightly. In the limit $b \to 0$ we studied various values of b and found marginal changes in the shape of the transition and in the slope of the resulting scale-free networks. The smallest value considered was $b = 10^{-7}$ and the change in the slope compared to $b = 1$ was < 0.05.

2.3.2 A Note on Entropy

The specific heat being the derivative of the internal energy (see inset to Figure 2.4a), can be used to compute the entropy of the system by

$$S(T) - S(T_0) = \int_{T_0}^{T} \mathrm{d}T' \frac{C(T')}{T'} \,. \tag{2.11}$$

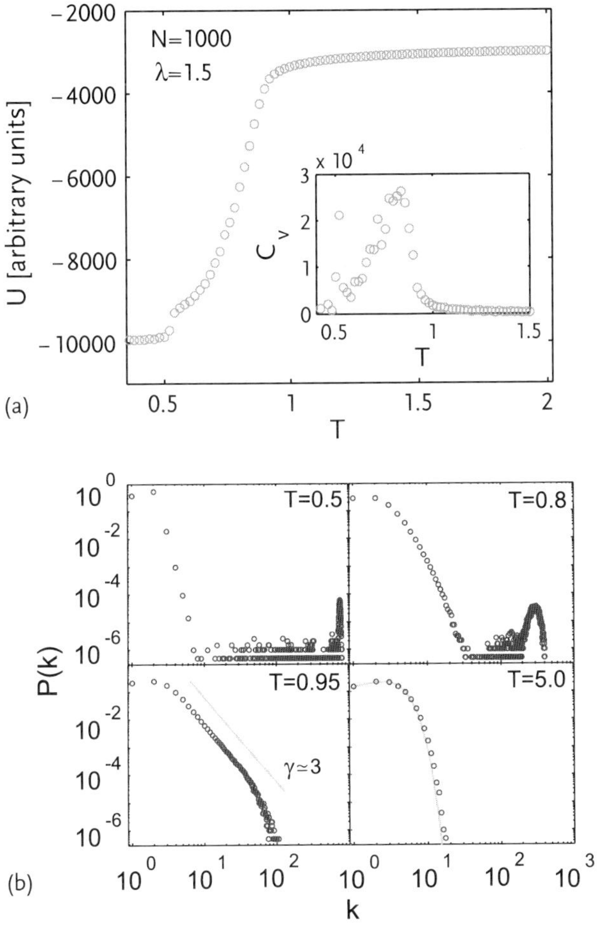

Figure 2.4 (a) Internal energy (utility) of the network system governed by the discussed Hamiltonian. The inset shows the specific heat. (b) Ensemble averages of degree distributions at different temperatures for $N = 10^3$, $\lambda = 3$, and $b = 5$. The line for $T = 5$ is the Poissonian $p(k) = \frac{e^{-\lambda}\lambda^k}{k!}$, expected for random graphs. Plot after [29].

Figure 2.5a shows the dependence of $S(T) - S(T_0)$ as a function of N. We fixed $T = 2$, $T_0 = 0.55$, and shifted by $S(T_0) = 750$. Note that this function is not strictly linear, as expected, and indicates nonextensive behavior. We compare this entropy with the microcanonical entropy S_c, from Equation (2.4) which is given by the solid line in Figure 2.5a for the same parameters L, N, and λ as used in the simulations. Note that the measured entropy is an empirical quantity, whereas S_c is a microscopic function of the distribution of microstates. The observed overlap is remarkable.

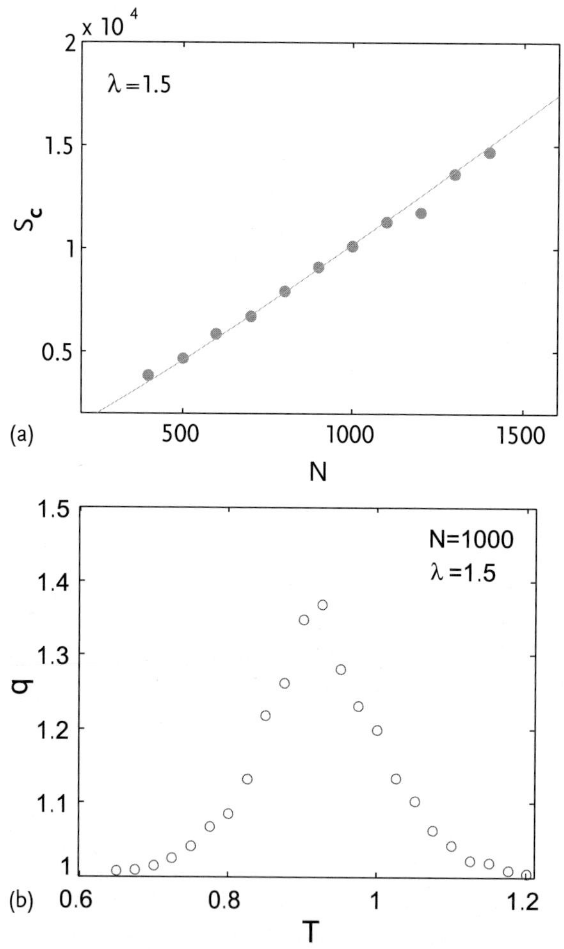

Figure 2.5 (a) Entropy (points) computed from specific heat as a function of N for $b = 5$; S_c (line) from Equation (2.4) is shown for comparison. (b) q-parameter from q-exponential fits to the degree distributions around the critical temperature. Plot after [30].

Within the temperature window $T \in [0.8, 1.1]$ degree distributions can be fitted reasonably well to q-exponentials [9, 31]. The temperature dependence of q is shown in Figure 2.5b around the critical point. Clearly, q has a maximum at the critical point and tends toward an ordinary exponential $(q \to 1)$, for high and low temperatures. The origin of these q-exponentials is not necessarily directly related to the nonextensiveness of the microcanonical density. Finally, it should be noted that scale-free networks were previously obtained by optimization with respect to different constraints (and by varying their relative weight) in [11, 12]. There, networks reach a stationary state after the optimization procedure and a scale-free region is found in a regime between random networks and stars.

2.4
Ensembles of Random Networks – Superstatistics

The simplest example for a (re-)wiring rule is the class of random networks introduced by Erdös and Rényi [1, 2], where N nodes are fully randomly connected by a set of L links. This corresponds to attaching a unique linking probability $\bar{p}$ to each node, i.e., $\bar{p}$ is the probability that any possible given pair of nodes is linked. The corresponding degree distribution is the binomial distribution

$$p(k) = \binom{N-1}{k} \bar{p}^k (1-\bar{p})^{N-1-k} , \qquad (2.12)$$

which in the large N limit reduces to the Poissonian distribution, $p(k) = \frac{e^{-\lambda}\lambda^k}{k!}$, where again $\lambda \equiv \bar{k} = \bar{p}(N-1) \sim L/N$. At this step one could introduce additional limitations on states, such as forbidding, e.g., self-linking, $c_{ii} = 0$. In the large N limit such limitations are of marginal importance.

In the Erdös–Rényi case each node has the same probability of being linked to any other node. In many realistic situations this is not the case and the linking probability of nodes is drawn from a distribution, $\Pi(p)$. For example, one could think of a social network of friends. The number of friends will depend on characteristics of the individual, such as being outgoing, rich, introverted, gregarious, or other factors that result in a linking probability. Assuming that the linking probability varies across nodes, and adopting a Gibbsean ensemble view on random networks, the degree distribution of Erdös–Rényi networks generalized to linking probabilities drawn from $\Pi(p)$, reads [22],

$$p(k) = \int_0^{\infty} d\lambda \, \Pi(\lambda) \frac{e^{-\lambda}\lambda^k}{k!} , \qquad (2.13)$$

for large enough N. This ensemble picture of generalized Erdös–Rényi networks is related to the concept of superstatistics [21], where the Boltzmann factor is generalized to a superposition of Boltzmann factors and where the inverse temperature is not unique across the system (or ensembles), but is represented by a distribution, $f(\beta)$. The generalized Boltzmann factor is

$$B(E) = \int_0^\infty d\beta f(\beta)\, e^{-\beta E}\,, \tag{2.14}$$

and with the partition function $Z \equiv \int_0^\infty dE B(E)$, one gets the probability $p(E) = \frac{1}{Z}B(E)$, given that the system relaxes much faster than β varies. The Boltzmann factor in Equation (2.14) is identified in our case with the Poisson factor in Equation (2.13). Equation (2.13) establishes a one-to-one relation of how linking probabilities that govern the linking dynamics are related to the resulting degree distributions. The problem left is to invert the relation to obtain the linking probability distribution, $\Pi(\lambda)$, which is sometimes called the *hidden variable distribution*, as a function of an observable degree distribution. In [22] this inversion was obtained exactly by using the methodology of quantum optics, which allows one to reduce the needed inverse "Poisson transform" to a 2-dimensional Fourier transform. The argument is briefly summarized here: recall the definition of an oscillator coherent state

$$|\alpha\rangle = \exp\left(-|\alpha|^2/2\right) \sum_{k=0}^{\infty} \frac{\alpha^k}{\sqrt{k!}}\, |k\rangle\,, \tag{2.15}$$

where the number states $|k\rangle$ form a Fock basis, $\left\{|k\rangle = (k!)^{-1/2}(\hat{a}^\dagger)^k|0\rangle\right\}_{k=0,1,2,\cdots}$, the ground state is defined by $\hat{a}|0\rangle \equiv 0$, and the usual oscillator algebra is given by $[\hat{a}, \hat{a}^\dagger] = 1$, $[\hat{a}, \hat{a}] = [\hat{a}^\dagger, \hat{a}^\dagger] = 0$. The number operator is $\hat{n} = \hat{a}^\dagger\hat{a}$, states are normalized, $\langle\alpha|\alpha'\rangle = \exp\left[-\frac{1}{2}\left(|\alpha|^2 + |\alpha'|^2\right) + \alpha'\alpha^*\right]$, and coherent states satisfy the overcompleteness relation, $\iint \frac{d^2\alpha}{\pi}|\alpha\rangle\langle\alpha| = 1$. This means that for the inner product of state and number state, $\langle k|\alpha\rangle = \exp\left(-|\alpha|^2/2\right)\frac{\alpha^k}{\sqrt{k!}}$, and for the phonon number $|\langle k|\alpha\rangle|^2 = \exp^{-|\alpha|^2}\frac{\left(|\alpha|^2\right)^k}{k!}$. The density operator is used in the Sudarshan–Glauber representation, $\hat{\varrho} = \iint d^2\alpha P(\alpha)|\alpha\rangle\langle\alpha|$, $\alpha P(\alpha) = 1$. Assuming that the density matrix depends on the number only, $\hat{\varrho} = f(\hat{n})$

$$f(k) = \langle k|\hat{\varrho}|k\rangle = \iint d^2\alpha P(\alpha)\, |\langle k|\alpha\rangle|^2 = \iint d^2\alpha P(\alpha)\frac{\exp^{-|\alpha|^2}\left(|\alpha|^2\right)^k}{k!}\,. \tag{2.16}$$

Since $\hat{\varrho}$ is a function of the number, $P(\alpha) = P(r)$ holds, and one gets $f(k) = \int_0^\infty d\lambda \Pi(\lambda)\frac{\lambda^k e^{-\lambda}}{k!}$, with $\Pi(\lambda) \equiv \pi P(|\alpha|)$ and $r^2 = \lambda$. This methodology now tells one how to perform the inversion of the problem: Compute

$$\langle -\beta|\hat{\varrho}|\beta\rangle = \iint d^2\alpha P(\alpha)\langle -\beta|\alpha\rangle\langle\alpha|\beta\rangle = e^{-|\beta|^2}\iint d^2\alpha P(\alpha)\, e^{-|\alpha|^2}\, e^{\beta\alpha^* - \beta^*\alpha} \tag{2.17}$$

and after an inverse Fourier transform get

$$P(\alpha) = e^{|\alpha|^2}\frac{1}{\pi^2}\iint d^2\beta\langle -\beta|\hat{\varrho}|\beta\rangle\, e^{|\beta|^2}\, e^{\alpha\beta^* - \alpha^*\beta}\,, \tag{2.18}$$

where $\Pi(\lambda) = \pi P(\alpha)$. The inversion problem thus reduces to a three-step process:

(1) Take a density matrix $\hat{\varrho}$ that corresponds to a given degree distribution.
(2) Compute $\langle -\beta|\hat{\varrho}|\beta\rangle$.
(3) Perform a double integral.

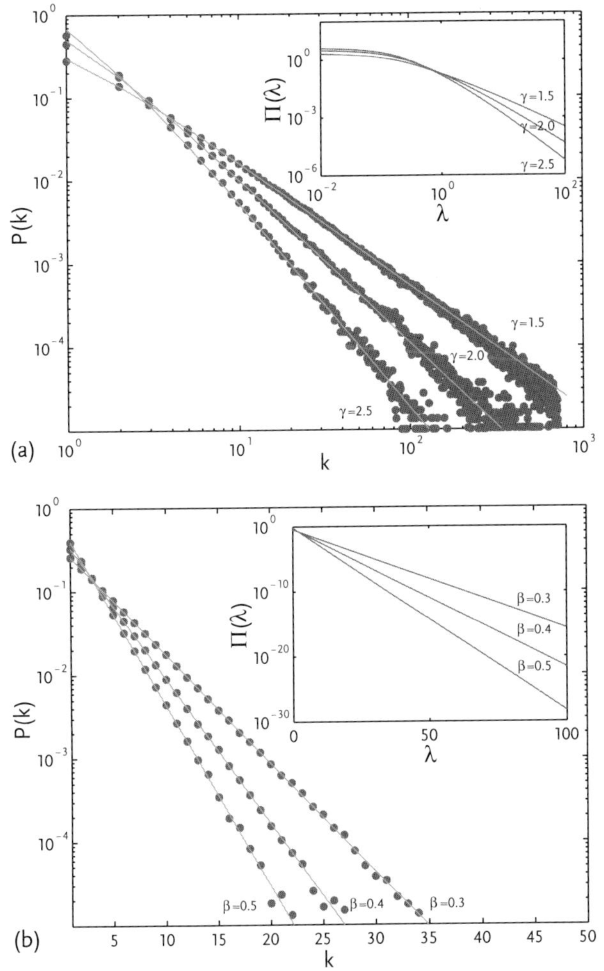

Figure 2.6 Degree distributions from a numerical simulation of networks where the linking probabilities for nodes are restricted to the distribution of Equation (2.20) (a) and an exponential distribution (b). Points are simulation results, lines correspond to theoretical predictions. Plot after [23].

As an example, the situation for scale-free networks was illustrated in [22]. Fix the density matrix to a q-exponential distribution, now for simplicity in other notation,

$$\hat{\varrho} = \frac{A}{(\hat{n} + k_0)^\gamma} \quad \langle k|\hat{\varrho}|k\rangle = \frac{A}{(k + k_0)^\gamma} = f(k) \equiv p(k) , \tag{2.19}$$

where $A^{-1} = \sum_{k=0}^{\infty} \frac{1}{(k+k_0)^\gamma} = \zeta(\gamma, k_0)$, and $\zeta(\gamma, k_0)$ is the Hurwitz generalized zeta function that is finite iff $\gamma > 1$. By use of the Mellin transformation and performing the above steps one gets the exact result

$$\Pi(\lambda) = \frac{A}{\Gamma(\gamma)} e^\lambda \int_0^\infty dt\, t^{\gamma-1} e^{(1-k_0)t-\lambda e^t} . \tag{2.20}$$

No closed form for this integral exists; however, from its moments the asymptotic behavior can be shown to be

$$\Pi(\lambda) \sim \frac{1}{\lambda^\gamma} , \tag{2.21}$$

i.e., the hidden variable distribution, Π, which is now *un-hidden*. The result is that it has the same asymptotic power as the degree distribution. For exponential networks the same procedure, starting from $\hat{\varrho} = A \exp(-\gamma \hat{n})$ and $\langle k|\hat{\varrho}|k\rangle = A \exp(-\gamma k)$, leads to an exact result for the hidden variable distribution $\Pi(\lambda) = A e^\gamma e^{-\lambda(e^\gamma - 1)}$. This further implies that a flat degree distribution ($\gamma = 0$) leads to flat linking probabilities. In Figure 2.6 we show the examples for the scale-free and the exponential cases.

Starting from Equation (2.13) the inversion problem can alternatively be solved by expanding

$$\Pi(\lambda) = \sum_{l=0}^{\infty} c_l L_l(\lambda) , \quad \lambda^k = k! \sum_{i=0}^{k} (-1)^i \binom{k}{i} L_i(\lambda) , \tag{2.22}$$

where L_i are Laguerre polynomials. The inversion becomes possible because the inverse of the Pascal triangle (in matrix notation) is again a Pascal triangle. The coefficients are then simply computed to be

$$c_l = \sum_{k=0}^{l} (-1)^k \binom{l}{k} p(k) . \tag{2.23}$$

This again establishes a direct link between a microscopic linking probability for the individual nodes and the bulk properties of the related networks.

2.5 Conclusion

We have discussed the question of how to formulate network theory – the theory of how a set of N nodes can get linked through a set of L links – in terms of a statistical mechanics approach. In particular we were interested in clarifying if and how it would be possible to relate microscopic linking rules between individual nodes to the bulk properties of networks, such as degree distributions or clustering. Further we search for the possibility to formulate a meaningful "thermodynamics" of networks. We focused our attention on stationary, nongrowing networks.

We addressed the question from three independent directions. First, starting from considerations of network entropies we discussed a wide class of network relinking models that maximize generalized entropies, such that the observed degree distributions follow certain functional forms. In particular we were interested in the frequently encountered q-exponential degree distributions. Second, we introduced an example of a network Hamiltonian that allowed us to introduce a "temperature" to the system. We demonstrated explicitly how the introduction of an energy dependence of states allows one to consistently use standard concepts of thermodynamics. With this framework we were able to look at thermodynamical quantities and extract a "thermodynamical" entropy, which we compared to the microscopic entropy definition. We found that networks can undergo a phase transition of degree distributions, from starlike networks at low temperatures to random graphs with Poissonian degree distributions at high temperatures. In the critical temperature region where the transition takes place, degree distributions seem to be again compatible with q-exponentials. Third, we discussed an ensemble view of networks where networks are seen as superpositions of random networks, in close analogy to superstatistics. These superpositions can lead to any desired degree distribution, or in other words, a given linking probability distribution for the individual nodes can be exactly linked to the emerging degree distribution.

In these approaches it was demonstrated explicitly that it is indeed reasonable to consider the possibility of a statistical mechanics – maybe even a thermodynamics – of networks.

I am deeply indebted to my colleagues Christoly Biely, Sumiyoshi Abe, Constantino Tsallis, and Fragiskos Kyriakopoulos, with whom I had the great pleasure of exploring these issues.

References

1 Erdös, P., Rényi, A., On random graphs, *Publ. Math. Debrecen* 6 (**1959**), 290.

2 On the evolution of random graphs, *Publ. Math. Inst. Hung. Acad. Sci.* 5 (**1960**), 17.

3 Barabási, A.-L., Statistical mechanics of complex networks, *Rev. Mod. Phys.* 74 (**2002**), 47.

4 Dorogovtsev, S., Mendes, J.F.F., *Evolution of Networks: From Biological Nets to the Internet and WWW*, Oxford University Press, Oxford, **2003**.

5 Barabási, A.-L., Albert R., Emergence of scaling in random networks, *Science* 286 (**1999**), 509–512.

6 Albert R., Barabási, A.-L., Topology of evolving networks: local events and universality, *Phys. Rev. Lett.* 85 (**2000**), 5234–5237.

7 Soares, D.J.B., Tsallis, C., Mariz, A.M., da Silva, L.R., Preferential attachment growth model and nonextensive statistical mechanics, *Europhys. Lett.* 70 (**2005**), 70–76.

8 Kim, B.J., Trusina, A., Minnhagen, P., Sneppen, K., Self organized scale-free networks from merging and regeneration, *Eur. Phy. J. B* 43 (**2005**), 369–372.

9 Thurner, S., Tsallis, C., Nonextensive aspects of self-organized scale-free gas-like networks, *Europhys. Lett.* 72 (**2005**), 197–203.

10 Parisi, G., Statistical mechanics of optimization problems, arxiv:cond-mat/0602350 (**2006**).

11 Valverde, S., Ferrer, R., Solé, R.V., Scale-free networks from optimal design, *Europhys. Lett.* 60 (**2002**), 512.

12 Ferrer, R., Solé, R.V., Optimization in complex networks, *Lect. Notes Phys.* 625 (**2003**), 114.

13 Rosvall, M., Sneppen, K., Modeling dynamics of information networks, *Phys. Rev. Lett.* 91 (**2003**), 178701.

14 Ramasco, J.J., Dorogovtsev, S.N., Pastor-Satorras, R., Self-organization of collaboration networks, *Phys. Rev. E* 70 (**2004**), 036106.

15 Caldarelli, G., Capocci, A., De Los Rios, P., Munoz, M.A., Scale-free networks from varying vertex intrinsic fitness, *Phys. Rev. Lett.* 89 (**2002**), 25.

16 Baiesi, M., Manna, S.S., Scale-free networks from a Hamiltonian dynamics, *Phys. Rev. E* 68 (**2003**), 047103.

17 Berg, J., Lässig, M., Correlated random networks, *Phys. Rev. Lett.* 89 (**2002**), 228701.

18 Park, J., Newman, M.E.J., The statistical mechanics of networks, *Phys. Rev. E* 70, (**2004**), 066117.

19 Farkas, I., Derenyi, I., Palla, G., Vicsek, T., Equilibrium statistical mechanics of network structures, *Lect. Notes Phys.* 650 (**2004**), 163.

20 Palla, G., Derenyi, I., Farkas, I., Vicsek, T., Statistical mechanics of topological phase transitions in networks, *Phys. Rev. E* 69 (**2004**), 046117.

21 Beck, C., Cohen, E.G.D., Superstatistics, *Physica A* 322 (**2003**), 267–275.

22 Abe, S., Thurner, S., Complex networks emerging from fluctuating random graphs: analytic formula for the hidden variable distribution, *Phys. Rev. E* 72 (**2005**), 036102.

23 Abe, S., Thurner, S., Hierarchical and mixing properties of static complex networks emerging from fluctuating classical random graphs, *Int. J. Phys. C* 17 (**2006**), 1303–1311.

24 Abe, S., Instability of q-expectation value, arxiv:0806.3934 (**2008**).

25 Clementi, F., Gallegati, M., Power law tails in the Italian personal income, *Physica A* 350 (**2005**), 427.

26 Souma, W., Nirei, M., Empirical study and model of personal income, arxiv:physics/0505173 (**2005**).

27 Ravasz, E., Barabasi, A.-L., Hierarchical organization in complex networks, *Phys. Rev. E* 67 (**2003**), 026112.

28 Watts, D.J., Strogatz, S.H., Collective dynamics of 'small-world' networks, *Nature* 393 (**1998**), 440.

29 Biely, C., Thurner, S., Statistical mechanics of scale-free networks at a critical point: complexity without irreversibility?, *Phys. Rev. E* 74 (**2006**), 066116.

30 Thurner, S., Biely, C., Two statistical mechanics aspects of complex networks, *Physica A* 372 (**2006**), 346–353.

31 Thurner, S., Nonextensive statistical mechanics and complex scale-free networks, *Eur. Phys. News* 36 (**2005**), 218.

32 Tsallis, C., Possible generalization of Boltzmann–Gibbs statistics, *J. Stat. Phys.* 52 (**1988**), 479–487.

33 Gell-Mann, M., Tsallis, C., *Nonextensive Entropy – Interdisciplinary Applications*, Oxford University Press, New York, **2004**.

34 Tsallis, C., Gell-Mann, M., Sato, Y., Asymptotically scale-invariant occupancy of phase space makes the entropy S_q extensive, *Proc. Natl. Acad. Sc. USA* 102 (**2005**), 15377–15382.

35 Tsallis, C., Mendes, R.S., Plastino, A.R., The role of constraints within generalized nonextensive statistics, *Physica A* 261 (**1998**), 534–554.

36 Hanel, R., Thurner, S., Generalized Boltzmann factors and the maximum entropy principle: entropies for complex systems, *Physica A* 380 (**2007**), 109–114.

37 Hanel, R., Thurner, S., Entropies for complex systems: generalized-generalized

entropies, in *Complexity, Metastability, and Nonextensivity*, AIP 965, **2007**, pp. 68–75.

38 Hanel, R., Thurner, S., Derivation of power-law distributions within standard statistical mechanics, *Physica A* 351 (**2005**), 260–268.

39 Thurner, S., Kyriakopoulos, F., Tsallis, C., Unified model for network dynamics exhibiting nonextensive statistics, *Phys. Rev. E* 76 (**2007**), 036111.

40 Ingersoll, J.E., *Theory of Financial Decision Making*, Rowman & Littlefield Publishers, **1987**.

41 von Neumann, J., Morgenstern, O., *Theory of Games and Economic Behavior*, Princeton University Press, Princeton, **1944**.

3
A Simple Integrated Approach to Network Complexity and Node Centrality

Danail Bonchev

3.1
Introduction

The traditional manner in which information theory [1–3] is applied to characterize the structure and complexity of graphs [4–9] is based on the partitioning of a certain set of graph elements N (vertices, edges, two-edge subgraphs, etc.) into k equivalence classes, having a cardinality of $N_1, N_2, \ldots, N_k$. This enables defining the probability of a randomly chosen element to belong to the class i as the ratio of the class cardinality and that of the entire set of graph elements, $p_i = N_i/N$. The Shannon equations for the mean and total information, $\bar{I}(\alpha)$ and $I(\alpha)$,

$$\bar{I}(\alpha) = -\sum_{i=1}^{k} p_i \log_2 p_i \,, \tag{3.1}$$

$$I(\alpha) = N\log_2 N - \sum_{i=1}^{k} N_i \log_2 N_i \tag{3.2}$$

are then directly applicable.

Depending on the selected criterion of equivalence α, different classes and correspondingly different information measures can be ascribed to the graph. The simplest example is the partitioning of the graph vertices according to the vertex degree (the number of the nearest neighboring vertices), the first subset of vertices including those of degree 1, then those of degree 2, 3, etc. Accounting for the second, third, and so forth neighbors [10, 11] is another option, as is the most rigorous definition of vertex equivalence, the equivalence classes being the orbits of the automorphisms group of the graph. The symmetry operation in the latter case is the interchange of vertices without breaking any adjacency relationship [12].

While useful for a variety of applications, this manner of applying information theory to graphs is not sensitive to structural details and frequently produces the same information content of nonisomorphic graphs. A viable alternative, offered by Bonchev and Trinajstić in 1977 [13–15], was to con-

Analysis of Complex Networks: From Biology to Linguistics. Edited by Matthias Dehmer and Frank Emmert-Streib

ISBN: 978-3-527-32345-6

sider weighted distributions of graph elements. In this case, one partitions not a set of graph elements but the overall value of some additive property of these elements. The simplest and most basic properties of graph vertices are their degree a_i, determining globally the graph total adjacency $A = \Sigma_i a_i$, and their distances d_i, determining the graph total distance D in the integer graph metric, $D = \Sigma_i d_i$. The vertex degree distribution $A\{a_1, a_2, \ldots, a_k\}$ and the distance degree distribution $D\{d_1, d_2, \ldots, d_k\}$ are the most basic graph distributions. Defining the probability $p(a_i)$ and $p(d_i)$ for a vertex to have a degree a_i and distance d_i, respectively, with $\Sigma_i p(a_i) = 1$ and $\Sigma_i p(a_i) = 1$, one enables the application of Shannon's equations. Two sensitive graph descriptors; the average information on the vertex degree distribution, $\bar{I}_{\mathrm{vd}}$, and the average information on the vertex distance distribution, $\bar{I}_{\mathrm{d}}$ (and their corresponding total descriptors) – are thus defined [13–15]:

$$\bar{I}_{\mathrm{vd}} = -\sum_{i=1}^{k} \frac{a_i}{A} \log_2 \frac{a_i}{A}\,, \tag{3.3}$$

$$\bar{I}_{\mathrm{d}} = -\sum_{i=1}^{k} \frac{d_i}{D} \log_2 \frac{d_i}{D}\,. \tag{3.4}$$

This approach to a more adequate characterization of the graph structure has found broad applications in studying molecular topology [16–21]. Recently, the expertise accumulated in chemical graph theory, along with similar contributions of graph theory to the social sciences [22–24] and computer sciences [25], have found broad applications to characterize complex nonrandom systems within the framework of network theory [26–28]. The major reason for these new avenues is the fact that complex systems are *relational*; their functioning as a whole is possible because of the relations existing between their basic elements. Networks (graphs) are the natural language to describe these complex dynamic systems, which promises to change important areas of science and technology. It is expected that during the next 10 to 15 years the major biological functions and human diseases will be redefined in terms of networks, and a similar paradigm change is expected in the area of drug design.

This new scientific revolution has prompted extensive contributions to network theory. Information theory descriptors like those given by Equations (3.3) and (3.4) have also been found useful in the characterization of network complexity [20, 29]. This paper reports another information measure of network complexity. The idea for it was generated under the influence of two major observations in complex networks. The most highly connected nodes in the network, termed "hubs" [30], are vital for the existence of the system described, e.g., the deletion of such a highly connected gene or protein in the living cell is lethal. More generally, the distribution of node degrees was shown to be scale-free, with few hubs, many sparsely connected nodes, and with all node degrees obeying in the majority of cases a power distribution law. The second general feature of all complex dynamic networks is their small size, the small degree of separation of any two nodes in the network. This property, termed "small-worldness" [30], indicates that the specific distribution of distances in networks is another network benchmark, along with the specific distribution of node degrees. It is therefore natural to try to integrate the two fundamental network features, connectivity and distances and their distributions, into single-number descriptors.

3.2 The Small-World Connectivity Descriptors (Bourgas Indices, BI)

The simplest way to integrate the information on network connectivity and distances is to use the ratio of the network total adjacency and total distance, A and D [29, 32]:

$$B1 = \frac{A}{D}. \tag{3.5}$$

The rationale for Equation (3.5) is straightforward: the graph (or network) complexity increases with the increase in the number of edges (links) E, where for undirected graphs $A = 2E$, and with the more compact "small-world" type of structure organization, that is with a smaller graph radius, and smaller total graph distance D.

The $B1$ index is a fast approximate measure of graph complexity, which will be illustrated by some examples in Figures 3.1 and 3.2. However, due to the fact that both A and D are additive functions, the same values could emerge from different vertex degree and vertex distance distributions. This partial degeneracy of $B1$ can be avoided (or at least very strongly reduced) if instead of using directly A and D, one proceeds with such vertex degree/vertex distance ratios b_i for each graph vertex and then sums up over all vertices to define the second small-world connectivity index $B2$:

$$b_i = \frac{a_i}{d_i}, \tag{3.6}$$

$$B2 = \sum_{i=1}^{k} b_i = \sum_{i=1}^{k} \frac{a_i}{d_i}. \tag{3.7}$$

The distribution of the vertex b_i descriptors, $B2\{b_1, b_2, b_3, \ldots, b_k\}$, can then be considered an important integrated distribution of vertex connectivity and distances. Moreover, when reordered by descending b_i values, this distribution will also show a new type of vertex-centric ordering (see next section). The information measure $B3$ defined on this distribution:

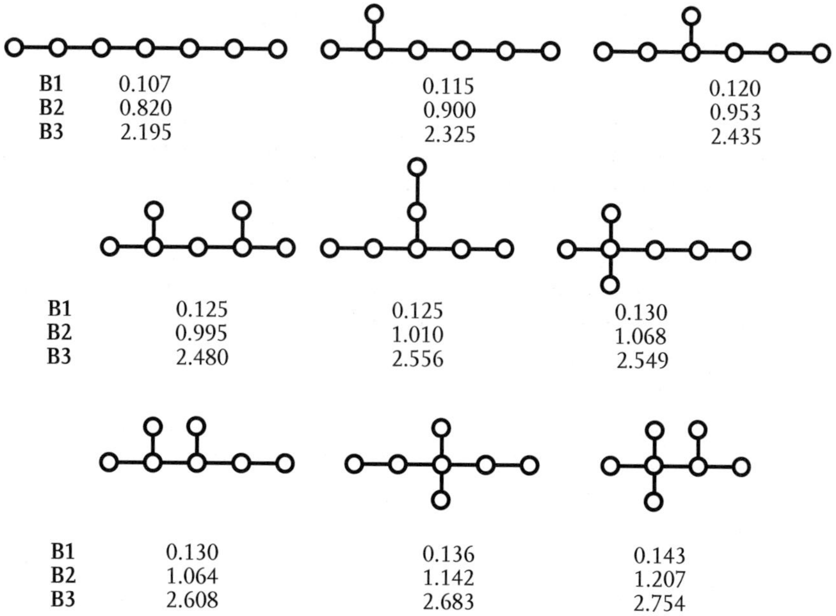

Figure 3.1 Nine acyclic graphs with seven nodes ordered according to their increasing complexity as characterized by the three Bourgas indices $B1$–$B3$.

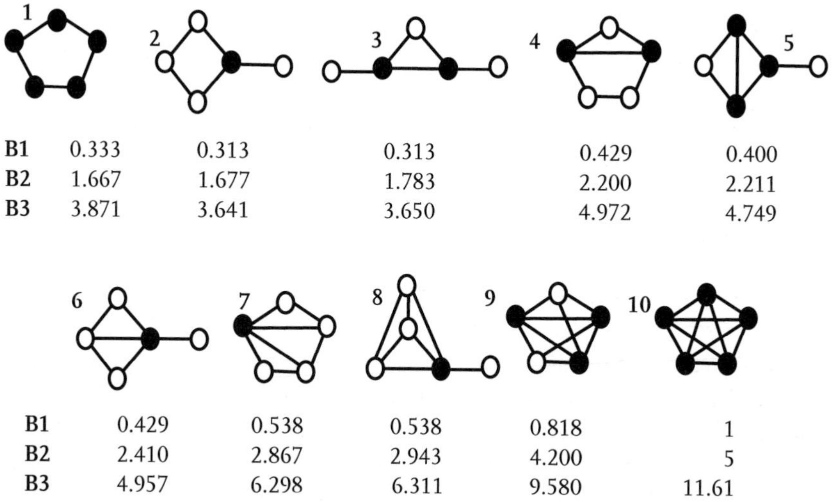

Figure 3.2 Ten cyclic graphs with five nodes ordered according to their increased complexity as reflected by the values of the three small-world connectivity indices. The black points mark the central vertices, as defined by the values of their b_i indices.

$$B3 = B2 \log_2 B2 - \sum_{i=1}^{k} b_i \log_2 b_i = B2 \log_2 B2 - \sum_{i=1}^{k} \frac{a_i}{d_i} \log_2 \frac{a_i}{d_i} \quad (3.8)$$

thus integrates connectivity, small-world, and centrality features of graphs and networks and might be expected to find application for characterizing network complexity. For brevity, the small-world connectivity indices $B1$–$B3$ were also named after the author's native city, "Bourgas indices".

Analysis of the topological descriptors $B1$ and $B2$, including derived formulas for the major classes of graphs (paths, stars, monocycles, and complete graphs), was presented in two preceding publications of the author [29, 32]. Equation (3.8) allows analyzing the basic trend of the integrated information index $B3$, as well. At a constant number of vertices, adding another edge increases the b_i terms of the vertices incident to that edge, not only because the corresponding vertex degrees a_i increase by one, but also because the distances d_i of the two vertices decrease, as a consequence of the formation of a new cycle.

Figure 3.1 shows all nine acyclic graphs of seven vertices, ordered according to the increasing number of complexifying elements. Thus the total number of branches increases from zero to three, and the number of branches incident to a vertex increases from zero to two, and the length of a branch increases from one to two. As previously mentioned, the $B1$ index cannot always distinguish all compared graphs, and it produces here two pairs of degenerate values (0.125 and 0.130). The $B2$ and $B3$ indices are nondegenerate and order the nine graphs almost identically, the only exception being the structures with $B1 = 0.130$, which have close but reversely ordered values of the $B2$ index.

Similar comparative analysis of ten cyclic graphs with the same number of vertices is presented in Figure 3.2. The graphs ordering provided by the increasing values of the $B1$–$B3$ indices again reflects the increase in the number of complexifying elements, the major factor being the number of cycles, followed at the same number of cycles by the number of branches, and the presence of vertex(es) of maximum degree. The $B1$ index produced pairs of degenerate values for graphs having the same number of cycles. It also frequently diverges from the very consistent complexity ordering produced by $B2$ and $B3$, which distinguish all ten graphs. One may conclude that while the $B1$ index can provide reasonably good approximate estimates of complexity for graphs with considerable structural differences, in cases of considerable graph similarity one should use only the $B2$ and $B3$ indices.

3.3
The Integrated Centrality Measure

The classical graph center [4] is the vertex(es) with the lowest eccentricity $e(i)$, that is, the lowest maximum vertex distance:

$$e(i) = \max d(i) = \min \,. \tag{3.9}$$

This definition often produces multiple central points and can also be considered as the first step in a hierarchical definition for a graph center [33, 34], the next hierarchical criterion (applied at the same minimal eccentricity) being the minimum vertex distance:

$$\text{for } e(i) = e(j), d(i) < d(j) \text{ for } i - \text{center}\,. \tag{3.10}$$

The vertex distance itself defines a different graph center called the *median* [35]. Its inverse has been used to define a centrality measure in the social sciences called the *closeness centrality, CC* [36]:

$$CC(i) = \frac{V-1}{d_i}\,, \tag{3.11}$$

where V is the number of graph vertices.

The *betweenness centrality* [37, 38] of a vertex i, $BC(i)$, is defined as the fraction of the shortest paths that traverse that vertex:

$$BC(i) = \frac{N_{\text{paths}}(i)}{N_{\text{paths}}}\,. \tag{3.12}$$

The idea behind this centrality concept is that vertices that occur more frequently on the shortest paths between other vertices have higher centrality than those that do not.

The simplest centrality measure is based on the vertex degree and is called the *degree centrality* [37]. The *integrated centrality* index $b(i) = a_i/d_i$ reported here combines in a single descriptor the properties of degree centrality and closeness centrality:

$$b(i) = \frac{a_i}{V-1} CC(i)\,. \tag{3.13}$$

Analysis has shown that for small graphs, the vertex distances in which are relatively close, the vertex degree is the dominant factor and the $b(i)$ index is more hub-oriented and closer to the degree centrality. In contrast, as the graph size increases, the distances become the dominant factor, and the integrated centrality index will behave more like the closeness centrality. In fact, the basic advantage of the new centrality measure is in the fact that it emerged within the framework of a more general approach aimed at describing graph complexity. Thus, the integrated centrality index reveals the centrality component of the graph and network complexity.

References

1 C. Shannon, W. Weaver, *Mathematical Theory of Communication*, University of Illinois Press, Urbana, IL, **1949**.

2 H. Kastler, Ed., *Essays on the Use of Information Theory in Biology*, University of Illinois Press, Urbana, IL, **1953**.

3 L. Brillouin, *Science and Information Theory*, Academic Press, New York, **1956**.

4 F. Harary, *Graph Theory*, 2nd ed., Addison-Wesley, Reading, MA, **1969**.

5 N. Trinajstić, *Chemical Graph Theory*, 2nd ed., CRC Press, Boca Raton, FL, **1992**.

6 N. Rashevsky, *Bull. Math. Biophys.* **1955**, 17, 229–235.

7 E. Trucco, *Bull Math. Biophys.* **1956**, 18, 129–135.

8 E. Trucco, *Bull. Math. Biophys.* **1956**, 18, 237–253.

9 A. Mowshowitz, *Bull. Math. Biophys.* **1968**, 30, 175–204.

10 H.L. Morgan, *J. Chem. Docum.* **1965**, 5, 107–113.

11 V.R. Magnuson, D.K. Harris, S.C. Basak, in: *ChemicalApplications of Topology and Graph Theory*, R.B. King (Ed.), Elsevier, Amsterdam, pp. 178–191, **1983**.

12 D. Bonchev, D. Kamenski, V. Kamenska, *Bull. Math. Biol.* **1976**, 38, 119–133.

13 D. Bonchev, N. Trinajstić, *J. Chem. Phys.* **1977**, 67, 4517–4533.

14 D. Bonchev, N. Trinajstić, *Intern. J. Quantum Chem. Symp.* **1982**, 16, 463–480.

15 D. Bonchev, *Information-Theoretic Indices for Characterization of Chemical Structures*. Research Studies Press, Chichester, U.K., **1983**.

16 D. Bonchev, O. Mekenyan, N. Trinajstić, *J. Comput. Chem.* **1981**, 2, 127–148.

17 D.H. Rouvray, R.B. King, Eds., *Topology in Chemistry. Discrete Mathematics of Molecules*, Horwood, Chichester, U.K., **2002**.

18 *Topological Indices and Related Descriptors*, J. Devillers and A.T. Balaban, Eds., Gordon and Breach, Reading, U.K., **1999**.

19 D. Bonchev, *Bulg. Chem. Commun.* **1995**, 28, 567–582.

20 D. Bonchev, in Mathematical Chemistry Series, Vol. 7, *Complexity in Chemistry*, D. Bonchev, D.H. Rouvray, Eds., Taylor & Francis, pp. 155–187, **2003**.

21 D. Bonchev, in *Encyclopedia of Complexity and System Science*, R. Meyers, Ed., Springer, Heidelberg, Germany, in press. 2

22 L. Freeman, *The Development of Social Network Analysis*, Empirical Press, Vancouver, **2006**.

23 S. Wasserman, K. Faust, *Social Network Analysis: Methods and Applications*. Cambridge University Press, Cambridge, U.K., **1994**.

24 B. Wellman, S.D. Berkowitz, Eds., *Social Structures: A Network Approach*. Cambridge University Press, Cambridge, U.K., **1988**.

25 A. Brandstädt, Van Bang Le, Eds., *Graph-Theoretic Concepts in Computer Science*, Lecture Notes in Computer Science , Vol. 2204, **2001**.

26 A.-L. Barabási, *Linked. The New Science of Networks*, Perseus, Cambridge, MA, **2002**.

27 S.N. Dorogovtsev, J.F.F. Mendes, *Adv. Phys.* **2002**, 51, 1079–1187.

28 M. Newman, A.-L. Barabási, D.J. Watts, Eds., *The Structure and Dynamics of Networks*, Princeton University Press, Princeton, NJ, **2006**.

29 D. Bonchev, G.A. Buck, in *Complexity in Chemistry, Biology and Ecology*, D. Bonchev, D.H. Rouvray, Eds., Springer, New York, pp. 191–235, **2005**.

30 A.-L Barabási, R. Albert, *Science* **1999**,286, 509–512.

31 D.J. Watts, S.H. Strogatz, *Nature* **1998**, 393, 440–442.

32 D. Bonchev, G. Buck, *J. Chem. Inform. Model.* **2007**, 47, 909–917.

33 D. Bonchev, A.T. Balaban, O. Mekenyan, *J. Chem. Inf. Comput. Sci.* **1980**, 20, 106–113.

34 D. Bonchev, *Theochem* **1989**, 185, 155–168.

35 O. Ore, *Theory of Graphs*, AMS Colloquium Publications 38, AMS, **1962**.

36 L.C. Freeman, *Social Networks* **1979**, 1, 215–239.

37 L.C. Freeman, *Sociometry* **1977**, 40, 35–41.

38 M.E.J. Newman, arXiv:cond-mat/0309045, **2003**.

4
Spectral Theory of Networks: From Biomolecular to Ecological Systems

Ernesto Estrada

4.1
Introduction

The best way to understand how things work is by understanding their structures [1]. Complex networks are not an exception [2]. To understand why some networks are more robust than others, or why the propagation of a disease is faster in one network than in another, it is necessary to understand how these networks are organized [3–5]. A complex network is a simplified representation of a complex system in which the entities of the system are represented by the nodes in the network and the interrelations between entities are represented by means of the links joining pairs of nodes [3–5]. In analyzing the architecture of a complex network we are concerned only with the topological organization of these nodes and links. That is to say, we are not concerned with any geometric characteristic of the systems we are representing by these networks but only with how the parts are organized or distributed to form the whole system. Some of these topological characteristics of a network can be evident by simple visual inspection. This is particularly easy when the networks (graphs) are small. For instance, the first two graphs displayed below do not contain cycles, i.e., they are *trees*. The first of them is simply a *linear chain* and the second a *star*. The third and fourth graphs are cyclic. The third graph is the cycle of four nodes, C_4, and the last is a graph having a connection between every pair of nodes, i.e., the complete graph K_4 [6]. All these graphs are connected, which means that we can travel from any given node to another in any graph.

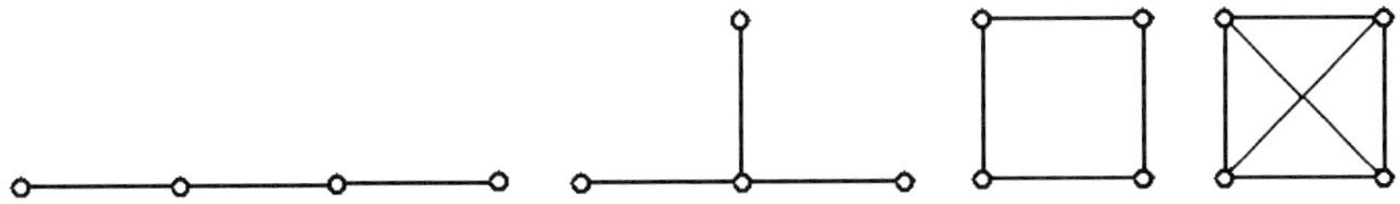

However, this visual analysis is not possible even for medium-sized networks. In addition, most real-world complex networks are very large and the questions we have to formulate to understand their structures and functioning are by far more complex [2–5]. To get a feel for how complex this prob-

Analysis of Complex Networks: From Biology to Linguistics. Edited by Matthias Dehmer and Frank Emmert-Streib

ISBN: 978-3-527-32345-6

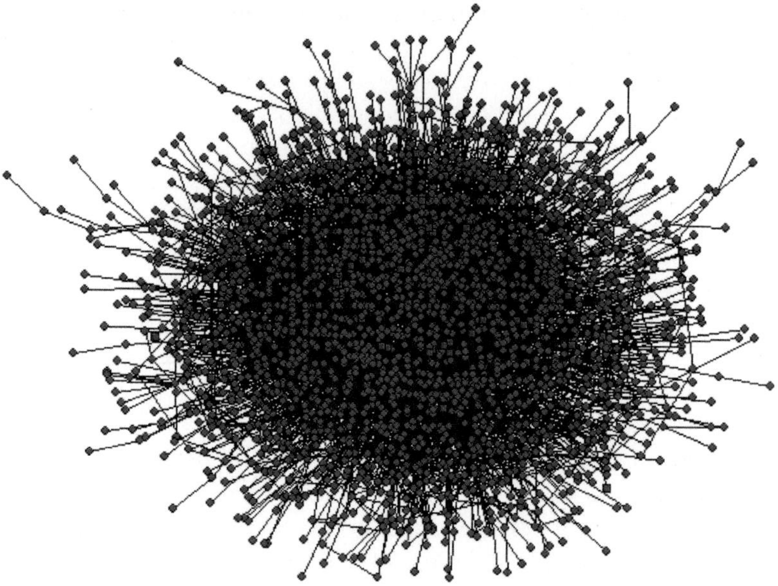

Figure 4.1 Representation of the human protein-protein interaction network. The proteins marked in red are those that have been identified as being involved in human diseases in the Online Mendelian Inheritance in Man (OMIM).

lem is we illustrate in Figure 4.1 the network of protein-protein interactions in human cells [7]. This network is far from being complete but it already contains more than 3000 nodes [7]. In Figure 4.1 we illustrate some proteins in red that have been identified as being responsible for hereditary diseases in humans [7].

It is evident that we need more sophisticated tools than visual inspection for analyzing the structure of complex networks. One of these tools is the *spectral graph theory* [8]. The spectrum of a graph (technically explained in the next section) can be considered the x-ray test for networks. Similarly to how we obtain information from x-ray spectroscopy about the internal structure of molecules, we can obtain information about the internal organization of complex networks with the use of spectral graph theory. This chapter is dedicated to the analysis of graph spectra to extract information about the architectural organization of real-world complex networks.

4.2 Background on Graph Spectra

A graph $G = (V, E)$ is a set of nodes V connected by means of the elements of the set of links E. Here we are dealing only with simple graphs [6], that is, an undirected graph without multiple links or self-loops. Thus, by graph we mean a simple graph. A node $v \in V$ is a terminal point of a link and represents an abstraction of an entity in a complex network such as a person, a city, a protein, an atom, etc. The links represent the relations between these entities.

A graph $G = (V, E)$ can be represented by different kinds of matrices [6]. The (ordinary) spectrum of a graph always refers to the spectrum of the adjacency matrix of the graph [9]. Thus, we will be concerned here only with this matrix. Excellent reviews about the Laplacian spectrum of graphs can be found in the literature [10]. The adjacency matrix $\mathbf{A} = A(G)$ of a graph $G = (V, E)$ is a symmetric matrix of order $n = |V|$, where $|\cdots|$ means the cardinality of the set and where $\mathbf{A}_{ij} = 1$ if there is a link between nodes i and j, and $\mathbf{A}_{ij} = 0$ otherwise.

The "spectrum" of a network is a listing of the *eigenvalues* of the adjacency matrix of such a network. It is well known that every $n \times n$ real symmetric matrix $\mathbf{A}$ has a spectrum of n orthonormal eigenvectors $\phi_1, \phi_2, \ldots, \phi_n$ with eigenvalues $\lambda_1 \geq \lambda_2 \geq \ldots \geq \lambda_n$ [11]. The largest eigenvalue of graph λ_1 is known as the principal eigenvalue, the spectral radius, or the Perron–Frobenius eigenvalue [11]. The eigenvector associated with this eigenvalue is also known as the principal eigenvector of a graph.

A walk of length l is any sequence of (not necessarily) different vertices $v_1, v_2, \ldots, v_k, v_{k+1}$ such that for each $i = 1, 2, \ldots, k$ there is an edge from v_i to v_{i+1}. A closed walk (CW) of length k is a walk in which $v_{k+1} = v_k$ [7]. The number of CWs of length μ_k is determined by the trace of the kth power of the adjacency matrix, $\mu_k = \operatorname{Tr} \mathbf{A}^k$. This number is also known as a spectral moment due to the following relationship with graph eigenvalues:

$$\mu_k = \sum_{j=1}^{n} \left(\lambda_j\right)^k . \tag{4.1}$$

The number of CWs of length k starting (and ending) at node p in the graph can also be expressed in terms of the graph eigenvalues and eigenvectors [12]:

$$\mu_k(p) = \sum_{j=1}^{n} \left[\phi_j(p)\right]^2 \left(\lambda_j\right)^k . \tag{4.2}$$

In a similar way, the number of walks of length k starting at node p and ending at node q are given by [12]

$$\mu_k(p, q) = \sum_{j=1}^{n} \phi_j(p)\, \phi_j(q) \left(\lambda_j\right)^k . \tag{4.3}$$

The spectrum of certain graphs is completely determined by the structure of the graph [9]. For instance, the complete graph, which is a graph in which every node is connected to every other node, has a spectrum $(n-1)^1$, $(-1)^{n-1}$. In the cycle graph, which is a graph on n nodes containing a single cycle through all nodes, the spectrum is given by $2\cos(2\pi j/n)$ $(j = 0, \ldots, n-1)$. The path or linear chain is also determined by its spectrum, which is given by $2\cos(2\pi j/n+1)$ $(j = 0, \ldots, n)$. The reader is referred to several books, such as [9,12,13], for a more thorough discussion and list of references to original papers.

4.3
Spectral Measures of Node Centrality

A local characterization of networks is made numerically by using one of several measures known as "centrality" [14]. One of the most used centrality measures is the "degree centrality" (DC) [15], which can be interpreted as a measure of immediate influence, as opposed to long-term effect, in a network [14]. Several other centrality measures have been introduced and studied for real-world networks, in particular for social networks. They account for the different node characteristics that permit them to be ranked in order of importance in the network. Betweenness centrality (BC) measures the number of times that a shortest path between nodes i and j travels through a node k whose centrality is being measured. The farness of a vertex is the sum of the lengths of the geodesics to every other vertex. The inverse of farness is closeness centrality (CC).

The first spectral measure of centrality was introduced by Bonacich in 1987 as the eigenvector centrality (EC) [16]. This centrality measure is not restricted to shortest paths [16]; it is defined as the principal or dominant eigenvector of the adjacency matrix A representing the connected subgraph or component of a network. It simulates a mechanism in which each node affects all of its neighbors simultaneously [17]. EC is better interpreted as a sort of extended-degree centrality that is proportional to the sum of the centralities of the node neighbors. Consequently, a node has a high value of EC either if it is connected to many other nodes or if it is connected to others that themselves have a high EC [18].

Here we designate the number of walks of length L starting at node i by $N_L(i)$ and the total number of walks of this length existing in the network by $N_L(G)$. The probability that a walk selected at random in the network has started at node i is simply

$$P_L(i) = \frac{N_L(i)}{N_L(G)} . \tag{4.4}$$

It is known that for a nonbipartite connected network with nodes $1, 2, \ldots, n$, for $L \to \infty$, the vector $[P_L(1)\ P_L(2)\ \ldots\ P_L(n)]$ tends toward the eigenvector corresponding to the largest eigenvalue of the adjacency matrix of the network. Consequently, the elements of EC represent the probabilities of selecting at random a walk of length L starting at node i when $L \to \infty$: $EC(i) = P_L(i)$ [12].

Another spectral measure of node centrality was introduced recently by Estrada as the *subgraph centrality* of vertex i in a network, which is given by [19]

$$SC(i) = \sum_{k=0}^{\infty} \frac{\mu_k(i)}{k!}, \tag{4.5}$$

where $\mu_k(i)$ are the number of closed walks of length k starting and ending at node i. The relation of this measure with the graph spectrum comes from the following results.

Let λ_1 be the principal eigenvalue of $\mathbf{A}$. For any nonnegative integer k and any $i \in \{1, \ldots, n\}$, $\mu_k(i) \le \lambda_1^k$, Series (4.2), whose terms are nonnegative, converges,

$$\sum_{k=0}^{\infty} \frac{\mu_k(i)}{k!} \le \sum_{k=0}^{\infty} \frac{\lambda_1^k}{k!} = e^{\lambda_1}. \tag{4.6}$$

Thus, the subgraph centrality of any vertex i is bounded above by $SC(i) \le e^{\lambda_1}$. The following result shows that the subgraph centrality can be obtained mathematically from the spectra of the adjacency matrix of the network.

Theorem 4.1 *[19]: Let $G = (V, E)$ be a simple graph of order n. Let $\varphi_1, \varphi_2, \ldots, \varphi_n$ be an orthonormal basis of R^n composed of eigenvectors of $\mathbf{A}$ associated to the eigenvalues $\lambda_1, \lambda_2, \ldots, \lambda_n$. Let $\varphi_j(i)$ denote the ith component of φ_j. For all $i \in V$, the subgraph centrality may be expressed as follows:*

$$SC(i) = \sum_{j=1}^{n} \left[\varphi_j(i)\right]^2 e^{\lambda_j}. \tag{4.7}$$

The sum of the subgraph centralities of all nodes in the network SC depends only on the eigenvalues of the adjacency matrix of the network [19]:

$$SC(G) = \sum_{i=1}^{n} SC(i) = \sum_{i=1}^{n} e^{\lambda_i}. \tag{4.8}$$

SC is also known as the Estrada index of a graph, and several mathematical results are available in the literature for this index [20–23]. Hereafter we will follow this designation and represent the subgraph centrality as $EE(G)$, or simply EE.

4.3.1 Subgraph Centrality as a Partition Function

To start with, let us now consider a network in which every pair of vertices is weighted by a parameter β. Let $\boldsymbol{B}$ be the adjacency matrix of this weighted network. It is obvious that $\boldsymbol{B} = \beta\boldsymbol{A}$ and $\mu_r(\boldsymbol{B}) = \mathrm{Tr}\,\boldsymbol{B}^r = \beta^r\,\mathrm{Tr}\,\boldsymbol{A}^r = \beta^r\mu_r$. In this case, the subgraph centrality can be generalized as follows [24]:

$$EE(G,\beta) = \sum_{r=0}^{\infty} \frac{\beta^r \mu_r}{r!} = \sum_{j=1}^{N} \mathrm{e}^{\beta\lambda_j}\,. \tag{4.9}$$

Alternatively, we can write $EE(G,\beta)$ as follows:

$$EE(G,\beta) = \mathrm{Tr} \sum_{r=0}^{\infty} \frac{\beta^r \boldsymbol{A}^r}{r!} = \mathrm{Tr}\,\mathrm{e}^{\beta A}\,. \tag{4.10}$$

It is straightforward to realize that the subgraph centrality is generalized to the partition function of the complex network in the form [25]

$$Z(G,\beta) \equiv EE(G,\beta) \equiv \mathrm{Tr}\,\mathrm{e}^{\beta A}\,, \tag{4.11}$$

where the Hamiltonian is $\boldsymbol{H} = -\boldsymbol{A}$ and β is the inverse temperature, that is, $\beta = 1/(k_B T)$. Note that β can be considered the "strength" of the interaction between a pair of vertices, assuming that every pair of vertices has the same interaction strength [25]. For instance, $\beta = 0$, which corresponds to the limit $T \to \infty$, corresponds to a graph with no links. This case is similar to a gas formed by monoatomic particles. On the other hand, very large values of β in the limit $T \to +0$ represents very large attractive interactions between pairs of bonded nodes in a similar manner to a solid. The "classical" subgraph centrality is the particular case where $\beta = 1$, i.e., the unweighted network.

Using this approach we can define the probability p_j that the system occupies a microstate j as follows [25]:

$$p_j = \frac{\mathrm{e}^{\beta\lambda_j}}{\sum_j \mathrm{e}^{\beta\lambda_j}} = \frac{\mathrm{e}^{\beta\lambda_j}}{EE(G,\beta)}\,. \tag{4.12}$$

Based on Equation (4.4) we can also define the information-theoretic entropy for the network using the Shannon expression [25]

$$S(G,\beta) = -k_B \sum_j \left[p_j\left(\beta\lambda_j - \ln EE\right)\right], \tag{4.13}$$

where we wrote $EE(G,\beta) = EE$. Then we can obtain the expressions for the total energy $H(G)$ and Helmholtz free energy $F(G)$ of the network [25]:

$$H(G, \beta) = -\sum_{j=1}^{n} \lambda_j p_j , \tag{4.14}$$

$$F(G, \beta) = -\beta^{-1} \ln EE . \tag{4.15}$$

These statistical mechanics functions of networks are bounded as follows [25]:

$$0 \leq S(G, \beta) \leq \beta \ln n , \tag{4.16}$$

$$-\beta (n-1) \leq H(G, \beta) \leq 0 , \tag{4.17}$$

$$-\beta (n-1) \leq F(G, \beta) \leq -\beta \ln n , \tag{4.18}$$

where the lower bounds are obtained for the complete graph as $n \to \infty$ and the upper bounds are reached for the null graph with n nodes.

4.3.2
Application

As a first illustration of the possibilities of the spectral measures of centrality we selected one example published recently by Choi *et al.* [26] in which the EC was used in comparing world city networks. The authors ranked the most central cities in the world by considering the Internet backbone and air-transport intercity linkages. When the authors considered only the number of direct links in the Internet backbone network, New York emerged as the most connected node, followed by London, Frankfurt, Tokyo, and Paris. However, when the EC was considered, the most central city was London, followed by New York, Paris, Frankfurt, and Amsterdam. In the network of air passengers the ranking according to the DC was dominated by London, followed by Frankfurt, Paris, New York, and Amsterdam. The use of the EC ranks London as the most central one, but changes the order of the other cities, Paris becomes the second most central followed by New York, Amsterdam and Frankfurt. The differences arise from the fact that in the EC a city that is connected to central cities has its own centrality boosted. Then, it is not only important to have a large number of connections but to have these connections with highly central nodes in the network.

To illustrate the characteristics of the subgraph centrality we selected an example from the collaboration network of *Computational Geometry* authors [19]. We selected at random two authors with the same degree and different subgraph centrality (Figure 4.2): Timothy M.Y. Chan and S.L. Abrams, both having $DC = 10$, but having $SC = 8.09 \cdot 10^9$ and $SC = 974.47$, respectively. Despite both authors' having the same number of coauthors, Chan is connected to five of the hubs of this collaboration network: Agarwal (98), Snoeyink (91), Sharir (87), Tamassia (79), and Yap (76) (DC is given in parenthesis). However, Abrams is connected to authors having lower numbers of

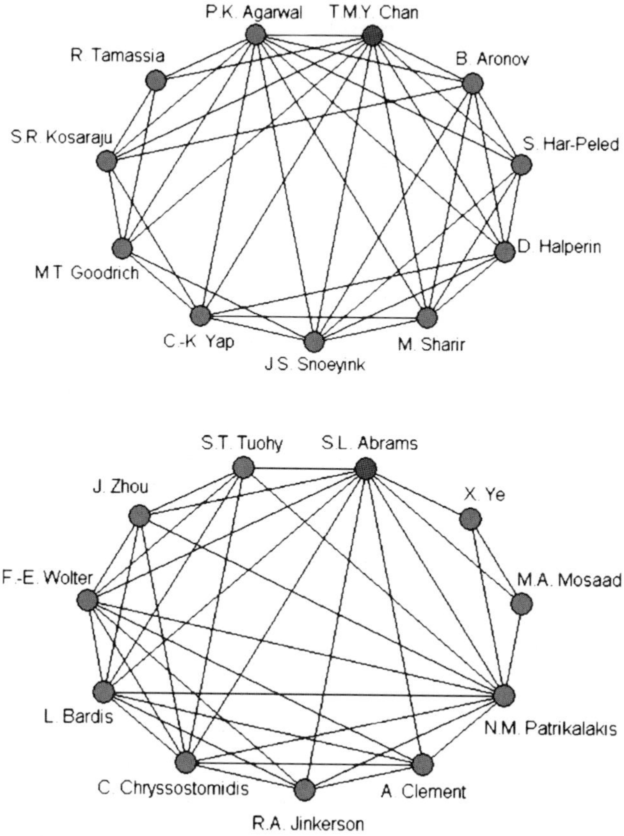

Figure 4.2 Part of the collaboration network in computational geometry for two authors with the same degree centrality but a different subgraph centrality: Timothy M.Y. Chan and S.L. Abrams and all their coworkers.

coworkers; e.g., Patrikalakis has 31 coauthors and the rest have only 5 to 16 collaborators. This simple difference means that Chan is separated from 623 other authors by a distance of only two; i.e., simply connected triplets, while this number is significantly lower for Abrams, i.e., only 116. The risk that Chan is "infected" with an idea circulating among the authors in this field of research is much higher than the risk with Abrams. This difference is accounted for by the subgraph centrality [19].

4.4 Global Topological Organization of Complex Networks

Our objective here is to give a characterization of the global organization of complex networks. The first step for analyzing the global architecture of a network is to determine whether the network is homogeneous or modular. In a homogeneous network, what you see locally is what you get globally from a topological point of view. However, in a modular network, the organization of certain modules or clusters would be different from one to another and to the global characteristics of the network [27–29].

Formally, we consider a network homogeneous if it has good expansion (GE) properties. A network has GE if every subset S of nodes ($S \leq 50\%$ of the nodes) has a neighborhood that is larger than some "expansion factor" Ω multiplied by the number of nodes in S. A neighborhood of S is the set of nodes that are linked to the nodes in S [30]. Formally, for each vertex $v \in V$ (where V is the set of nodes in the network), the neighborhood of v, denoted as $\Gamma(v)$, is defined as $\Gamma(v) = \{u \in V | (u, v) \in E\}$ (where E is the set of links in the network). Then, the neighborhood of a subset $S \subseteq V$ is defined as the union of the neighborhoods of the nodes in S: $\Gamma(S) = \bigcup_{v \in S} \Gamma(v)$, and the network has GE if $\Gamma(v) \geq \Omega |S| \; \forall S \subseteq V$.

Consequently, in a homogenous network we should expect that some local topological properties scale as a power law of global topological properties. A power-law relationship between a two variables x and y of the network is known by the term scaling and refers to the relationship [31]

$$y = Ax^{\eta} , \tag{4.19}$$

where A and η are constant. The existence of a scaling law reveals that the phenomenon under study reproduces itself on different time and/or space scales. That is, it has self-similarity [31]. Then, if x and y are variables representing some topological features of the network at the local and the global scale, the existence of such scaling implies that the network is topologically self-similar and *what we see locally is what we get globally*, which means that the network is homogeneous. In the following section we develop an approach to account for such scaling.

4.4.1 Spectral Scaling Method

Our first task here is to find a couple of appropriate topological variables for a network that characterize the local and global environment around a node. As for the local property we consider the subgraph centrality. As we already noted, this spectral measure characterizes the local cliquishness around a node because it gives larger weights to the participation of a node in smaller subgraphs. It should be noted that $EE(i)$ counts all CWs in the network, which can be of even or odd length. CWs of even length might be trivial on moving back and forth in acyclic subgraphs, i.e., those that do not contain cycles, while odd CWs do not contain contributions from acyclic subgraphs.

It is easy to show [32] that

$$EE(i) = \sum_{j=1}^{N} \left[\varphi_j(i)\right]^2 \cosh\left(\lambda_j\right) + \sum_{j=1}^{N} \left[\varphi_j(i)\right]^2 \sinh\left(\lambda_j\right) = EE_{\text{even}}(i) + EE_{\text{odd}}(i)\,, \tag{4.20}$$

which means that the term $EE_{\text{odd}}(i)$ only accounts for subgraphs containing at least one odd cycle. In this way, $EE_{\text{odd}}(i)$ can be considered a topological property of local organization in networks that characterize the odd-cyclic wiring of a typical neighborhood. We consider the EC a global topological characterization of the environment around a node. We have already shown that the EC represents the probability of selecting at random a walk of length L starting at node i when $L \to \infty$ [12]. Due to the infinite length of the walk we are considering, such a walk visits all nodes and links of the network obtaining a global picture of the topological environment around the corresponding node.

Now, we can establish the relationship between the local and global spectral properties of a network. To start with, we consider a graph with GE properties. Then, it is known that for a network to have good expansion, the gap between the first and second eigenvalues of the adjacency matrix ($\Delta\lambda = \lambda_2 - \lambda_1$) needs to be sufficiently large. For instance, what follows is a well-known result in the field of expander graphs [33–35].

Theorem 4.2 *(Alon–Milman): Let G be a regular graph with spectrum $\lambda_1 \geq \lambda_2 \geq \ldots \geq \lambda_n$. Then, the expansion factor is bounded as*

$$\frac{\lambda_1 - \lambda_2}{2} \leq \phi(G) \leq \sqrt{2\lambda_1(\lambda_1 - \lambda_2)}.$$

Then, let us write $EE_{odd}(i)$ as follows:

$$EE_{\text{odd}}(i) = \left[EC(i)\right]^2 \sinh(\lambda_1) + \sum_{j=2} \left[\varphi_j(i)\right]^2 \sinh\left(\lambda_j\right)\,, \tag{4.21}$$

where $EC(i)$ is the EC (the principal eigenvector $\varphi_1(i)$) and λ_1 is the principal eigenvalue of the network. Then, let us assume that $\lambda_1 \gg \lambda_2$ in such a way that we can consider that $\left[EC(i)\right]^2 \sinh(\lambda_1) \gg \sum_{j=2} \left[\varphi_j(i)\right]^2 \sinh\left(\lambda_j\right)$. Consequently, we can write the odd-subgraph centrality as

$$EE_{\text{odd}}(i) \approx \left[EC(i)\right]^2 \sinh(\lambda_1)\,, \tag{4.22}$$

and the principal eigenvector of the network is directly related to the subgraph centrality in GENs according to the following spectral scaling relationship [36, 37]:

$$EC(i) \propto A\left[EE_{\text{odd}}(i)\right]^{\eta}\,, \tag{4.23}$$

which corresponds to the power-law relationship between $EC(i)$ and $EE_{odd}(i)$ for GENs, which is similar to the one given by Equation (4.19), where x and y are variables representing some topological features of the network at the local and the global scale. Here, $A \approx [\sinh(\lambda_1)]^{-0.5}$ and $\eta \approx 0.5$. This expression can be written in a log-log scale as [36, 37]

$$\log [EC(i)] = \log A + \eta \log [EE_{odd}(i)] . \quad (4.24)$$

Consequently, a log-log plot of $EC(i)$ vs. $EE_{odd}(i)$ in a homogeneous network has to show a linear fit with slope $\eta \approx 0.5$ and intercept $\log A$ for GENs.

4.4.2
Universal Topological Classes of Networks

There are several classification schemes grouping networks according to their structures. For instance, complex networks can be classified according to the existence or not of the "small-world" property [38, 39] or according to their degree distribution. The last classification permits us to classify networks as "scale-free" [40] if their node degree distribution decays as a power law, "broad-scale" networks, which are characterized by a connectivity distribution that has a power-law regime followed by a sharp cutoff, or "single-scale" networks in which degree distribution displays a fast decaying tail [41]. Even scale-free networks have been classified into two different subclasses according to their exponent in the power-law distribution of the betweenness centrality [42].

Each of these classification schemes reproduces different characteristics of complex networks. "Small-worldness" [38] and "scale-freeness" [40] reflect global organizational principles of complex systems. The first characterizes the relatively small separation among pairs of nodes and the high cliquishness of some real-world networks [38]. The second reproduces the presence of a few highly connected hubs that keep glued together the vast majority of poorly connected nodes in certain networks [40]. Both properties are of great relevance in analyzing other important properties of complex networks, such as disease propagation [43–45] or robustness against targeted or random attacks [46–48]. However, there are important organizational principles of complex networks that escape the analysis of these global network characteristics.

The theoretical approach we presented in the previous section permits the classification of complex networks into two groups: homogeneous (GEN) and nonhomogeneous networks. Here we are interested in identifying the topological differences existing among the nonhomogeneous networks in such a way that permit us to classify them into some universal classes.

Let us consider the ideal case in which a network displays perfect spectral scaling, such that we can calculate the eigenvector centrality by using the following expression:

$$\log EC^{\text{Ideal}}(i) = 0.5 \log EE_{\text{odd}}(i) - 0.5 \log\left[\sinh(\lambda_1)\right] . \tag{4.25}$$

Now, let us consider the deviations from the ideal behavior represented by Equation (4.22) in nonhomogeneous networks. We can account for these deviations from the *ideal* by measuring the departure of the points from the perfect straight line with respect to $\log EC^{\text{Ideal}}(i)$:

$$\Delta \log EC(i) = \log \frac{EC(i)}{EC^{\text{Ideal}}(i)} = \log \left\{ \frac{\left[EC(i)\right]^2 \sinh(\lambda_1)}{EE_{\text{odd}}(i)} \right\}^{0.5} . \tag{4.26}$$

Then, according to the values of $\Delta \log EC(i)$, there are four different classes of complex networks. These classes are as follows [49].

- **Class I**: networks displaying perfect spectral scaling:

$$\Delta \log EC(i) \approxeq 0, \ \forall i \in V \Rightarrow \left[EC(i)\right]^2 \sinh(\lambda_1) \approxeq EE_{\text{odd}}(i) . \tag{4.27}$$

- **Class II**: networks displaying spectral scaling with negative deviations:

$$\Delta \log EC(i) \le 0 \Rightarrow \left[EC(i)\right]^2 \sinh(\lambda_1) \le EE_{\text{odd}}(i), i \in V . \tag{4.28}$$

- **Class III**: networks displaying spectral scaling with positive deviations:

$$\Delta \log EC(i) \ge 0 \Rightarrow \left[EC(i)\right]^2 \sinh(\lambda_1) \ge EE_{\text{odd}}(i), i \in V . \tag{4.29}$$

- **Class IV**: networks displaying spectral scaling with mixed deviations:

$$\Delta \log EC(p) \le 0, p \in V \quad \text{and} \quad \Delta \log EC(q) > 0, q \in V . \tag{4.30}$$

We previously showed that the first of such classes corresponds to networks displaying good expansion properties, that is, networks in which nodes and links are homogeneously distributed through the network in such a way that there are not structural bottlenecks. The other three classes correspond to different organizations of the community structure in the networks. Class II corresponds to networks in which there are two or more communities of highly interconnected nodes, which display low intermodule connectivity. These kind of networks look like networks containing holes in their structures. In class III the networks display a typical "core-periphery" structure characterized by a highly interconnected central core surrounded by a sparser periphery of nodes. Finally, class IV networks display a combination of highly connected groups (quasi-cliques) and some groups of nodes partitioned into disjoint subsets (quasi-bipartite), without a predominance of either structure. In Figure 4.3 we illustrate the main structural

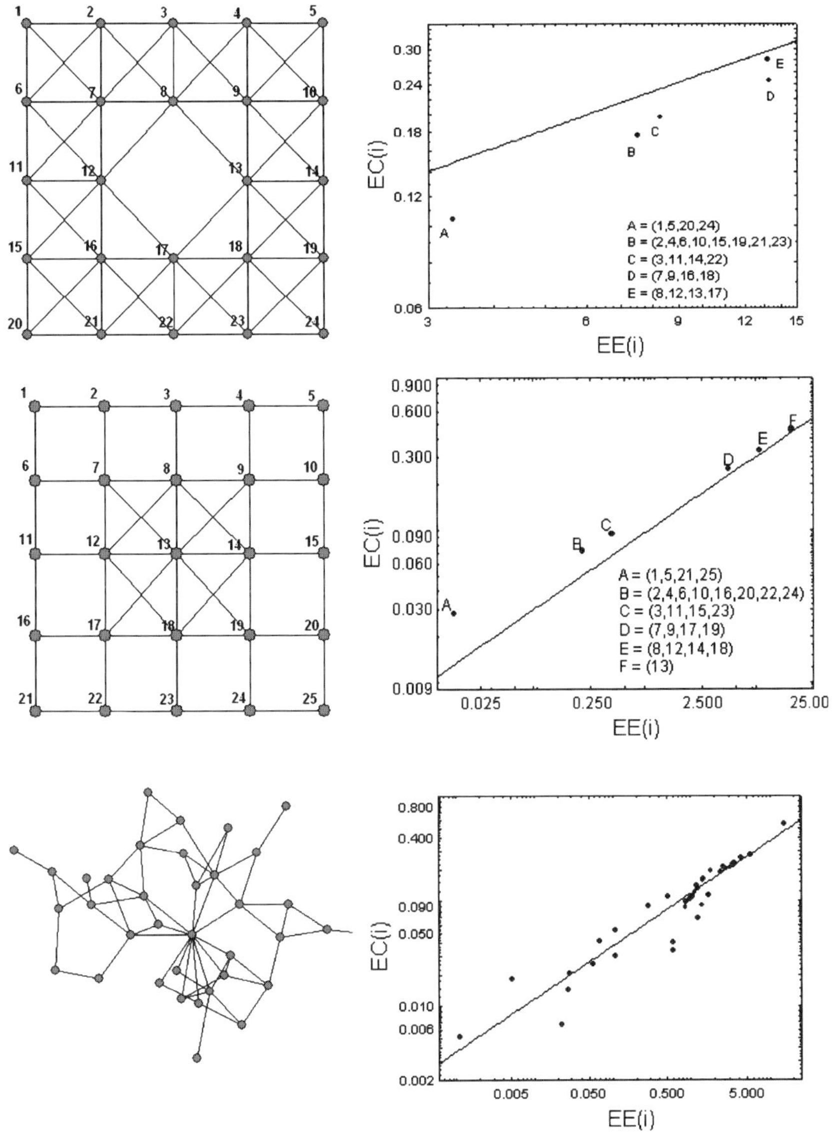

Figure 4.3 Models of networks in classes II, III, and IV according to the classification of the spectral scaling approach. On the left-and side we show the corresponding spectral scaling for these networks.

properties of nonhomogeneous networks and their respective spectral scaling plots.

To quantify the degree of deviation of the nodes from the ideal spectral scaling, we account for the mean square error of all points with positive and negative deviations in the spectral scaling, respectively [49]:

$$\xi^{+} = \sqrt{\frac{1}{N_{+}} \sum_{+} \left(\log \frac{EC(i)}{EC^{\text{Ideal}}(i)} \right)} \quad \text{and} \quad \xi^{-} = \sqrt{\frac{1}{N_{-}} \sum_{-} \left(\log \frac{EC(i)}{EC^{\text{Ideal}}(i)} \right)} .$$

where $\sum_{+}$ and $\sum_{-}$ are the sums carried out for the N_{+} points having $\Delta \log EC(i) > 0$ and for the N_{-} having $\Delta \log EC(i) < 0$, respectively.

4.4.3 Applications

We have studied 61 real-world complex networks accounting for ecological, biological, protein secondary structures, informational, technological, and social systems [49]. Using the values of ξ^{-} and ξ^{+} we have classified these networks into the four different classes that are predicted to exist from a theoretical point of view. We have carried out a canonical discriminant analysis (CDA) [44] for the 61 networks studied using $\log\left(\xi^{-} + 10^{-3}\right)$ and $\log\left(\xi^{+} + 10^{-3}\right)$ as classifiers, where the sum of the constant 10^{-3} is necessary to avoid indeterminacies due to zero values. In Figure 4.4 we can see the main factors (roots) that perfectly separate the networks studied into the four different structural classes.

Consequently, we have identified the existence of the four classes of networks in real-world systems by studying a large pool of networks representing ecological, biological, informational, technological, and social systems. While classes I, II, and IV are equally populated, each having about 32% of the total networks, class III is less frequent and only appeared in two ecological networks. In general, most ecological networks correspond to class I (70%), and they represent the only systems in which the four classes of networks are represented. Most biological networks studied correspond to class IV (67%), while all protein secondary structure networks correspond to class II. Informational networks are mainly classified into two classes: class I (50%) and class II (33.3%). On the other hand, technological networks are mainly in class IV (64%), while 27% correspond to class I. Social networks also display great homogeneity in their structural classes as they correspond mainly to classes II and IV (91%) [49].

We finally have explored the possible growing mechanisms determining the structural classes observed in this work. We found that a random growing mechanism giving rise to uniform distributions of node degrees and the preferential attachment mechanism of Barabási-Albert reproduces very well the characteristics of networks in class I when the average degree is larger than 5. For sparser networks, such as those having an average degree of less than 3, both mechanisms reproduce the characteristics of networks in class IV. However, neither growing mechanism is able to reproduce the topo-

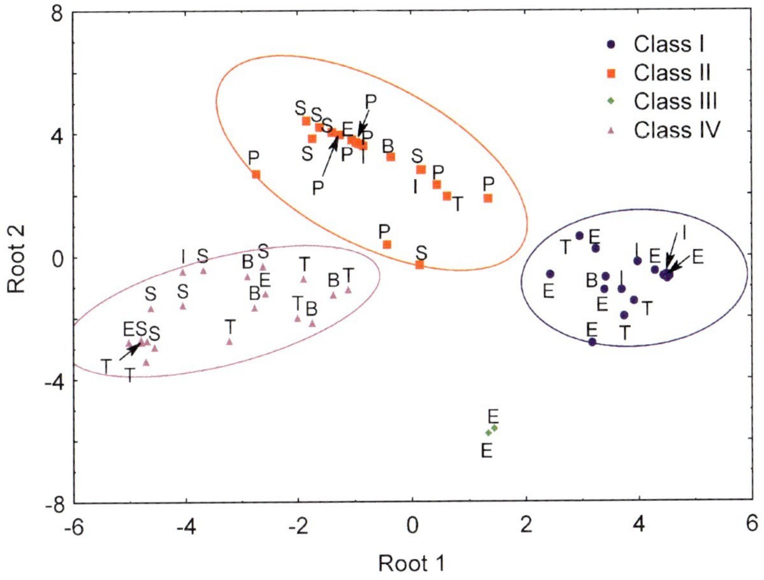

Figure 4.4 Plot of the two principal roots obtained in the canonical discriminant analysis (CDA) of the 61 networks classified into 4 different structural classes. Ellipses correspond to 95% of confidence in the CDA.

logical organization of networks in classes II and III [49]. Similar results are obtained when generating random networks with the same degree sequence as real-world networks. Our results confirm previous findings about the necessity of investigating new growing mechanisms for generating networks to model real-world systems [50].

4.5 Communicability in Complex Networks

The *communicability* between a pair of nodes in a network is usually considered as taking place through the shortest path connecting both nodes. However, it is known that communication between a pair of nodes in a network does not always take place through the shortest paths but it can follow other nonoptimal walks [51–53]. Then, we can consider a communicability measure that accounts for a weighted sum of all walks connecting two nodes in the network. We can design our measure in such a way that the shortest path connecting these two nodes always receives the largest weight. Then, if $P_{pq}^{(s)}$ is the number of shortest paths between nodes p and q having length s and $W_{pq}^{(k)}$ is the number of walks connecting p and q of length $k > s$, we propose to consider the quantity [54]

$$G_{pq} = \frac{1}{s!} P_{pq} + \sum_{k>s} \frac{1}{k!} W_{pq}^{(k)} . \tag{4.31}$$

In fact, Equation (4.31) can be written as the sum of the p, q entry of the different powers of the adjacency matrix:

$$G_{pq} = \sum_{k=0}^{\infty} \frac{\left(A^k\right)_{pq}}{k!} , \tag{4.32}$$

which converges to [54]

$$G_{pq} = \left(e^A\right)_{pq} = \sum_{j=1}^{n} \phi_j (p)\, \phi_j (q)\, e^{\lambda_j} . \tag{4.33}$$

We call G_{pq} the *communicability* between nodes p and q in the network. The communicability should be minimum between the end nodes of a chain, where it vanishes as the length of the chain is increased. On the other hand, the communicability between an arbitrary pair of nodes in a complete graph diverges as the size of the graph is increased because the oscillation is greatly amplified by the infinitely many walks between the nodes. Thus, the communicability between a pair of nodes in a network is bounded between zero and infinity, which are obtained for the two end nodes of an infinite linear chain and for a pair of nodes in an infinite complete graph. For the linear chain P_n the value of G_{pq} is equal to [54]

$$G_{pq} = \frac{1}{n+1} \sum_{j} \left(\cos \frac{j\pi (p-q)}{n+1} - \cos \frac{j\pi (p+q)}{n+1} \right) e^{2\cos\left(\frac{j\pi}{n+1}\right)} . \tag{4.34}$$

Let P_∞ be a chain of infinite length. It is straightforward to realize by simple substitution in Equation (4.34) that $G_{1,\infty} = 0$ for the end nodes $p = 1$ and $q = \infty$. For the complete graph we have that [54]

$$G_{pq} = \frac{e^{n-1}}{n} + e^{-1} \sum_{j=2}^{n} \phi_j(p)\phi_j(q) = \frac{e^{n-1}}{n} - \frac{1}{ne} = \frac{1}{ne}\left(e^n - 1\right) , \tag{4.35}$$

and it is easy to see that $G_{pq} \to \infty$ as $n \to \infty$ for K_n.

A physical interpretation of the communicability can be done by considering a continuous-time quantum walk on the network. Take a quantum-mechanical wave function $|\psi (t)\rangle$ at time t. It obeys the Schrödinger equation [55]

$$i\hbar \frac{d}{dt} |\psi (t)\rangle = -A |\psi (t)\rangle , \tag{4.36}$$

where we use the adjacency matrix as the negative Hamiltonian.

Assuming from now on that $\hbar = 1$ we can write down the solution of the time-dependent Schrödinger Equation (4.33) in the form $|\psi(t)\rangle = e^{iAt}|\psi(0)\rangle$. The final state $e^{iAt}|q\rangle$ is a state of the graph that results after time t from the initial state $|q\rangle$. The "particle" that resided on the node q at time $t = 0$ diffuses for the time t because of the quantum dynamics. Then, we can obtain the amplitude that the "particle" ends up in at node p of the network by computing the product $\langle p| e^{iAt}|q\rangle$. By continuation from real time t to imaginary time, we have the thermal Green's function defined as $G_{pq} = \langle p| e^{A}|q\rangle$, which is the communicability between nodes p and q in the network as defined in this work [54]. Consequently, the communicability between nodes p and q in the network represents the probability that a particle starting from node p ends up at node q after wandering on the complex network due to the thermal fluctuation. By regarding the thermal fluctuation as some form of random noise, we can identify the particle as an information carrier in a society or a needle in a drug-user network.

4.5.1
Communicability and Network Communities

Many complex networks in the real world are not homogeneous, as we have already seen in this chapter. Instead, the nodes in most networks appear to group into subgraphs in which the density of internal connections is larger than the connections with the rest of the nodes in the network. This notion was first introduced by Girvan and Newman [56] and it is known as the *community structure* of complex networks [57–61]. In the language of communicability we are using in this section, we can say that a community is a group of nodes having larger communicability among them than with the rest of the nodes in the graph. Later on we will give a more formal definition of community.

To perform further analysis, we now use the spectral decomposition of the Green's function [62]. Imagine that the network has a spring on each link. Each eigenvector indicates a mode of oscillation of the entire network and its eigenvalue represents the weight of the mode. It is known that the eigenvector of the largest eigenvalue λ_1 has elements of the same sign. This means that the most important mode is the oscillation where all nodes move in the same direction at one time.

The second largest eigenvector ϕ_2 has both positive and negative elements. Suppose that a network has two clusters connected through a bottleneck but each cluster is closely connected within. The second eigenvector represents the mode of oscillation where the nodes of one cluster move coherently in one direction and the nodes of the other cluster move coherently in the opposite direction. Then the sign of the product $\phi_2(p)\,\phi_2(q)$ tells us whether nodes p and q are in the same cluster or not.

The same analysis can be applied to the rest of the eigenvalues of the network. The third eigenvector, ϕ_3, which is orthonormal to the first two eigenvectors, has a different pattern of signs, dividing the network into three different blocks after appropriate arrangement of the nodes. In general, the second eigenvector divides the graph into biants, the third divides it into triants, the fourth into quadrants, and so forth, but these clusters are not necessarily independent of each other.

According to this pattern of signs, we have the following decomposition of the thermal Green's function [54]:

$$G_{pq} = \left[\phi_1(p)\,\phi_1(q)\,e^{\lambda_j}\right] + \left[\sum_{j=2}^{++} \phi_j^+(p)\,\phi_j^+(q)\,e^{\lambda_j} + \sum_{j=2}^{--} \phi_j^-(p)\,\phi_j^-(q)\,e^{\lambda_j}\right]$$
$$+ \left[\sum_{j=2}^{+-} \phi_j^+(p)\,\phi_j^-(q)\,e^{\lambda_j} + \sum_{j=2}^{-+} \phi_j^-(p)\,\phi_j^+(q)\,e^{\lambda_j}\right], \tag{4.37}$$

where ϕ_j^+ and ϕ_j^- refer to the eigenvector components with positive and negative signs, respectively. According to the partitions made by the pattern of signs of the eigenvectors in a graph, two nodes have the same sign in an eigenvector if they can be considered as being in the same partition of the network, while those pairs having different signs correspond to nodes in different partitions. Thus, the first bracket in Equation (4.34) represents the background mode of translational movement. The second bracket represents the *intracluster communicability* between nodes in the network, and the third bracket represents the *intercluster communicability* between nodes [54].

The above consideration motivates us to define a quantity ΔG_{pq} by subtracting the contribution of the largest eigenvalue λ_1 from Equation (4.34) [54]:

$$\Delta G_{pq} = \sum_{j=2}^{\text{intracluster}} \phi_j(p)\,\phi_j(q)\,e^{\lambda_j} + \sum_{j=2}^{\text{intercluster}} \phi_j(p)\,\phi_j(q)\,e^{\lambda_j}. \tag{4.38}$$

By focusing on the sign of ΔG_{pq}, we can unambiguously define a community for a group of nodes. If ΔG_{pq} for a pair of nodes p and q have a positive sign, then they are in the same community. If ΔG_{pq} the two nodes have a negative sign, then they are in different clusters [54].

Definition 4.1 A community in a network is a group of nodes $U \subseteq V$ for which the intracluster communicability is larger than the intercluster communicability, i.e., $\Delta G_{pq} > 0, \forall\,(p,q) \in U$.

4.5.2
Detection of Communities: The Communicability Graph

To start with, we represent the values of ΔG_{pq} in the form of a matrix $\boldsymbol{\Delta}$. $\boldsymbol{\Delta}$ is a matrix whose nondiagonal entries are given by the values of ΔG_{pq} and zeroes in the main diagonal. Now, let us introduce a Heaviside step function:

$$\Theta(x) = \begin{cases} 1 & \text{if } x > 0 \\ 0 & \text{if } x \leq 0 \end{cases} \tag{4.39}$$

Let $\Theta(\boldsymbol{\Delta})$ be the result of applying the Heaviside step function in an elementwise way to matrix $\boldsymbol{\Delta}$. Then, in the resulting matrix $\Theta(\boldsymbol{\Delta})$ a pair of nodes p and q is connected if, and only if, they have $\Delta G_{pq} > 0$. Then let us define the following graph [63].

Definition 4.2 The *communicability graph* $\Theta(G)$ is a graph having adjacency matrix $\Theta(\boldsymbol{\Delta})$.

In such a graph two nodes are connected if they have $\Delta G_{pq} > 0$. That is to say, the nodes forming a community in the original graph are connected in the communicability graph. Now, suppose that there is a link between nodes p and q and there are also links between them and a third node r. This means that $\Delta G_{pq} > 0$, $\Delta G_{pr} > 0$, and $\Delta G_{qr} > 0$. Consequently, the three nodes form a positive subgraph C. As we want to detect the largest subset of nodes connected to this triple, we have to search for the nodes s for which $\Delta G_{is} > 0 \, \forall i \in C$. Using the *communicability graph,* this search is reduced to finding the cliques in a simple graph, $\Theta(\boldsymbol{\Delta})$. These cliques correspond to the communities of the network. A clique is a maximum complete subgraph in a graph, that is, a maximum subgraph in which every pair of nodes is connected.

Finding the cliques in a graph is a classical problem in combinatorial optimization that has found applications in diverse areas [64]. Here we use a well-known algorithm developed by Bron and Kerbosch [65], which is a depth-first search for generating all cliques in a graph. This algorithm consumes a time per clique that is almost independent of the graph size for random graphs, and for the Moon–Moser graphs of n vertices the total time is proportional to $(3.14)^{n/3}$. The Moon–Moser graphs have the largest number of maximal cliques possible among all n-vertex graphs regardless of the number of edges in the graph [66].

4.5.3
Application

As an example of a real-world network, we consider a friendship network known as the Zachary karate club, which has 34 members (nodes) with some friendship relations (links). The members of the club, after some entanglement, were eventually divided into two groups, one formed by the followers of the instructor and the other formed by the followers of the administrator [67]. This network has been analyzed in practically every paper consider-

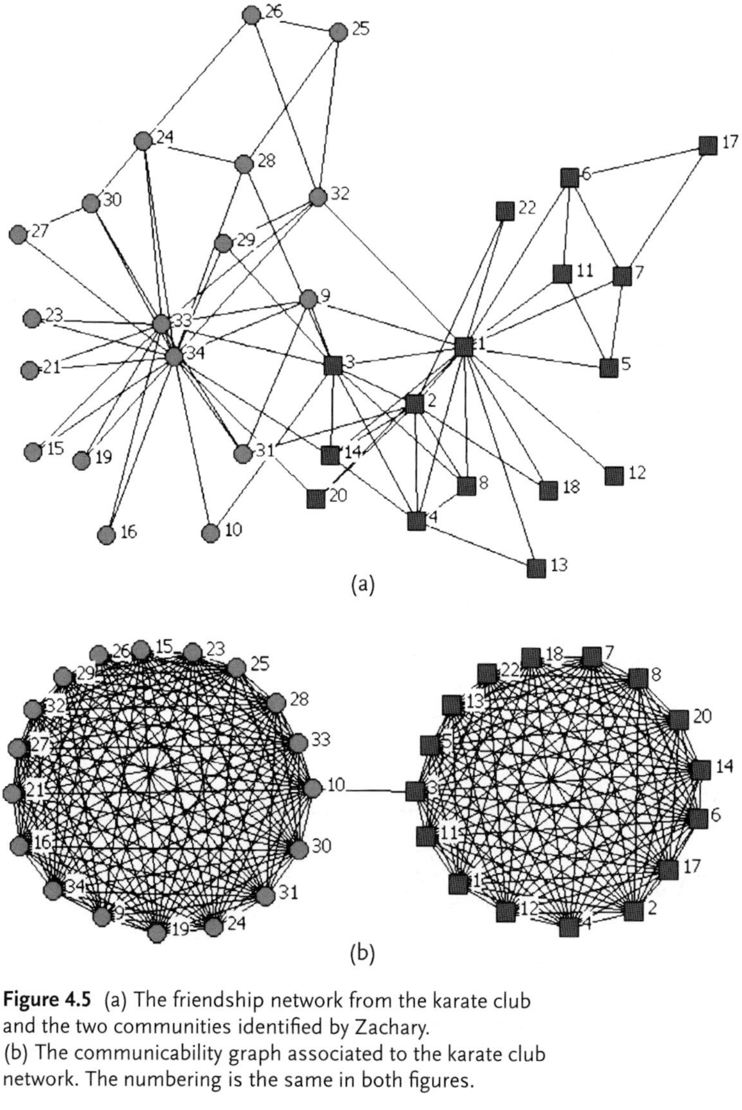

Figure 4.5 (a) The friendship network from the karate club and the two communities identified by Zachary.
(b) The communicability graph associated to the karate club network. The numbering is the same in both figures.

ing the problem of community identification in complex networks. In Figure 4.5a we illustrate the Zachary network in which the nodes are divided into the two classes observed by Zachary on the basis of the friendship relationships among the members of the club.

In the Figure 4.5b we illustrate the communicability graph $\Theta(G)$ of the Zachary network. As can be seen, $\Theta(G)$ correctly divides the network into two groups. There is very high internal communicability among the members of the respective groups, but there is almost no communicability between the groups. In fact, node 3 is correctly included in the group of the instructor (node 1).

An analysis of the cliques in the communicability graph reveals a more detailed view of the community structure of this network. Accordingly, there are five different cliques representing five overlapping communities in the network. These communities are given below, where the numbers correspond to the labels of the nodes in Figure 4.5a:

$1: \{10, 15, 16, 19, 21, 23, 24, 26, 27, 28, 29, 30, 31, 32, 33, 34\}$

$2: \{9, 10, 15, 16, 19, 21, 23, 24, 27, 28, 29, 30, 31, 32, 33, 34\}$

$3: \{10, 15, 16, 19, 21, 23, 24, 25, 26, 27, 28, 29, 30, 32, 33, 34\}$

$4: \{1, 2, 3, 4, 5, 6, 7, 8, 11, 12, 13, 14, 17, 18, 20, 22\}$

$5: \{3, 10\}$

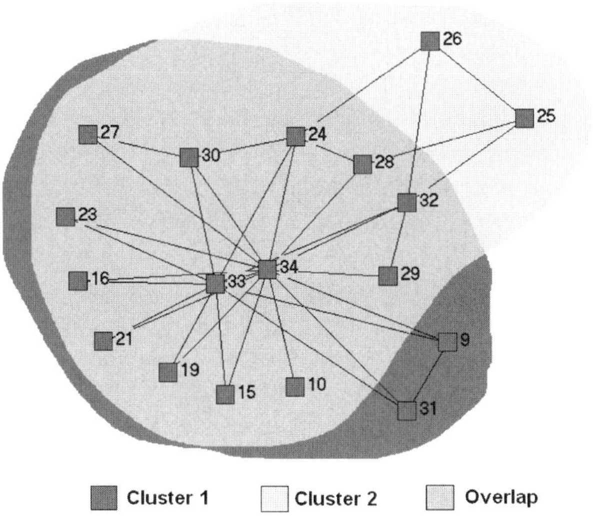

Figure 4.6 Illustration of the overlapping between two groups or neighborhoods formed among the followers of the administrator (node 34) in the Zachary karate club network.

As can be seen, the first three communities, which correspond to the group of the administrator (node 34), are formed by 16 members each and display an overlap of about 94% (Figure 4.6). The fourth community corresponds to that of the instructor (node 1) and also has 16 members. The last community is formed by nodes 3 and 10 only. This community displays overlaps with the communities of the administrator as well as with that of the instructor. In fact, node 10 appears in communities 1 to 4, and node 3 appears in communities 4 and 5.

4.6 Network Bipartivity

There are numerous natural systems that can be modeled by making a partition of the nodes into two disjoint sets [68,69]. For instance, in a network representing heterosexual relationships, one set of nodes corresponds to female and the other to male partners. In some trade networks, one set of nodes can represent buyers and the other sellers, and so forth. These networks are called bipartite networks or graphs and are formally defined below [6].

Definition 4.3 A network (graph) $G = (V, E)$ is called *bipartite* if its vertex set V can be partitioned into two subsets V_1 and V_2 such that all edges have one endpoint in V_1 and the other in V_2.

Now, let us consider the case in which some connections between the nodes in the same set of a formerly bipartite network are allowed. Strictly speaking these networks are not bipartite, but we can consider them loosely as *almost bipartite* networks. For instance, if we consider a sexual relationships network in which not only heterosexual but also some homosexual relations are present, the network is not bipartite, but it could be almost bipartite if the number of homosexual relations is low compared to the number of heterosexual ones. It is known that the transmission rates for homosexual and heterosexual contacts differ [69]. Consequently, the transmission of this disease will depend on how bipartite the corresponding network is. In other words, having an idea of the bipartivity of sexual networks we will have an idea on the rate of spreading of a sexually transmitted disease.

The following is a well-known result due to König that permits us to characterize bipartite graphs [70].

Theorem 4.3 *(König): A graph is bipartite if and only if all its cycles are even.*

We will make use of this result in order to characterize the bipartivity of a network. To start with, we consider the subgraph centrality of the whole graph defined by Equation (4.5). We can express this index as the sum of two contributions, one coming from odd and the other from even CWs [32]:

$$EE(G) = \sum_{j=1}^{n} \left[\cosh(\lambda_j) + \sinh(\lambda_j)\right] = EE_{\text{even}} + EE_{\text{odd}} . \tag{4.40}$$

If $G(V, E)$ is bipartite, then according to the theorem of König [70]: $EE_{\text{odd}} = \sum_{j=1}^{n} \sinh(\lambda_j) = 0$ because there are no odd CWs in the network [32]. Therefore:

$$EE(G) = EE_{\text{even}} = \sum_{j=1}^{n} \cosh(\lambda_j) . \tag{4.41}$$

Consequently, the proportion of even CWs to the total number of CWs is a measure of the network bipartivity [32]:

$$b(G) = \frac{EE_{\text{even}}}{EE(G)} = \frac{EE_{\text{even}}}{EE_{\text{even}} + EE_{\text{odd}}} = \frac{\sum_{j=1}^{n} \cosh(\lambda_j)}{\sum_{j=1}^{n} e^{\lambda_j}} . \tag{4.42}$$

It is evident that $b(G) \le 1$ and $b(G) = 1$ if and only if G is bipartite, i.e., $EE_{\text{odd}} = 0$. Furthermore, as $0 \le EE_{\text{odd}}$ and $\sinh(\lambda_j) \le \cosh(\lambda_j)$, $\forall \lambda_i$, then $\frac{1}{2} < b(G)$ and $\frac{1}{2} < b(G) \le 1$. The lower bound is reached for the least possible bipartite graph with n nodes, which is the complete graph K_n. As the eigenvalues of K_n are $n-1$ and -1 (with multiplicity $n-1$), then $b(G) \to \frac{1}{2}$ when $n \to \infty$ in K_n.

Then $b(G)$ represents a quantitative characterization of the bipartivity of a complex network. Now we have a quantitative measure that permits us to discern between quasi-bipartite networks as well as to differentiate them from bipartite and not bipartite graphs. However, an open question remains: Can we identify the bipartite subgraphs existing in a network?

4.6.1
Detecting Bipartite Substructures in Complex Networks

It is known that the eigenvectors corresponding to positive eigenvalues give a partition of a network into clusters of tightly connected nodes [71, 72]. In contrast, the eigenvectors corresponding to negative eigenvalues make partitions in which nodes are not close to those with which they are linked, but rather with those with which they are not linked [71, 72]. Then we can make use of the communicability function to identify the bipartite structures in complex networks. In general, we can say that a positive (negative) value of β in the communicability function (30) increases the contribution of the positive (negative) eigenvalues to the communicability function. Then, if we write the communicability function as [73]

$$G_{pq}(\beta) = \sum_{\lambda_j<0} \phi_j(p)\phi_j(q)\,\mathrm{e}^{\beta\lambda_j} + \sum_{\lambda_j=0} \phi_j(p)\phi_j(q)\,\mathrm{e}^{\beta\lambda_j} + \sum_{\lambda_j>0} \phi_j(p)\phi_j(q)\,\mathrm{e}^{\beta\lambda_j}\,, \tag{4.43}$$

we have that

$$G_{pq}(\beta > 0) \approx \sum_{\lambda_j>0}^{n} \phi_j(p)\phi_j(q)\,\mathrm{e}^{\beta\lambda_j}\,, \tag{4.44}$$

$$G_{pq}(\beta < 0) \approx \sum_{\lambda_j<0}^{n} \phi_j(p)\phi_j(q)\,\mathrm{e}^{-|\beta|\lambda_j}\,. \tag{4.45}$$

In other words, $G_{pq}(\beta > 0)$ determines a partition of the network into clusters of tightly connected nodes, which corresponds to the network communities. On the other hand, for $G_{pq}(\beta < 0)$ the network is partitioned in such a way that the nodes are close to other nodes that have similar patterns of connections with other sets of nodes, i.e., nodes to which they are structurally equivalent. In the first case, we say that the nodes corresponding to larger components tend to form *quasi-cliques*, that is, clusters in which every two nodes tend to interact with each other. In the second case, the nodes tend to form *quasi-bipartites*, i.e., nodes are partitioned into *almost* disjoint subsets with high connectivity between sets but low internal connectivity.

Let us consider a bipartite graph and let p and q be nodes that are in two different disjoint sets of the graph. Then, there are no walks of even length between p and q in the graph and [73]

$$G_{pq}(\beta = -1) = [-\sinh(A)]_{pq} < 0\,. \tag{4.46}$$

However, if p and q are nodes in the same disjoint set, then there is no walk of odd length connecting them due to the lack of odd cycles in the bipartite graph, which yields

$$G_{pq}(\beta = -1) = [\cosh(A)]_{pq} > 0\,. \tag{4.47}$$

The above argument shows that, in general, the sign of the communicability at a negative temperature, $G_{pq}(\beta = -1) = \left(\mathrm{e}^{-A}\right)_{pq}$, gives an indication as to how the nodes can be separated into disjoint sets.

Our strategy for detecting quasi-bipartite clusters in complex networks is as follows. First we start by calculating $\exp(-\boldsymbol{A})$, whose (p, q)-entry gives the communicability between the nodes p and q in the network. Then we introduce the following definition [73].

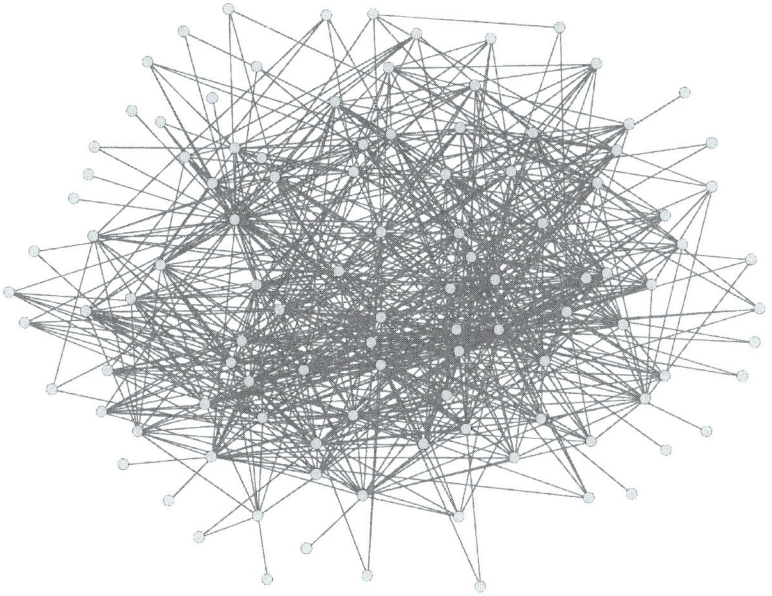

(a)

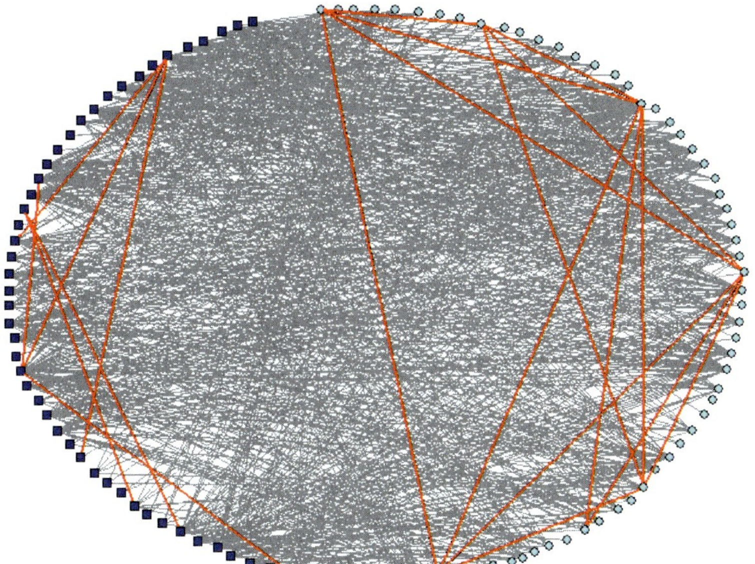

(b)

Figure 4.7 (a) Network representation of the food web of Canton Creek. (b) Bipartite structure of this network as found by the method explained here. Nodes in each quasibipartite cluster are represented by squares and circles of two different colors. The thick red lines represent the intracluster connections and the gray lines the intercluster links.

Definition 4.4 The *node-repulsion graph* is a graph whose adjacency matrix is given by $\Theta\,[\exp(-A)]$, which results from the elementwise application of the function $\Theta(x)$ to the matrix $\exp(-A)$. A pair of nodes p and q in the node-repulsion graph $\Theta\,[\exp(-A)]$ is connected if and only if they have $G_{pq} > 0$. The Heaviside function $\Theta\,(x)$ was already introduced in Equation (4.39).

Using the *node-repulsion graph*, the search for quasi-bipartite subgraphs in a complex network is reduced to finding the cliques in a simple graph, $\Theta\,[\exp(-A)]$. These cliques correspond to the quasi-bipartite clusters of the network [73].

4.6.2 Application

Here we study the food web of Canton Creek, which consists primarily of invertebrates and algae in a tributary, surrounded by pasture, of the Taieri River in the South Island of New Zealand [74]. This network consists of 108 nodes (species) and 707 links (trophic relations). Using our current approach, we find that this network can be divided into two almost-bipartite clusters, one having 66 nodes and the other 42. Only 20 links connect nodes in the same clusters, 13 of them connect nodes in the set containing 66 nodes and the other 7 connect nodes in the set of 42 nodes. Thus 97.2% of links are connections between the two almost-bipartite clusters and only 2.8% links are intracluster connections [73]. In Figure 4.7, we illustrate the network and its quasi-bipartite clusters as found in the current work. The value of the bipartivity measure for this network $b(G) = 0.775$ indicates that the network in general is not bipartite but that important bipartite and quasi-bipartite structures are present in the graph, which is corroborated by our algorithm for finding such structures [73].

4.7 Conclusion

The discovery of X-rays more than a century ago has increased our knowledge in many fields, such as the structure of matter, cosmology, security in technology, and X-ray diagnostics, among others. The existence of a tool, like X-rays and other spectroscopic techniques, permits us to understand the internal structure of the systems under study from molecules and materials to the human body. In a similar way, spectral graph theory is the X-ray machine for studying complex networks. As we have shown here, the use of graph spectral techniques permits us to analyze the local and global structure of complex networks.

Using graph spectral theory it is possible to "see" how central a node is based on its weighted participation in all substructures present in the graph. The same techniques permit us to analyze whether a network is homogeneous or modular. In the last case it permits us to classify their structures according to certain universal structural classes, regardless of whether it is representing a cell or a society. In addition, the spectral techniques explained in this chapter permit us to identify the communities existing in a complex network, as well as the bipartivity structure of certain substructures present in such systems. Many other characteristics of complex networks could be investigated using the spectra of graphs. Some of them have already been described by the scientists working in this field, others are still waiting for the development of the appropriate tools. I hope this chapter helps to inspire the development of new spectral measures for characterizing the structure and functioning of complex networks.

Acknowledgment

This chapter was written between the University of Santiago de Compostela, Spain, the University of Strathclyde, UK, and the University of Tokyo, Japan. The author would like to thank the program "Ramón y Cajal", Spain, and acknowledge the support of the Royal Society of Edinburgh and the Edinburgh Mathematical Society during a visit to the Department of Mathematics, Strathclyde University (March 2008), to the IIS, University of Tokyo, for a fellowship as Research Visitor during April–June 2008. The author is grateful to both institutions for their warm hospitality.

References

1 Bar-Yam, Y., *Making Things Work: Solving Complex Problems in a Complex World.* NECSI Knowledge Press, Cambridge, MA, **2004**.
2 da F. Costa, L., Rodríguez, F.A., Travieso, G., Villa Boas, P.R., *Adv. Phys.* 56 (**2007**) p. 167.
3 Albert, R., Barabási, A.-L., *Rev. Mod. Phys.* 74 (**2002**) p. 47.
4 Newman, M.E.J., *SIAM Rev.* 45 (**2003**) p. 167.
5 Boccaletti, S., Latora, V., Moreno, Y., Chavez, M., Hwang, D.-U., *Phys. Rep.* 424 (**2006**) p. 175.
6 Harary, F. *Graph Theory*. Addison-Wesley, Reading, MA, **1994**.
7 Rual, J.-F. et al., *Nature* 437 (**2005**) p. 1173.
8 Biggs, N.L., *Algebraic Graph Theory*. Cambridge University Press, Cambridge, **1993**.
9 Cvetković, D., Doob, M., Sach, H., *Spectra of Graphs*. Academic Press, NY, **1980**.
10 Chung, F.R.K., *Spectral Graph Theory*. CBMS 92, AMS, Providence, RI, **1997**.
11 Horn, R., Johnson, C.R., *Matrix Analysis*. Cambridge University Press, Cambridge, **1985**.
12 Cvetković, D., Rowlinson, P., Simiæ, S., *Eigenspaces of Graphs*. Cambridge University Press, Cambridge, **1997**.
13 Cvetković, D., Doob, M., Gutman, I., Torgašev, A. *Recent Results in the Theory of Graph Spectra*. North-Holland, Amsterdam, **1988**.

14 Freeman, L.C., *Social Networks* 1 (**1979**) p. 215.

15 Albert, R., Jeong, H., Barabási, A.-L., Nature 401 (**1999**) p. 130.

16 Bonacich, P., *J. Math. Sociol.* 2 (**1972**) p. 113.

17 Bonacich, P., *Am. J. Sociol.* 92 (**1987**) p. 1170.

18 Newman, M.E.J., *Phys. Rev. E* 70 (**2004**) p. 056131.

19 Estrada, E., Rodríguez-Velázquez, J.A., *Phys. Rev. E* 71 (**2005**) p. 056103.

20 de la Peña, J.A., Gutman, I. Rada, J., *Lin. Algebra Appl.* 427 (**2007**) p. 70.

21 Gutman, I. Graovac, A., *Chem. Phys. Lett.* 436 (**2007**) p. 294.

22 Ginosar, Y., Gutman, I., Mansour, T., Schork, M., *Chem. Phys. Lett.* 454 (**2008**) p. 145.

23 Carbó-Dorca, R., *J. Math. Chem.* 44 (**2008**) p. 373.

24 Rodríguez-Velázquez, J.A., Estrada, E., Gutiérrez, A., *Lin. Multil. Algebra* 55 (**2007**) 293.

25 Estrada, E., Hatano, N., *Chem. Phys. Lett.* 439 (**2007**) 247.

26 Choi, J.H., Barnett, G.A., Chon, B.-S., *Global Networks* 6 (**2006**) p. 81.

27 Variano, E.A., McCoy, J.H., *Phys. Rev. Lett.* 92 (**2004**) p. 188701.

28 Newman, M.J.E., *Proc. Natl. Acad. Sci USA* 103 (**2006**) p. 8577.

29 Wagner, G.P., Pavlicev, M., Cheverud, J.M., *Nature Rev. Genet.* 8 (**2007**) p. 921.

30 Hoory, S., Linial, N., Wigderson, A., *Bull. Am. Math. Soc.* 43 (**2006**) 439.

31 Barenblatt, G.I., *Scaling*. Cambridge University Press, Cambridge, **2003**.

32 Estrada, E., Rodríguez-Velázquez, J.A., *Phys. Rev. E* 72 (**2005**) 046105.

33 Dodziuk, J., *Trans. Amer. Math. Soc.* 284 (**1984**) p. 787.

34 Alon, N., Milman, V.D., *J. Combin. Theory Ser. B*, 38 (**1985**) p. 73.

35 Alon, N., *Combinatorica*, 6 (**1986**) p. 83.

36 Estrada, E., *Europhys. Lett.* 73 (**2006**) p. 649.

37 Estrada, E., *Eur. Phys. J. B* 52 (**2006**) p. 563.

38 Watts, D.J., Strogatz, S.H., *Nature* 393 (**1998**) p. 440.

39 Dunne, J.A., Williams, R.J., Martinez, N.D., *Proc. Natl. Acad. Sci. USA* 99 (**2002**) p. 12917.

40 Barabási, A.-L., Albert, R., *Science* 286 (**1999**) p. 509.

41 Amaral, L.A.N., Scala, A., Barthélémy, M., Stanley, H.E., *Proc. Natl. Acad. Sci. USA* 97 (**2000**) p. 11149.

42 Goh, K.-I., Oh, E., Jeong, H., Kahng, B., Kim, D., *Proc. Natl. Acad. Sci. USA* 99 (**2002**) p. 12583.

43 Pastor-Satorras, R., Vespignani, A., *Phys. Rev. Lett.* 86 (**2001**) p. 3200.

44 Newman, M.E.J., *Phys. Rev. E* 66 (**2002**) p. 016128.

45 May, R.M., Lloyd, A.L., *Phys. Rev. E* 64 (**2001**) p. 066112.

46 Albert, R., Jeong, H., Barabási, A.-L., *Nature* 406 (**2000**) p. 378.

47 Paul, G., Tanizawa, T., Havlin, S., Stanley, H.E., *Eur. Phys. J. B* 38 (**2004**) p. 187.

48 Balthrop, J., Forrest, S., Newman, M.J.E., Williamson, M.M., *Science* 304 (**2004**) p. 527.

49 Estrada, E., *Phys. Rev. E* 75 (**2007**) p. 016103.

50 Jungsbluth, M., Burghardt, B., Hartmann, A.K., *Physica A* 381 (**2007**) p. 444.

51 Borgatti, S.P., *Social Networks* 27 (**2005**) p. 55.

52 Hromkovic, J., Klasing, R., Pelc, A., Ruzicka, P., Unger, W., *Dissemination of Information in Communication Networks: Broadcasting, Gossiping, Leader Election, and Fault Tolerance*. Springer, Berlin, **2005**.

53 Shi, T.J., Mohan, G., *Comput. Comm.* 29 (**2006**) p. 1284.

54 Estrada, E., Hatano, N., *Phys. Rev. E* 77 (**2008**) p. 036111.

55 Morse, P.M., Feshbach, H., *Methods of Theoretical Physics*. McGraw-Hill, New York, **1953**.

56 Girvan, M., Newman, M.E.J., *Proc. Natl. Acad. Sci. USA* 99 (**2002**) p. 7821.

57 Radicchi, F., Castellano, C., Cecconi, F., Loreto, V., Parisi, D., *Proc. Natl. Acad. Sci. USA* 101 (**2004**) p. 2658.

58 Newman, M.E.J., *Eur. Phys. J. B* 38 (**2004**) p. 321.

59 Newman, M.E.J., *Phys. Rev. E* 69 (**2004**) p. 066133.

60 Palla, G., Derényi, I., Farkas, I., Vicsek, T., *Nature* 435 (**2005**) p. 814.

61 Newman, M.E.J., *Phys. Rev. E* 74 (**2006**) p. 036104.

62 Amaral, L.A.N., Ottino, J., *Eur. Phys. J. B* 38 (**2004**) p. 147.

63 Estrada, E., Hatano, N., submitted (**2008**).

64 Bomze, I.M., Budinich, M., Pardalos, P.M., Pelillo, M., in: Du, D.-Z., Pardalos, P.M. (Eds.), *Handbook of Combinatorial Optimization, Supplement Vol. A*, pp. 1–74. Kluwer Academic Publishers, Dordrecht, **1999**.

65 Bron, C., Kerbosch, J., *Comm. ACM* 16 (**1973**) 575.

66 Moon, J.W., Moser, L., *Israel J. Math.* 3 (**1965**) 23.

67 Zachary, W.W., *J. Anthropol. Res.* 33 (**1977**) 452.

68 Guillaume, J.-L., Lapaty, M., *Inform. Proces. Lett.* 90 (**2004**) 215.

69 Holme, P., Liljeros, F., Edling, C.R., Kim, B.J., *Phys. Rev. E.* 68 (**2003**) p. 056107.

70 König, D., *Theorie der endlichen und unendlichen Graphen*. Leipzig, **1936**.

71 Seary, A.J., Richards, Jr., W.D., in: *Proceedings of the International Conference on Social Networks*, London, edited by M.G. Everett and K. Rennolds. Greenwich University Press, London, **1995**, Vol. 1, p. 47.

72 Xiao, T. Gong, S., *Pattern Recog.* 41 (**2008**) 1012.

73 Estrada, E., Higham, D.J., Hatano, N., *Physica A* 388 (**2009**) 264.

74 Townsend, C., Thompson, R.M., McIntosh, A.R., Kilroy, C., Edwards, E., Scarsbrook, M.R., *Ecol. Lett.* 1 (**1998**) 200.

5
On the Structure of Neutral Networks of RNA Pseudoknot Structures

Christian M. Reidys

5.1
Motivation and Background

Random induced subgraphs arise in the context of molecular folding maps [24] where the neutral networks of molecular structures can be modeled as random induced subgraphs of n-cubes [17]. They also occur in the context of neutral evolution of populations (i.e., families of Q_2^n-vertices) consisting of erroneously replicating RNA strings. Here, one works in Q_4^n since we have for RNA the nucleotides $\{\mathbf{A}, \mathbf{U}, \mathbf{G}, \mathbf{C}\}$. Random induced subgraphs of n-cubes have had an impact on the conceptual level [23] and led to experimental work identifying sequences that realize two distinct ribozymes [22]. A systematic computational analysis of neutral networks of molecular folding maps can be found in [11, 12]. An RNA structure, $\mathfrak{s}$, is a graph over $[n]$ having vertex degree ≤ 1 and whose arcs are drawn in the upper half-plane (Figure 5.1). The set of $\mathfrak{s}$-compatible sequences, $C[\mathfrak{s}]$, consists of all sequences that have at any two paired positions one of the 6 nucleotide pairs $(\mathbf{A}, \mathbf{U})$, $(\mathbf{U}, \mathbf{A})$, $(\mathbf{G}, \mathbf{U})$, $(\mathbf{U}, \mathbf{G})$, $(\mathbf{G}, \mathbf{C})$, $(\mathbf{C}, \mathbf{G})$. The structure $\mathfrak{s}$ gives rise to a new adjacency relation

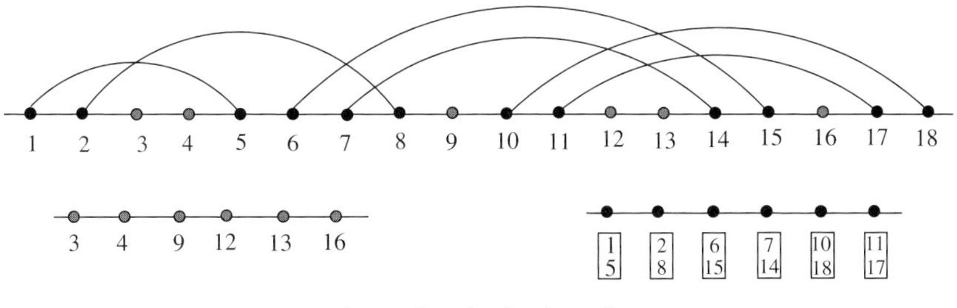

Figure 5.1 An RNA structure and its induced subcubes $Q_4^{n_u}$ and $Q_6^{n_p}$: a structure allows one to rearrange its compatible sequences into unpaired and paired segments. The former is a sequence over the original alphabet $\mathbf{A}$, $\mathbf{U}$, $\mathbf{G}$, $\mathbf{C}$ and for the latter we derive a sequence over the alphabet of base pairs, $(\mathbf{A}, \mathbf{U})$, $(\mathbf{U}, \mathbf{A})$, $(\mathbf{G}, \mathbf{U})$, $(\mathbf{U}, \mathbf{G})$, $(\mathbf{G}, \mathbf{C})$, $(\mathbf{C}, \mathbf{G})$.

Analysis of Complex Networks: From Biology to Linguistics. Edited by Matthias Dehmer and Frank Emmert-Streib

ISBN: 978-3-527-32345-6

within $C[\mathfrak{s}]$. Indeed, we can reorganize a sequence $(x_1, \ldots, x_n)$ into the tuple

$$\left((u_1, \ldots, u_{n_\mathrm{u}}), (p_1, \ldots, p_{n_\mathrm{p}})\right) , \tag{5.1}$$

where u_j denotes the unpaired nucleotides and $p_j = (x_i, x_k)$ all base pairs, respectively (Figure 5.1). We can view $v_\mathrm{u} = (u_1, \ldots, u_{n_\mathrm{u}})$ and $v_\mathrm{p} = (p_1, \ldots, p_{n_\mathrm{p}})$ as elements of the cubes $Q_4^{n_\mathrm{u}}$ and $Q_6^{n_\mathrm{p}}$, implying the new adjacency relation for elements of $C[\mathfrak{s}]$. That is, $C[\mathfrak{s}]$ carries the natural graph structure $Q_4^{n_\mathrm{u}} \times Q_6^{n_\mathrm{p}}$, where "×" denotes the direct product of graphs. The neutral network of $\mathfrak{s}$ is the set of all sequences that fold into $\mathfrak{s}$ and are contained in the set of compatible sequences. Whether or not some compatible sequence is contained in the neutral network can be decided by independently selecting vertices v_u and v_p with probability λ_u and λ_p, respectively [17]. This modeling ansatz leads accordingly to random induced subgraphs of n-cubes. Note that the probabilities λ_u and λ_p are easily measured *locally* via RNA computer folding maps: they coincide with the average fraction of neutral neighbors within the compatible neighbors. Explicitly, λ_u is the percentage of sequences that differ by a neutral mutation in an unpaired position, while λ_p corresponds to the percentage of neutral sequences that are compatible via a base pair mutation (for instance, $(\mathbf{A}, \mathbf{U}) \mapsto (\mathbf{G}, \mathbf{C})$) (Figure 5.2). One particularly fascinating property of random induced is the existence of unique, large components. Its existence has a profound impact on the evolutionary optimization process since it shows that large Hamming distances can be traversed in the course of neutral evolution by successively performing point mutations. Burtin was

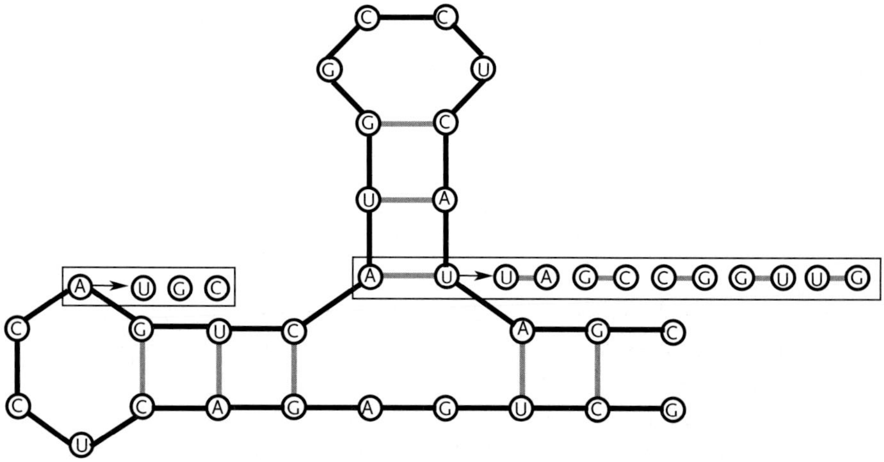

Figure 5.2 Compatible mutations: representing a secondary structure differently, the red edges correspond to the arcs in the upper half-plane of its diagram representation. We illustrate the different alphabets for compatible mutations in unpaired and paired positions.

the first [8] to study the connectedness of random subgraphs of n-cubes, Q_2^n, obtained by selecting all Q_2^n-edges independently (with probability p_n). He proved that a.s. all such subgraphs are connected for $p > 1/2$ and are disconnected for $p < 1/2$. Erdős and Spencer [10] refined Burtin's result and, more importantly in our context, they conjectured that there exists a.s. a giant component for $p_n = \frac{1+\varepsilon}{n}$ and $\varepsilon > 0$. Their conjecture was proved by Ajtai, Komlós, and Szemerédi [1], who established the existence of a giant component for $p_n = \frac{1+\varepsilon}{n}$. Key ingredients in their proof are Harper's isoperimetric inequality [13] and a two-round randomization, used for showing the nonexistence of certain splits. Considerably less is known for random induced subgraphs of the n-cube obtained by independently selecting each Q_2^n-vertex with probability λ_n. One main result here is the paper of Bollobás *et al.*, who have shown in [5] for constant χ that $C_n^{(1)} = (1+o(1))\kappa\chi\frac{1+\chi}{n}2^n$. Recently their result has been improved [20]: for $\chi_n \geq n^{-\frac{1}{3}+\delta}$, where $\delta > 0$, a unique largest component of specific size exists. For $\chi_n = \varepsilon$ this result (combined with a straightforward argument for $\lambda_n \leq \frac{1-\varepsilon}{n}$) implies the analog of Ajtai *et al.*'s [1] result for random induced subgraphs. In this contribution we will introduce the key concepts and ideas in the context of connectivity and large components in binary and generalized n-cubes.

5.1.1
Notation and Terminology

The n-cube, Q_α^n, is a combinatorial graph with vertex set $\mathbb{A}^n$, where $\mathbb{A}$ is some finite alphabet of size $\alpha \geq 2$. W.l.o.g. we will assume $\mathbb{F}_2 \subset \mathbb{A}$ and call Q_2^n the binary n-cube. In an n-cube two vertices are adjacent if they differ in exactly one coordinate. Let $d(v, v')$ be the number of coordinates by which v and v' differ. We set

$$\forall A \subset \mathbb{A}^n, j \leq n; \quad \mathsf{B}(A,j) = \{v \in \mathbb{A}^n \mid \exists \alpha \in A;\ d(v,\alpha) \leq j\} \tag{5.2}$$

$$\mathsf{S}(A,j) = \{v \in \mathbb{A}^n \mid \exists \alpha \in A;\ d(v,\alpha) = j\} \tag{5.3}$$

$$\forall A \subset \mathbb{A}^n; \quad \mathsf{d}(A) = \{v \in \mathbb{A}^n \setminus A \mid \exists \alpha \in A;\ d(v,\alpha) = 1\} \tag{5.4}$$

and call $\mathsf{B}(A,j)$ and $\mathsf{d}(A)$ the ball of radius j around A and the vertex boundary of A in Q_α^n, respectively. If $A = \{v\}$, we simply write $\mathsf{B}(v,j)$. Let A, $B \subset \mathbb{A}^n$; we call A ℓ-dense in B if $\mathsf{B}(v,\ell) \cap A \neq \varnothing$ for $v \in B$.

Q_2^n can be viewed as the Cayley graph $\mathsf{Cay}(\mathbb{F}_2^n, \{e_i \mid i = 1, \ldots, n\})$, where e_i is the canonical base vector. We will view $\mathbb{F}_2^n$ as a $\mathbb{F}_2$-vector space and denote the linear hull over $\{v_1, \ldots, v_h\}$, $v_j \in \mathbb{F}_2^n$ by $\langle v_1, v_2, \ldots, v_h\rangle$. There exists a natural linear order $\leq$ over Q_2^n given by

$$v \leq v' \iff (d(v,0) < d(v',0)) \ \vee\ (d(v,0) = d(v',0) \ \wedge\ v <_{\text{lex}} v')\,, \tag{5.5}$$

where $<_{\text{lex}}$ denotes the lexicographical order. Any notion of minimal element or smallest element in $A \subset Q_2^n$ is considered w.r.t. the linear order $\le$ of Equation (5.5).

Each $A \subset \mathbb{A}^n$ induces a unique induced subgraph in Q_α^n, denoted by $Q_\alpha^n[A]$, in which $a_1, a_2 \in A$ are adjacent if a_1, a_2 are adjacent in Q_α^n. Let Q_{α,λ_n}^n be a random graph consisting of Q_α^n-subgraphs, Γ_n, induced by selecting each Q_α^n-vertex with independent probability λ_n. Q_{α,λ_n}^n is the finite probability space $(\{Q_\alpha^n[A] \mid A \subset \mathbb{A}^n\}, \mathbb{P})$, with the probability measure $\mathbb{P}(A) = \lambda_n^{|A|}(1-\lambda_n)^{\alpha^n - |A|}$. A property M is a subset of induced subgraphs of Q_α^n closed under graph isomorphisms. The terminology "M holds a.s." is equivalent to $\lim_{n\to\infty}\mathbb{P}(\mathsf{M}) = 1$. A component of Γ_n is a maximal connected induced Γ_n-subgraph, C_n. The largest Γ_n-component is denoted by $C_n^{(1)}$. It is called a giant component if and only if

$$\exists\,\kappa > 0, \quad |C_n^{(1)}| \ge \kappa\,|\Gamma_n|\,, \tag{5.6}$$

and $x_n \sim y_n$ is equivalent to (a) $\lim_{n\to\infty} x_n/y_n$ exists and (b) $\lim_{n\to\infty} x_n/y_n = 1$. Let $Z_n = \sum_{i=1}^n \xi_i$ be a sum of mutually independent indicator random variables (r.v.), ξ_i having values in $\{0, 1\}$. Then we have, [9], for $\eta > 0$ and $c_\eta = \min\{-\ln(e^\eta[1+\eta]^{-[1+\eta]}), \frac{\eta^2}{2}\}$

$$\mathsf{Prob}\,(\,|\,Z_n - \mathbb{E}[Z_n]\,| > \eta\,\mathbb{E}[Z_n]\,) \le 2\,e^{-c_\eta \mathbb{E}[Z_n]}\,. \tag{5.7}$$

n is always assumed to be sufficiently large and ε is a positive constant satisfying $0 < \varepsilon < \frac{1}{3}$. We use the notation $B_m(\ell, \lambda_n) = \binom{m}{\ell}\lambda_n^\ell(1-\lambda_n)^{m-\ell}$ and write $g(n) = O(f(n))$ and $g(n) = o(f(n))$ for $g(n)/f(n) \longrightarrow \kappa$ as $n \longrightarrow \infty$ and $g(n)/f(n) \longrightarrow 0$ as $n \longrightarrow \infty$, respectively.

5.2
Preliminaries

In this section we present three theorems, instrumental for our analysis of connectivity, large components, and distances in n-cubes. The first result is due to [4] used for Sidon sets in groups in the context of Cayley graphs. In what follows G denotes a finite group and M a finite set acted upon by G.

Proposition 5.1 *Suppose G acts transitively on M and let $A \subset M$. Then we have*

$$\frac{1}{|G|}\sum_{g\in G}|A \cap gA| = |A|^2/|M|\,. \tag{5.8}$$

Proof. We prove Equation (5.8) by induction on $|A|$. For $A = \{x\}$ we derive $\frac{1}{|G|}\sum_{gx=x} 1 = |G_x|/|G|$, since $|M| = |G|/|G_x|$. We next prove the induction step. We write $A = A_0 \cup \{x\}$ and compute

$$\begin{aligned}\frac{1}{|G|}\sum_g |A \cap gA| &= \frac{1}{|G|}\sum_g \big(|A_0 \cap gA_0| + |\{gx\} \cap A_0| + |\{x\} \cap gA_0| \\ &\quad + |\{gx\} \cap \{x\}| \\ &= \frac{1}{|G|}\left(|A_0|^2|G_x| + 2|A_0||G_x| + |G_x|\right) \\ &= \frac{1}{|G|}\left((|A_0|+1)^2\,|G_x|\right) = \frac{|A|^2}{|M|}\,.\end{aligned}$$

Aldous [2, 3] observed how to use Proposition 5.1 for deriving a very general lower bound for vertex boundaries in Cayley graphs:

Theorem 5.1 *Suppose G acts transitively on M and let $A \subset M$, and let S be a generating set of the Cayley graph* Cay(G, S), *where $|S| = n$. Then we have*

$$\exists s \in S; \quad |sA \setminus A| \geq \frac{1}{n}|A|\left(1 - \frac{|A|}{|M|}\right)\,. \tag{5.9}$$

Proof. We compute

$$|A| = \frac{1}{|G|}\sum_g(|gA \setminus A| + |A \cap gA|) = \frac{1}{|G|}\sum_g |gA \setminus A| + |A|\frac{|A|}{|M|}\,, \tag{5.10}$$

and hence $|A|\left(1 - \frac{|A|}{|M|}\right) = \frac{1}{|G|}\sum_g |gA \setminus A|$. From this we can immediately conclude

$$\exists g \in G; \quad |gA \setminus A| \geq |A|\left(1 - \frac{|A|}{|M|}\right)\,.$$

Let $g = \prod_{j=1}^k s_j$. Since each element of $gA \setminus A$ is contained in at least one set $s_jA \setminus A$, we obtain

$$|gA \setminus A| \leq \sum_{j=1}^{k} |s_jA \setminus A|\,.$$

Hence there exists some $1 \leq j \leq k$ such that $|s_jA \setminus A| \geq \frac{1}{k}|gA \setminus A|$, and the lemma follows.

Let us next recall some basic facts about branching processes [14, 16]. Suppose ξ is a random variable and $\left(\xi_i^{(t)}\right)$, $i, t \in \mathbb{N}$ counts the number of offspring of the ith individual at generation $t-1$. We consider the family of r.v. $(Z_i)_{i \in \mathbb{N}_0}$: $Z_0 = 1$ and $Z_t = \sum_{i=1}^{Z_{t-1}} \xi_i^{(t)}$ for $t \geq 1$ and interpret Z_t as the number of individuals "alive" in generation t. We will be interested in the limit probability $\lim_{t \to \infty} \text{Prob}(Z_t > 0)$, i.e., the probability of infinite survival. We have the following theorem.

Theorem 5.2 *Let* $u_n = n^{-\frac{1}{3}}$, $\lambda_n = \frac{1+\chi_n}{n}$, $m = n - \lfloor \frac{3}{4} u_n n \rfloor$, *and* $\mathrm{Prob}(\xi = \ell) = B_m(\ell, \lambda_n)$. *Then for* $\chi_n = \varepsilon$ *the r.v.* ξ *becomes asymptotically Poisson, i.e.,* $\mathbb{P}(\xi = \ell) \sim \frac{(1+\varepsilon)^\ell}{\ell!} e^{-(1+\varepsilon)}$ *and*

$$0 < \lim_{t \to \infty} \mathrm{Prob}(Z_t > 0) = \alpha(\varepsilon) < 1 . \tag{5.11}$$

For $o(1) = \chi_n \geq n^{-\frac{1}{3}+\delta}$, $\delta > 0$, *we have*

$$\lim_{t \to \infty} \mathrm{Prob}(Z_t > 0) = (2 + o(1))\, \chi_n . \tag{5.12}$$

We proceed by stating Janson's inequality [15]. It is the key tool for proving Theorem 5.4 in Section 5.3 and Theorem 5.7 in Section 5.5. Intuitively, Janson's inequality can be viewed as a large deviation result in the presence of correlation.

Theorem 5.3 *Let* R *be a random subset of some set* $[V] = \{1, \ldots, V\}$ *obtained by selecting each element* $v \in V$ *independently with probability* λ. *Let* $S_1, \ldots, S_s$ *be subsets of* $[V]$ *and* X *be the r.v. counting the number of* S_i *for which* $S_i \subset R$. *Let furthermore*

$$\Omega = \sum_{(i,j);\, S_i \cap S_j \neq \emptyset} \mathbb{P}(S_i \cup S_j \subset R) , \tag{5.13}$$

where the sum is taken over all ordered pairs (i, j). *Then for any* $\gamma > 0$, *we have*

$$\mathbb{P}(X \leq (1 - \gamma)\mathbb{E}[X]) \leq e^{-\frac{\gamma^2 \mathbb{E}[X]}{2+2\Omega/\mathbb{E}[X]}} . \tag{5.14}$$

5.3
Connectivity

As already mentioned, the connectivity property of neutral networks of RNA structures has a profound impact on our picture of evolutionary optimization. It is closely related to the connectivity of the two subcubes induced by the unpaired and paired nucleotides. We present the combinatorial, constructive proof that localizes the threshold value for generalized n-cubes due to [18]. The particular construction has led to several computational studies on the connectivity of neutral networks [11, 12].

Lemma 5.1 *Let* Q_α^n *be a generalized* n*-cube,* $\lambda > 1 - \sqrt[\alpha-1]{\alpha^{-1}}$, *and* Γ_n *an induced* Q_α^n*-subgraph obtained by selecting each* Q_α^n*-vertex with independent probability* λ. *Then we have*

$$\lim_{n \to \infty} \mathbb{P}(\forall v, v' \in \Gamma_n,\ d_{Q_\alpha^n}(v, v') = k;\ v \text{ is connected to } v') = 1 . \tag{5.15}$$

Proof. *Claim* 1. Suppose $\lambda > 1 - \sqrt[\alpha-1]{\alpha^{-1}}$. Then for arbitrary $\ell \in \mathbb{N}$, Γ_n contains a.s. exclusivcly vertices of degree $\geq \ell$.

To prove the claim we first observe that $\lambda > 1 - \sqrt[\alpha-1]{\alpha^{-1}}$ is equivalent to $(1-\lambda)^{(\alpha-1)}\alpha < 1$. We fix $\ell \in \mathbb{N}$. Using the linearity of expectation, the expected number of vertices of degree $\leq \ell$ is given by

$$\begin{aligned}\alpha^n \sum_{i=0}^{\ell} \binom{(\alpha-1)n}{i} \lambda^i (1-\lambda)^{(\alpha-1)n-i} &\leq \ell\left((\alpha-1)n\right)^{\ell} \alpha^n (1-\lambda)^{(\alpha-1)n-\ell} \\ &= c' n^{\ell} \left[\alpha(1-\lambda)^{(\alpha-1)}\right]^n, \quad c' > 0 \\ &\sim e^{-cn}, \quad c > 0\,.\end{aligned}$$

Since we have for any r.v. X with positive integer values: $\mathbb{E}(X) \geq \mathbb{P}(X > 0)$, Claim 1 follows.

According to Claim 1 we can now choose for $v, v' \in \Gamma_n$ with $d(v, v') = k$ and $\ell \in \mathbb{N}$ the two sets of neighbors $\{v^{(j_h)} \mid 1 \leq h \leq \ell\}$ and $\{v'^{(i_h)} \mid 1 \leq h \leq \ell\}$. W.l.o.g. we may assume that $\{j_h\} = \{1, \ldots, \ell\}$ and $\{i_h\} = \{\ell+1, \ldots, 2\ell\}$ and that v, v' differ exactly in the positions $2\ell+1, \ldots, 2\ell+k$. Furthermore we may assume that v, v' and $v^{(i)}, v'^{(\ell+i)}$ differ by 0 and 1 entries, i.e., are of the form

$$v = (\underbrace{0, \ldots, 0}_{\ell}, \underbrace{0, \ldots, 0}_{\ell}, \underbrace{0, \ldots, 0}_{k}, x_{2\ell+k+1}, \ldots, x_n) \tag{5.16}$$

$$v' = (\underbrace{0, \ldots, 0}_{\ell}, \underbrace{0, \ldots, 0}_{\ell}, \underbrace{1, \ldots, 1}_{k}, x_{2\ell+k+1}, \ldots, x_n) \tag{5.17}$$

$$v^{(i)} = (\underbrace{0, \ldots, 1, 0, \ldots, 0}_{1 \text{ in } i\text{th position}}, \underbrace{0, \ldots, 0}_{\ell}, \underbrace{0, \ldots, 0}_{k}, x_{2\ell+k+1}, \ldots, x_n) \tag{5.18}$$

$$v'^{(\ell+i)} = (\underbrace{0, \ldots, 0}_{\ell}, \underbrace{0, \ldots, 1, 0, \ldots, 0}_{1 \text{ in } (\ell+i)\text{th position}}, \underbrace{1, \ldots, 1}_{k}, x_{2\ell+k+1}, \ldots, x_n)\,. \tag{5.19}$$

For each pair of elements $(v^{(i)}, v'^{(\ell+i)})$ with $1 \leq i \leq \ell$ we consider the sets $\mathsf{B}^{n-(2\ell+k)}(v^{(i)}, 1)$ and $\mathsf{B}^{n-(2\ell+k)}(v'^{(i+\ell)}, 1)$, where

$$\mathsf{B}^{n-(2\ell+k)}(w, 1) = \{e_h + w \mid 2\ell + k < h \leq n\}\,. \tag{5.20}$$

$(v^{(i)}, v'^{(i+\ell)})$ is connected by the Q^n_α-path

$$\gamma_i = \left(v^{(i)}, e_{i+\ell}, \underbrace{e_{2\ell+1}, \ldots, e_{2\ell+k}}_{k}, e_i, v'^{(i+\ell)}\right), \quad 1 \leq i \leq \ell\,. \tag{5.21}$$

γ_i is contained in Γ_n with a probability of at least λ^{k+2}. Since all neighbors of v and v' are of the form $v^{(i)}$, for $1 \leq i \leq \ell$ and $v'^{(i+\ell)}$ for $\ell + 1 \leq i + \ell \leq 2\ell$, for $i \neq j$ any two paths

$$\gamma_i = \left(v^{(i)}, e_{i+\ell}, e_{2\ell+1}, \ldots, e_{2\ell+k}, e_i, v'^{(\ell+i)}\right) \tag{5.22}$$

$$\gamma_j = \left(v^{(j)}, e_{j+\ell}, e_{2\ell+1}, \ldots, e_{2\ell+k}, e_j, v'^{(\ell+j)}\right) \tag{5.23}$$

are vertex disjoint. The probability of selecting a pair of vertices $(v^{(i)}+e_h, v'^{(i+\ell)}+e_h)$ is λ^2. We have the pairs

$$\left(v^{(i)} + e_h, v'^{(i+\ell)} + e_h\right), \quad \left(v^{(j)} + e_{h'}, v'^{(j+\ell)} + e_{h'}\right), \quad 1 \le i,j \le \ell,\ h \neq h'$$

and for $i = j$ we have the vertex disjoint paths $\gamma_i + e_h, \gamma_i + e_{h'}$ since $h, h' > 2\ell + (k)$. Two paths $\gamma_i + e_h$ and $\gamma_j + e_h$ of two pairs

$$\left(v^{(i)} + e_h, v'^{(i+\ell)} + e_h\right) \quad \text{and} \quad \left(v^{(j)} + e_h, v'^{(j+\ell)} + e_h\right)$$

are, in view of Equations (5.22) and (5.23), also disjoint (Figure 5.3). The probability that for all pairs $(v^{(i)} + e_h, v'^{(i+\ell)} + e_h)$, where $1 \le i \le \ell$, $2\ell + k < h$ we select none of the $\gamma_i + e_h$-paths is by linearity of expectation less than

$$\alpha^n n^{k+2}\left(1 - \lambda^2\lambda^{2+k}\right)^{\ell(n-(2\ell+k))} = \alpha^n n^{k+2}\left(1 - \lambda^{4+k}\right)^{-\ell(2\ell+k)}\left(1 - \lambda^{4+k}\right)^{\ell n}.$$

By choosing ℓ large enough we can satisfy

$$\left(1 - \lambda^{4+k}\right)^{\ell} < (1 - \lambda)^{(\alpha-1)}, \tag{5.24}$$

whence

$$n^{k+2}(1 - \lambda^{4+k})^{-\ell(2\ell+k)}\left[\alpha(1 - \lambda)^{(\alpha-1)}\right]^n,$$

which obviously tends to zero. Accordingly, there exists a.s. at least one path of the form $\gamma_i + e_h$ (Equation (5.21)) that connects v and v' in Γ_n, and the proof of the lemma is complete.

Theorem 5.4 *Let Q^n_α be a generalized n-cube and $\mathbb{P}$ the probability $\mathbb{P}(\Gamma_n) = \lambda^{|\Gamma_n|}(1 - \lambda)^{\alpha^n - |\Gamma_n|}$. Then the following assertions hold:*

$$\lim_{n\to\infty} \mathbb{P}(\Gamma_n \text{ is connected}) = \begin{cases} 0 & \text{for } \lambda < 1 - \sqrt[\alpha-1]{\alpha^{-1}} \\ 1 & \text{for } \lambda > 1 - \sqrt[\alpha-1]{\alpha^{-1}} \end{cases}. \tag{5.25}$$

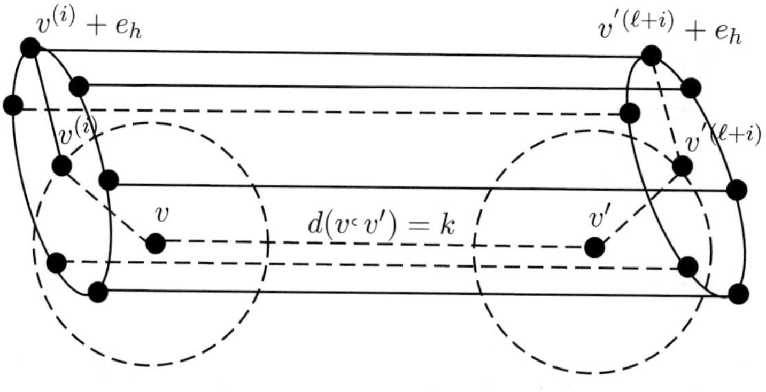

Figure 5.3 Illustration of the proof idea: constructing the independent paths.

Proof. Suppose first we have $\lambda > 1 - \sqrt[\alpha-1]{\alpha^{-1}}$. For any two vertices $w, w' \in \Gamma_n$ we fix a shortest Q_α^n-path $\gamma_{w,w'}$ connecting them. Let Z be the r.v. counting the isolated vertices in Γ_n and w_{i_j} the vertex of the jth step of $\gamma_{w,w'}$. Since

$$\mathbb{P}(\mathsf{B}(w_{i_j}, 1) \cap \Gamma_n = \varnothing, 1 \le j \le s) \le \mathbb{E}(Z) = \alpha^n \lambda (1-\lambda)^{(\alpha-1)n}, \tag{5.26}$$

we observe that for all $1 \le j \le s$, $\mathsf{B}(w_{i_j}, 1) \cap \Gamma_n \neq \varnothing$ holds. Let $a_j \in \mathsf{B}(w_{i_j}, 1) \cap \Gamma_n$. All pairs (a_j, a_{j+1}) have distance $d(a_j, a_{j+1}) \le 3$ and are by Lemma 5.1 a.s. connected. We can therefore select a Γ_n-path, γ_j, connecting a_j and a_{j+1}. Concatenating all paths γ_j produces a Γ_n-path connecting w and w', whence for $\lambda > 1 - \sqrt[\alpha-1]{\alpha^{-1}}$ Γ_n is a.s. connected.

Claim 2. For $\lambda < 1 - \sqrt[\alpha-1]{\alpha^{-1}}$ the random graph Γ_n contains a.s. isolated points. We consider $\mathsf{B}(v, 1) \subset Q_\alpha^n$ and define I_v as the indicator r.v. of the event

$$\{\Gamma_n \mid v \in \Gamma_n, \mathsf{S}(v, 1) \cap \Gamma_n = \varnothing\} \tag{5.27}$$

and set

$$\Omega = \sum_{\{(v,v') \mid v \neq v', \mathsf{B}(v,1) \cap \mathsf{B}(v',1) \neq \varnothing\}} \mathbb{P}(I_v \cdot I_{v'} = 1). \tag{5.28}$$

Suppose for $v \neq v'$, $\mathsf{B}(v, 1) \cap \mathsf{B}(v', 1) \neq \varnothing$. Then either $d(v, v') = 1$ and $|\mathsf{B}(v, 1) \cap \mathsf{B}(v', 1)| = \alpha$ or $d(v, v') = 2$ and $|\mathsf{B}(v, 1) \cap \mathsf{B}(v', 1)| = 2$. Therefore,

$$\mathsf{B}(v, 1) \cap \mathsf{B}(v', 1) \neq \varnothing \quad \Rightarrow \quad \mathbb{P}(I_v \cdot I_{v'} = 1) \le \lambda^2 (1-\lambda)^{2(\alpha-1)n-\alpha}.$$

Clearly we have $Z = \sum_{v \in Q_\alpha^n} I_v$ and $\mathbb{E}(Z) = \alpha^n \lambda (1-\lambda)^{(\alpha-1)n}$. Since $\lambda < 1 - \sqrt[\alpha-1]{\alpha^{-1}}$, we have $\mathbb{E}(Z) \sim e^{cn}$, for $c > 0$. We next compute

$$\begin{aligned}
\Omega &\le \alpha^n (\alpha-1)^2 \binom{n}{2} \lambda^2 (1-\lambda)^{-\alpha} \left[(1-\lambda)^{(\alpha-1)n}\right]^2 \\
&= (\alpha-1)^2 \binom{n}{2} (1-\lambda)^{-\alpha} \lambda (1-\lambda)^{(\alpha-1)n}\, \mathbb{E}(Z) \\
&\sim e^{-c'n} \mathbb{E}(Z), \quad c' > 0.
\end{aligned}$$

Janson's inequality (Theorem 5.3) guarantees

$$\mathbb{P}(Z \le (1-\gamma)\mathbb{E}[Z]) \le e^{-\frac{\gamma^2 \mathbb{E}[Z]}{2+2\Omega/\mathbb{E}[Z]}}. \tag{5.29}$$

Therefore, Γ_n contains a.s. isolated points, which proves that Γ_n is not connected for $\lambda < 1 - \sqrt[\alpha-1]{\alpha^{-1}}$.

5.4
The Largest Component

In this section we assume $\alpha = 2$, i.e., we work in binary n-cubes. All results and proofs easily extend to arbitrary alphabets. The analysis of large components presented here follows [20]. Section 5.3 has shown that the connectivity threshold $1-\sqrt[\alpha-1]{\alpha^{-1}}$ reflects the disappearance of isolated vertices. Therefore, it seems natural to ask which probabilities we find a unique large component for. In addition, one would like to know its size. Note that the connectivity property alone is not sufficient for understanding the neutral evolution. The relevant property will be identified in Section 5.5 and is related to the emergence of short paths in n-cubes. Intuitively the largest component is in its "early" stage locally "treelike" and therefore not suited for preserving sequence-specific information. We set

$$\pi(\chi_n) = \begin{cases} \alpha(\varepsilon) & \text{for} \quad \chi_n = \varepsilon \\ 2(1+o(1))\chi_n & \text{for} \quad o(1) = \chi_n \geq n^{-\frac{1}{3}+\delta} \end{cases} \tag{5.30}$$

and

$$\nu_n = \left\lfloor \frac{1}{2k(k+1)} u_n n \right\rfloor, \quad \iota_n = \left\lfloor \frac{k}{2(k+1)} u_n n \right\rfloor, \quad \text{and} \quad z_n = k\nu_n + \iota_n . \tag{5.31}$$

We write a Q_2^n-vertex $v = (x_1, \ldots, x_n)$ as

$$(\underbrace{x_1^{(1)}, \ldots, x_{\nu_n}^{(1)}}_{\nu_n \text{ coordinates}}, \underbrace{x_1^{(2)}, \ldots, x_{\nu_n}^{(2)}}_{\nu_n \text{ coordinates}}, \ldots, \underbrace{x_1^{(k+1)}, \ldots, x_{\iota_n}^{(k+1)}}_{\iota_n \text{ coordinates}}, \underbrace{x_{z_n+1}, \ldots, x_n}_{\substack{n-z_n \geq \\ n-\lfloor \frac{1}{2} u_n n \rfloor \text{ coordinates}}}) . \tag{5.32}$$

For any $1 \leq s \leq \nu_n$, $r = 1, \ldots, k$ we set $e_s^{(r)}$ to be the $s + (r-1)\nu_n$th-unit vector, i.e., $e_s^{(r)}$ has exactly one 1 at its $(s + (r-1)\nu_n)$th coordinate. Similarly, let $e_s^{(k+1)}$, $1 \leq s \leq \iota_n$, denote the $(s + k\nu_n)$th-unit vector. We use the standard notation for the $z_n + 1 \leq t \leq n$ unit vectors, i.e., e_t is the vector where $x_t = 1$, and $x_j = 0$, otherwise.

Let us outline the strategy of the proof. First we prove Lemma 5.2, which generates small subcomponents of size $\geq \lfloor \frac{1}{4} u_n n \rfloor$. The size is small enough to assure that they exist for any Q_2^n-vertex with probability $\pi(\chi_n)$ (Theorem 5.2). In Lemma 5.3 we build on these subcomponents, proving that they can, for certain λ_n, be extended to size $\geq n^h$, $h \in \mathbb{N}$. We next prove Lemma 5.4, which shows that the number of vertices not contained in subcomponents of size $\geq n^h$ is concentrated. We integrate our results thus far in Lemma 5.5, showing that the number of vertices contained in subcomponents of size $\geq n^h$ is concentrated at $\pi(\chi_n)\,|\Gamma_n|$. The idea is now to prove that exactly the latter merge into the unique large component. To show this, we first prove Lemma 5.6, a technical prerequisite for Lemma 5.7 that will be instrumental in proving that the vertices contained in subcomponents of size $\geq n^h$ merge into the unique largest component.

Lemma 5.2 *Suppose* $\lambda_n = \frac{1+\chi_n}{n}$ *and* $\varepsilon \geq \chi_n \geq n^{-\frac{1}{3}+\delta}$, *where* $\delta > 0$. *Then each* Γ_n*-vertex is contained in a* Γ_n*-subcomponent of size* $\lfloor \frac{1}{4} u_n n \rfloor$ *with probability at least* $\pi(\chi_n)$.

Proof. We consider the following branching process in the subcube $Q_2^{n-z_n}$ (Equation (5.32)). W.l.o.g. we initialize the process at $v = (0, \dots, 0)$ and set $E_0 = \{e_{z_n+1}, \dots, e_n\}$ and $L_0[0] = \{(0, \dots, 0)\}$. We consider the $n - \lfloor \frac{3}{4} u_n n \rfloor$ smallest neighbors of v. Starting with the smallest we select each of them with independent probability $\lambda_n = \frac{1+\chi_n}{n}$. Suppose $v + e_j$ is the first one selected. Then we set $E_1 = E_0 \setminus \{e_j\}$ and $L_1[0] = L_0[0] \cup \{e_j\}$ and proceed inductively, setting $E_s = E_{s-1} \setminus \{e_w\}$ and $L_t[0] = L_{t-1}[0] \cup \{e_w\}$ for each neighbor $v + e_w$ being selected subject to the condition $|E_s| > n - (\lfloor \frac{3}{4} u_n n \rfloor - 1)$. This procedure generates the set containing all selected 0-neighbors and 0 itself, which we denote by $N_*[0]$. We consider $L_*[0] = N_*[0] \setminus \{0\}$. If $\varnothing \neq L_*[0]$, we proceed by choosing its smallest element, v_1^*. By construction, v_1^* has at least $n - \lfloor \frac{3}{4} u_n n \rfloor$ neighbors of the form $v_1^* + e_r$, where $e_r \in E_s$. We iterate the process selecting from the smallest $n - \lfloor \frac{3}{4} u_n n \rfloor$ neighbors of v_1^* and set $L_*[1] = (N_*[1] \cup L_*[0]) \setminus \{v_1^*\}$. We proceed inductively, setting $L_*[r] = (N_*[r] \cup L_*[r-1]) \setminus \{v_r^*\}$. This process constructs an induced subtree of $Q_2^{n-z_n}$. It stops if we have $L_*[r] = \varnothing$ for some $r \geq 1$ or

$$|E_s| = n - \left(\left\lfloor \frac{3}{4} u_n n \right\rfloor - 1 \right),$$

in which case $\lfloor \frac{1}{4} u_n n \rfloor - 1$ vertices have been connected. Theorem 5.2 guarantees that this $Q_2^{n-z_n}$-tree has size $\lfloor \frac{1}{4} u_n n \rfloor$ with probability at least $\pi(\chi_n)$.

We refer to the particular branching process used in Lemma 5.2 as a γ-process (Figure 5.4). The γ-process produces a subcomponent of size $\lfloor \frac{1}{4} u_n n \rfloor$, which we refer to as γ-(sc). The γ-process employed in Lemma 5.2 did not by construction involve the first z_n coordinates. Following our outline, in the following lemma we will use the first $k\, v_n$ of them in order to build inductively larger subcomponents (sc).

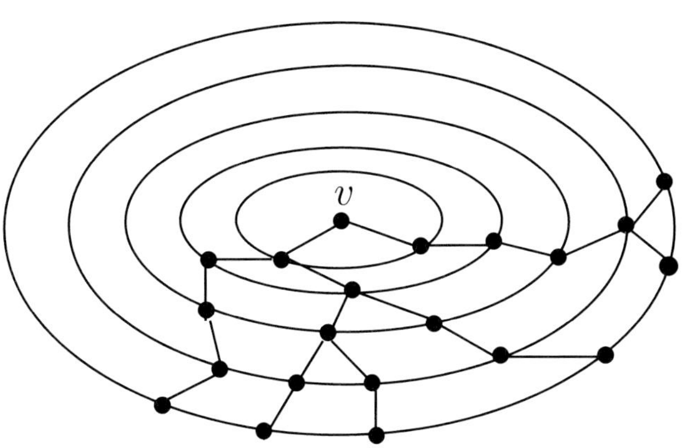

Figure 5.4 Lemma 5.2: The induced tree of size $\leq \lfloor \frac{1}{4} u_n n \rfloor$.

Lemma 5.3 *Let $k \in \mathbb{N}$ be arbitrary but fixed, $\lambda_n = \frac{1+\chi_n}{n}$, $\nu_n = \lfloor \frac{u_n n}{2k(k+1)} \rfloor$, and $\varphi_n = \pi(\chi_n)\nu_n(1 - e^{-(1+\chi_n)u_n/4})$. Then there exists $\varrho_k > 0$ such that each Γ_n-vertex is with probability at least*

$$\pi_k(\chi_n) = \pi(\chi_n)\ (1 - e^{-\varrho_k \varphi_n}) \tag{5.33}$$

contained in a Γ_n-subcomponent of size at least $c_k\,(u_n n)\varphi_n^k$, where $c_k > 0$.

Lemma 5.3 allows us to introduce the induced subgraph $\Gamma_{n,k} = Q_2^n[A]$, where

$$A = \left\{ v \mid v \text{ is contained in a } \Gamma_n\text{-(sc) of size} \geq c_k\,(u_n n)\varphi_n^k,\ c_k > 0 \right\}\ . \tag{5.34}$$

In case of $\varepsilon \geq \chi_n \geq n^{-\frac{1}{3}+\delta}$, we have $1 - e^{-\frac{1}{4}(1+\chi_n)u_n} \geq u_n/4$, and consequently $\varphi_n \geq c'\,(1 + o(1))\chi_n u_n^2\, n \geq c_0\, n^\delta$ for some $c', c_0 > 0$. Furthermore

$$\left\lfloor \frac{1}{4} u_n n \right\rfloor\ \varphi_n^k \geq c_k\, n^{\frac{2}{3}} n^{k\delta}, \qquad c_k > 0\,. \tag{5.35}$$

Accordingly, choosing k sufficiently large, each Γ_n-vertex is contained in a (sc) of arbitrary polynomial size with probability at least

$$\pi(\chi_n)\ \left(1 - e^{-\varrho_k n^\delta}\right), \qquad 0 < \delta,\ 0 < \varrho_k\,. \tag{5.36}$$

Proof. Since all translations are Q_2^n-automorphisms, we can w.l.o.g. assume that $v = (0, \ldots, 0)$. We use the notation of Equation (5.32) and recruit the $n - z_n$-unit vectors e_t for a γ-process. The γ-process of Lemma 5.2 yields a γ-(sc), $C(0)$, of size $\lfloor \frac{1}{4} u_n n \rfloor$ with probability $\geq \pi(\chi_n)$. We consider for $1 \leq i \leq k$ the sets of ν_n elements $B_i = \{e_1^{(i)}, \ldots, e_{\nu_n}^{(i)}\}$ and set $H = \langle e_{z_n+1}, \ldots, e_n \rangle$. By construction we have

$$\langle B_i \cup \langle \bigcup_{1\leq j\leq i-1} B_j \rangle \oplus H \rangle = \langle B_i \rangle \oplus \langle \bigcup_{1\leq j\leq i-1} B_j \rangle \oplus H\,. \tag{5.37}$$

In particular, for any $1 \leq s < j \leq \nu_n$: $e_s^{(1)} - e_j^{(1)} \in H$ is equivalent to $e_s^{(1)} = e_j^{(1)}$. Since all vertices are selected independently and $|C(0)| = \lfloor \frac{1}{4} u_n n \rfloor$, for fixed $e_s^{(1)} \in B_1$ the probability of not selecting a vertex $v' \in e_s^{(1)} + C(0)$ is given by

$$\mathbb{P}\left(\{e_s^{(1)} + \xi \mid \xi \in C(0)\} \cap \Gamma_n = \varnothing\right) = \left(1 - \frac{1 + \chi_n}{n}\right)^{\lfloor \frac{1}{4} u_n n \rfloor} \sim e^{-(1+\chi_n)\frac{1}{4}u_n}\,. \tag{5.38}$$

We set $\mu_n = (1 - e^{-(1+\chi_n)\frac{1}{4}u_n})$, i.e., $\mu_n = \mathbb{P}((e_s^{(1)} + C(0)) \cap \Gamma_n \neq \varnothing)$, and introduce the r.v.

$$X_1 = \left|\{e_s^{(1)} \in B_1 \mid \exists \xi \in C(0);\ e_s^{(1)} + \xi \in \Gamma_n\}\right|\,. \tag{5.39}$$

Obviously, $\mathbb{E}(X_1) = \mu_n\nu_n$, and, using the large deviation result of Equation (5.7), we can conclude that

$$\exists \varrho > 0; \quad \mathbb{P}\left(X_1 < \frac{1}{2}\mu_n\nu_n\right) \le e^{-\varrho\mu_n\nu_n} . \tag{5.40}$$

Suppose for $e_s^{(1)}$ there exists some $\xi \in C(0)$ such that $e_s^{(1)} + \xi \in \Gamma_n$ (that is, $e_s^{(1)}$ is counted by X_1). We then select the smallest element of the set $\{e_s^{(1)} + \xi \mid \xi \in C(0)\}$, say $e_s^{(1)} + \xi_0$ and initiate a γ-process using the $n - z_n$ elements $\{e_{z_n+1}, \ldots, e_n\}$ at $e_s^{(1)} + \xi_0$. The process yields a γ-(sc) of size $\lfloor \frac{1}{4}u_n n \rfloor$ with probability at least $\pi(\chi_n)$. For any two elements $e_s^{(1)}, e_j^{(1)}$ with $e_s^{(1)} + \xi(e_s^{(1)}), e_j^{(1)} + \xi(e_j^{(1)}) \in \Gamma_n$ the respective sets are vertex disjoint since $\langle B_1 \cup H\rangle = \langle B_1\rangle \oplus H$. Let $\tilde{X}_1$ be the random variable counting the number of these new, pairwise vertex disjoint sets of γ-(sc) of size $\lfloor \frac{1}{4}u_n n \rfloor$. By construction each of them is connected to $C(0)$. We immediately observe $\mathbb{E}(\tilde{X}_1) \ge \pi(\chi_n)\mu_n\nu_n$ and set $\varphi_n = \pi(\chi_n)\mu_n\nu_n$. Using the large deviation result in Equation (5.7) we derive

$$\exists \varrho_1 > 0; \quad \mathbb{P}\left(\tilde{X}_1 < \frac{1}{2}\varphi_n\right) \le e^{-\varrho_1\varphi_n} . \tag{5.41}$$

We proceed by proving that for each $1 \le i \le k$ there exists a sequence of r.v.s. $(\tilde{X}_1, \tilde{X}_2, \ldots, \tilde{X}_i)$, where $\tilde{X}_i$ counts the number of pairwise disjoint sets of γ-(sc) added at step $1 \le j \le i$ such that:

(a) all sets, $C_\tau^{(j)}$, $1 \le j \le i$, added until step i are pairwise vertex disjoint and are of size $\lfloor \frac{1}{4}u_n n \rfloor$;
(b) all sets added until step i are connected to $C(0)$ and

$$\exists \varrho_i > 0; \quad \mathbb{P}\left(\tilde{X}_i < \frac{1}{2^i}(\varphi_n)^i\right) \le e^{-\varrho_i\varphi_n} , \text{ where } \varphi_n = \pi(\chi_n)\mu_n\nu_n . \tag{5.42}$$

We prove the assertion by induction on i. Indeed in our construction of $\tilde{X}_1$ we have already established the induction basis. To define $\tilde{X}_{i+1}$, we use the set $B_{i+1} = \{e_1^{(i+1)}, \ldots, e_{\nu_n}^{(i+1)}\}$. For each $C_\tau^{(i)}$ counted by $\tilde{X}_i$ (i.e., the vertices that were connected in step i) we form the set $e_s^{(i+1)} + C_\tau^{(i)}$. By induction hypothesis two different $C_\tau^{(i)}, C_{\tau'}^{(i)}$, counted by $\tilde{X}_i$, are vertex disjoint and connected to $C(0)$. Since $\langle B_{i+1}\rangle \bigoplus \langle \bigcup_{1\le j\le i} B_j\rangle \bigoplus H$ are disjoint, we can conclude

$$(s \ne s' \ \vee \ \tau \ne \tau') \quad \Rightarrow \quad (e_s^{(i+1)} + C_\tau^{(i)}) \cap (e_{s'}^{(i+1)} + C_{\tau'}^{(i)}) = \varnothing ,$$

and the probability that we have for fixed $C_\tau^{(i)}$: $(e_s^{(i+1)} + C_\tau^{(i)}) \cap \Gamma_n = \varnothing$ for some $e_s^{(i+1)} \in B_{i+1}$ is exactly as in Equation (5.38)

$$\mathbb{P}\left((e_s^{(i+1)} + C_\tau^{(i)}) \cap \Gamma_n = \varnothing\right) = \left(1 - \frac{1+\chi_n}{n}\right)^{\lfloor \frac{1}{4}u_n n\rfloor} \sim e^{-(1+\chi_n)\frac{1}{4}u_n} .$$

As for the induction basis, $\mu_n = (1 - e^{-(1+\chi_n)\frac{1}{4}u_n})$ is the probability that $(e_s^{(i+1)} + C_\tau^{(i)}) \cap \Gamma_n \neq \varnothing$. We proceed by defining the r.v.

$$X_{i+1} = \sum_{C_\tau^{(i)}} \left| \left\{ e_s^{(i+1)} \in B_{i+1} \mid \exists \xi \in C_\tau^{(i)};\ e_s^{(i+1)} + \xi \in \Gamma_n \right\} \right| . \tag{5.43}$$

X_{i+1} counts the number of events where $(e_s^{(i+1)} + C_\tau^{(i)}) \cap \Gamma_n \neq \varnothing$ for each $C_\tau^{(i)}$. For fixed $C_\tau^{(i)}$ and fixed $e_s^{(i+1)} \in B_{i+1}$ we choose the minimal element

$$e_s^{(i+1)} + \xi_{0,\tau} \in \left\{ e_s^{(i+1)} + \xi_\tau \mid \xi_\tau \in C_\tau^{(i)},\ e_s^{(i+1)} + \xi_\tau \in \Gamma_n \right\} .$$

Then X_{i+1} counts exactly the minimal elements $e_s^{(i+1)} + \xi_{0,\alpha}, e_{s'}^{(i+1)} + \xi_{0,\tau'}, \ldots$ for all $C_\tau^{(i)}, C_{\tau'}^{(i)}, \ldots$, and any two can be used to construct pairwise vertex disjoint γ-(sc) of size $\lfloor \frac{1}{4} u_n n \rfloor$. We next define $\tilde{X}_{i+1}$ to be the r.v. counting the number of events in which the γ-process in H initiated at the $e_s^{(i+1)} + \xi_{0,\tau} \in \Gamma_n$ yields a γ-(sc) of size $\lfloor \frac{1}{4} u_n n \rfloor$. By construction each of these is connected to a unique $C_\tau^{(i)}$. Since $\langle B_{i+1} \rangle \bigoplus \langle \bigcup_{1 \le j \le i} B_j \rangle \bigoplus H$, all newly added sets are pairwise vertex disjoint to all previously added vertices. We derive

$$\begin{aligned}\mathbb{P}\left(\tilde{X}_{i+1} < \frac{1}{2^{i+1}} \varphi_n^{i+1}\right) &\le \underbrace{\mathbb{P}\left(\tilde{X}_i < \frac{1}{2^i} \varphi_n^i\right)}_{\text{failure at step } i} + \underbrace{\mathbb{P}\left(\tilde{X}_{i+1} < \frac{1}{2^{i+1}} \varphi_n^{i+1} \wedge \tilde{X}_i \ge \frac{1}{2^i} \varphi_n^i\right)}_{\text{failure at step } i+1 \text{ conditional to } \tilde{X}_i \ge \frac{1}{2^i}\varphi_n^i} \\ &\le e^{-\varrho_i \varphi_n} + e^{-\varrho\, \varphi_n^{i+1}} (1 - e^{-\varrho_i \varphi_n}), \quad \varrho > 0 \\ &\le e^{-\varrho_{i+1} \varphi_n} .\end{aligned}$$

Therefore, each Γ_n-vertex is with probability at least $\pi(\chi_n)\,(1 - e^{-\varrho_k \varphi_n})$ contained in a Γ_n-(sc) of size at least $c_k\,(\chi_n n) \varphi_n^k$, for $c_k > 0$, and the lemma is proved.

We next prove a technical lemma that will be instrumental for the proof of Lemma 5.5. We show that the number of vertices not contained in $\Gamma_{n,k}$ is sharply concentrated, using a strategy similar to that in Bollobás *et al.* [6]. Let U_n denote the complement of $\Gamma_{n,k}$ in Γ_n.

Lemma 5.4 *Let $k \in \mathbb{N}$ and $\lambda_n = \frac{1+\chi_n}{n}$, where $\varepsilon \ge \chi_n \ge n^{-\frac{1}{3}+\delta}$. Then we have*

$$\mathbb{P}\left(\big| |U_n| - \mathbb{E}[|U_n|] \big| \ge \frac{1}{n} \mathbb{E}[|U_n|] \right) = o(1) . \tag{5.44}$$

Proof. Let C be a Q_2^n-component of size strictly smaller than $\tau = c_k\,(u_n n) \varphi_n^k$ and let v be a fixed C-vertex. We shall denote the ordered pair (C, v) by C_v and the indicator variable of the pair C_v by X_{C_v}. Clearly, we have

$$|U_n| = \sum_{C_v} X_{C_v},$$

where the summation is taken over all ordered pairs (C, v) with $|C| < \tau$. Considering isolated points, we immediately obtain $\mathbb{E}[U_n] \geq c|\Gamma_n|$ for some $1 \geq c > 0$.

Claim. The random variable $|U_n|$ is sharply concentrated.
We prove the claim by estimating $\mathbb{V}[|U|]$ via computing the correlation terms $\mathbb{E}[X_{C_v} X_{D_{v'}}]$ and applying Chebyshev's inequality. Suppose $C_v \neq D_{v'}$. There are two ways by which $X_{C_v}, X_{D_{v'}}$ viewed as r.v. over Q^n_{2,λ_n}, can be correlated. First v, v' can belong to the same component, i.e. $C = D$, in which case we write $C_v \sim_1 D_{v'}$. Clearly,

$$\sum_{C_v \sim_1 D_{v'}} \mathbb{E}[X_{C_v} X_{D_{v'}}] \leq \tau \ \mathbb{E}[|U_n|]. \tag{5.45}$$

Second, correlation arises when v, v' belong to two different components C_v, $D_{v'}$ having minimal distance 2 in Q^n_2. In this case we write $C_v \sim_2 D_{v'}$. Then there exists some Q^n_2-vertex, w, such that $w \in \mathrm{d}(C_v) \cap \mathrm{d}(D_{v'})$ and we derive

$$\begin{aligned} \mathbb{P}(d(C_v, D_{v'}) = 2) &= \frac{1-\lambda_n}{\lambda_n} \ \mathbb{P}(C_v \cup D_{v'} \cup \{w\} \text{ is a } \Gamma_n\text{-component}) \\ &\leq n \ \mathbb{P}(C_v \cup D_{v'} \cup \{w\} \text{ is a } \Gamma_n\text{-component}). \end{aligned}$$

We can now immediately give the upper bound

$$\sum_{C_v \sim_2 D_{v'}} \mathbb{E}[X_{C_v} X_{D_{v'}}] \leq n\,(2\tau + 1)^3 \ |\Gamma_n|. \tag{5.46}$$

The uncorrelated pairs $(X_{C_v}, X_{D_{v'}})$, writing $C_v \not\sim D_{v'}$, can easily be estimated by

$$\sum_{C_v \not\sim D_{v'}} \mathbb{E}[X_{C_v} X_{D_{v'}}] = \sum_{C_v \not\sim D_{v'}} \mathbb{E}[X_{C_v}]\mathbb{E}[X_{D_{v'}}] \leq \mathbb{E}[|U_n|]^2. \tag{5.47}$$

Consequently we arrive at

$$\begin{aligned} \mathbb{E}[|U_n|(|U_n|-1)] &= \sum_{C_v \sim_1 D_{v'}} \mathbb{E}[X_{C_v} X_{D_{v'}}] + \sum_{C_v \sim_2 D_{v'}} \mathbb{E}[X_{C_v} X_{D_{v'}}] + \sum_{C_v \not\sim D_{v'}} \mathbb{E}[X_{C_v} X_{D_{v'}}] \\ &\leq \tau \ \mathbb{E}[|U_n|] + n\,(2\tau+1)^3 |\Gamma_n| + \mathbb{E}[|U_n|]^2. \end{aligned}$$

Using $\mathbb{V}[|U_n|] = \mathbb{E}[|U_n|(|U_n|-1)] + \mathbb{E}[|U_n|] - \mathbb{E}[|U_n|]^2$ and $\mathbb{E}[U_n] \geq c\,|\Gamma_n|$ we obtain

$$\frac{\mathbb{V}[|U_n|]}{\mathbb{E}[|U_n|]^2} \leq \frac{c_k\,(u_n n)\varphi_n^k + \frac{1}{c}\,n\,\left(2c_k\,(u_n n)\varphi_n^k + 1\right)^3 + 1}{|\mathbb{E}[U_n]|} = o\left(\frac{1}{n^2}\right).$$

Chebyshev's inequality guarantees $\mathbb{P}(||U_n| - \mathbb{E}[|U_n|]| \geq \frac{1}{n}\,\mathbb{E}[|U_n|]) \leq n^2\,\frac{\mathbb{V}[|U_n|]}{\mathbb{E}[|U_n|]^2}$, whence the claim and the lemma follows.

Lemma 5.5 *Let $\lambda_n = \frac{1+\chi_n}{n}$, where $\varepsilon \geq \chi_n \geq n^{-\frac{1}{3}+\delta}$. Then we have for sufficiently large $k \in \mathbb{N}$*

$$(1 - o(1))\,\pi(\chi_n)\,|\Gamma_n| \leq |\Gamma_{n,k}| \leq (1 + o(1))\,\pi(\chi_n)\,|\Gamma_n| \qquad a.s. \tag{5.48}$$

In view of Lemma 5.3 the crucial part is to show that there are sufficiently many Γ_n-vertices contained in Γ_n-(sc) of size $< c_k\, u_n n \varphi_n^k$. For this purpose we use a strategy introduced by Bollobás *et al.* [6] and consider the n-regular rooted tree T_n. Let v^* denote the root of T_n. Then v^* has n descendents and all other T_n-vertices have $n-1$. Selecting the T_n-vertices with independent probability λ_n we obtain the probability space T_{n,λ_n} whose elements, A_n, are random induced subtrees. We will be interested in the A_n-component that contains the root, denoted by C_{v^*}. Let ξ_{v^*} and ξ_v, for $v \neq v^*$, be two r.v. such that $\mathsf{Prob}(\xi_{v^*} = \ell) = B_n(\ell, \lambda_n)$ and $\mathsf{Prob}(\xi_v = \ell) = B_{n-1}(\ell, \lambda_n)$, respectively. We assume that ξ_{v^*} and ξ_v count the offspring produced at v^* and $v \neq v^*$. Then the induced branching process initialized at v^*, $(Z_i)_{i\in\mathbb{N}_0}$, constructs C_{v^*}. Let $\pi_0(\chi)$ denote its survival probability; then we have, in view of Theorem 5.2 and [6], Corollary 6:

$$\pi_0(\chi_n) = (1 + o(1))\,\pi(\chi_n)\,. \tag{5.49}$$

Proof. Claim 1. $|\Gamma_{n,k}| \geq \big((1 - o(1))\,\pi(\chi_n)\big)\,|\Gamma_n|$ a.s.
According to Lemma 5.3 we have $\mathbb{E}[|U_n|] < (1 - \pi_k(\chi_n))\,|\Gamma_n|$, and we can conclude using Lemma 5.4 and $\mathbb{E}[|U_n|] = O(|\Gamma_n|)$ that

$$|U_n| < \left(1 + O\left(\frac{1}{n}\right)\right)\,\mathbb{E}[|U_n|] < \left(1 - \left(\pi_k(\chi_n) - O\left(\frac{1}{n}\right)\right)\right)\,|\Gamma_n| \quad \text{a.s.} \tag{5.50}$$

In view of Equation (5.33) and $\chi_n \geq n^{-\frac{1}{3}+\delta}$ we have for arbitrary but fixed k

$$\pi_k(\chi_n) - O(\frac{1}{n}) = (1 - o(1))\,\pi(\chi_n)\,.$$

Therefore, we derive

$$|\Gamma_{n,k}| \geq (1 - o(1))\,\pi(\chi_n)\,|\Gamma_n| \quad \text{a.s.}\,, \tag{5.51}$$

and Claim 1 follows.

Claim 2. For sufficiently large k, $|\Gamma_{n,k}| \leq \big((1 + o(1))\,\pi(\chi_n)\big)\,|\Gamma_n|$ a.s. holds.
For any fixed Q_2^n-vertex, v, we have the inequality

$$\mathbb{P}\left(|C_{v^*}| \leq \ell\right) \leq \mathbb{P}\left(|C_v| \leq \ell\right)\,. \tag{5.52}$$

Indeed we can obtain C_v by inductively constructing a spanning tree as follows. Suppose the set of all C_v-vertices at distance h is $M_h^{C_v}$. Starting with the smallest $w \in M_h^{C_v}$ $(h \geq 1)$ there are at most $n-1$ w-neighbors contained in $M_{h+1}^{C_v}$ that are not neighbors for some smaller $w' \in M_h^{C_v}$. Hence for any $w \in M_h^{C_v}$ at most $n-1$ vertices have to be examined. The A_n-component C_{v^*} is generated by the same procedure. Then for each $w \in M_h^{C_{v^*}}$ there are exactly $n-1$ neighbors in $M_{h+1}^{C_{v^*}}$. Since the process adds at each stage less or equally many vertices for C_v, we have by construction $|C_v| \leq |C_{v^*}|$. Standard estimates for binomial coefficients allow one to estimate the numbers of T_n-subtrees containing the root [6], Corollary 3. Since vertex boundaries in T_n are easily obtained, we can accordingly compute $\mathbb{P}(|C_{v^*}| = \ell)$. Choosing k sufficiently large, the estimates in [6], Lemma 22, guarantee

$$\mathbb{P}\left(|C_{v^*}| < c_k\, u_n n\, \varphi_n^k\right) = (1 - \pi_0(\chi_n)) + o(e^{-n})\,. \tag{5.53}$$

In view of $\mathbb{P}(|C_{v^*}| \leq \ell) \leq \mathbb{P}(|C_v| \leq \ell)$ and Equation (5.49), we can conclude from Equation (5.53) that

$$(1 - (1 + o(1))\pi(\chi_n))\,|\Gamma_n| + o(1) \leq \mathbb{E}[|U_n|]\,. \tag{5.54}$$

According to Lemma 5.4 we have $(1 - O(\frac{1}{n}))\,\mathbb{E}[|U_n|] < |U_n|$ a.s., and therefore

$$(1 - (1 + o(1) + O\left(\frac{1}{n}\right))\,\pi(\chi_n))\,|\Gamma_n| \leq |U_n| \qquad \text{a.s.} \tag{5.55}$$

Equations (5.51) and (5.55) imply

$$(1 - o(1))\,\pi(\chi_n)|\Gamma_n| \leq |\Gamma_{n,k}| \leq (1 + o(1))\,\pi(\chi_n)|\Gamma_n| \qquad \text{a.s.}\,, \tag{5.56}$$

whence the lemma.

Finally, we show that $\Gamma_{n,k}$ is a.s. 2-dense in Q_2^n with the exception of $2^n\, e^{-\tilde{\Delta}\, n^{\delta}}$ vertices. Accordingly, $\Gamma_{n,k}$ is uniformly distributed in Γ_n. The lemma will allow us to establish via Lemma 5.7 the existence of many vertex disjoint short paths between certain splits of the $\Gamma_{n,k}$-vertices.

Lemma 5.6 *Let $k \in \mathbb{N}$ and $\lambda_n = \frac{1+\chi_n}{n}$ and $\varepsilon \geq \chi_n \geq n^{-\frac{1}{3}+\delta}$. Then we have*

$$\exists \Delta > 0;\ \forall v \in Q_2^n,\ \mathbb{P}\left(|\mathsf{S}(v,2) \cap \Gamma_{n,k}| < \frac{1}{2}\left(\frac{k}{2(k+1)}\right)^2 n^{\delta}\right) \leq e^{-\Delta\, n^{\delta}}\,. \tag{5.57}$$

Let $D_\delta = \{v \mid |\mathsf{S}(v,2) \cap \Gamma_{n,k}| < \frac{1}{2}\left(\frac{k}{2(k+1)}\right)^2 n^{\delta}\}$; then

$$|D_\delta| \leq 2^n\, e^{-\tilde{\Delta}\, n^{\delta}} \qquad \text{a.s., where} \quad \Delta > \tilde{\Delta} > 0\,. \tag{5.58}$$

Proof. To prove the lemma we use the last (Equation (5.32)) $\iota_n = \lfloor \frac{k}{2(k+1)} u_n n \rfloor$ elements $e_1^{(k+1)}, \ldots, e_{\iota_n}^{(k+1)}$. We consider for arbitrary $v \in Q_2^n$

$$\mathsf{S}^{(k+1)}(v,2) = \left\{ v + e_i^{(k+1)} + e_j^{(k+1)} \mid 1 \le i < j \le \iota_n, \right\} . \tag{5.59}$$

Clearly, $|\mathsf{S}^{(k+1)}(v,2)| = \binom{\iota_n}{2}$ holds. By construction, the Γ_n-(sc) of size $\ge c_k\,(u_n n)\varphi_n^k$ of Lemma 5.3 are vertex disjoint for any two vertices in $\mathsf{S}^{(k+1)}(v,2) \cap \Gamma_n$ and each Γ_n-vertex belongs to $\Gamma_{n,k}$ with probability $\ge \pi_k(\chi_n)$. Let Z be the r.v. counting the number of vertices in $\mathsf{S}^{(k+1)}(v,2) \cap \Gamma_{n,k}$. Then we have $\mathbb{E}[Z] \ge \left(\frac{k}{2(k+1)}\right)^2 \frac{u_n^2}{2}\, n\, \pi(\chi_n)$. Equation (5.57) follows now from Equation (5.7), $u_n^2 n \chi_n \ge n^{\delta}$, and $\mathbb{P}(|\mathsf{S}(v,2) \cap \Gamma_{n,k}| < \eta) \le \mathbb{P}(|\mathsf{S}^{(k+1)}(v,2) \cap \Gamma_{n,k}| < \eta)$. Now let $D_\delta = \{v \mid |\mathsf{S}(v,2) \cap \Gamma_{n,k}| < \frac{1}{2}\left(\frac{k}{2(k+1)}\right)^2 n^{\delta}\}$. By linearity of expectation $\mathbb{E}(|D_\delta|) \le 2^n \mathrm{e}^{-\Delta n^{\delta}}$ holds, and using Markov's inequality, $\mathbb{P}(X > t\mathbb{E}(X)) \le 1/t$ for $t > 0$, we derive that $|D_\delta| \le 2^n \mathrm{e}^{-\tilde{\Delta} n^{\delta}}$ a.s. for any $0 < \tilde{\Delta} < \Delta$.

The next lemma proves the existence of many vertex disjoint paths connecting the boundaries of certain splits of $\Gamma_{n,k}$-vertices. The lemma is related to a result in [7] but is much stronger since the actual length of these paths is ≤ 3. The shortness of these paths results from the 2-density of $\Gamma_{n,k}$ (Lemma 5.6) and is a consequence of our particular construction of small subcomponents in Lemma 5.3.

Lemma 5.7 *Suppose $\lambda_n = \frac{1+\chi_n}{n}$, where $\varepsilon \ge \chi_n \ge n^{-\frac{1}{3}+\delta}$. Let (A, B) be a split of the $\Gamma_{n,k}$-vertex set with the properties*

$$\exists\, 0 < \sigma_0 \le \sigma_1 < 1; \quad \frac{1}{n^2}\, 2^n \le |A| = \sigma_0 |\Gamma_{n,k}| \quad \text{and} \quad \frac{1}{n^2}\, 2^n \le |B| = \sigma_1 |\Gamma_{n,k}| \,. \tag{5.60}$$

Then there exists some $t > 0$ such that a.s. $\mathsf{d}(A)$ is connected to $\mathsf{d}(B)$ in Q_2^n via at least

$$\frac{t}{n^4}\, 2^n / \binom{n}{7} \tag{5.61}$$

vertex disjoint (independent) paths of length ≤ 3.

Proof. We consider $\mathsf{B}(A,2)$ and distinguish the cases

$$|\mathsf{B}(A,2)| \le \frac{2}{3}\, 2^n \quad \text{and} \quad |\mathsf{B}(A,2)| > \frac{2}{3}\, 2^n \,. \tag{5.62}$$

Suppose first $|\mathsf{B}(A,2)| \le \frac{2}{3}\, 2^n$ holds. According to Theorem 5.1 and Equation (5.60) we have

$$\exists\, d_1 > 0; \quad |\mathsf{d}(\mathsf{B}(A,2))| \ge \frac{d_1}{n^3}\, 2^n \,, \tag{5.63}$$

and Lemma 5.6 guarantees that a.s. all except of at most $2^n\, e^{-\tilde{\Delta} n^{\delta}}$ Q_2^n-vertices are within distance 2 to some $\Gamma_{n,k}$-vertex. Hence there exist at least $\frac{d}{n^3}\, 2^n$ vertices of $\mathsf{d}(\mathsf{B}(A,2))$ (which by definition are not contained in $\mathsf{B}(A,2)$) contained in $\mathsf{B}(B,2)$ i.e.,

$$|\mathsf{dB}(A,2) \cap \mathsf{B}(B,2)| \geq \frac{d}{n^3}\, 2^n \quad \text{a.s.} \tag{5.64}$$

For each $\beta_2 \in \mathsf{d}(\mathsf{B}(A,2)) \cap \mathsf{B}(B,2)$ there exists a path $(\alpha_1,\alpha_2,\beta_2)$, starting in $\mathsf{d}(A)$ with terminus β_2. In view of $\mathsf{B}(B,2) = \mathsf{d}(\mathsf{B}(B,1))\dot{\cup}\mathsf{B}(B,1)$, we distinguish the following cases:

$$|\mathsf{d}(\mathsf{B}(A,2)) \cap \mathsf{d}(\mathsf{B}(B,1))| \geq \frac{1}{n^3} d_{2,1}\, 2^n \quad \text{and} \quad |\mathsf{d}(\mathsf{B}(A,2)) \cap \mathsf{B}(B,1)| \geq \frac{1}{n^3}\, d_{2,2}\, 2^n\,. \tag{5.65}$$

Suppose we have $|\mathsf{d}(\mathsf{B}(A,2)) \cap \mathsf{d}(\mathsf{B}(B,1))| \geq \frac{1}{n^3} d_{2,1}\, 2^n$. For each $\beta_2 \in \mathsf{d}(\mathsf{B}(B,1))$ we select some element $\beta_1(\beta_2) \in \mathsf{d}(B)$ and set $B^* \subset \mathsf{d}(B)$ to be the set of these endpoints. Clearly, at most n elements in $\mathsf{B}(B,2)$ can produce the same endpoint, whence

$$|B^*| \geq \frac{1}{n^4} d_{2,1}\, 2^n\,.$$

Let $B_1 \subset B^*$ be maximal subject to the condition that for any pair of B_1-vertices (β_1, β_1') we have $d(\beta_1,\beta_1') > 6$. Then we have $|B_1| \geq |B^*|/\binom{n}{7}$ since $|\mathsf{B}(v,7)| = \binom{n}{7}$. Any two of the paths from $\mathsf{d}(A)$ to $B_1 \subset \mathsf{d}(B)$ are of the form $(\alpha_1,\alpha_2,\beta_2,\beta_1)$ and vertex disjoint since each of them is contained in $\mathsf{B}(\beta_1,3)$. Therefore, there are a.s. at least

$$\frac{1}{n^4} d_{2,1}\, 2^n \Big/ \binom{n}{7} \tag{5.66}$$

vertex disjoint paths connecting $\mathsf{d}(A)$ and $\mathsf{d}(B)$. Suppose next $|\mathsf{d}(\mathsf{B}(A,2)) \cap \mathsf{B}(B,1)| \geq \frac{1}{n^3}\, d_{2,2}\, 2^n$. We conclude in complete analogy that there exist a.s. at least

$$\frac{1}{n^3} d_{2,2}\, 2^n \Big/ \binom{n}{5} \tag{5.67}$$

vertex disjoint paths of the form $(\alpha_1,\alpha_2,\beta_2)$ connecting $\mathsf{d}(A)$ and $\mathsf{d}(B)$. It remains to consider the case $|\mathsf{B}(A,2)| > \frac{2}{3} 2^n$. By construction both A and B satisfy Equation (5.60), whence it suffices to assume that also $|\mathsf{B}(B,2)| > \frac{2}{3} 2^n$ holds. In this case we have

$$|\mathsf{B}(A,2) \cap \mathsf{B}(B,2)| > \frac{1}{3}\, 2^n\,,$$

and to each $\alpha_2 \in \mathsf{B}(A,2) \cap \mathsf{B}(B,2)$ we select $\alpha_1 \in \mathsf{d}(A)$ and $\beta_1 \in \mathsf{d}(B)$. We derive in analogy to the previous arguments that there exist a.s. at least

$$\frac{1}{n^2} d_2\, 2^n \Big/ \binom{n}{5} \tag{5.68}$$

pairwise vertex disjoint paths of the form $(\alpha_1, \alpha_2, \beta_1)$, and the proof of the lemma is complete.

Theorem 5.5 *[20] Let Q^n_{2,λ_n} be a random graph consisting of Q^n_2-subgraphs, Γ_n, induced by selecting each Q^n_2-vertex with independent probability λ_n. Suppose $\lambda_n = \frac{1+\chi_n}{n}$, where $\varepsilon \geq \chi_n \geq n^{-\frac{1}{3}+\delta}$, $\delta > 0$. Then we have*

$$\lim_{n\to\infty} \mathbb{P}\left(|C_n^{(1)}| \sim \pi(\chi_n)\,\frac{1+\chi_n}{n}\,2^n \text{ and } C_n^{(1)} \text{ is unique} \right) = 1\,. \tag{5.69}$$

Proof. Claim. We have $|C_n^{(1)}| \sim |\Gamma_{n,k}|$ a.s.
To prove the claim we use an idea introduced by Ajtai *et al.* [1] and select Q^n_2-vertices in two rounds. First we select Q^n_2-vertices with independent probability $\frac{1+\chi_n/2}{n}$ and subsequently with $\frac{\chi_n}{2n}$. The probability for some vertex not to be chosen in both randomizations is $(1-\frac{1+\chi_n/2}{n})(1-\frac{\chi_n/2}{n}) = 1 - \frac{1+\chi_n}{n} + \frac{(1+\chi_n/2)\chi_n/2}{n^2} \geq 1 - \frac{1+\chi_n}{n}$. Hence selecting first with probability $\frac{1+\chi_n/2}{n}$ (first round) and then with $\frac{\chi_n/2}{n}$ (second round) a vertex is selected with probability less than $\frac{1+\chi_n}{n}$ (all preceding lemmas hold for the first randomization $\frac{1+\chi_n/2}{n}$). We now select in our first round each Q^n_2-vertex with probability $\frac{1+\chi_n/2}{n}$. According to Lemma 5.5

$$|\Gamma_{n,k}| \sim \pi(\chi_n)\,|\Gamma_n| \quad \text{a.s.} \tag{5.70}$$

Suppose $\Gamma_{n,k}$ contains a component, A, such that

$$\frac{1}{n^2}\,2^n \leq |A| \leq (1-b)\,|\Gamma_{n,k}|, \quad b > 0\,;$$

then there exists a split of $\Gamma_{n,k}$, (A, B) satisfying the assumptions of Lemma 5.7 (and $\mathsf{d}(A) \cap \mathsf{d}(B) = \varnothing$). We now observe that Lemma 5.3 limits the number of ways these splits can be constructed. In view of

$$\left\lfloor \frac{1}{4} u_n n \right\rfloor \varphi_n^k \geq c_k\, n^{\frac{2}{3}} n^{k\delta}, \quad c_k > 0\,, \tag{5.71}$$

each A-vertex is contained in a component of size at least $c_k\, n^{\frac{2}{3}} n^{k\delta}$. Therefore, there are at most

$$2^{\left(2^n / \left(c_k\, n^{\frac{2}{3}} n^{k\delta}\right)\right)} \tag{5.72}$$

ways to choose A in such a split. According to Lemma 5.7, there exists $t > 0$ such that a.s. $\mathsf{d}(A)$ is connected to $\mathsf{d}(B)$ in Q^n_2 via at least $\frac{t}{n^4}\,2^n/\binom{n}{7}$ vertex disjoint paths of length ≤ 3. We now select Q^n_2-vertices with probability $\frac{\chi_n/2}{n}$. None of the above $\geq \frac{t}{n^4}\,2^n/\binom{n}{7}$ paths can be selected during this process. Since any two paths are vertex disjoint, the expected number of such splits is less than

$$2^{\left(2^n / \left(c_k\, n^{\frac{2}{3}} n^{k\delta}\right)\right)} \left(1 - \left(\frac{\chi_n/2}{n}\right)^4\right)^{\frac{t}{n^4} 2^n/\binom{n}{7}} \sim 2^{\left(2^n / \left(c_k\, n^{\frac{2}{3}} n^{k\delta}\right)\right)}\, e^{-\frac{t\chi_n^4}{2^4 n^8}\,2^n/\binom{n}{7}}\,. \tag{5.73}$$

Hence choosing k sufficiently large, we can conclude that a.s. there cannot exist such a split. Therefore, $|C_n^{(1)}| \sim |\Gamma_{n,k}|$, a.s. and the claim is proved. According to Lemma 5.5, we therefore have $|C_n^{(1)}| \sim \pi(\chi_n)\,|\Gamma_n|$. In particular, for $\chi_n = \varepsilon$, Theorem 5.2 ($0 < \alpha(\varepsilon) < 1$) implies that there exists a giant component. It remains to prove that $C_n^{(1)}$ is unique. By construction, any large component, C_n', is necessarily contained in $\Gamma_{n,k}$. In the proof of the claim we have shown that a.s. there cannot exist a component C_n' in Γ_n with the property $|C_n'| \geq \frac{1}{n^2}\,|\Gamma_n|$. Therefore, $C_n^{(1)}$ is unique, and the proof of the theorem is complete.

Theorem 5.6 below is the analog of Ajtai *et al.*'s result [1] (for random subgraphs of n-cubes obtained by selecting Q_2^n-edges independently).

Theorem 5.6 *Let Q_{2,λ_n}^n be a random graph consisting of Q_2^n-subgraphs, Γ_n, induced by selecting each Q_2^n-vertex with independent probability λ_n. Then*

$$\lim_{n\to\infty} \mathbb{P}(\Gamma_n \textit{ has an unique giant component}) = \begin{cases} 1 & \textit{for } \lambda_n \geq \frac{1+\varepsilon}{n} \\ 0 & \textit{for } \lambda_n \leq \frac{1-\varepsilon}{n}\,. \end{cases} \tag{5.74}$$

Proof. We proved the first assertion in Theorem 5.5. It remains to consider the case $\lambda_n = \frac{1-\varepsilon}{n}$.

Claim. Suppose $\lambda_n = \frac{1-\varepsilon}{n}$; then there exists $\kappa' > 0$ such that $|C_n^{(1)}| \leq \kappa' n$ holds. The expected number of components of size ℓ is less than

$$\frac{1}{\ell}\, 2^n\, n^{\ell-1} \left(\frac{1-\varepsilon}{n}\right)^{\ell} = \frac{1}{\ell\, n}\, 2^n\, (1-\varepsilon)^{\ell} \tag{5.75}$$

since there are 2^n ways to choose the first element and at most n-vertices to choose from subsequently. This component is counted ℓ times corresponding to all ℓ choices for the "first" vertex. Let $X_{\kappa' n}$ be the r.v. counting the number of components of size $\geq \kappa' n$. Choosing κ' such that $(1-\varepsilon)^{\kappa'} < 1/4$ we obtain

$$\mathbb{E}(X_{\kappa' n}) \leq \sum_{\ell \geq \kappa' n} \frac{1}{\ell\, n}\, 2^n\, (1-\varepsilon)^{\ell} \leq \frac{1}{n^2} 2^n (1-\varepsilon)^{\kappa' n} \sum_{\ell\geq 0} (1-\varepsilon)^{\ell} < \frac{1}{n^2} \left(\frac{1}{2}\right)^n \frac{1}{1-(1-\varepsilon)}\,, \tag{5.76}$$

whence the claim, and the proof of the theorem is complete.

5.5 Distances in n-Cubes

In this section we analyze for which probabilities, λ_n, random induced subgraphs of n-cube exhibit "short" distances. To be precise, we ask for which λ_n

does there exist some constant Δ such that

$$(\dagger) \quad \exists \Delta > 0; \quad d_{\Gamma_n}(v, v') \le \Delta \, d_{Q_2^n}(v, v') \quad \text{a.s. provided } v, v' \text{ are in } \Gamma_n . \qquad (5.77)$$

Before we give the main result of this section we prove a combinatorial lemma [21]. A weaker version of this lemma is proved in [6].

Lemma 5.8 *Let $d \in \mathbb{N}$, $d \ge 2$, and let v, v' be two Q_2^n-vertices where $d(v, v') = d$. Then any Q_2^n-path from v to v' has length $2\ell + d$, and there are at most*

$$\binom{2\ell + d}{\ell + d} \binom{\ell + d}{\ell} n^{\ell} \, \ell! \, d! \qquad (5.78)$$

Q_2^n-paths from v to v' of length $2\ell + d$.

Proof. W.l.o.g. we can assume $v = (0, \ldots, 0)$ and $v' = (x_i)_i$, where $x_i = 1$ for $1 \le i \le d$, and $x_i = 0$ otherwise. Each path of length m induces the family of steps $(\varepsilon_s)_{1 \le s \le m}$, where $\varepsilon_s \in \{e_j \mid 1 \le j \le n\}$. Since the path ends at v', we have for fixed $1 \le j \le n$

$$\sum_{\{\varepsilon_s | \varepsilon_s = e_i\}} \varepsilon_s = \begin{cases} 1 & \text{for } 1 \le i \le d \\ 0 & \text{otherwise} \end{cases} . \qquad (5.79)$$

Hence the families induced by our paths contain necessarily the set $\{e_1, \ldots, e_d\}$. Let $(\varepsilon'_s)_{1 \le s \le m'}$ be the family obtained from $(\varepsilon_s)_{1 \le s \le m}$ by removing the steps $e_1, \ldots, e_d$ at the smallest index at which they occur. Then $(\varepsilon'_s)_{1 \le s \le m'}$ represents a cycle starting and ending at v. Furthermore, we have for all j; $\sum_{\{\varepsilon'_s | \varepsilon'_s = e_i\}} \varepsilon'_s = 0$, i.e., all steps must come in (e_j, e_j), that is, as (up-step,down-step) pairs. As a result, we derive $m = 2\ell + d$, and there are exactly ℓ steps of the form e_j that can be freely chosen (free up-steps). We now count the number of $2\ell + d$-tuples $(\varepsilon_s)_{1 \le s \le 2\ell + d}$. There are exactly $\binom{2\ell+d}{\ell+d}$ ways to select the $(\ell + d)$ indices for the up-steps within the set of all $2\ell + d$ indices. Furthermore, there are $\binom{\ell+d}{\ell}$ ways to select the ℓ positions for the free up-steps and at most n^{ℓ} ways to choose the free up-steps themselves. Since a free up-step is paired with a unique down-step reversing it, the ℓ free up-steps determine all ℓ down-steps. Clearly, there are at most $\ell!$ ways to assign the down steps to their ℓ indices. Finally, there are at most $d!$ ways to assign the fixed up-steps, and the lemma follows.

The following theorem [21] establishes the threshold value for the existence of the above constant Δ. The result is of relevance in the context of local connectivity of neutral networks, a structural property that allows populations of RNA strings to preserve sequence-specific information.

Theorem 5.7 *Let $0 < \delta \le \frac{1}{2}$ and v, v' be arbitrary but fixed Q_2^n-vertices having distance $d_{Q_2^n}(v, v') = d$, $d \ge 2$, $d \in \mathbb{N}$. Let Γ_n denote the random subgraph of Q_2^n obtained by independently selecting Q_2^n-vertices with probability λ_n. Suppose v, v' are contained in Γ_n. Then we have*

(a) Suppose $\lambda_n < n^{\delta-\frac{1}{2}}$ for any $\delta > 0$. Then there exists a.s. no $\Delta > 0$ satisfying

$$d_{\Gamma_n}(v, v') \le \Delta\, d_{Q_2^n}(v, v') . \tag{5.80}$$

(b) Suppose $\lambda_n \ge n^{\delta-\frac{1}{2}}$ for some $\delta > 0$. Then there exists a.s. some finite $\Delta = \Delta(\delta) > 0$ such that

$$d_{\Gamma_n}(v, v') \le \Delta\, d_{Q_2^n}(v, v') . \tag{5.81}$$

Proof. Suppose $d = d(v, v')$ and $\Delta > 0$ are fixed. Let $Z = Z(d, \Delta)$ be the r.v. counting the paths of length $\le \Delta\, d$ from v to v'. According to Lemma 5.8 we have

$$\mathbb{E}[Z] \le \sum_{2\ell+d\le\Delta\, d} \binom{2\ell+d}{\ell+d}\binom{\ell+d}{\ell} n^{\ell}\,\ell!\,d!\,\lambda_n^{2\ell+d-1} . \tag{5.82}$$

Since $\lambda_n < n^{\delta-\frac{1}{2}}$ for any $\delta > 0$, we obtain

$$\sum_{2\ell+d\le\Delta\, d} \binom{2\ell+d}{\ell+d}\binom{\ell+d}{\ell} n^{\ell}\,\ell!\,d!\,\lambda_n^{2\ell+d-1}$$
$$\le \sum_{2\ell+d\le\Delta\, d} \binom{2\ell+d}{\ell+d}\binom{\ell+d}{\ell} \ell!\,d!\,n^{\delta\,2\ell}\left[\frac{1}{n^{\frac{1}{2}-\delta}}\right]^{d-1} . \tag{5.83}$$

For given $d \ge 2$ and Δ, the quantity ℓ is bounded, and choosing δ sufficiently small we derive the upper bound

$$\mathbb{E}[Z] \le O(n^{-\mu}) \quad \text{for some} \quad \mu > 0 , \tag{5.84}$$

and assertion (a) is proved.
To prove (b) we consider the subset of paths $\mathfrak{A}_\sigma$, where σ is some permutation of $d-1$ elements. $\mathfrak{A}_\sigma$-elements are called $\mathfrak{a}$-paths and given by the following data:

(I) some family $(e_{j_1}, \ldots, e_{j_\ell})$, where $d-1 \le j_i \le n$ and $|\{j_i \mid 1 \le i \le \ell\}| = \ell$;
(II) the fixed family $(e_{\sigma(1)}, \ldots, e_{\sigma(d-1)})$; and, finally,
(III) the family $(e_{j_\ell}, \ldots, e_{j_1})$, i.e., the mirror image of the family chosen in (I).

Let $X_{\mathfrak{a}}$ be the indicator r.v. for the event "$\mathfrak{a}$ is a path in Γ_n." Clearly, $A = \sum_{\mathfrak{a}\in\mathfrak{A}_\sigma} X_{\mathfrak{a}}$ is the r.v. counting the number of $\mathfrak{a}$-paths contained in Γ_n. Let $n' = n-(d-1)$. By construction of $\mathfrak{a}$-paths and the linearity of expectation, our first observation is

$$\mathbb{E}[A] = \ell!\binom{n'}{\ell}\lambda_n^{2\ell+(d-1)} = (n')_\ell\,\lambda_n^{2\ell+(d-1)} , \tag{5.85}$$

where $(n)_\ell = n(n-1)\cdots(n-(\ell-1))$. Since $\lambda_n \geq n^{-\frac{1}{2}+\delta}$ for some $0 < \delta < \frac{1}{2}$,

$$\mathbb{E}[A] \geq \left[\frac{(n'-\ell)}{n}\right]^{\ell} n^{2\ell\delta}\left[n^{-\frac{1}{2}+\delta}\right]^{d-1} . \tag{5.86}$$

The idea is to use Janson's inequality (Theorem 5.3) to show that a.s. at least one $\mathfrak{a}$-path is contained in Γ_n. For this purpose we estimate the correlation between the indicator r.v. $X_{\mathfrak{a}}$ and $X_{\mathfrak{a}'}$. The key term we have to analyze is $\Omega = \sum_{\mathfrak{a}\in\mathfrak{A}_\sigma}\sum_{\substack{\mathfrak{a}'\in\mathfrak{A}_\sigma;\\ \mathfrak{a}'\cap\mathfrak{a}\neq\varnothing}} \mathbb{E}[X_{\mathfrak{a}}X_{\mathfrak{a}'}]$. Let $u_s = v + (\sum_{i=1}^{s} e_{j_i})$, where $s \leq \ell$. Since the sequence given in (III) represents the mirror image of the sequence $(e_{j_1},\ldots,e_{j_\ell})$, we inspect

$$|\mathfrak{a}\cap\mathfrak{a}'| = 2\,|\{u_h \in \mathfrak{a}\cap\mathfrak{a}'\}| + \begin{cases} d-1 & \text{if } u_\ell \in \mathfrak{a}\cap\mathfrak{a}' \\ 0 & \text{otherwise.}\end{cases} \tag{5.87}$$

Indeed, only if $\mathfrak{a}$ and $\mathfrak{a}'$ intersect at u_ℓ do the subsequent $d-1$ steps of (II) coincide. In view of Equation (5.87) we distinguish the cases

$$\text{(i)}\ \ u_\ell \notin \mathfrak{a}\cap\mathfrak{a}' \quad \text{and} \quad \text{(ii)}\ \ u_\ell \in \mathfrak{a}\cap\mathfrak{a}' . \tag{5.88}$$

Ad (i): then we have $|\mathfrak{a}\cap\mathfrak{a}'| = 2h$, where $1 \leq h \leq \ell-1$. For fixed h there are exactly $\binom{\ell-1}{h}$ ways to select the h vertices where $\mathfrak{a}$ and $\mathfrak{a}'$ intersect. For each such selection there are at most $h!\,(n'-h)_{\ell-h}$ paths $\mathfrak{a}'$, whence

$$|\{\mathfrak{a}' \mid |\mathfrak{a}'\cap\mathfrak{a}| = 2h\}| \leq \binom{\ell-1}{h} h!\,(n'-h)_{\ell-h} . \tag{5.89}$$

The probability for choosing a correlated $\mathfrak{a}'$-path is given by $\lambda_n^{2[2\ell+(d-1)]-2h}$, and we compute

$$\begin{aligned}
\sum_{\mathfrak{a}\in\mathfrak{A}_\sigma}\sum_{\substack{\mathfrak{a}'\in\mathfrak{A}_\sigma;\\ u_\ell\notin\mathfrak{a}'\cap\mathfrak{a}\neq\varnothing}} \mathbb{E}[X_{\mathfrak{a}}X_{\mathfrak{a}'}] &= \mathbb{E}[A]\sum_{h=1}^{\ell-1} |\{\mathfrak{a}' \mid |\mathfrak{a}'\cap\mathfrak{a}| = 2h\}|\,\lambda_n^{[2\ell+(d-1)]-2h} \\
&\leq \mathbb{E}[A]\sum_{h=1}^{\ell-1} h!\binom{\ell-1}{h}(n'-h)_{\ell-h}\lambda_n^{[2\ell+(d-1)]-2h} \\
&= \mathbb{E}[A]^2\sum_{h=1}^{\ell-1} h!\binom{\ell-1}{h}(n')_h^{-1}\,\lambda_n^{-2h} \\
&\leq \mathbb{E}[A]^2\sum_{h=1}^{\ell-1} h!\binom{\ell-1}{h}\frac{n^h}{(n')_h}\,n^{-2h\delta} ,
\end{aligned}$$

where the last inequality is implied by $\lambda_n \geq n^{-\frac{1}{2}+\delta}$. In view of Equation (5.85) we have for sufficiently large n

$$\sum_{h=1}^{\ell-1} h!\binom{\ell-1}{h}\frac{n^h}{(n')_h}\,n^{-2h\delta} = \underbrace{(\ell-1)\frac{n}{n'}\,n^{-2\delta}}_{h=1} + \underbrace{O(n^{-4\delta})}_{h>1} . \tag{5.90}$$

Consequently we can give the following upper bound for case (i):

$$\sum_{\mathfrak{a}\in\mathfrak{A}_\sigma} \sum_{\substack{\mathfrak{a}'\in\mathfrak{A}_\sigma;\\ u_\ell\notin\mathfrak{a}'\cap\mathfrak{a}\neq\varnothing}} \mathbb{E}[X_\mathfrak{a} X_{\mathfrak{a}'}] \le \left[(\ell-1)\,\frac{n}{n'}\,n^{-2\delta} + O\left(n^{-4\delta}\right)\right]\,\mathbb{E}[A]^2\,. \tag{5.91}$$

Ad (ii): the key observation is that for fixed $\mathfrak{a}$ there are at most $\ell!$ paths $\mathfrak{a}'$ that intersect $\mathfrak{a}$ at least in u_ℓ. Each of these appears with probability of at most 1, whence

$$\sum_{\mathfrak{a}\in\mathfrak{A}_\sigma} \sum_{\substack{\mathfrak{a}'\in\mathfrak{A}_\sigma;\\ u_\ell\in\mathfrak{a}'\cap\mathfrak{a}\neq\varnothing}} \mathbb{E}[X_\mathfrak{a} X_{\mathfrak{a}'}] \le \ell!\,\mathbb{E}[A]\,. \tag{5.92}$$

Equations (5.91) and (5.92) guarantee

$$\Omega \le \left(\underbrace{(\ell-1)\,\frac{n}{n'}\,n^{-2\delta} + O\left(n^{-4\delta}\right)}_{\text{(i)}} + \underbrace{\frac{\ell!}{\mathbb{E}[A]}}_{\text{(ii)}}\right)\mathbb{E}[A]^2\,. \tag{5.93}$$

According to Theorem 5.3 we have $\mathbb{P}(A \le (1-\gamma)\mathbb{E}[A]) \le e^{-\frac{\gamma^2\mathbb{E}[A]}{2+2\Omega/\mathbb{E}[A]}}$, i.e.,

$$\mathbb{P}(A \le (1-\gamma)\mathbb{E}[A]) \le \exp\left[-\frac{\gamma^2}{2/\mathbb{E}[A] + 2\left((\ell-1)\,\frac{n}{n'}\,n^{-2\delta} + O\left(n^{-4\delta}\right) + \frac{\ell!}{\mathbb{E}[A]}\right)}\right]. \tag{5.94}$$

In view of $\mathbb{E}[A] \ge \left[\frac{(n'-\ell)}{n}\right]^\ell n^{2\ell\delta}\left[n^{-\frac{1}{2}+\delta}\right]^{d-1}$, we observe

$$\left[\frac{\gamma^2}{2/\mathbb{E}[A] + 2\left((\ell-1)\,\frac{n}{n'}\,n^{-2\delta} + O\left(n^{-4\delta}\right) + \frac{\ell!}{\mathbb{E}[A]}\right)}\right] = O\left(n^{2\delta}\right)\,, \tag{5.95}$$

for sufficiently large ℓ. Setting $\gamma = 1$, Equation (5.94) becomes

$$\mathbb{P}(A = 0) \le e^{-c' n^{2\delta}} \quad \text{for some } c' > 0\,. \tag{5.96}$$

Since an $\mathfrak{a}$-path has length $2\ell + d$, Equation (5.96) proves (b), and the proof of the theorem is complete.

Corollary 5.1 *Let $\alpha, d \in \mathbb{N}$, $\alpha, d \ge 2$, v, v' be arbitrary but fixed Q_α^n-vertices having distance $d_{Q_\alpha^n}(v, v') = d$ and $n' = n - (d-1)$. Suppose we select Q_α^n-vertices with the probability $0 < \lambda < 1$. Then there exists a Γ_n-path connecting v and v' of length exactly $2 + d_{Q_\alpha^n}(v, v')$ with a probability of at least*

$$\sigma_{\lambda,d}^{[\alpha]}(n) = 1 - \exp\left(-\frac{(\alpha-1)\,n'\,\lambda^{2+(d-1)}}{4}\right)\,, \tag{5.97}$$

provided v, v' are contained in Γ_n.

Proof. The expected number of $\mathfrak{a}$-paths is, according to Theorem 5.7,

$$\mathbb{E}[A] = (\alpha - 1)\,(n - (d-1))\,\lambda^{2+(d-1)} = (\alpha - 1)\,n'\,\lambda^{d+1}\,.$$

For $\ell = 1$ we have only type (ii) correlation, given by Equation (5.92). In this case any two correlated paths necessarily coincide, whence

$$\sum_{\mathfrak{a}\in\mathfrak{A}_\sigma}\ \sum_{\substack{\mathfrak{a}'\in\mathfrak{A}_\sigma;\\ u_\ell\in\mathfrak{a}'\cap\mathfrak{a}\neq\varnothing}} \mathbb{E}[X_\mathfrak{a}X_{\mathfrak{a}'}] = (\alpha - 1)\,n'\,\lambda^{d+1}\,.$$

Consequently Equation (5.94) becomes

$$\mathbb{P}(A = 0) \le \exp\left[-\frac{\mathbb{E}[A]}{4}\right] = \exp\left[-(\alpha - 1)\,(n - (d-1))\,\lambda^{d+1}/4\right]\,.$$

5.6 Conclusion

We began by showing how RNA sequence-structure relations give rise to particular subcubes within a sequence space. These subcubes reduce many questions arising in the context of a neutral evolution of RNA sequences to structural properties of random induced subgraphs of n-cubes. The first and probably best known property is the connectivity of n-cubes in Section 5.3. Here we give a constructive proof that shows how the actual paths can be obtained. In Section 5.4 we discuss the largest component in n-cubes. We prove an extension of Ajtai *et al.*'s [1] result for random graphs in which edges are selected with independent probability. We adopt an "algorithmic" approach and try to give constructive proofs of our results; see, for instance, Lemma 5.3. Only the argument given in Theorem 5.5, where we show that the small subcomponents constructed in Lemma 5.3 have to "melt," does not indicate how to obtain the largest component constructively. The existence of the largest component is of vital importance for neutral evolution. It represents the structural prerequisite for changing the nucleotides of a sequence by successive local "computations" while remaining on the neutral network of a given structure. Upon closer inspection, however, additional properties for neutral evolution are needed [21]. The neutral network has to have many "short" paths whose lengths scale with the Hamming distance of the sequences. This led in Section 5.5 to the analysis of the local connectivity of n-cubes [21]. Local connectivity is a scaling property that reflects a structural relation between the neutral network and sequence space itself. Since local connectivity is a monotone graph property (i.e., once Γ_n is locally connected increasing the probability λ_n does not change the local connectedness), there exists a threshold value. In Theorem 5.7 we localized this threshold value. If locally connected, neutral networks can be viewed as Δ-dilated n-cubes. In particular, a small Hamming distance for two sequences on the neutral network implies the existence of a short neutral path connecting them. We have studied the "algorithmic" perspective, i.e., how to obtain such short paths, in Corollary 5.1.

Acknowledgment

We thank Rita R. Wang for her help. This work was supported by the 973 Project, the PCSIRT Project of the Ministry of Education, the Ministry of Science and Technology, and the National Science Foundation of China.

References

1 M. Ajtai, J. Komlós, and E. Szemerédi. Largest random component of a k-cube. *Combinatorica*, 2:1–7, 1982.

2 D. Aldous and P. Diaconis. Strong uniform times and finite random walks. *Adv. Appl. Math.*, 2:69–97, 1987.

3 L. Babai. Local expansion of vertex transitive graphs and random generation in finite groups. *Proceedings of the 23rd ACM Symposium on Theory of Computing (ACM, New York)*, 1:164–174, 1991.

4 L. Babai and V.T. Sos. Sidon sets in groups and induced subgraphs of cayley graphs. *Eur. J. Combin.*, 1:1–11, 1985.

5 B. Bollobás, Y. Kohayakawa, and T. Luczak. On the evolution of random boolean functions. In: P. Frankl, Z. Furedi, G. Katona, D. Miklos (eds.) *Extremal Problems in Finite Sets*, Hungary Janos Bolyai Mathematical Society, pp. 137–156, 1994.

6 B. Bollobás, Y. Kohayakawa, and T. Luczak. The evolution of random subgraphs of the cube. *Random Struct. Algorithms*, 3:55–90, 1992.

7 C. Borgs, J.T. Chayes, H. Remco, G. Slade, and J. Spencer. Random subgraphs of finite graphs: III. The phase transition for the n-cube. *Combinatorica*, 26:395–410, 2006.

8 J.D. Burtin. The probability of connectedness of a random subgraph of an n-dimensional cube. *Problems Inf. Transmiss.*, 13:147–152, 1977.

9 H. Chernoff. A measure of the asymptotic efficiency for tests of a hypothesis based on the sum of observations. *Ann. Math. Stat.*, 23:493–509, 1952.

10 P. Erdős and J. Spencer. The evolution of the n-cube. *Comput. Math. Appl.*, 5:33–39, 1979.

11 U. Goebel and C.V. Forst. RNA Pathfinder–global properties of neutral betworks, *Z. Phys. Chem.*, 216, 2002.

12 W. Grüner, R. Giegerich, D. Strothmann, C.M. Reidys, J. Weber, Hofacker I.L., Stadler P.F., and Schuster P. Analysis of RNA sequence structure maps by exhaustive enumeration I. Neutral networks. *Chem. Monthly*, 127:355–374, 1996.

13 L.H. Harper. Minimal numberings and isoperimetric problems on cubes. *Theory of Graphs, International Symposium, Rome*, 1966.

14 T.E. Harris. *The Theory of Branching Processes*. Springer, Berlin, 1963.

15 S. Janson. Poisson approximation for large deviations. *Random Struct. Algorithms*, 1:221–229, 1990.

16 V.F. Kolchin. *Random Mappings*. Optimization Software, New York, 1986, 206 pp.

17 C.M. Reidys, P.F. Stadler, and P.K. Schuster. Generic properties of combinatory maps and neutral networks of RNA secondary structures. *Bull. Math. Biol.*, 59(2):339–397, 1997.

18 C.M. Reidys. Random induced subgraphs of generalized n-cubes. *Adv. Appl. Math.*, 19:360–377, 1997.

19 C.M. Reidys. Distance in Random induced subgraphs of generalized

n-cubes. *Combinator. Probabil. Comput.*, 11:599–605, 2002.

20 C.M. Reidys. Large components in random induced subgraphs of n-cubes. *Discrete Math.*, 2009, in press.

21 C.M. Reidys. Local connectivity of neutral networks. *Bull. Math. Biol.*, 71:265–290, 2009.

22 E.A. Schultes and D.P. Bartel. One sequence, two ribozymes: implications for the emergence of new ribozyme folds. *Science*, 289(5478):448–452, 2000.

23 P. Schuster. A testable genotype–phenotype map: modeling evolution of RNA molecules. M. Laessig and A. Valeriani (eds.), Springer, 2002.

24 P. Schuster, W. Fontana, P.F. Stadler, and I.L. Hofacker. From sequences to shapes and back: a case study in RNA secondary structures. *Proc. Roy. Soc. B* 255:279–284, 1994.

6
Graph Edit Distance – Optimal and Suboptimal Algorithms with Applications

Horst Bunke and Kaspar Riesen

6.1
Introduction

In recent years, the use of graph-based data structures has gained popularity in various fields of computer science. Informally, a graph is a set of entities often referred to as nodes connected by links termed edges. The edges represent binary relationships that might exist between pairs of nodes. In general, both nodes and edges can be labeled by one or several attribute values describing their respective properties. Due to the ability of graphs to represent properties of entities as well as binary relations at the same time, graphs have found widespread applications in science and engineering [13, 28].

In the fields of bioinformatics and chemoinformatics, for instance, graph-based representations have been intensively used [5, 33, 41]. In [5] graphs are used to model proteins for protein function prediction. In [33, 41] graphs serve for molecular structure-activity relationship analysis. Another field of research where graphs are studied with emerging interest is that of web content mining. In [48] it is described how graphs can be used to model relational information that is often not present in a vectorial representation of the underlying web document. Image analysis is another field of research where graph-based representation has attracted attention [23, 30, 32, 35]. The basic idea in [23, 30] is to represent color images by means of region adjacency graphs where the nodes are labeled according to RGB color information. In [32] corner points of 2D views of 3D objects are used to construct Delaunay graphs. Graph-based fingerprint classification in the Henry scheme is studied in [35]. The idea here is to use the directional variance in order to extract regions from fingerprints and convert them subsequently into attributed graphs. Finally, graphs have been used to detect network anomalies and to predict abnormal events in computer networks [9].

From an algorithmic perspective, graphs are the most general data structures, as all common data types are simple instances of graphs [4]. Vectors

Analysis of Complex Networks: From Biology to Linguistics. Edited by Matthias Dehmer and Frank Emmert-Streib

ISBN: 978-3-527-32345-6

or strings, for instance, can be seen as simple graphs. In this case each node represents the value of a specific data item and edges connect each item with its successor. Yet, in contrast with the high representational power of graphs as well as their high degree of flexibility, we observe that many operations on graphs, though conceptually simple, are computationally expensive. Consider, for instance, the comparison of objects, i.e., the computation of a similarity or dissimilarity value of a pair of objects. In the case of feature vectors, this operation is linear in the number of data items. Using strings the comparison becomes quadratic in the length of the underlying strings [58], and for graphs the same operation is exponential in the number of nodes [55].

In fact, a dissimilarity computation is needed in various applications (e.g., data mining, machine learning, pattern recognition, etc.). Hence, when graphs are used as the basic data structure, an adequate dissimilarity model for graphs has to be defined. In contrast to vectorial data structures, where the efficiently computable Euclidean distance is widely accepted as a natural distance model, no universally accepted metric on graphs is available. However, in the last three decades quite a large number of different graph distance measures have been proposed in the literature [13] ranging from spectral decompositions of graph matrices over tree search procedures to the training of neural networks.

It turns out that the concept of graph edit distance [7,47] is one of the most flexible graph distance measures, as it can be applied to arbitrary graphs (labeled, unlabeled, directed, undirected) with unconstrained labels on both nodes and edges. The basic idea of the graph edit distance is to define the dissimilarity of two graphs by the minimum amount of distortion that has to be applied to transfrom one graph into the other. Yet, the major problem of the graph edit distance is its computational complexity, which is intractable for large graphs. Typically, the computation of an exact graph edit distance is limited to graphs with at most a few tens of nodes. In order to overcome this severe limitation, different approximative approaches to graph edit distance have been proposed. In the present chapter one particular methodology is presented, allowing one to approximate the edit distance of graphs in cubic time [42].

In the context of the work presented in this chapter, graphs and (suboptimal) graph edit distance are used for the classification of structured data. Classification refers to the process of assigning an unknown input object to one of a given set of classes. It is a common task in the areas of pattern recognition, machine learning, and data mining. Applications of classification can be found in biometric person identification, optical character recognition, automatic protein prediction, medical diagnosis, and other domains. Usually, a classifier is built on the basis of a training set of objects on which the classification rule is learned, based on some underlying mathematical model. Though graph edit distance provides us with a general dissimilarity model in the graph domain, this is typically not sufficient for many standard classification algorithms. In fact, edit-distance-based graph classification is basically limited to nearest-neighbor classification (e.g., [48]). Recently, however, a novel approach to graph-based object classification has emerged [10,45]. In this particular method the graph edit distance is used to transform graphs explicitly into feature vectors. This transformation implies that all standard methods in object classification [17] – originally developed for vectors – become applicable to graphs.

The remainder of this chapter is organized as follows. In Section 6.2 the concept of graph edit distance is introduced in detail. Then, Section 6.3 addresses the problem of optimal and suboptimal graph edit distance computation. In Section 6.4 four different graph data sets with quite different characteristics comming from various applications are reviewed, and moreover, two different approaches to edit-distance-based object classification are discussed. An experimental evaluation of (suboptimal) graph edit distance in conjunction with the two classification scenarios is conducted in Section 6.5. Finally, in Section 6.6 the chapter is summarized and conclusions are drawn.

6.2
Graph Edit Distance

In this section the basic notation for graphs and graph edit distance (GED) is introduced. Let L_V and L_E be a finite or infinite set of labels for nodes and edges, respectively.

Definition 6.1 A graph g is a four-tuple $g = (V, E, \mu, \nu)$, where V is the finite set of nodes, $E \subseteq V \times V$ the set of edges, $\mu : V \longrightarrow L_V$ the node-labeling function, and $\nu : E \longrightarrow L_E$ the edge-labeling function.

This definition allows us to handle arbitrary graphs with unconstrained labeling functions. For example, the labels can be given by the set of integers, the vector space $\mathbb{R}^n$, or a set of symbolic labels $L = \{\alpha, \beta, \gamma, \ldots\}$. Moreover, unlabeled graphs are obtained as a special case by assigning the same label l to all nodes and edges. Edges are given by pairs of nodes (u, v), where $u \in V$ denotes the source node and $v \in V$ the target node of a directed edge. Undirected graphs can be modeled by inserting a reverse edge $(v, u) \in E$ for each edge $(u, v) \in E$ with $\nu(u, v) = \nu(v, u)$.

Graph matching refers to the task of evaluating the similarity of graphs. There are two major versions of this task: exact and error-tolerant graph matching. The aim of exact graph matching is to determine whether two graphs or parts of them are identical in terms of structure and labels. Compared to vectors, where the determination of identical parts is trivial, exact matching of graphs is substantially more difficult [55]. Generally, there is no canonical ordering for nodes and edges in a graph, which makes the comparison of two sets of nodes and edges quite complex.

Based on the exact graph matching paradigms of maximum common subgraph and minimum common supergraph, a few graph similarity measures have been proposed [11, 18, 59]. However, the main restriction of exact graph matching and related similarity measures is the requirement that a significant part of the topology together with the corresponding node and edge labels in two graphs have to be identical to obtain a high similarity value. In fact, this is not realistic for many applications. Especially if the node or edge label alphabet L_V or L_E, respectively, is given by the n-dimensional vector space $\mathbb{R}^n$, the exact matching paradigm is too restrictive. In order to make graph matching better applicable to real-world problems, several error-tolerant, or inexact, graph matching methods have been proposed.

One class of error-tolerant graph matching methods employs artificial neural networks. In two seminal papers [19, 52] it is shown that neural networks can be used to classify directed acyclic graphs. Further examples of graph matching methods based on neural networks can be found in [2, 26]. Another class of error-tolerant graph matching procedures is based on relaxation labeling techniques for structural matching. In [12, 61], for instance, the graph matching problem is stated as a labeling problem. The spectral decomposition of graphs is another approach to graph matching [32, 56]. The basic idea here is to represent graphs by the eigendecomposition of their adjacency or Laplacian matrix. Other examples of graph matching algorithms are based on graduated assignment [21] or random walks [22].

A common problem of the above-mentioned methods for error-tolerant graph matching is that they are often restricted to special classes of graphs. One of the most flexible methods for error-tolerant graph matching that does not suffer from this restriction is based on the edit distance of graphs [7, 47]. Originally, the edit distance was proposed in the context of string matching [58]. Procedures for edit distance computation aim at deriving a dissimilarity measure from the number of distortions one has to apply to transform one pattern into another. The concept of edit distance has been extended from strings to trees [50] and eventually to graphs [7, 47]. Similarly to string edit distance, the key idea of GED is to define the dissimilarity, or distance, of graphs by the minimum amount of distortion that is needed to transform one graph into another.

A standard set of distortion operations is given by *insertions*, *deletions*, and *substitutions* of both nodes and edges. Other operations, such as *merging* and *splitting* of nodes [1], can be useful in certain applications but are not considered in the remainder of this chapter. We denote the substitution of two nodes u and v by $(u \rightarrow v)$, the deletion of node u by $(u \rightarrow \varepsilon)$, and the insertion of node v by $(\varepsilon \rightarrow v)$. For edges we use a similar notation. Given two graphs,

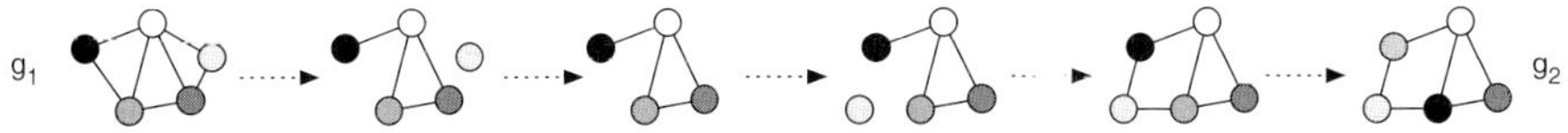

Figure 6.1 A possible edit path between graph g_1 and g_2 (node labels are represented by different shades of gray).

the source graph g_1 and the target graph g_2, the idea of GED is to delete some nodes and edges from g_1, relabel (substitute) some of the remaining nodes and edges, and insert some nodes and edges in g_2, such that g_1 is finally transformed into g_2. A sequence of edit operations $e_1, \ldots, e_k$ that transform g_1 into g_2 is called an *edit path* between g_1 and g_2. Figure 6.1 gives an example of an edit path between two graphs g_1 and g_2. This edit path consists of three edge deletions, one node deletion, one node insertion, two edge insertions, and two node substitutions.

Obviously, for every pair of graphs (g_1, g_2) there exist a number of different edit paths transforming g_1 into g_2. Let $Y(g_1, g_2)$ denote the set of all such edit paths. To find the most suitable edit path out of $Y(g_1, g_2)$, one introduces a cost for each edit operation, measuring the strength of the corresponding operation. The idea of such a cost function is to define whether or not an edit operation represents a strong modification of the graph. Obviously, the cost function is defined with respect to the underlying node or edge labels. Consequently the method is versatile, i.e., it is possible to integrate domain-specific knowledge about object similarity, if available, when defining the costs of the elementary edit operations. Hence GED can be made more discriminative by tuning graph similarity to the specific application area. However, automatic procedures for learning the edit costs from a set of sample graphs are available as well [36, 37].

Clearly, between two similar graphs, there should exist an inexpensive edit path, representing low-cost operations, while for dissimilar graphs an edit path with high costs is needed. Consequently, the *edit distance* of two graphs is defined by the minimum cost edit path between two graphs.

Definition 6.2 Let $g_1 = (V_1, E_1, \mu_1, \nu_1)$ be the source and $g_2 = (V_2, E_2, \mu_2, \nu_2)$ the target graph. The *graph edit distance* between g_1 and g_2 is defined by

$$d(g_1, g_2) = \min_{(e_1,\ldots,e_k) \in Y(g_1,g_2)} \sum_{i=1}^{k} c(e_i)\,,$$

where $Y(g_1, g_2)$ denotes the set of edit paths transforming g_1 into g_2, and c denotes the cost function measuring the strength $c(e_i)$ of edit operation e_i.

6.3 Computation of GED

Unlike exact graph matching methods, the edit distance for graphs allows every node of a graph to be matched to every node of another graph. This flexibility makes GED particularly appropriate for noisy data but is, on the other hand, computationally more expensive than simpler graph matching models. One can distinguish two computation paradigms for GED, viz. *optimal* and *approximate* or *suboptimal* GED algorithms. The former approach always finds a solution that represents the minimum cost edit path between two given graphs. Consequently, the time and space complexity of optimal GED is exponential in the number of nodes of the two involved graphs. That is, for large graphs the computation of edit distance is intractable. The latter computation paradigm addresses the GED problem by only ensuring to find a local minimum of the matching costs. Often this minimum is not very far from the global one, but this property cannot be guaranteed. If such an approximation of the edit distance is acceptable in a certain application, then the subopotimality can be traded for a shorter, usually polynomial, matching time [13].

6.3.1 Optimal Algorithms

The computation of the exact edit distance is usually carried out by means of a tree search algorithm that explores the space of all possible mappings of the nodes and edges of the first graph to the nodes and edges of the second graph. A widely used method is based on the A* algorithm [24], which is a best-first search algorithm. The basic idea is to organize the underlying search space as an ordered tree. The root node of the search tree represents the starting point of our search procedure, inner nodes of the search tree correspond to partial solutions, and leaf nodes represent complete – not necessarily optimal – solutions. Such a search tree is constructed dynamically at runtime by iteratively creating successor nodes linked by edges to the currently considered node in the search tree. In order to determine the most promising node in the current search tree, i.e., the node that will be used for further expansion of the desired mapping in the next iteration, a heuristic function is usually used. Formally, for a node p in the search tree, we use $g(p)$ to denote the cost of the optimal path from the root node to the current node p, i.e., $g(p)$ is set equal to the cost of the partial edit path accumulated so far, and we use $h(p)$ for denoting the estimated cost from p to a leaf node. The sum $g(p) + h(p)$ gives the total cost assigned to an open node in the search tree. One can show that, given that the estimation of the future cost $h(p)$ is lower than, or equal to, the real cost, the algorithm is admissible,

Algorithm 6.1 Graph edit distance algorithm.

Input: Nonempty graphs $g_1 = (V_1, E_1, \mu_1, \nu_1)$ and $g_2 = (V_2, E_2, \mu_2 \nu_2)$, where $V_1 = \{u_1, \ldots, u_{|V_1|}\}$ and $V_2 = \{v_1, \ldots, v_{|V_2|}\}$

Output: A minimum cost edit path from g_1 to g_2, e.g., $p_{\min} = \{u_1 \to v_3, u_2 \to \varepsilon, \ldots, \varepsilon \to v_2\}$

1: Initialize *OPEN* to the empty set $\{\}$
2: For each node $w \in V_2$, insert the substitution $\{u_1 \to w\}$ into *OPEN*
3: Insert the deletion $\{u_1 \to \varepsilon\}$ into *OPEN*
4: **loop**
5:   Remove $p_{\min} = \text{argmin}_{p \in OPEN}\{g(p) + h(p)\}$ from *OPEN*
6:   **if** $p_{\min}$ is a complete edit path
7:     Return $p_{\min}$ as the solution
8:   **else**
9:     Let $p_{\min} = \{u_1 \to v_{i1}, \cdots, u_k \to v_{ik}\}$
10:     **if** $k < |V_1|$ **then**
11:       For each $w \in V_2 \setminus \{v_{i1}, \cdots, v_{ik}\}$, insert $p_{\min} \cup \{u_{k+1} \to w\}$ into *OPEN*
12:       Insert $p_{\min} \cup \{u_{k+1} \to \varepsilon\}$ into *OPEN*
13:     **else**
14:       Insert $p_{\min} \cup \bigcup_{w \in V_2 \setminus \{v_{i1}, \cdots, v_{ik}\}} \{\varepsilon \to w\}$ into *OPEN*
15:     **end if**
16:   **end if**
17: **end loop**

i.e., an optimal path from the root node to a leaf node is guaranteed to be found [24].

In Algorithm 6.1 the A*-based method for optimal GED computation is given. The nodes of the source graph are processed in the order $(u_1, u_2, \ldots)$. The deletion (line 12) or the substitution of a node (line 11) are considered simultaneously, which produces a number of successor nodes in the search tree. If all nodes of the first graph have been processed, the remaining nodes of the second graph are inserted in a single step (line 14). The set *OPEN* of partial edit paths contains the search tree nodes to be processed in the next steps. The most promising partial edit path $p \in OPEN$, i.e., the one that minimizes $g(p) + h(p)$, is always chosen first (line 5). This procedure guarantees that the complete edit path found by the algorithm first is always optimal, i.e., has minimal costs among all possible competing paths (line 7).

Note that edit operations on edges are implied by edit operations on their adjacent nodes, i.e., whether an edge is substituted, deleted, or inserted depends on the edit operations performed on its adjacent nodes. Formally, let $u, u' \in V_1 \cup \{\varepsilon\}$ and $v, v' \in V_2 \cup \{\varepsilon\}$, and assume that the two node operations $(u \to v)$ and $(u' \to v')$ have been executed. We distinguish three cases.

1. Assume there are edges $e_1 = (u, u') \in E_1$ and $e_2 = (v, v') \in E_2$ in the corresponding graphs g_1 and g_2. Then the edge substitution $(e_1 \to e_2)$ is implied by the node operations given above.
2. Assume there is an edge $e_1 = (u, u') \in E_1$ but there is no edge $e_2 = (v, v') \in E_2$. Then the edge deletion $(e_1 \to \varepsilon)$ is implied by the node operations

given above. Obviously, if $v = \varepsilon$ or $v' = \varepsilon$ there cannot be any edge $(v, v') \in E_2$ and thus an edge deletion $(e_1 \rightarrow \varepsilon)$ has to be performed.

3. Assume there is no edge $e_1 = (u, u') \in E_1$ but an edge $e_2 = (v, v') \in E_2$. Then the edge insertion $(\varepsilon \rightarrow e_2)$ is implied by the node operations given above. Obviously, if $u = \varepsilon$ or $u' = \varepsilon$, there cannot be any edge $(u, u') \in E_1$. Consequently, an edge insertion $(\varepsilon \rightarrow e_2)$ has to be performed.

Obviously, the implied edge operations can be derived from every partial or complete edit path during the search procedure given in Algorithm 6.1. The costs of these implied edge operations are dynamically added to the corresponding paths in *OPEN*.

In order to integrate more knowledge about partial solutions in the search tree, it has been proposed to use heuristics [24]. Basically, such heuristics for a tree search algorithm aim at the estimation of a lower bound $h(p)$ of the future costs. In the simplest scenario this lower bound estimation $h(p)$ for the current node p is set to zero for all p, which is equivalent to using no heuristic information about the present situation at all. The other extreme would be to compute for a partial edit path the actual optimal path to a leaf node, i.e., perform a complete edit distance computation for each node of the search tree. In this case, the function $\left(h(p)\right)$ is not a lower bound, but the exact value of the optimal costs. Of course, the computation of such a perfect heuristic is both unreasonable and intractable. Somewhere in between the two extremes one can define a function $h(p)$ evaluating how many edit operations have to be performed in a complete edit path at least [7]. One possible function of this type is described in the next paragraph.

Let us assume that a partial edit path at a position in the search tree is given, and let the number of unprocessed nodes of the first graph g_1 and second graph g_2 be n_1 and n_2, respectively. For an efficient estimation of the remaining optimal edit operations, we first attempt to perform as many node substitutions as possible. To this end, we potentially substitute each of the n_1 nodes from g_1 with any of the n_2 nodes from g_2. To obtain a lower bound of the exact edit cost, we accumulate the costs of the $\min\{n_1, n_2\}$ least expensive of these node substitutions, and the costs of $\max\{0, n_1 - n_2\}$ node deletions and $\max\{0, n_2 - n_1\}$ node insertions. Any of the selected substitutions that is more expensive than a deletion followed by an insertion operation is replaced by the latter. The unprocessed edges of both graphs are handled analogously. Obviously, this procedure allows multiple substitutions involving the same node or edge and, therefore, it possibly represents an invalid way to edit the remaining part of g_1 into the remaining part of g_2. However, the estimated cost certainly constitutes a lower bound of the exact cost, and thus an optimal edit path is guaranteed to be found [24]. We refer to this method for GED computation as *Heuristic-A**.

6.3.2 Suboptimal Algorithms

The method described in the previous section finds an optimal edit path between two graphs. Unfortunately, the computational complexity of the edit distance algorithm, whether or not a heuristic function $h(p)$ is used to govern the tree traversal process, is exponential in the number of nodes of the involved graphs. This means that the running time and space complexity may be huge even for rather small graphs.

In recent years, a number of methods addressing the high computational complexity of GED computation have been proposed. A common way to make graph matching more efficient is to restrict considerations to special classes of graphs. Examples include the classes of planar graphs [25], bounded-valence graphs [31], trees [54], and graphs with unique node labels [15]. Recently, a suboptimal edit distance algorithm has been proposed [35] that requires the nodes of graphs to be planarly embedded, which is satisfied in many, but not all, computer vision applications of graph matching. In some approaches, the basic idea is to perform a local search to solve the graph matching problem, that is, to optimize local criteria instead of global or optimal ones [3]. In [27], a linear programming method for computing the edit distance of graphs with unlabeled edges is proposed. The method can be used to derive lower and upper edit distance bounds in polynomial time. A number of graph matching methods based on genetic algorithms have been proposed [14]. Genetic algorithms offer an efficient way to cope with large search spaces, but they are nondeterministic.

In [39] a simple variant of an optimal edit distance algorithm based on Heuristic-A* is proposed. Instead of expanding all sucessor nodes in the search tree, only a fixed number s of nodes to be processed are kept in the *OPEN* set at all times. Whenever a new partial edit path is added to *OPEN* in Algorithm 6.1, only the s partial edit paths p with the lowest costs $g(p) + h(p)$ are kept, and the remaining partial edit paths in *OPEN* are removed. Obviously, this procedure corresponds to a pruning of the search tree during the search procedure, i.e., not the full search space is explored, but only those nodes are expanded that belong to the most promising partial matches. This method with parameter s is referred to as *Beamsearch((s))*, or *Beam((s))* for short.

6.3.2.1 Bipartite Graph Matching

Another approach to solving the problem of GED computation is introduced in [42]. In this approach the GED is approximated by finding an optimal match between nodes of two graphs together with their local structure. The computation of GED is then reduced to the *assignment problem*. The assignment problem considers the task of finding an optimal assignment of the elements of a set A to the elements of a set B, where A and B have the same cardinality. Assuming that numerical costs are given for each assignment pair, an optimal assignment is one that minimizes the sum of the assignment costs. Formally, the assignment problem can be defined as follows.

Definition 6.3 Let us assume there are two sets A and B together with an $n \times n$ cost matrix $\mathbf{C}$ of real numbers given, where $|A| = |B| = n$. The matrix elements $\mathbf{C}_{ij}$ correspond to the costs of assigning the ith element of A to the jth element of B. The assignment problem can be stated as finding a permutation $p = p_1, \ldots, p_n$ of the integers $1, 2, \ldots, n$ that minimizes $\sum_{i=1}^{n} \mathbf{C}_{ip_i}$.

The assignment problem can be reformulated as finding an optimal matching in a complete bipartite graph and is therefore also referred to as a bipartite graph matching problem. Solving the assignment problem in a brute force manner by enumerating all permutations and selecting the one that minimizes the objective function leads to an exponential complexity that is unreasonable, of course. However, there exists an algorithm that is known as Munkres' algorithm [34][1)] that solves the bipartite matching problem in polynomial time. In Algorithm 6.2 Munkres' method is described in detail. The assignment cost matrix $\mathbf{C}$ given in Definition 6.3 is the algorithms' input, and the output corresponds to the optimal permutation, i.e., the assignment pairs resulting in the minimum cost. In the description of Munkres' method in Algorithm 6.2, some lines (rows or columns) of the cost matrix $\mathbf{C}$ and some zero elements are distinguished. They are termed *covered* or *uncovered* lines and *starred* or *primed* zeros, respectively. In the worst case the maximum number of operations needed by the algorithm is $O(n^3)$. Note that the $O(n^3)$ complexity is much smaller than the $O(n!)$ complexity required by a brute force algorithm.

Let us assume a source graph $g_1 = (V_1, E_1, \mu_1, \nu_1)$ and a target graph $g_2 = (V_2, E_2, \mu_2, \nu_2)$ of equal size, i.e., $|V_1| = |V_2|$, are given. One can use Munkres' algorithm in order to map the nodes of V_1 to the nodes of V_2 such that the resulting node substitution costs are minimal, i.e., we solve the assignment problem of Definition 6.3 with $A = V_1$ and $B = V_2$. In our solution we define the cost matrix $\mathbf{C}$ such that entry $\mathbf{C}_{i,j}$ corresponds to the cost of substituting the ith node of V_1 with the jth node of V_2. Formally, $\mathbf{C}_{i,j} = c(u_i \rightarrow v_j)$, where $u_i \in V_1$ and $v_j \in V_2$, for $i, j = 1, \ldots, |V_1|$.

The constraint that both graphs to be matched are of equal size is too restrictive since it cannot be expected that all graphs in a specific problem domain will always have the same number of nodes. However, one can define a quadratic cost matrix $\mathbf{C}$ that is more general in the sense that we allow insertions and/or deletions to occur in both graphs under consideration.

1) Munkres' algorithm is a refinement of an earlier version by Kuhn [29] and is also referred to as Kuhn–Munkres, or Hungarian algorithm.

Algorithm 6.2 Munkres' algorithm for the assignment problem.

Input: A cost matrix **C** with dimensionality n
Output: The minimum-cost node or edge assignment

```
For each row r and column c in C, subtract its smallest element from every element in r
  and c, respectively.
For all zeros z_i in C, mark z_i with a star if there is no starred zero in its row or column
STEP 1:
Cover each column containing a starred zero
if n columns are covered then GOTO DONE else GOTO STEP 2 end if
STEP 2:
if C contains an uncovered zero then
   Find an arbitrary uncovered zero Z_0 and prime it
   if There is no starred zero in the row of Z_0 then
      GOTO STEP 3
   else
      Cover this row, and uncover the column containing the starred zero GOTO STEP 2.
   end if
else
   Save the smallest uncovered element e_min GOTO STEP 4
end if
STEP 3: Construct a series S of alternating primed and starred zeros as follows:
Insert Z_0 into S
while In the column of Z_0 there exists a starred zero Z_1
   Insert Z_1 into S
   Replace Z_0 with the primed zero in the row of Z_1. Insert Z_0 into S
end while
Unstar each starred zero in S and replace all primes with stars. Erase all other primes
  and uncover every
  line in C GOTO STEP 1
STEP 4: Add e_min to every element in covered rows and subtract it from every element
  in uncovered columns. GOTO STEP 2
DONE: Assignment pairs are indicated by the positions of starred zeros in the cost matrix.
```

Moreover, matrix **C** is now by definition quadratic, regardless of the size of the underlying graphs. That is, $|V_1| \neq |V_2|$ is explicitly allowed.

Definition 6.4 Let $g_1 = (V_1, E_1, \mu_1, \nu_1)$ be the source and $g_2 = (V_2, E_2, \mu_2, \nu_2)$ the target graph with $V_1 = \{u_1, \ldots, u_n\}$ and $V_2 = \{v_1, \ldots, v_m\}$, respectively. The cost matrix **C** is defined as

$$
\mathbf{C} = \left[\begin{array}{cccc|cccc}
c_{1,1} & c_{1,2} & \cdots & c_{1,m} & c_{1,\varepsilon} & \infty & \cdots & \infty \\
c_{2,1} & c_{2,2} & \cdots & c_{2,m} & \infty & c_{2,\varepsilon} & \ddots & \vdots \\
\vdots & \vdots & \ddots & \vdots & \vdots & \ddots & \ddots & \infty \\
c_{n,1} & c_{n,2} & \cdots & c_{n,m} & \infty & \cdots & \infty & c_{n,\varepsilon} \\
\hline
c_{\varepsilon,1} & \infty & \cdots & \infty & 0 & 0 & \cdots & 0 \\
\infty & c_{\varepsilon,2} & \ddots & \vdots & 0 & 0 & \ddots & \vdots \\
\vdots & \ddots & \ddots & \infty & \vdots & \ddots & \ddots & 0 \\
\infty & \cdots & \infty & c_{\varepsilon,m} & 0 & \cdots & 0 & 0
\end{array}\right]
$$

where $c_{i,j}$ denotes the cost of a node substitution, $c_{i,\varepsilon}$ the cost of a node deletion $c(u_i \rightarrow \varepsilon)$, and $c_{\varepsilon,j}$ the costs of a node insertion $c(\varepsilon \rightarrow v_j)$.

Obviously, the upper left corner of the cost matrix represents the costs of all possible node substitutions, the diagonal of the upper right corner the costs of all possible node deletions, and the diagonal of the bottom left corner the costs of all possible node insertions. Note that each node can be deleted or inserted at most once. Therefore, any nondiagonal element of the upper right and lower left part is set to ∞. The bottom right corner of the cost matrix is set to zero since substitutions of the form $(\varepsilon \rightarrow \varepsilon)$ should not incur any costs.

On the basis of the new cost matrix **C** defined above, Munkres' algorithm [34] can be executed (Algorithm 6.2). This algorithm finds the optimal, i.e., the minimum cost, permutation $p = p_1, \ldots, p_{n+m}$ of the integers $1, 2, \ldots, n + m$ that minimizes $\sum_{i=1}^{n+m} C_{ip_i}$. Obviously, this is equivalent to the minimum cost assignment of the nodes of g_1 represented by the rows to the nodes of g_2 represented by the columns of matrix **C**. That is, Munkres' algorithm indicates the minimum cost assignment pairs with starred zeros in the transformed cost matrix **C**. These starred zeros are independent, i.e., each row and each column of **C** contains exactly one starred zero. Consequently, each node of graph g_1 is either uniquely assigned to a node of g_2 (upper left corner of **C**) or to the deletion node ε (upper right corner of **C**). Conversely, each node of graph g_2 is either uniquely assigned to a node of g_1 (upper left corner of **C**) or to the insertion node ε (bottom left corner of **C**). The ε-nodes in g_1 and g_2 corresponding to rows $n+1, \ldots, n+m$ and columns $m+1, \ldots, m+n$ in **C** that are not used cancel each other out without any costs (bottom right corner of **C**).

So far the proposed algorithm considers the nodes only and takes no information about the edges into account. In order to achieve a better approximation of the true edit distance, it would be highly desirable to involve edge operations and their costs in the node assignment process as well. In order to achieve this goal, an extension of the cost matrix is needed. To each entry $c_{i,j}$, i.e., to each cost of a node substitution $c(u_i \rightarrow v_j)$, the minimum sum of edge edit operation costs, implied by node substitution $u_i \rightarrow v_j$, is added. Formally, assume that node u_i has adjacent edges E_{ui} and node v_j has adjacent edges E_{vj}. With these two sets of edges, E_{ui} and E_{vj}, an individual cost matrix similar to Definition 6.4 can be established and an optimal assignment of elements E_{ui} to elements E_{vj} according to Algorithm 6.2 performed. Clearly, this procedure leads to the minimum sum of edge edit costs implied by the given node substitution $u_i \rightarrow v_j$. These edge edit costs are added to the entry $c_{i,j}$. Clearly, to the entry $c_{i,\varepsilon}$, which denotes the cost of a node deletion, the cost of the deletion of all adjacent edges of u_i is added, and to the entry $c_{\varepsilon,j}$, which denotes the cost of a node insertion, the cost of all insertions of the adjacent edges of v_j is added.

Note that Munkres' algorithm used in its original form is optimal for solving the assignment problem, but it provides us with a suboptimal solution for the GED problem only. This is due to the fact that each node edit operation is considered individually (considering the local structure only), such that no implied operations on the edges can be inferred dynamically. The result returned by Munkres' algorithm corresponds to the minimum cost mapping, according to matrix $\mathbf{C}$, of the nodes of g_1 to the nodes of g_2. Given this mapping, the implied edit operations of the edges are inferred, and the accumulated costs of the individual edit operations on both nodes and edges can be computed. The approximate edit distance values obtained by this procedure are equal to, or larger than, the exact distance values, since it finds an optimal solution in a subspace of the complete search space. We refer to this method as *Bipartite*, or *BP* for short.

6.4 Applications

The intention of this section is twofold. First, four different graph data sets are discussed in order to give an exemplary insight into how graphs can be used to model objects in certain applications. Secondly, two different approaches to graph-based object classification in conjunction with GED are discussed. First we make use of the edit distance for a direct classification by means of a nearest-neighbor classifier. The second approach uses the GED in order to transform graphs into feature vectors. The classification process is then carried out in the target vector space.

6.4.1 Graph Data Sets

In this section four different graph data sets with quite different characteristics are presented. They represent line drawings, grayscale images, HTML web sites, and molecular compounds. The graph data sets emerged in the context of the authors' recent work on graph kernels [38] and graph embedding [10, 45]. All graph data sets discussed in the present paper are publicly available or will be made available in the near future [43].

Letter Database The first graph data set involves graphs that represent distorted letter drawings. We consider the 15 capital letters of the Roman alphabet that consist of straight lines only (*A, E, F, H, I, K, L, M, N, T, V, W, X, Y, Z*). For each class, a prototype line drawing is manually constructed. These prototype drawings are then converted into prototype graphs by representing lines by undirected edges and ending points of lines by nodes. Each node is labeled with a two-dimensional attribute giving its position relative to a reference coordinate system. Edges are unlabeled. The graph database consists of a training set, a validation set, and a test set of size 750 each. In order to test classifiers under different conditions, distortions are applied on the prototype graphs with three different levels of strength, viz. *low*, *medium*, and *high*. Hence, our experimental data set comprises 6,750 graphs altogether. Figure 6.2 illustrates the prototype graph and a graph instance for each distortion level representing the letter *A*.

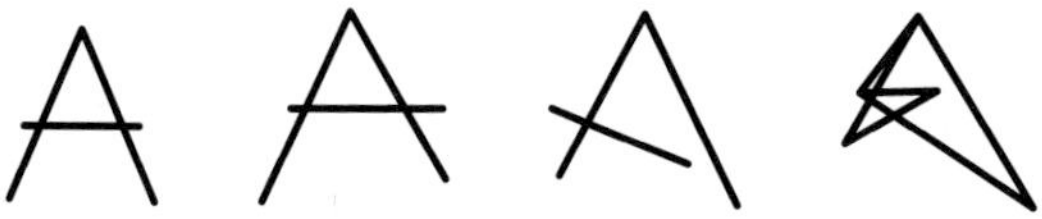

(a) (b) (c) (d)

Figure 6.2 Instances of letter *A*: Original and distortion levels *low*, *medium* and *high* (from left to right).

Fingerprint Database Fingerprints are converted into graphs by filtering the images and extracting regions that are relevant [35]. In order to obtain graphs from fingerprint images, the relevant regions are binarized and a noise removal and thinning procedure is applied. This results in a skeletonized representation of the extracted regions. Ending points and bifurcation points of the skeletonized regions are represented by nodes. Additional nodes are inserted in regular intervals between ending points and bifurcation points. Finally, undirected edges are inserted into link nodes that are directly connected through a ridge in the skeleton. Each node is labeled with a two-dimensional attribute giving its position. The edges are attributed with an angle denoting the orientation of the edge with respect to the horizontal direction.

The fingerprint database used in our experiments is based on the NIST-4 reference database of fingerprints [60]. It consists of a training set of size 500,

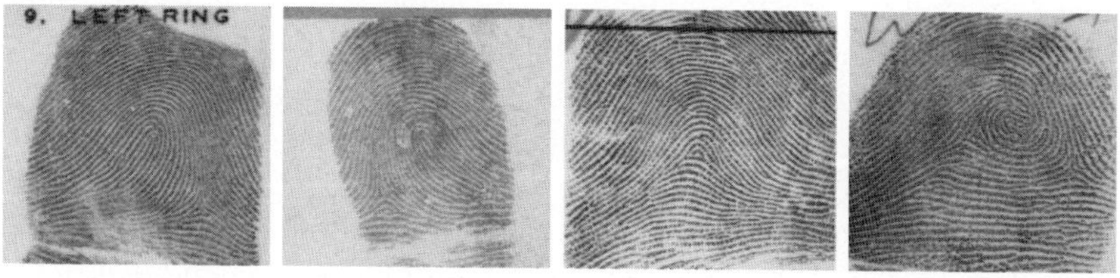

(a) *Left* (b) *Right* (c) *Arch* (d) *Whorl*

Figure 6.3 Fingerprint examples from the four classes.

a validation set of size 300, and a test set of size 2000. Thus, there are a total of 2800 fingerprint images from the four classes *arch, left, right,* and *whorl* of the Galton–Henry classification system. For examples of these fingerprint classes, see Figure 6.3.

AIDS Database The AIDS data set consists of graphs representing molecular compounds. We construct graphs from the AIDS Antiviral Screen Database of Active Compounds [16]. This data set consists of two classes (*active, inactive*), which represent molecules with activity against HIV or not. The molecules are converted into graphs in a straightforward manner by representing atoms as nodes and the covalent bonds as edges. Nodes are labeled with the number of the corresponding chemical symbol and edges by the valence of the linkage. Figure 6.4 illustrates one molecular compound from each class. Note that different shades of gray represent different chemical symbols, i.e., node labels. We use a training set and a validation set of size 250 each, and a test set of size 1500. Thus, there are 2000 elements in all (1600 inactive and 400 active elements).

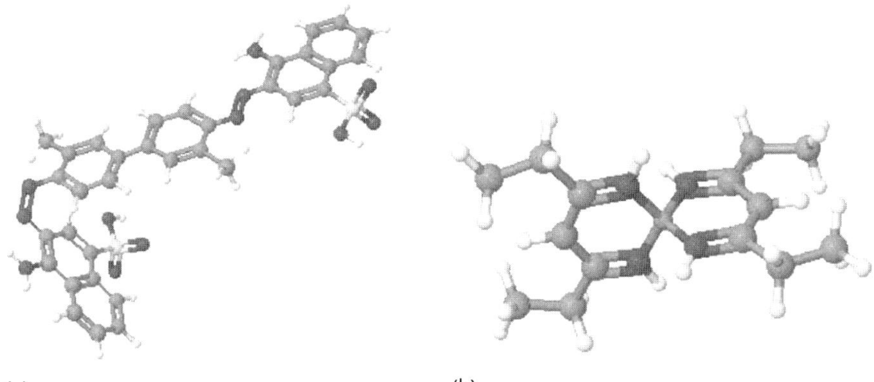

(a) (b)

Figure 6.4 A molecular compound of both classes: (a) active, (b) inactive.

Web Page Database In [48] several methods for creating graphs from web documents are introduced. For the graphs used in the experiments, the following method was applied. First, all words occuring in the web document – except for stop words, which contain little information – are converted into nodes in the resulting web graph. We attribute the nodes with the corresponding word and its frequency, i.e., even if a word appears more than once in the same web document, we create only one unique node for it and store its total frequency as an additional node attribute. Next, different sections of the web document are investigated individually. These sections are *title*, which contains the text related to the document's title, *link*, which is a text in a clickable hyperlink, and *text*, which comprises any of the readable text in the web document. If word w_i immediately precedes word w_{i+1} in any of the sections *title*, *link*, or *text*, a directed edge from the node corresponding to word w_i to the node corresponding to the word w_{i+1} is inserted in our web graph. The resulting edge is given the appropriate section label. Although word w_i might immediately precede word w_{i+1} in more than one section, only one edge is inserted. That is, an edge is possibly labeled with more than one section label. Finally, only the most frequently used words (nodes) are kept in the graph and the terms are conflated to the most frequently occurring form.

In our experiments we make use of a data set that consists of 2340 documents from 20 categories (*Business, Health, Politics, Sports, Technology, Entertainment, Art, Cable, Culture, Film, Industry, Media, Multimedia, Music, Online, People, Review, Stage, Television,* and *Variety*). The last 14 categories are subcategories related to entertainment. These web documents were originally hosted at Yahoo as news pages (http://www.yahoo.com). The database is split into a training, a validation, and a test set of equal size (780).

In contrast to the other data sets, these graphs are characterized by the existence of unique node labels. This is interesting since it implies that whenever two graphs are being matched with each other, each node has at most one uniquely defined candidate for possible assignment in the other graph. Hence, the computationally expensive step in graph matching, i.e., the exploration of all possible mappings between the nodes of the two graphs under consideration, is no longer needed [6].

Note that the graph data sets are of a quite different nature, coming from a variety of applications. Furthermore, the graph sets differ in their characteristics, such as the number of available graphs ($|G|$), the number of different classes ($|\Omega|$), and the average and maximum number of nodes and edges per graph ($\varnothing|V|$, $\varnothing|E|$, $\max|V|$, $\max|E|$). Table 6.1 gives a summary of all graph data sets and their corresponding characteristics.

Table 6.1 Graph data set characteristics.

Database	$\|G\|$	$\|\Omega\|$	$\varnothing\|V\|$	$\varnothing\|E\|$	$\max\|V\|$	$\max\|E\|$
Letter	6750	15	4.7	4.5	9	9
Fingerprint	2800	4	5.4	4.4	26	24
AIDS	2000	2	9.5	10.0	85	328
Web page	2340	20	186.1	104.6	834	596

6.4.2
GED-Based Nearest-Neighbor Classification

A traditional approach to addressing the classification problem in a graph space is to apply a k-nearest-neighbor (k-NN) classifier in conjunction with edit distance. Given a labeled set of training graphs, an unknown graph is assigned to the class that occurs most frequently among the k nearest graphs (in terms of edit distance) from the training set. Formally, let us assume that a graph space $\mathcal{G}$, a space of class labels $\mathcal{Y}$, and a labeled training set of patterns $\{(g_i, y_i)\}_{i=1,\ldots,N} \subseteq \mathcal{G} \times \mathcal{Y}$ are given. If $\{(g_{(1)}, y_{(1)}), \ldots, (g_{(k)}, y_{(k)})\} \subseteq \{g_i, y_i)\}_{i=1,\ldots,N}$ are the k patterns that have the smallest distance $d(g, g_{(i)})$ to a test pattern g, then the k-NN classifier $f\colon \mathcal{G} \rightarrow \mathcal{Y}$ can be defined by

$$f(g) = \operatorname*{argmax}_{y \in \mathcal{Y}} |\{(g_{(i)}, y_{(i)}) : y_{(i)} = y\}| .$$

If $k = 1$, the k-NN classifier's decision is based on just one element from the training set, regardless of whether this element is an outlier or a true class representative. Obviously, a choice of parameter $k > 1$ reduces the influence of outliers by evaluating which class occurs most frequently in a neighborhood around the test pattern.

Classifiers of the nearest-neighbor type in conjunction with GED have been succesfully applied to various classification problems. In [48], for instance, this concept is used for web content mining, which involves the clustering and classification of web documents based on their textual substance. The automatic identification of diatoms using GED and k-NN classifiers is conducted in [1]. Diatoms are unicellular algae found in humid places where light provides the basis for photosynthesis. The classification of diatoms is useful for various applications such as environmental monitoring or forensic medicine. The huge estimated number of more than 10,000 diatom classes makes the classification of diatoms very difficult. Finally, in [38] nearest-neighbor classification based on GED is applied to various classification problems, viz. fingerprint, color image, and molecule classification.

6.4.3
Dissimilarity-Based Embedding Graph Kernels

Although GED and related similarity measures allow us to compute distances between general graphs, this is not sufficient for most standard pattern recognition algorithms. In fact, edit-distance-based graph matching that can be applied directly in the domain of graphs is limited to nearest-neighbor classification described in Section 6.4.2.

A promising direction to overcome the lack of algorithmic tools for graph classification is graph embedding. Basically, an embedding of graphs into a vector space establishes access to the rich repository of algorithmic tools for pattern analysis. Examples of graph embeddings can be found in [32, 46, 62]. Recently a new class of graph embedding procedures that are based on prototype selection and GED computation has emerged. Originally the idea was proposed in [40] in order to map feature vectors into dissimilarity spaces. This idea was generalized first to strings [53] and eventually to the domain of graphs [45]. The key idea of this approach is to use the distances of an input graph to a number of training graphs, termed prototype graphs, as a vectorial description of the input graph. That is, we use a dissimilarity representation rather than the original graph representation for pattern recognition tasks.

This graph embedding procedure makes use of GED. Consequently, it can be applied to both directed and undirected graphs, as well as to graphs without and with labels on their nodes and/or edges. In case there are labels on the nodes and/or edges, these labels can be of any nature (discrete symbols, the set of integer or real numbers, or whole attribute vectors). Even hypergraphs can be embedded with this embedding method [8]. Hence, the proposed embedding approach is more general than other graph embedding techniques where (sometimes quite severe) restrictions on the type of underlying graph are imposed.

General Embedding Procedure Assume we have a labeled set of sample graphs, $\mathcal{G} = \{g_1, \ldots, g_N\}$. After having selected a set $\mathcal{P} = \{p_1, \ldots, p_n\} \subseteq \mathcal{G}$, we compute the GED of a given input graph g to each prototype $p \in \mathcal{P}$. Note that g can be an element of $\mathcal{G}$ or any other graph. This leads to n dissimilarities, $d_1 = d(g, p_1), \ldots, d_n = d(g, p_n)$, which can be arranged in an n-dimensional vector $(d_1, \ldots, d_n)$. In this way we can transform any graph from the training as well as any other graph set (for instance a validation or a test set of a classification problem) into a vector of real numbers.

Definition 6.5 Let $\mathcal{G}$ be a finite or infinite set of graphs and $\mathcal{P} = \{p_1, \ldots, p_n\} \subseteq \mathcal{G}$ a set of prototypes. Then, the mapping $\varphi_n^{\mathcal{P}} : \mathcal{G} \to \mathbb{R}^n$ is defined as the function

$$\varphi_n^P(g) = (d(g, p_1), \ldots, d(g, p_n)) ,$$

where $d(g, p_i)$ is the edit distance between graph g and the ith prototype[2)].

The complexity of the embedding procedure in conjunction with GED is exponential since the exact computation of GED is exponential in the number of nodes for general graphs. However, as mentioned above, there exist a number of efficient approximation algorithms for GED computation

2) Note that any other graph dissimilarity measure can be used as well.

(e.g., the bipartite matching approach [42] with cubic time complexity described in the present chapter). Consequently, given n predefined prototypes, the embedding of one particular graph is established by means of n distance computations with polynomial time.

Prototype Selection One crucial question about the proposed graph embedding is how to define a set $\mathcal{P}$ of prototypes that lead to a good performance of the classifier in the feature space. Often, the prototype set $\mathcal{P}$ is defined as a subset of the training set of graphs $\mathcal{T}$, i.e., $\mathcal{P} \subseteq \mathcal{T}$ [45]. In [40, 45, 53] different prototype selection algorithms are discussed. These prototype selection strategies use some heuristics based on the underlying dissimilarities in the original graph domain. It is shown that none of them is globally best, i.e., the quality of the selected prototypes and in particular their number depends on the underlying data set. Thus, both the selection strategy and dimensionality are determined with the target classifier on a validation set.

An alternative approach is to use all available elements from the training set as prototypes, i.e., $\mathcal{P} = \mathcal{T}$, and subsequently apply dimensionality reduction methods. This process is more principled and allows us to completely avoid the problem of finding the optimal prototype selection strategy. For dimensionality reduction, for instance, the well-known principal component analysis (PCA) and Fisher's linear discriminant analysis (LDA) [17] can be applied. This approach (using all available elements from the training set as prototypes) has been applied to graph embeddings in [44]. Note that in this approach the vector space embedded graphs of dimensionality N are mapped into another vector space of dimensionality $n \leq N$. For the sake of simplicity, however, we also denote these embedded, and eventually transformed, graphs with $\varphi_n^P(g)$.

Relationship to Graph Kernel Methods Another idea to overcome the lack of algorithmic tools for graph classification, which is closely related to graph embedding procedures, is kernel methods [49, 51, 57]. In recent years, kernel methods have become one of the most rapidly emerging subfields in intelligent information processing. The vast majority of work on kernel methods is concerned with transforming a given feature space into another one of higher dimensionality without computing the transformation explicitly for each individual feature vector. As a fundamental extension the existence of kernels for symbolic data structures, especially for graphs, has been shown [20]. By means of suitable kernel functions, graphs can be implicitly mapped into vector spaces. Consequently, a large class of kernel machines for classification, most of them originally developed for feature vectors, become applicable to graphs.

Definition 6.6 Let $\mathcal{G}$ be a finite or infinite set of graphs, $g_i, g_j \in \mathcal{G}$, and $\varphi : \mathcal{G} \to \mathbb{R}^n$ a function with $n \in \mathbb{N}$. A graph kernel function is a mapping $\kappa : \mathcal{G} \times \mathcal{G} \to \mathbb{R}$ such that $\kappa(g_i, g_j) = \langle \varphi(g_i), \varphi(g_j) \rangle$.

According to this definition a graph kernel function takes two graphs g_i and g_j as arguments and returns a real number that is equal to the result achieved by first mapping the two graphs by a function φ to a vector space and then computing the dot product $\langle \varphi(g_i), \varphi(g_j) \rangle$ in the feature space. The kernel function $\kappa(g_i, g_j)$ provides us with a shortcut (kernel trick) that eliminates the need for computing $\varphi(.)$ explicitly. It is well-known that many classification algorithms can be kernelized, i.e., formulated in such a way that only scalar products of vectors rather than the vectors of individual objects are needed. Such algorithms are commonly referred to as *kernel machines*. Hence, applying a graph kernel provides us access to all these algorithms.

Based on the graph embedding $\varphi_n^{\mathcal{P}}$ established above, one can define a valid graph kernel κ by computing the standard dot product of two graph maps in the resulting vector space

$$\kappa_{\langle\rangle}(g_i, g_j) = \langle \varphi_n^{\mathcal{P}}(g_i), \varphi_n^{\mathcal{P}}(g_j) \rangle .$$

Of course, not only the standard dot product can be used but any valid kernel function defined for vectors, e.g., an RBF kernel function

$$\kappa_{\mathrm{RBF}}(g_i, g_j) = \exp\left(-\gamma ||\varphi_n^{\mathcal{P}}(g_i) - \varphi_n^{\mathcal{P}}(g_j)||^2\right)$$

where $\gamma > 0$. We denote this graph kernel as *dissimilarity embedding graph kernel*.

In a recent book, graph kernels were proposed that directly use GEDs [38]. This approach turns the existing dissimilarity measure (GED) into a similarity measure by mapping low distance values to high similarity values and vice versa. To this end, a simple monotonically decreasing transformation is applied to the GED. Note the fundamental difference between such an approach and our embedding procedure. While in the former methodology the existing dissimilarity measure is turned into a similarity measure (i.e., a kernel value) and subsequently plugged into a kernel machine, the latter uses the dissimilarities to n prototypes as features for a new description of the underlying object. Therefore, not only kernel machines but also other nonkernelizable algorithms can be applied in conjunction with the proposed graph embedding method.

6.5 Experimental Evaluation

Two different experiments are described in this section in order to demonstrate the feasibility of (suboptimal) GED for graph-based object classification. First, GED is computed with optimal and suboptimal methods. A comparison of the computation time and the classification accuracy using a nearest-neighbor classifier is carried out. In the second experiment the suboptimal GEDs as computed for the first experiment are used for the dissimilarity embedding graph kernel described in Section 6.4.3.

6.5.1
Optimal vs. Suboptimal Graph Edit Distance

In our experiments we divide each database into three disjoint subsets, viz. the training, the validation, and the test set. The elements of the training set are used as prototypes in the NN classifier. The validation set is used to determine the values of the meta parameters τ_{node}, which correspond to the cost of a node deletion or insertion, and τ_{edge}, which corresponds to the costs of an edge deletion or insertion. For all considered graph data sets node and edge labels are integer numbers, real numbers, real vectors, or strings, and the substitution cost of a pair of labels is given by a suitable distance measure (Euclidean distance or string edit distance [58]). Finally, the parameter pair that leads to the highest classification accuracy on the validation set is used on the independent test set to perform GED computation and NN classification.

We use three algorithms for GED computation, viz. Heuristic-A*, Beam, and BP, which are all described in detail above. Besides the computation time (t) and classification accuracy of the NN classifier, we are interested in other indicators. In particular, we compute the correlation (ϱ) between exact and suboptimal distances. Clearly, the correlation coefficient between exact and suboptimal distances is an indicator of how good a suboptimal method approximates the exact edit distance.

In Table 6.2 the results achieved on all data sets except the Webpage data are given.[3)] The computation time in Table 6.2 corresponds to the total time elapsed while performing all GED computations on a given data set. Missing table entries correspond to cases where a time limit was exceeded and the computation was aborted. We observe that exact edit distance computation by means of Heuristic-A* is feasible for the Letter graphs only. The graphs of the remaining data sets are too complex, too large, or too dense to compute exact GEDs.

Comparing the runtime of the suboptimal method, BP, with the other systems we observe a massive speedup. On the Letter data at the lowest distortion level, for instance, the novel bipartite edit distance algorithm is about

3) Due to unique node labels, no approximation is needed for matching these graphs. Consequently, a comparison on this data set is omitted.

Table 6.2 Accuracy of a NN classifier (NN), correlation (ϱ), and time (t)

Database	Heuristic-A*			Beam(10)			BP		
	NN	ϱ	t	NN	ϱ	t	NN	ϱ	t
Letter L	91.0	1.00	649′05″	91.1	0.95	107′52″	91.1	0.98	7′52″
Letter M	77.9	1.00	2061′29″	78.5	0.93	120′04″	77.6	0.93	6′11″
Letter H	63.0	1.00	4914′45″	63.9	0.93	149′47″	61.6	0.97	8′48″
Fingerprint	–	–	–	84.6	–	166′05″	78.7	–	2′09″
AIDS	–	–	–	96.2	–	1047′55″	97.0	–	4′17″

81 times faster than the exact algorithm (Heuristic-A*) and about 13 times faster than the second system (Beam). On the Letter graphs at medium and high distortion levels the corresponding speedups of the novel algorithm are even higher. On the two data sets Fingerprint and AIDS exact computation of the edit distance is not possible within a reasonable amount of time. The runtime of BP on these data sets is about 83 (Fingerprint) and 260 (Molecules) times faster than the other suboptimal algorithm Beam.

From Table 6.2 a significant speedup of the novel bipartite method for GED computation compared to the exact procedure is evident. However, the question remains whether the approximate edit distances found by BP are accurate enough for pattern recognition tasks. As mentioned before, the distances found by BP are equal to, or larger than, the exact GED. In fact, this can be seen in the correlation scatter plots in Figure 6.5. These scatter plots give us a visual representation of the accuracy of the suboptimal methods Beam and BP on the Letter data at the lowest distortion level. We plot for each pair consisting of one test and one training graph its exact (horizontal axis) and approximate (vertical axis) distance value.

Based on the scatter plots given in Figure 6.5 we find that Beam approximates small distance values accurately, i.e., all small approximate distances are equal to the exact distances. On the other hand, large distance values are overestimated quite strongly. The mean and the standard deviation of the difference between the approximate and exact distances are 0.23 and 0.59, respectively. Based on the fact that graphs within the same class usually have a smaller distance than graphs belonging to two different classes, this means that the suboptimality of Beam mainly increases interclass distances, while intraclass distances are not strongly affected.

A similar conclusion can be drawn for the suboptimal algorithm BP. Many of the small distance values are not overestimated, while higher distance values are increased due to the suboptimal nature of the novel approach. In contrast with the suboptimal Beam method, where the level of overestimation increases with larger distance values, the distance values are better bounded by BP, i.e., large distance values are likewise not strongly overestimated. That is, both the mean (0.16) and the standard deviation (0.27) of the difference

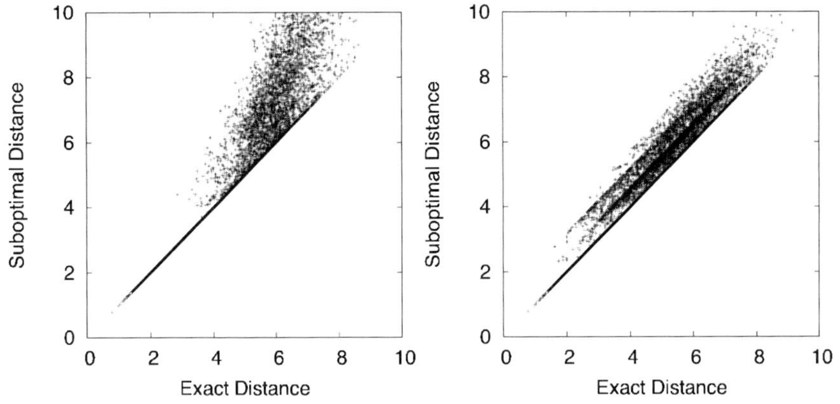

(a) Beam(10) (b) BP

Figure 6.5 Scatter plots of the exact edit distances (x-axis) and the suboptimal edit distances (y-axis).

between the approximate and exact distances are smaller than with Beam. On all other data sets, we obtain scatter plots similar to those in Figure 6.5 and can draw the same conclusions.

BP considers the local edge structure of the graphs only. Hence, in comparison with Heuristic-A*, the novel algorithm BP might find an optimal node mapping that eventually causes additional edge operations. These additional edge edit operations are often deletions or insertions. This leads to additional costs of a multiple of the edit cost parameter τ_{edge}. Obviously, this explains the accumulation of points in the two linelike areas parallel to the diagonal in the distance scatter plot in Figure 6.5b.

Based on the scatter plots and the high correlation coefficient ϱ reported in Table 6.2, one can presume that the classification results of an NN classifier will not be negatively affected when we substitute the exact edit distances by approximate ones. In fact, this can be observed in Table 6.2 for all data sets. On Letter L the classification accuracy is improved, while on Letters M and H it drops compared to the exact algorithm. Note that the only difference that is statistically significant (using a Z-test at the 95% level) is the deterioration on Letter H from 63.0% to 61.6%. On the AIDS data set we observe that BP outperforms Beam (with statistical significance), while on Fingerprints the accuracy drops statistically significantly.

From the results reported in Table 6.2 we can conclude that in general the classification accuracy of the NN classifier is not negatively affected by using the approximate rather than the exact edit distances. This is due to the fact that most of the overestimated distances belong to interclass pairs of graphs, while intraclass distances are not strongly affected. Obviously, intraclass distances are of much higher importance for a distance-based classifier than interclass distances. In other words, through the approximation of the edit distances, the graphs are rearranged with respect to each other such that a better classification becomes possible. Graphs that belong to the same class (according to the ground truth) often remain near each other, while graphs from different classes are pulled apart from each other. Obviously, if the approximation is too inaccurate, the similarity measure and the underlying classifier will be unfavorably disturbed.

6.5.2
Dissimilarity Embedding Graph Kernels Based on Suboptimal Graph Edit Distance

The purpose of the experiments described in this section is to employ the suboptimal GEDs computed by BP for the dissimilarity embedding graph kernel. In an experimental evaluation, the classification accuracy of a kernel machine in conjunction with the dissimilarity embedding graph kernel is compared to a k-NN classifier in the original graph domain. The kernel machine used is a support vector machine (SVM). The basic idea of SVM is to separate classes of patterns by hyperplanes. The sum of distances from the hyperplane to the closest pattern of each class is commonly termed *margin*. The SVM is characterized by the property that it finds the maximum-margin hyperplane, which is expected to perform best on an independent test set. This type of classifier has proven very powerful in various applications and has become one of the most popular classifiers in machine learning, pattern recognition, and related areas recently.

For graph embedding we use all available elements from the training set as prototypes, i.e., $\mathcal{P} = \mathcal{T}$, and subsequently apply dimensionality reduction methods [44]. For dimensionality reduction, we make use of the well-known PCA and Fisher's LDA [17].

PCA [17] is a linear transformation. It seeks the projection that best represents the data. PCA is based on the observation that the first principal component points in the direction of the highest variance of the underlying data and, therefore, includes the most information about the data. The second principal component is perpendicular to the first principal component and points in the direction of the second highest variance and so on. For reducing the dimensionality of the transformed data we retain only the $n \leq N$ principal components with the highest variance. The data are then represented in a new coordinate system defined by these n principal components.

Fisher's LDA [17] is a linear transformation as well. In contrast to PCA, LDA takes class label information into account. In its original form, LDA can be applied to two-class problems only. However, we make use of a generalization, called multiple discriminant analysis (MDA), which can cope with more than two classes. In MDA, we are seeking the projection of the data that best separates the classes from each other. In this approach the transformed data points have a maximal dimensionality of $c - 1$, where c is the number of classes.

In Figure 6.6a the SVM parameter validation is illustrated on the Webpage data set for one fixed value of the dimensionality. The SVM, in conjunction with the dissimilarity embedding graph kernel used in this chapter, has parameters C and γ, where C corresponds to the weighting factor for misclassification and γ is used in the RBF kernel function. For each dimensionality and each possible value of C and γ an SVM is trained and its performance is evaluated on the validation set. In Figure 6.6a the classification accuracy on the validation set is plotted as a function of C and γ.

Figure 6.6b (solid line) displays the best classification results achieved on the validation set with the best (C, γ) pair for various dimensionality values. Together with the accuracy, another curve that corresponds to the fraction of the variance is shown in the same figure (dashed line). This curve displays

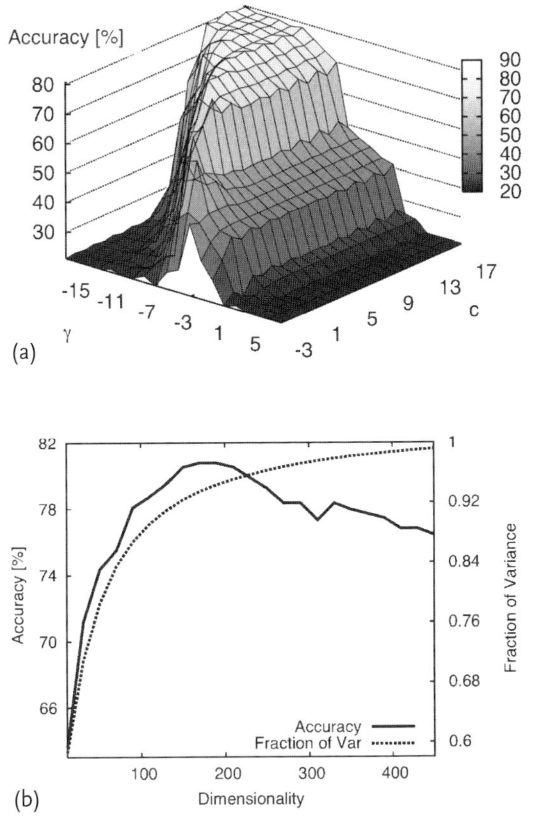

Figure 6.6 Validation of meta parameters of dissimilarity embedding graph kernel. (a) Optimizing C and γ for a specific dimensionality. (b) Validation of the PCA space dimensionality.

the fraction of the variance kept in the PCA-reduced vectors as a function of the dimensionality. As one expects, the fraction monotonically increases with the number of principal components retained. However, there is no clear cutoff point in the fraction curve. This is why we use a validation set to find the optimal number of dimensions. The parameter values (C, γ, and dimensionality) that result in the lowest classification error on the validation set are applied to the independent test set.

The procedure for the MDA-transformed data differs from the validation on PCA data in that no validation set is used. There are two reasons for not using a validation set. First, as the number of dimensions is limited by the number of classes minus one, we always use the maximum possible value. Second, for MDA it is more important to provide a large training set for transformation than optimizing the SVM parameter values. Hence, for MDA transformation we merge the validation and training set to one large set. The MDA transformation is applied on this new set.

In our experiments we use both nonnormalized and normalized data. Normalized data are obtained by linearly scaling all individual feature values to the range $[-1, 1]$. Of course, the whole optimization process of the meta parameters is independently carried out for both nonnormalized and normalized data. In Table 6.3 all classification accuracies on all data sets are shown. In the first column the reference systems' accuracy is given (k-NN). In the remaining columns the results of the dissimilarity embedding graph kernel in conjunction with PCA- and MDA-reduced data are given. The classifiers applied to normalized data are marked with an asterisk (PCA* and MDA*).

The dissimilarity space embedding graph kernel in conjunction with PCA outperforms the reference system on all data sets. Note that all improvements are statistically significant. Comparing the use of normalized vs. nonnormalized data, we note that three times we obtain further improve-

Table 6.3 Classification accuracy on the graph data sets.

	Reference	Embedding kernel			
DB	k-NN	PCA	PCA*	MDA	MDA*
Letter L	91.1	92.1 ∘	92.7 ∘	88.7 •	89.8
Letter M	77.6	81.4 ∘	81.1 ∘	58.0 •	68.5 •
Letter H	61.6	73.8 ∘	73.3 ∘	60.2	60.5
Fingerprint	80.6	81.9 ∘	83.1 ∘	71.5 •	74.7 •
AIDS	97.1	98.3 ∘	98.2 ∘	94.6 •	95.4 •
Webpage	77.4	81.5 ∘	84.0 ∘	87.9 ∘	82.6 ∘

∘ Statistically significant improvement over the reference system (Z-test, $\alpha = 0.05$)
• Statistically significant deterioration over the reference system (Z-test, $\alpha = 0.05$)

ments by means of normalization. Two of these improvements are statistically significant (Fingerprint and Webpage) while none of the deteriorations (Letter M, Letter H, AIDS) are significant.

Regarding the embedding kernel applied to MDA reduced data, we observe that the recognition rates are lower than those of the PCA in general. This can be explained by the fact that the maximum possible dimensionality is restricted by the number of classes minus one. On five data sets, the results achieved with an MDA are even worse than those of the reference system. However, on the Webpage data set the MDA-based method achieves the best result among all classifiers with 87.9%.

Comparing MDA and MDA* we observe that the graph kernel applied to normalized data outperforms the kernel applied on the raw data five times. Four of these improvements (Letters L and M, Fingerprint, AIDS) and the deterioration (Webpage) are statistically significant. The problem of too low dimensionalities also arises on MDA*, of course.

6.6 Summary and Conclusions

Graphs, which can be seen as the most general data structure in computer science, are used for object representation in the present chapter. As a basic dissimilarity measure for graphs, the concept of edit distance is applied. The edit distance of graphs is a powerful and flexible concept that has found various applications. A serious drawback is, however, the exponential complexity of GED computation. Hence using optimal, or exact, algorithms restricts the applicability of the edit distance to graphs of rather small size.

In the current chapter a suboptimal approach to GED computation is introduced (BP) that is based on Munkres' algorithm for solving the assignment problem. The proposed solution allows for the insertion, deletion, and substitution of both nodes and edges, but considers these edit operations in a rather independent fashion from each other. Therefore, while Munkres' algorithm returns the optimal solution to the assignment problem, BP yields only a suboptimal, or approximate, solution to the GED problem. However, the time complexity is only cubic in the number of nodes of the two underlying graphs. In the experimental section, optimal and approximative edit distance is computed on different graph data sets. The first finding of our work is that, although Beam, which is another suboptimal algorithm developed previously, achieves remarkable speedups compared to the exact method, BP is still much faster. Furthermore, the new approach makes GED feasible for graphs with up to 130 nodes.

The second finding is that suboptimal GED need not necessarily lead to a deterioration of the classification accuracy of a distance-based classifier. An experimental analysis has shown that the larger the true distances are, the larger is their overestimation. In other words, smaller distances are computed more accurately than larger distances by the suboptimal algorithm. However, for a NN classifier, small distances have more influence on the decision than large distances. Hence no serious deterioration of the classification accuracy occurs when the proposed suboptimal algorithm is used instead of an exact method.

In a second application, suboptimal distances computed by BP are used for the dissimilarity space embedding graph kernel. This graph kernel computes the GED of a sample graph to a training set of size N. As a result, we obtain N real numbers, which are used as a high dimensional vectorial description of the given graph. Previous work on graph embedding depends on the selection of suitable prototypes. However, the way of selecting these prototypes is a critical issue. With the method described in this chapter the difficult task of prototype selection is avoided by taking all available graphs from the training set as prototypes, then reducing the dimensionality by applying the mathematically well-founded dimensionality reduction algorithms PCA and MDA. With several experimental results obtained on four databases with quite different characteristics we show that the performance of a k-NN classifier in the graph domain, used as a reference system, can be outperformed with statistical significance. The SVM in the optimized PCA space outperforms the reference system on all data sets. One of the findings of the experiments is that the PCA-based system is superior compared to the MDA-based classifier in most cases.

References

1 R. Ambauen, S. Fischer, and H. Bunke. Graph edit distance with node splitting and merging and its application to diatom identification. In E. Hancock and M. Vento, (eds.), *Proceedings of the 4th International Workshop on Graph Based Representations in Pattern Recognition*, LNCS 2726, pp. 95–106. Springer, Berlin, 2003.

2 M. Bianchini, M. Gori, L. Sarti, and F. Scarselli. Recursive processing of cyclic graphs. *IEEE Transactions on Neural Networks*, 17(1):10–18, January 2006.

3 M.C. Boeres, C.C. Ribeiro, and I. Bloch. A randomized heuristic for scene recognition by graph matching. In C.C. Ribeiro and S.L. Martins, (eds.), *Proceedings of the 3rd Workshop on Efficient and Experimental Algorithms*, LNCS 3059, pp. 100–113. Springer, Berlin, 2004.

4 K. Borgwardt. *Graph Kernels*. PhD thesis, Ludwig-Maximilians-Universität München, 2007.

5 K. Borgwardt, C. Ong, S. Schönauer, S. Vishwanathan, A. Smola, and H.-P. Kriegel. Protein function prediction via graph kernels. *Bioinformatics*, 21(1):47–56, 2005.

6 H. Bunke. Graph-based tools for data mining and machine learning. In P. Perner and A. Rosenfeld, (eds.), *Proceedings of the 3rd International Conference on Machine Learning and Data Mining in Pattern Recognition*, LNAI 2734, pp. 7–19. Springer, Berlin, 2003.

7 H. Bunke and G. Allermann. Inexact graph matching for structural pattern

recognition. *Pattern Recognition Letters*, 1:245–253, 1983.

8 H. Bunke, P. Dickinson, and M. Kraetzl. Theoretical and algorithmic framework for hypergraph matching. In F. Roli and S. Vitulano, (eds.), *Proceedings of the 13th International Conference on Image Analysis and Processing – ICIAP 2005*, LNCS 3617, pp. 463–470. Springer, Berlin, 2005.

9 H. Bunke, P.J. Dickinson, M. Kraetzl, and W.D. Wallis. *A Graph-Theoretic Approach to Enterprise Network Dynamics*, Vol. 24 of *Progress in Computer Science and Applied Logic (PCS)*. Birkhäuser, 2007.

10 H. Bunke and K. Riesen. A family of novel graph kernels for structural pattern recognition. In L. Rueda, D. Mery, and J. Kittler, (eds.), *Proceedings of the 12th Iberoamerican Congress on Pattern Recognition*, LNCS 4756, pp. 20–31. Springer, Berlin, 2007. Invited Paper for the 12th Iberoamerican Congress on Pattern Recognition, CIARP.

11 H. Bunke and K. Shearer. A graph distance metric based on the maximal common subgraph. *Pattern Recognition Letters*, 19(3):255–259, 1998.

12 W.J. Christmas, J. Kittler, and M. Petrou. Structural matching in computer vision using probabilistic relaxation. *IEEE Transactions on Pattern Analysis and Machine Intelligence*, 17(8):749–764, 1995.

13 D. Conte, P. Foggia, C. Sansone, and M. Vento. Thirty years of graph matching in pattern recognition. *International Journal of Pattern Recognition and Artificial Intelligence*, 18(3):265–298, 2004.

14 A. Cross, R. Wilson, and E. Hancock. Inexact graph matching using genetic search. *Pattern Recognition*, 30(6):953–970, 1997.

15 P.J. Dickinson, H. Bunke, A. Dadej, and M. Kraetzl. Matching graphs with unique node labels. *Pattern Analysis and Applications*, 7(3):243–254, 2004.

16 Development Therapeutics Program DTP. AIDS antiviral screen, 2004. dtp.nci.nih.gov/docs.

17 R. Duda, P. Hart, and D. Stork. *Pattern Classification*. Wiley-Interscience, 2nd Edition, 2000.

18 M.-L. Fernandez and G. Valiente. A graph distance metric combining maximum common subgraph and minimum common supergraph. *Pattern Recognition Letters*, 22(6–7):753–758, 2001.

19 P. Frasconi, M. Gori, and A. Sperduti. A general framework for adaptive processing of data structures. *IEEE Transactions on Neural Networks*, 9(5):768–786, 1998.

20 T. Gärtner, J. Lloyd, and P. Flach. Kernels and distances for structured data. *Machine Learning*, 57(3):205–232, 2004.

21 S. Gold and A. Rangarajan. A graduated assignment algorithm for graph matching. *IEEE Transactions on Pattern Analysis and Machine Intelligence*, 18(4):377–388, 1996.

22 M. Gori, M. Maggini, and L. Sarti. Exact and approximate graph matching using random walks. *IEEE Transactions on Pattern Analysis and Machine Intelligence*, 27(7):1100–1111, 2005.

23 Z. Harchaoui and F. Bach. Image classification with segmentation graph kernels. In *IEEE Conference on Computer Vision and Pattern Recognition*, pp. 1–8, 2007.

24 P.E. Hart, N.J. Nilsson, and B. Raphael. A formal basis for the heuristic determination of minimum cost paths. *IEEE Transactions of Systems, Science, and Cybernetics*, 4(2):100–107, 1968.

25 J.E. Hopcroft and J. Wong. Linear time algorithm for isomorphism of planar graphs. In *Proceedings of the 6th Annual ACM Symposium on Theory of Computing*, pp. 172–184, 1974.

26 B. Jain and F. Wysotzki. Solving inexact graph isomorphism problems using neural networks. *Neurocomputing*, 63:45–67, 2005.

27 D. Justice and A. Hero. A binary linear programming formulation of the graph edit distance. *IEEE Transactions on Pattern Analysis ans Machine Intelligence*, 28(8):1200–1214, 2006.

28 A. Kandel, H. Bunke, and M. Last, (eds.). *Applied Graph Theory in Computer Vision and Pattern Recognition*, Vol. 52 of *Studies in Computational Intelligence*. Springer, Berlin, 2007.

29 H.W. Kuhn. The Hungarian method for the assignment problem. *Naval Research Logistic Quarterly*, 2:83–97, 1955.

30 B. Le Saux and H. Bunke. Feature selection for graph-based image classifiers. In J. Marques, N. Perez de Blanca, and P. Pina, (eds.), *Proceedings of the 2nd Iberian Conference on Pattern Recognition and Image Analysis, Part II*, LNCS 3523, pp. 147–154. Springer, Berlin, 2005.

31 E.M. Luks. Isomorphism of graphs of bounded valence can be tested in ploynomial time. *Journal of Computer and Systems Sciences*, 25:42–65, 1982.

32 B. Luo, R. Wilson, and E.R. Hancock. Spectral embedding of graphs. *Pattern Recognition*, 36(10):2213–2223, 2003.

33 P. Mahé, N. Ueda, and T. Akutsu. Graph kernels for molecular structures – activity relationship analysis with support vector machines. *Journal of Chemical Information and Modeling*, 45(4):939–951, 2005.

34 J. Munkres. Algorithms for the assignment and transportation problems. In *Journal of the Society for Industrial and Applied Mathematics*, Vol. 5, pp. 32–38, March 1957.

35 M. Neuhaus and H. Bunke. An error-tolerant approximate matching algorithm for attributed planar graphs and its application to fingerprint classification. In A. Fred, T. Caelli, R. Duin, A. Campilho, and D. de Ridder, (eds.), *Proceedings of the 10th International Workshop on Structural and Syntactic Pattern Recognition*, LNCS 3138, pp. 180–189. Springer, Berlin, 2004.

36 M. Neuhaus and H. Bunke. Self-organizing maps for learning the edit costs in graph matching. *IEEE Transactions on Systems, Man, and Cybernetics (Part B)*, 35(3):503–514, 2005.

37 M. Neuhaus and H. Bunke. Automatic learning of cost functions for graph edit distance. *Information Sciences*, 177(1):239–247, 2007.

38 M. Neuhaus and H. Bunke. *Bridging the Gap Between Graph Edit Distance and Kernel Machines.* World Scientific, Singapore, 2007.

39 M. Neuhaus, K. Riesen, and H. Bunke. Fast suboptimal algorithms for the computation of graph edit distance. In C.C. Ribeiro and S.L. Martins, (eds.), *Proceedings of the 11th International Workshop on Strucural and Syntactic Pattern Recognition*, LNCS 3059, pp. 163–172. Springer, Berlin 2006.

40 E. Pekalska and R. Duin. *The Dissimilarity Representation for Pattern Recognition: Foundations and Applications.* World Scientific, Singapore, 2005.

41 L. Ralaivola, S.J. Swamidass, H Saigo, and P. Baldi. Graph kernels for chemical informatics. *Neural Networks*, 18(8):1093–1110, 2005.

42 K. Riesen and H. Bunke. Approximate graph edit distance computation by means of bipartite graph matching. *Image and Vision Computing*, 2008. Accepted for publication.

43 K. Riesen and H. Bunke. IAM graph database, 2008. www.iam.unibe.ch/fki/databases.

44 K. Riesen and H. Bunke. Reducing the dimensionality of dissimilarity space embedding graph kernels. *Engineering Applications of Artificial Intelligence Engineering Applications of Artificial Intelligence*, 2008. Accepted for publication.

45 K. Riesen, M. Neuhaus, and H. Bunke. Graph embedding in vector spaces by means of prototype selection. In F. Escolano and M. Vento, (eds.), *Proc. 6th Int. Workshop on Graph Based Representations in Pattern Recognition*, LNCS 4538, pp. 383–393. Springer, Berlin, 2007.

46 A. Robles-Kelly and E.R. Hancock. A Riemannian approach to graph embedding. *Pattern Recognition*, 40:1024–1056, 2007.

47 A. Sanfeliu and K.S. Fu. A distance measure between attributed relational graphs for pattern recognition. *IEEE Transactions on Systems, Man, and Cybernetics (Part B)*, 13(3):353–363, 1983.

48 A. Schenker, H. Bunke, M. Last, and A. Kandel. *Graph-Theoretic Techniques for Web Content Mining.* World Scientific, Singapore, 2005.

49 B. Schölkopf and A. Smola. *Learning with Kernels.* MIT Press, Cambridge, MA, 2002.

50 S.M. Selkow. The tree-to-tree editing problem. *Information Processing Letters*, 6(6):184–186, 1977.

51 J. Shawe-Taylor and N. Cristianini. *Kernel Methods for Pattern Analysis.* Cambridge University Press, Cambridge, UK, 2004.

52 A. Sperduti and A. Starita. Supervised neural networks for the classification of structures. *IEEE Transactions on Neural Networks*, 8(3):714–735, 1997.

53 B. Spillmann, M. Neuhaus, H. Bunke, E. Pekalska, and R. Duin. Transforming strings to vector spaces using prototype selection. In Dit-Yan Yeung, J.T. Kwok, A. Fred, F. Roli, and D. de Ridder, (eds.), *Proceedings of the 11th International Workshop on Strucural and Syntactic Pattern Recognition*, LNCS 4109, pp. 287–296. Springer, Berlin, 2006.

54 A. Torsello, D. Hidovic-Rowe, and M. Pelillo. Polynomial-time metrics for attributed trees. *IEEE Transactions on Pattern Analysis and Machine Intelligence*, 27(7):1087–1099, 2005.

55 J.R. Ullman. An algorithm for subgraph isomorphism. *Journal of the Association for Computing Machinery*, 23(1):31–42, 1976.

56 S. Umeyama. An eigendecomposition approach to weighted graph matching problems. *IEEE Transactions on Pattern Analysis and Machine Intelligence*, 10(5):695–703, 1988.

57 V. Vapnik. *Statistical Learning Theory*. John Wiley & Sons, New York, 1998.

58 R.A. Wagner and M.J. Fischer. The string-to-string correction problem. *Journal of the Association for Computing Machinery*, 21(1):168–173, 1974.

59 W.D. Wallis, P. Shoubridge, M. Kraetzl, and D. Ray. Graph distances using graph union. *Pattern Recognition Letters*, 22(6):701–704, 2001.

60 C.I. Watson and C.L. Wilson. *NIST Special Database 4, Fingerprint Database*. National Institute of Standards and Technology, March 1992.

61 R.C. Wilson and E. Hancock. Structural matching by discrete relaxation. *IEEE Transactions on Pattern Analysis and Machine Intelligence*, 19(6):634–648, 1997.

62 R.C. Wilson, E.R. Hancock, and B. Luo. Pattern vectors from algebraic graph theory. *IEEE Transactions on Pattern Analysis and Machine Intelligence*, 27(7):1112–1124, 2005.

7
Graph Energy

Ivan Gutman, Xueliang Li, and Jianbin Zhang

7.1
Introduction

In this chapter we are concerned with the eigenvalues of graphs and some of their chemical applications. Let G be a (simple) graph, with vertex set $V(G)$ and edge set $E(G)$. The number of vertices of G is n, and its vertices are labeled by $v_1, v_2, \ldots, v_n$. The adjacency matrix $A(G)$ of the graph G is a square matrix of order n, whose (i,j)-entry is equal to 1 if the vertices v_i and v_j are adjacent and equal to zero otherwise.

The eigenvalues $\lambda_1, \lambda_2, \ldots, \lambda_n$ of the adjacency matrix $A(G)$ are said to be the *eigenvalues of the graph* G and to form its *spectrum*. Details of the spectral theory of graphs can be found in the seminal monograph [1].

The characteristic polynomial of the adjacency matrix, i.e., $\det(\lambda I_n - A(G))$, where I_n is the unit matrix of order n, is said to be the *characteristic polynomial of the graph* G and will be denoted by $\phi(G, \lambda)$. From linear algebra it is known that the graph eigenvalues are just the solutions of the equation $\phi(G, \lambda) = 0$.

One of the most remarkable chemical applications of graph theory is based on the close correspondence between the graph eigenvalues and the molecular orbital energy levels of π-electrons in conjugated hydrocarbons. For details, see [2–4]. If G is a molecular graph of a conjugated hydrocarbon with n vertices and $\lambda_1, \lambda_2, \ldots, \lambda_n$ are its eigenvalues, then in the so-called Hüchkel molecular orbital (HMO) approximation [3,5], the energy of the ith molecular orbital is given by

$$E_i = \alpha + \lambda_i \beta \,,$$

where α and β are pertinent constants. In order to simplify the formalism, it is customary to set $\alpha = 0$ and $\beta = 1$, in which case the π-electron orbital energies and the graph eigenvalues coincide.

Analysis of Complex Networks: From Biology to Linguistics. Edited by Matthias Dehmer and Frank Emmert-Streib

ISBN: 978-3-527-32345-6

The total π-electron energy (E) is equal to the sum of the energies of all π-electrons that are present in the respective molecule, i.e., $E = \sum_{i=1}^{n} g_i\, E_i = \sum_{i=1}^{n} g_i\, \lambda_i$, where g_i is the number of electrons in the ith molecular orbital (whose energy is E_i). Because of restrictions coming from the Pauli exclusion principle [5], g_i is 2, 1, or 0. In the majority of chemically relevant cases, $g_i = 2$ whenever $\lambda_i > 0$ and $g_i = 0$ whenever $\lambda_i < 0$, implying $E = 2\sum_{+} \lambda_i$ with $\sum_{+}$ indicating the summation over positive eigenvalues. Because the sum of all eigenvalues is zero, one immediately arrives at

$$E = E(G) = \sum_{i=0}^{n} |\lambda_i| \,. \tag{7.1}$$

The total π-electron energy and, in particular, the right-hand side of Equation 7.1 was studied already in the pioneering days of quantum chemistry (see, e.g., [6]). In the 1970s one of the present authors [7] came to the idea of defining the *energy of a graph* G as the sum of the absolute values of its eigenvalues. By this, Equation 7.1 could now be viewed as the definition of a graph invariant (that in the case of some special graphs has a chemical interpretation) that is applicable to all graphs. This seemingly insignificant change of the approach to $E(G)$ eventually resulted in the development of an entirely new *theory of graph energy*. In this chapter we outline its main results, especially those obtained in the last decade. For earlier mathematical results on graph energy see the review [8]; for its chemical aspects see [9, 10]

Although put forward already in the 1970s [7], and having much older roots in theoretical chemistry [6], the concept of graph energy has for a long time failed to attract the attention of mathematicians and mathematical chemists. However, around the year 2000, research on graph energy suddenly became a very popular topic, resulting in numerous significant discoveries and in a remarkable number of publications. Since 2001 over 100 mathematical papers on E were produced, more than one per month.

This chapter has six sections, followed by a detailed (yet far from complete) bibliography on graph energy. In the second section numerous upper and lower bounds for graph energy are given, and in many cases the graphs achieving these bounds are characterized. The third section is concerned with hyperenergetic ($E > 2n - 2$) and hypoenergetic ($E < n$) graphs, as well as with pairs of equienergetic graphs ($E(G_1) = E(G_2)$). The fourth section outlines some selected (of very many existing) results on graphs extremal with regard to energy. In the fifth section we briefly state a few results on graph energy that could not be included in the preceding three sections. Concluding remarks are given in the last section.

7.2 Bounds for the Energy of Graphs

Let G be a graph possessing n vertices and m edges. We say that G is an (n, m)-graph.

For any (n, m)-graph [1], $\sum_{i=1}^{n} \lambda_i^2 = 2m$.
In what follows we assume that the graph eigenvalues are labeled in a non-increasing manner, i.e., that

$$\lambda_1 \geq \lambda_2 \geq \cdots \geq \lambda_n .$$

If G is connected, then $\lambda_1 > \lambda_2$ [1]. Because $\lambda_1 \geq |\lambda_i|$, $i = 2, \ldots, n$, the eigenvalue λ_1 is referred to as the *spectral radius* of graph G.

Some of the simplest and longest standing [8] bounds for the energy of graphs are given below.

Theorem 7.1 *[11] For an (n, m)-graph G,*

$$E(G) \leq \sqrt{2mn}$$

with equality if and only if G is either an empty graph (with $m = 0$, i.e., $G \approx \overline{K_n}$) or a regular graph of degree 1, i.e., $G \approx (n/2)K_2$.

Theorem 7.2 *[12] For a graph G with m edges,*

$$2\sqrt{m} \leq E(G) \leq 2m .$$

Equality $E(G) = 2\sqrt{m}$ holds if and only if G consists of a complete bipartite graph $K_{a,b}$, such that $a \cdot b = m$, and arbitrarily many isolated vertices. Equality $E(G) = 2m$ holds if and only if G consists of m copies of K_2 and arbitrarily many isolated vertices.

7.2.1 Some Upper Bounds

Using

$$\sum_{i=2}^{n} \lambda_i^2 = 2m - \lambda_1^2$$

together with the Cauchy–Schwarz inequality, applied to the $(n-1)$-dimensional vectors $(|\lambda_2|, \ldots, |\lambda_n|)$ and $(1, \ldots, 1)$, we obtain the inequality

$$\sum_{i=2}^{n} |\lambda_i| \leq \sqrt{(n-1)(2m - \lambda_1^2)} .$$

Thus, we have

$$E(G) \le \lambda_1 + \sqrt{(n-1)(2m-\lambda_1^2)}\,.$$

Since $F(x) := x + \sqrt{(n-1)(2m-x^2)}$ is a decreasing function in the variable x and the spectral radius obeys the inequality $\lambda_1 \ge 2m/n$ [1], we have the following theorem.

Theorem 7.3 *[13] Let G be an (n, m)-graph. If $2m \ge n$, then*

$$E(G) \le \frac{2m}{n} + \sqrt{(n-1)\left[2m - \left(\frac{2m}{n}\right)^2\right]}\,. \tag{7.2}$$

Moreover, equality holds in (7.2) if and only if G consists of $n/2$ copies of K_2, or $G \approx K_n$, or G is an noncomplete connected strongly regular graph with two nontrivial eigenvalues both having absolute values equal to $\sqrt{(2m-(2m/n)^2)/(n-1)}$.
If $2m \le n$, then the inequality

$$E(G) \le 2m \tag{7.3}$$

holds. Moreover, equality holds in (7.3) if and only if G is a disjoint union of edges and isolated vertices.

Recall [1] that a graph G that is neither complete nor empty is said to be *strongly regular* with parameters (n, k, a, c) if it has n vertices, it is regular of degree k, every pair of its adjacent vertices has a common neighbors, and every pair of its nonadjacent vertices has c common neighbors. A strongly regular graph with parameters (n, k, a, c) has only three distinct eigenvalues and the eigenvalues of G that are different from k are the zeros of the quadratic polynomial $x^2 - (a-c)x - (k-c)$. Denote these eigenvalues by s and t, and let m_s and m_t be, respectively, their multiplicities. Since k has a multiplicity equal to one, and the sum of all the eigenvalues is 0, we have $m_s + m_t = n - 1$ and $m_s\, s + m_t\, t = -k$.

Using routine calculus, it can be shown that the left-hand side of Inequality 7.2 becomes maximal when $m = (n^2 + n\sqrt{n})/4$. It thus follows:

Theorem 7.4 *[13] Let G be a graph on n vertices. Then*

$$E(G) \le \frac{n}{2}(\sqrt{n}+1) \tag{7.4}$$

with equality if and only if G is a strongly regular graph with parameters

$$\left(n\,,\ \frac{n+\sqrt{n}}{2}\,,\ \frac{n+2\sqrt{n}}{4}\,,\ \frac{n+2\sqrt{n}}{4}\right).$$

Obviously, if such a graph with property $E = n(\sqrt{n} + 1)/2$ does exist, then n must be a square of a positive integer. Very recently, Haemers [14] conjectured that $n = p^2$ is necessary and sufficient for the existence of such graphs. He also tried to construct such strongly regular graphs, and proved:

Theorem 7.5 *[14] There are strongly regular graphs with parameters*

$$\left(n\,,\ \frac{n+\sqrt{n}}{2}\,,\ \frac{n+2\sqrt{n}}{4}\,,\ \frac{n+2\sqrt{n}}{4}\right)$$

for (i) $n = 4^p$, $p \geq 1$; *(ii)* $n = 4^p q^4$, $p, q \geq 1$; *(iii)* $n = 4^{p+1} q^2$, $p \geq 1$ *and* $4q - 1$ *is a prime power, or* $2q - 1$ *is a prime power, or* q *is a square, or* $q < 167$.

As explained above, the graphs specified in Theorem 7.5 have maximum energy. Haemers also found that for $n = 4, 16, 36$ the above extremal graphs are unique, whereas for $n = 64, 100, 144$, these are not unique.

Earlier, McCelland [11] showed that $E(G) \leq \sqrt{2mn}$ (Theorem 7.1). It is easy to demonstrate [15] that Inequality 7.2, and therefore also (7.4), improve this bound.

For special classes of graphs one can obtain better bounds.

Theorem 7.6 *[16] Let G be a bipartite graph on* $n > 2$ *vertices. Then*

$$E(G) \leq \frac{n}{\sqrt{8}}(\sqrt{n} + \sqrt{2}) \tag{7.5}$$

with equality if and only if $n = 2v$ *and G is the incidence graph of a 2-*$\left(v, \frac{v+\sqrt{v}}{2}, \frac{v+2\sqrt{v}}{4}\right)$*-design.*

Recall [17] that a *2-*(v, k, λ)*-design* is a collection of k-subsets or blocks of a set of v points such that each 2-set of points lies in exactly λ blocks. The incident matrix B of a 2-(v, k, λ)-design is the $v \times b$ matrix defined so that for each point x and block S, $B_{x,S} = 0$ if $x \in S$ and $B_{x,S} = 1$ otherwise.

A graph is said to be *semiregular bipartite* if it is bipartite and each vertex in the same part of bipartition has the same degree.

Among known bounds for λ_1, we need here the following [18]:

$$\lambda_1 \geq \sqrt{\frac{1}{n}\sum_{i=1}^{n} d_i^2}\,,$$

where $d_1, d_2, \ldots, d_n$ is the degree sequence of the underlying graph G. Equality holds if and only if G is either regular or semiregular bipartite.

Theorem 7.7 *[19] If G is an (n, m)-graph with degree sequence $d_1, d_2, \ldots, d_n$, then*

$$E(G) \le \sqrt{\frac{1}{n}\sum_{i=1}^{n} d_i^2} + \sqrt{(n-1)\left[2m - \left(\sqrt{\frac{1}{n}\sum_{i=1}^{n} d_i^2}\right)^2\right]} .$$

Equality holds if and only if G is either $(n/2)K_2$ (if $m = n/2$), or K_n (if $m = n(n-1)/2$), or a noncomplete connected strongly regular graph with two nontrivial eigenvalues both having absolute value $\sqrt{(2m-(2m/n)^2)/(n-1)}$, or $n\,K_1$ (if $m = 0$).

Since

$$4m^2 = \left(\sum_{i=1}^{n} d_i\right)^2 \le n \sum_{i=1}^{n} d_i^2$$

and $F(x) = x + \sqrt{(n-1)(2m-x^2)}$ decreases for $\sqrt{2m/n} \le x \le \sqrt{2m}$, it follows that the upper bound of Theorem 7.8 is better than that of Theorem 7.6.

Theorem 7.8 *[19] If G is a bipartite (n, m)-graph, $n > 2$, with degree sequence $d_1, d_2, \ldots, d_n$, then*

$$E(G) \le 2\sqrt{\frac{1}{n}\sum_{i=1}^{n} d_i^2} + \sqrt{(n-2)\left[2m - \frac{2}{n}\sum_{i=1}^{n} d_i^2\right]} .$$

Equality holds if and only if G is $(n/2)K_2$, or a complete bipartite graph, or the incidence graph of a symmetric 2-(v, k, λ)-design with $k = 2m/n$ and $\lambda = k(k-1)/(v-1)$, $(n = 2v)$, or $n\,K_1$.

An extension of Theorem 7.8, for the case where the number of zero eigenvalues is known, was reported in [20].

For $v_i \in V(G)$, the *2-degree* of v_i, denoted by t_i, is the sum of degrees of the vertices adjacent to v_i. We call $\frac{t_i}{d_i}$ the *average degree* of v_i. The *average 2-degree* of v_i, denoted by m_i, is the average of the degrees of the vertices adjacent to v_i. Then $t_i = d_i m_i$. Furthermore, denote by σ_i the sum of the 2-degrees of the vertices adjacent to v_i. A graph G is called *p-pseudoregular* if there is a constant p such that each vertex of G has an average degree equal to p. A bipartite graph $G = (X, Y)$ is said to be *(p_x, p_y)-pseudo-semiregular* if there are two constants p_x and p_y such that each vertex in X has an average degree p_x and each vertex in Y has an average degree p_y.

Theorem 7.9 *[21] Let G be an (n, m)-graph, $m > 0$, with degree sequence $d_1, d_2, \ldots, d_n$ and 2-degree sequence $t_1, t_2, \ldots, t_n$. Let*

$$D_2 = \sum_{i=1}^{n} d_i^2 \quad \textit{and} \quad T_2 = \sum_{i=1}^{m} t_i^2 .$$

Then

$$E(G) \leq 2\sqrt{T_2/D_2} + \sqrt{(n-1)(2m - T_2/D_2)}\,.$$

Equality holds if and only if $G \approx (n/2)K_2$ *or* $G \approx K_n$ *or G is a nonbipartite connected p-pseudoregular graph with three distinct eigenvalues* p, $\sqrt{(2m-p^2)/(n-1)}$, *and* $-\sqrt{(2m-p^2)/(n-1)}$, *provided* $p > \sqrt{2m/n}$.

Theorem 7.10 *[21] Let G be a bipartite* (n, m)*-graph,* $m > 0$. *Using the same notation as in Theorem 7.9 we have*

$$E(G) \leq 2\sqrt{T_2/D_2} + \sqrt{(n-2)(2m - 2\,T_2/D_2)}\,.$$

Equality holds if and only if $G \approx (n/2)K_2$, *or* $G \approx K_{r_1,r_2} \cup (n-r_1-r_2)K_1$, *where* $r_1 r_2 = m$, *or G is a connected* (p_x, p_y)*-pseudo-semiregular bipartite graph with four distinct eigenvalues* $\sqrt{p_x p_y}$, $\sqrt{(2m-2p_x p_y)/(n-2)}$, $-\sqrt{(2m-2p_x p_y)/(n-2)}$, *and* $-\sqrt{p_x p_y}$, *provided* $p_x p_y > \sqrt{2m/n}$.

Theorem 7.11 *[22] Let G be an* (n, m)*-graph,* $m > 0$ *with degree sequence* $d_1, d_2, \cdots, d_n$ *and 2-degree sequence* $t_1, t_2, \cdots, t_n$. *Let*

$$S_2 = \sum_{i=1}^{n} \sigma_i^2\,,$$

and let the other symbols be the same as in Theorem 7.9. Then

$$E(G) \leq 2\sqrt{S_2/T_2} + \sqrt{(n-1)(2m - S_2/T_2)}\,.$$

Equality holds if and only if $G \approx (n/2)K_2$, *or* $G \approx K_n$, *or G is a nonbipartite connected graph satisfying* $\sigma_1/t_1 = \sigma_2/t_2 = \cdots = \sigma_n/t_n = p$ *and has three distinct eigenvalues* p, $\sqrt{(2m-p^2)/(n-1)}$, *and* $-\sqrt{(2m-p^2)/(n-1)}$, *provided* $p > \sqrt{2m/n}$.

Theorem 7.12 *[22] Let G be a bipartite* (n, m)*-graph and everything else the same as in Theorem 7.11. Then*

$$E(G) \leq 2\sqrt{S_2/T_2} + \sqrt{(n-2)(2m - 2\,S_2/T_2)}\,.$$

Equality holds if and only if $G \approx (n/2)K_2$, *or* $G \approx K_{r_1,r_2} \cup (n - r_1 - r_2)K_1$, *where* $r_1 r_2 = m$, *or G is a connected bipartite graph with* $V = \{v_1, v_2, \ldots, v_s\} \cup \{v_{s+1}, v_{s+2}, \ldots, v_n\}$ *such that* $\sigma_1/t_1 = \cdots = \sigma_s/t_s = p_x$ *and* $\sigma_{s+1}/t_{s+1} = \cdots = \sigma_n/t_n = p_y$ *and has four distinct eigenvalues* $\sqrt{p_x p_y}$, $\sqrt{(2m-2p_x p_y)/(n-2)}$, $-\sqrt{(2m-2p_x p_y)/(n-2)}$, *and* $-\sqrt{p_x p_y}$, *provided* $p_x p_y > \sqrt{2m/n}$.

For $v \in V(G)$, the *k-degree* $d_k(v)$ of v is the number of walks of length k of G, starting at v.

Theorem 7.13 *[21] Let G be an (n, m)-graph, m > 0. Then*

$$E(G) \leq \sqrt{\frac{\sum_{v\in V(G)} d_2^2(v)}{\sum_{v\in V(G)} d^2(v)}} + \sqrt{(n-1)\left(2m - \frac{\sum_{v\in V(G)} d_2^2(v)}{\sum_{v\in V(G)} d^2(v)}\right)}\,.$$

Equality holds if and only if $G \approx (n/2)K_2$, *or* $G \approx K_n$, *or G is a non-bipartite connected p-pseudoregular graph with three distinct eigenvalues p,* $\sqrt{(2m-p^2)/(n-1)}$, *and* $-\sqrt{(2m-p^2)/(n-1)}$, *provided* $p > \sqrt{2m/n}$.

Theorem 7.14 *[23] Let G be a connected (n, m)-graph. Then*

$$E(G) \leq \sqrt{\frac{\sum_{v\in V(G)} d_{k+1}^2(v)}{\sum_{v\in V(G)} d_k^2(v)}} + \sqrt{(n-1)\left(2m - \frac{\sum_{v\in V(G)} d_{k+1}^2(v)}{\sum_{v\in V(G)} d_k^2(v)}\right)}\,.$$

Equality holds if and only if G is either the complete graph K_n *or G is a strongly regular graph with two nontrivial eigenvalues both having absolute value equal to* $\sqrt{[2m-(2m/n)^2]/(n-1)}$.

Theorem 7.15 *[23] Let G be a connected bipartite (n, m)-graph,* $n \geq 2$. *Then*

$$E(G) \leq 2\sqrt{\frac{\sum_{v\in V(G)} d_{k+1}^2(v)}{\sum_{v\in V(G)} d_k^2(v)}} + \sqrt{(n-2)\left(2m - 2\frac{\sum_{v\in V(G)} d_{k+1}^2(v)}{\sum_{v\in V(G)} d_k^2(v)}\right)}\,.$$

Equality holds if and only if G is either the complete bipartite graph or G is the incidence graph of a symmetric 2-(v,k,λ)*-design with* $v = n/2$, $k = 2m/n$, *and* $\lambda = k(k-1)/(v-1)$.

More upper bounds of the same kind can be found in [24, 25].

It is well known [1] that the eigenvalues of a bipartite graph G on $n = 2N$ vertices occur in pairs: $\pm\lambda_1, \pm\lambda_2, \cdots, \pm\lambda_N$, where $\lambda_1 \geq \lambda_2 \geq \cdots \geq \lambda_N$. Then the energy of G is given by

$$E(G) = 2(\lambda_1 + \lambda_2 + \cdots + \lambda_N)$$

and

$$\sum_{i=1}^{N} \lambda_i^2 = m\,.$$

Let $q = \sum_{i=1}^{N} \lambda_i^4$. By the Cauchy–Schwarz inequality, $m^2 \leq Nq$.

Theorem 7.16 *[26] Let G be a bipartite graph on 2N vertices. Then the following holds. (i)* $m^2 = Nq$ *if and only if* $G \approx N K_2$*. (ii)* $m^2 = q$ *if and only if G is the direct sum of h isolated vertices and a copy of a complete bipartite graph* $K_{r,s}$*, such that* $rs = m$ *and* $h + r + s = 2N$*. (iii) If* $1 < m^2/q < N$*, then*

$$E(G) \le \frac{2}{\sqrt{N}} \left[\left(m - \sqrt{(N-1)Q} \right) + (N-1) \left(m - \sqrt{Q/(N-1)} \right) \right] , \qquad (7.6)$$

where $Q = Nq - m^2$*. Equality holds if G is the graph of a symmetric BIBD. Conversely, if the equality holds and G is regular, then G is the graph of a symmetric BIBD.*

Recall [17] that a *balanced incomplete block design* (BIBD) is a family of b blocks of a set of v elements, such that (i) each element is contained in r blocks, (ii) each block contains k elements, and (iii) each pair of elements is simultaneously contained in λ blocks. The integers (v, b, r, k, λ) are called the parameters of the design. In the particular case $r = k$, the design is said to be symmetric. The graph of a design is formed in the following way: the $b + v$ vertices of the graph correspond to the blocks and elements of the design with two vertices adjacent if and only if one corresponds to a block and the other corresponds to an element contained in that block.

Theorem 7.17 *[26] Let G be a bipartite graph on 2N+1 vertices. Then the following holds. (i)* $Q \ge 0$ *and the equality is obeyed if and only if G is the direct sum of an isolated vertex with* $N K_2$*. (ii) Inequality 7.6 remains true if* $q < m^2 < Nq$*, and the equality holds if G consists of an isolated vertex and a copy of the graph of a symmetric BIBD.*

If $n = 2N$ and $m \ge N$, then the upper bound of Theorem 7.3 is

$$E_*(N, m) = \frac{2m}{N} + 2\sqrt{(N-1)\left[m - \left(\frac{m}{N}\right)^2\right]} .$$

Theorem 7.18 *[26] If* $N^3 q \ge m^4$*, then* $E(G) \le E_*(N, m)$*.*

Therefore, if $N^3 q \le m^4$, then the bound of Theorem 7.16 improves that of Theorem 7.3.

Ending this subsection, we state one of the several bounds for energy obtained by Morales [27–29]. Let G be a bipartite graph on $2N$ vertices. Then

$$E(G) \le 2\sqrt{m(N-1) + \sqrt{\frac{N(m^2 - q)}{N-1}}} .$$

7.2.2 Some Lower Bounds

In [30] it was shown that for all regular graphs G with degree $k > 0$, the energy is not less than the number of vertices, $E(G) \geq n$. Equality is attained if G consists of $n/(2p)$ components isomorphic to the complete bipartite graph $K_{p,p}$.

Eventually several other classes of graphs were characterized for which $E \geq n$ holds [31]. Among these are the hexagonal systems (representing benzenoid hydrocarbons [32]).

A lover bound for E was obtained by McClelland [11]. Start with

$$\left(\sum_{i=1}^{n} |\lambda_i|\right)^2 = \sum_{i=1}^{n} \lambda_i^2 + \sum_{i \neq j} |\lambda_i||\lambda_j| .$$

Since the geometric mean of positive numbers is not greater than their arithmetic mean,

$$\frac{1}{n(n-1)} \sum_{i \neq j} |\lambda_i||\lambda_j| \geq \prod_{i \neq j} (|\lambda_i||\lambda_j|)^{1/n(n-1)} = \prod_{i=1}^{n} (|\lambda_i|)^{2/n} = |\det(A)|^{2/n} .$$

Hence,

$$E(G)^2 \geq \sum_{i=1}^{n} \lambda_i^2 + n(n-1)|\det(A)|^{2/n} .$$

Theorem 7.19 *[11]* $E(G) \geq \sqrt{2m + n(n-1)|\det A|^{2/n}}$.

If $\det A \neq 0$, which is equivalent to the condition that no graph eigenvalue is equal to zero, then from Theorem 7.19 it follows that $E(G) \geq n$.
For bipartite graphs a similar argument yields [33]

$$E(G) \geq \sqrt{4m + n(n-2)|\det A|^{2/n}} .$$

There are some other lower bounds:

Theorem 7.20 *[26] (i) Let G be a bipartite graph with $2N$ vertices. Then*

$$E(G) \geq 2m\sqrt{\frac{m}{q}} . \tag{7.7}$$

Equality holds if and only if either $G = N K_2$ or G is the direct sum of isolated vertices and complete bipartite graphs $K_{r_1,s_1}, \ldots, K_{r_j,s_j}$ such that $r_1 s_1 = \cdots = r_j s_j$.

(ii) If G is a bipartite graph with $2N + 1$ vertices, then Inequality 7.7 remains true. Moreover, the equality holds if and and only if G is the direct sum of isolated vertices and complete bipartite graphs $K_{r_1,s_1}, \ldots, K_{r_j,s_j}$ such that $r_1 s_1 = \cdots = r_j s_j$.

Theorem 7.21 *[34] Let G be a bipartite graph with at least one edge and let r, s, t be positive integers, such that $4r = s + t + 2$. Then*

$$E(G) \geq M_r(G)^2[M_s(G)M_t(G)]^{-1/2}, \tag{7.8}$$

where $M_k = M_k(G) = \sum_{i=1}^{n}(\lambda_i)^k$ is the kth spectral moment of the graph G.

For a bipartite graph, the odd spectral moments are necessarily zero. In order to overcome this limitation we define the momentlike quantities

$$M_k^* = M_k^*(G) = \sum_{i-1}^{n} |\lambda_i|^k .$$

Then we have the following theorem.

Theorem 7.22 *[35] Let G be a graph with at least one edge and let r, s, t be nonnegative real numbers such that $4r = s + t + 2$. Then*

$$E(G) \geq M_r^*(G)^2[M_s^*(G)M_t^*(G)]^{-\frac{1}{2}} \tag{7.9}$$

with equality if and only if the components of graph G are isolated vertices and complete bipartite graphs $K_{p_1,q_1}, \ldots, K_{p_k,q_k}$ for some $k \geq 1$ such that $p_1 q_1 = \cdots = p_k q_k$.

From [11] we know that $E(G) \leq \sqrt{2mn}$ holds for all graphs. There exists a constant g such that $g\sqrt{2mn}$ is a lower bound for $E(G)$.

For a quadrangle-free (n, m)-graph G with maximum vertex degree 2, and no isolated vertices, we have [36]

$$E(G) > \frac{4}{5}\sqrt{2mn}.$$

If the maximum vertex degree is 3, then [36]

$$E(G) > \frac{2\sqrt{6}}{7}\sqrt{2mn}.$$

Some other lower bounds of this type are found in [37–41]. Of these we state here:

Theorem 7.23 *[41] Let G be a quadrangle-free (n, m)-graph with minimum vertex degree $\delta \geq 1$ and maximum vertex degree Δ. Then*

$$E(G) > \frac{2\sqrt{2\delta\Delta}}{2(\delta+\Delta)-1}\sqrt{2mn}. \tag{7.10}$$

The authors of [13] expressed the opinion that for a given $\varepsilon > 0$ and almost all $n \geq 1$, there exists a graph G on n vertices for which $E(G) \geq (1-\varepsilon)(n/2)(\sqrt{n}+1)$. Nikiforov [42,43] arrived at a stronger statement for sufficiently large n.

Theorem 7.24 *[42] (i) For all sufficiently large n, there exists a graph G of order n with* $E(G) \geq \frac{1}{2}n^{3/2} - n^{11/10}$. *(ii) For almost all graphs*

$$\left(\frac{1}{4}+o(1)\right)n^{3/2} < E(G) < \left(\frac{1}{2}+o(1)\right)n^{3/2} .$$

7.3
Hyperenergetic, Hypoenergetic, and Equienergetic Graphs

7.3.1
Hyperenergetic Graphs

The energy of the n-vertex complete graph K_n is equal to $2(n-1)$. We call an n-vertex graph G *hyperenergetic* if $E(G) > 2(n-1)$. From Nikiforov's Theorem 7.24 we see that almost all graphs are hyperenergetic. Therefore, any search for hyperenergetic graphs appears nowadays a futile task. Yet, before Theorem 7.24 was discovered, a number of such results were obtained. We outline here some of them.

In [7] it was conjectured that the complete graph K_n had greatest energy among all n-vertex graphs. This conjecture was soon shown to be false [44].

The first systematic construction of hyperenergetic graphs was proposed by Walikar, Ramane, and Hampiholi [45], who showed that the line graphs of K_n, $n \geq 5$, and of $K_{n/2,n/2}$, $n \geq 8$, are hyperenergetic. These results were eventually extended to other graphs with a large number of edges [46,47].

Hou and Gutman [48] showed that the line graph of any (n,m)-graph, $n \geq 5$, $m \geq 2n$, is hyperenergetic. Also, the line graph of any bipartite (n,m)-graph, $n \geq 7$, $m \geq 2(n-1)$, is hyperenergetic. Some classes of circulant graphs [49–51] as well as Kneser graphs and their complements [52] are hyperenergetic. In fact, almost all circulant graphs are hyperenergetic [49].

Graphs on n vertices with fewer than $2n-1$ edges are not hyperenergetic [53,54]. This, in particular, implies that Hückel graphs (graphs representing conjugated molecules [2–4], in which the vertex degrees do not exceed 3) cannot be hyperenergetic.

7.3.2 Hypoenergetic Graphs

A graph on n vertices, whose energy is less than n is said to be *hypoenergetic*. In what follows, for obvious reasons we assume that the graphs considered have no isolated vertices.

Studies of hypoenergetic graphs started only quite recently [31, 55], and until now very few results on such graphs have been known.

There are reasons to believe (cf. Theorem 7.24) that there are few hypoenergetic graphs.

Theorem 7.25 *[56] (i) There exist hypoenergetic trees of order n with maximum vertex degree $\Delta \leq 3$ only for $n = 1, 3, 4, 7$ (a single such tree for each value of n; see Figure 7.1). (ii) If $\Delta = 4$, then there exist hypoenergetic trees for all $n \geq 5$, such that $n \equiv k$ (mod 4) $k = 0, 1, 3$. (iii) If $\Delta \geq 5$, then there exist hypoenergetic trees for all $n \geq \Delta + 1$.*

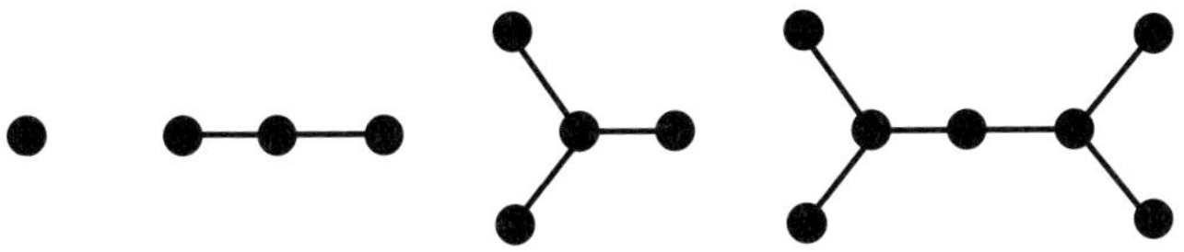

Figure 7.1 The only four hypoenergetic trees with maximum vertex degree not exceeding 3.

Independently of the paper [56], and almost at the same time, Nikiforov [57] arrived at results essentially the same as Theorem 7.25, (i).

Computer search indicates that there exist hypoenergetic trees with $\Delta = 4$ also for $n \equiv 2$ (*mod* 4). The existence of these kinds of trees is still under our consideration.

7.3.3 Equienergetic Graphs

Two nonisomorphic graphs are said to be *equienergetic* if they have the same energy. There exist numerous pairs of graphs with identical spectra, so-called cospectral graphs [1]. In a trivial manner such graphs are equienergetic.

Therefore, in what follows we will be interested only in noncospectral equienergetic graphs.

It is also trivial that the graphs G and $G \cup \overline{K_p}$ (which are not cospectral) are equienergetic. Namely, the spectrum of the graph whose components are G and additional p isolated vertices consists of the eigenvalues of G and of p zeros.

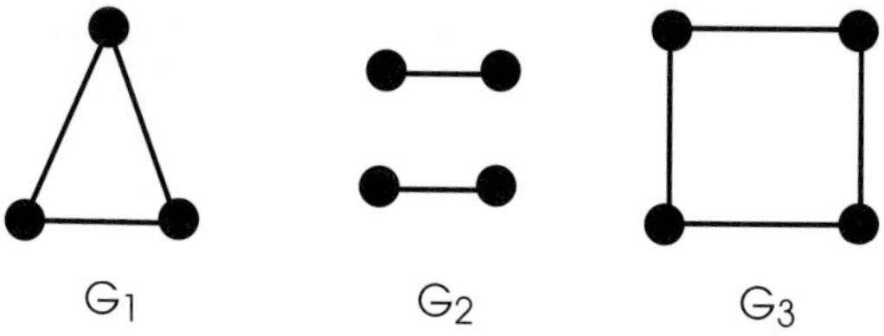

Figure 7.2 Three noncospectral equienergetic graphs with $E = 4$. Note that $Sp(G_1) = \{2, -1, -1\}$, $Sp(G_2) = \{1, 1, -1, -1\}$, and $Sp(G_3) = \{2, 0, 0, -2\}$.

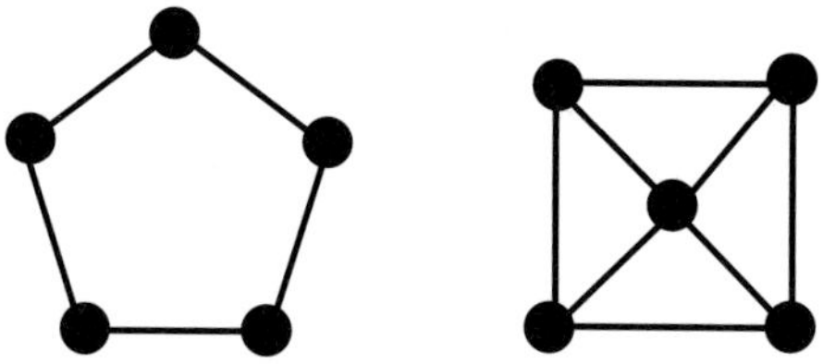

Figure 7.3 The smallest pair of connected equienergetic graphs with an equal number of vertices.

The smallest triplet of nontrivial equienergetic graphs (all having $E = 4$) is shown in Figure 7.2. The smallest pair of equienergetic noncospectral connected graphs with an equal number of vertices is shown in Figure 7.3. These examples indicate that there exist many (nontrivial) families of equienergetic graphs, and that the construction/finding of such families will not be particularly difficult.

The concept of equienergetic graphs was put forward independently and almost simultaneously by Brankov, Stevanović, and Gutman [58] and Balakrishnan [59]. Since 2004 a plethora of papers has been published on equienergetic graphs [60–72]. In what follows we state some of the results obtained along these lines.

Let G be a graph on n vertices and let $V(G) = \{v_1, v_2, \ldots, v_n\}$. Take another set of vertices $U = \{u_1, u_2, \ldots, u_n\}$. Define a graph DG whose vertex set is $V(HDG) = V(G) \cup U$ and whose edge set consists only of the edges joining u_i to the neighbors of v_i in G, for $i = 1, 2, \ldots, n$. The resulting graph DG is called the identity duplication graph of G [64, 73].

With the same notation as above, let $u_1, u_2, \ldots, u_n$ be vertices of another copy of G. Make u_i adjacent to the neighbors of v_i in G, for $i = 1, 2, \ldots, n$. The resulting graph [64] is denoted by D_2G.

The adjacency matrix of DH is

$$A(DG) = \begin{bmatrix} 0 & A(G) \\ A(G) & 0 \end{bmatrix} = A \bigotimes \begin{bmatrix} 0 & 1 \\ 1 & 0 \end{bmatrix} .$$

Thus if $spec(G) = \{\lambda_i\ ,\ i = 1, \ldots, n\}$, then $spec(DH) = \{\lambda_i, \lambda_i\ ,\ i = 1, \ldots, n\}$. The adjacency matrix of D_2H is

$$A(D_2G) = \begin{bmatrix} A(G), & A(G) \\ A(G), & A(G) \end{bmatrix} = A \bigotimes \begin{bmatrix} 1, & 1 \\ 1, & 1 \end{bmatrix},$$

and therefore $spec(D_2G) = \{2\lambda_1, 2\lambda_2 \ldots, 2\lambda_n, 0, 0, \ldots, 0\}$. We thus have the following theorem.

Theorem 7.26 *[64] DG and D_2G are a pair of equienergetic graphs.*

Let G be an r-regular graph on n vertices, and $V(G) = \{v_1, \ldots, v_n\}$. Introduce a set of n isolated vertices $\{u_1, u_2, \ldots, u_n\}$ and make each u_i adjacent to the neighbors of v_i in G for every i. Then introduce a set of k, $(k \geq 0)$, isolated vertices and make all of them adjacent to all vertices of G. The resultant graph is denoted by H.

By direct computation it follows that

$$E(H) = \sqrt{5}\left[E(G) + \sqrt{r^2 + \frac{4}{5} n\, k} - r\right].$$

Combining this and Theorem 7.26 one arrives at:

Theorem 7.27 *[64] There exists a pair of n-vertex noncospectral equienergetic graphs for $n = 6, 14, 18$ and $n \geq 20$.*

Ramane and Walikar [65] recently obtained a stronger result:

Theorem 7.28 *[65] There exists a pair of connected noncospectral equienergetic n-vertex graphs for all $n \geq 9$.*

If G is a graph and $L(G) = L^1(G)$ its line graph, then $L^k(G), k = 2, 3, \ldots,$ defined recursively via $L^k(G) = L(L^{k-1}(G))$, are the iterated line graphs of G.

If G is an r-regular graph with n vertices and m edges, then the characteristic polynomials of G and $L(G)$ are related as [1]

$$\phi(L(G), x) = (x + 2)^{m-n} \phi(G, x - r + 2).$$

If $spec(G) = \{r, \lambda_2, \ldots, \lambda_n\}$, then $spec(L(G)) = \{r + r - 2, \lambda_2 + r - 2, \ldots, \lambda_n + r - 2, -2, \ldots, -2\}$ and $spec(L^2(G)) = \{2r-6, \ldots, 2r-6, r+3r-6, \lambda_2+3r-6, \ldots, \lambda_n + 3r - 6, -2, \ldots, -2\}$. Now, because the eigenvalues of any r-regular graph G obey the condition $|\lambda_i| \leq r$, we see that the only negative eigenvalues of $L^2(G)$ are those equal to -2, whose multiplicity is equal to $nr(r-2)/2$. Consequently,

$$E(L^2(G)) = 2 \times 2 \times \frac{nr(r-2)}{2} = 2nr(r-2).$$

In a similar manner, also $E(L^k(G))$, $k > 2$, depends solely on n and r.

Theorem 7.29 *[62] Let G_1 and G_2 be two noncospectral regular graphs of the same order and of the same degree $r \geq 3$. Then, for $k \geq 2$, the iterated line graphs $L^k(G_1)$ and $L^k(G_2)$ form a pair of noncospectral equienergetic graphs of equal order and with the same number of edges. If, in addition, G_1 and G_2 are chosen to be connected, then also $L^k(G_1)$ and $L^k(G_2)$ are connected.*

Let G_1 and G_2 be two r-regular graphs of order n. From [61] we know that $\overline{L^2(G_1)}$ and $\overline{L^2(G_2)}$ are also equienergetic, and $E(\overline{L^2(G_1)}) = E(\overline{L^2(G_2)}) = (nr-4)(2r-3)-2$, where $\overline{G}$ denotes the complement of graph G.

Let G be a simple graph with vertex set $V = \{v_1, v_2, \ldots, v_n\}$. The extended double cover of G, denoted by G^*, is the bipartite graph with bipartition (X, Y), where $X = \{x_1, x_2, \ldots, x_n\}$ and $Y = \{y_1, y_2, \ldots, y_n\}$, in which x_i and y_i are adjacent if and only if either $i = j$ or v_i and v_j are adjacent in G. Further, G^* is regular of degree $r+1$ if and only if G is regular of degree r. Then we have:

Theorem 7.30 *[66] Let G_1, G_2 be two r-regular graphs of order n. Then*

(i) $(L^2(G_1))^*$ *and* $(L^2(G_2))^*$ *are equienergetic bipartite graphs, and*

$$E((L^2(G_1))^*) = E((L^2(G_2))^*) = nr(3r-5)\,.$$

(ii) $(\overline{L^2(G_1)})^*$ *and* $(\overline{L^2(G_2)})^*$ *are equienergetic bipartite graphs, and*

$$E((\overline{L^2(G_1)})^*) = E((\overline{L^2(G_2)})^*) = (5nr-16)(r-2) + nr - 8\,.$$

(iii) $\overline{(L^2(G_1))^*}$ *and* $\overline{(L^2(G_2))^*}$ *are equienergetic bipartite graphs, and*

$$E(\overline{(L^2(G_1))^*}) = E(\overline{(L^2(G_2))^*}) = (2nr-4)(2r-3)-2\,.$$

A computer search showed that there are numerous pairs of noncospectral equienergetic trees [58]. Some of these are depicted in Figure 7.4.

Numerical calculations, no matter how accurate, cannot be considered as proof that two graphs are equienergetic. In the case of equienergetic trees this problem can sometimes be overcome as in the following example.

Consider the trees T_A, T_B, and T_C, depicted at the bottom of Figure 7.4. Using standard recursive methods [1,4], one can compute their characteristic polynomials as:

$$\phi(T_A, \lambda) = \lambda^{18} - 17\lambda^{16} + 117\lambda^{14} - 421\lambda^{12} + 853\lambda^{10} - 973\lambda^8 + 588\lambda^6 - 164\lambda^4 + 16\lambda^2$$

$$\phi(T_B, \lambda) = \lambda^{18} - 17\lambda^{16} + 117\lambda^{14} - 421\lambda^{12} + 853\lambda^{10} - 973\lambda^8 + 588\lambda^6 - 164\lambda^4 + 16\lambda^2$$

$$\phi(T_C, \lambda) = \lambda^{18} - 17\lambda^{16} + 111\lambda^{14} - 359\lambda^{12} + 632\lambda^{10} - 632\lambda^8 + 359\lambda^6 - 111\lambda^4 + 17\lambda^2 - 1$$

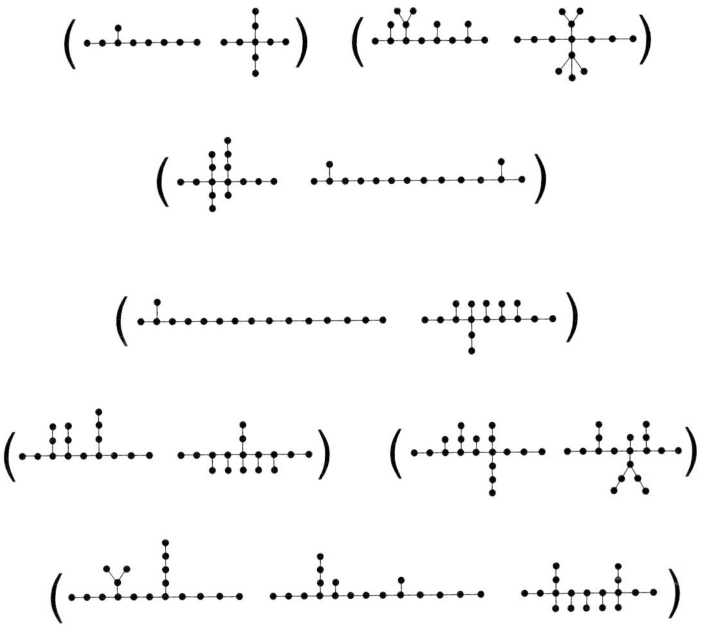

Figure 7.4 Equienergetic trees [58]. Of the three 18-vertex trees at the bottom of this figure, the first two are cospectral, but not cospectral with the third tree.

Trees T_A and T_B have identical characteristic polynomials and, consequently, are cospectral. The characteristic polynomial of T_C is different, implying that T_C is not cospectral with T_A and T_B.

Now, if we are lucky, the above characteristic polynomials can be factored. In this particular case we are lucky, and by easy calculation we find that:

$$\phi(T_A, \lambda) = \lambda^2 (\lambda^2 - 1)(\lambda^2 - 2)^2 (\lambda^2 - 4)(\lambda^4 - 3\lambda^2 + 1)(\lambda^4 - 5\lambda^2 + 1)$$

$$\phi(T_C, \lambda) = (\lambda^2 - 1)^3 (\lambda^4 - 3\lambda^2 + 1)(\lambda^4 - 5\lambda^2 + 1)(\lambda^4 - 6\lambda^2 + 1)\,.$$

It is now an elementary exercise in algebra to verify that

$$E(T_A) = E(T_B) = E(T_C) = 6 + 4\sqrt{2} + 2\sqrt{5} + 2\sqrt{7}\,.$$

If, however, the characteristic polynomials cannot be properly factored, then at the present moment there is no way to prove that the underlying trees are equienergetic. Note that until now no general method (different from computer search) for finding equienergetic trees has been discovered.

7.4
Graphs Extremal with Regard to Energy

One of the fundamental questions that is encountered in the study of graph energy is which graphs (from a given class) have the greatest and smallest E-values. The first such result was obtained for trees [74], when it was demonstrated that the star has minimum and the path maximum energy. In the meantime, a remarkably large number of papers were published on such extremal problems: for general graphs [13, 14, 16, 75–78], trees and chemical trees [79–93], unicyclic [94–107], bicyclic [108–114], tricyclic [115, 116], and tetracyclic graphs [117], as well as for benzenoid and related polycyclic systems [118–122].

In this section we state a few of these results, selecting those that can be formulated in a simple manner.

We first present elementary results.

The n-vertex graph with minimum energy is $\overline{K_n}$, the graph consisting of isolated vertices. Its energy is zero.

The minimum-energy n-vertex graph without isolated vertices is the complete bipartite graph $K_{n-1,1}$, also known as the star [12]. Its energy is equal to $2\sqrt{n-1}$; cf. Theorem 7.2.

Finding the maximum-energy n-vertex graph(s) is a much more difficult task, and a complete solution of this problem is not known. For some results along these lines see Theorem 7.5.

Let G be a graph on n vertices and $A(G)$ its adjacency matrix. As before, let the characteristic polynomial of G be

$$\phi(G,\lambda) = \det(\lambda I_n - A(G)) = \sum_{k=0}^{n} a_k \lambda^{n-k} .$$

A classical result of the theory of graph energy is [6, 8] that $E(G)$ can be computed from the characteristic polynomial of G by means of

$$E(G) = \frac{1}{\pi} \int_{-\infty}^{+\infty} \left[n - \frac{ix\,\phi'(G,ix)}{\phi(G,ix)} \right] \mathrm{d}x ,$$

where $\phi'(G,\lambda)$ denotes the first derivative of $\phi(G,\lambda)$, and where $i = \sqrt{-1}$. More on the Coulson integral formula can be found elsewhere [4, 123, 124].

Another way to write the Coulson integral formula is [74]

$$E(G) = \frac{1}{\pi} \int_{-\infty}^{+\infty} \frac{1}{x^2} \ln \left[\left(\sum_{k \ge 0} (-1)^k a_{2k} x^{2k} \right)^2 + \left(\sum_{k \ge 0} (-1)^k a_{2k+1} x^{2k+1} \right)^2 \right] \mathrm{d}x . \tag{7.11}$$

If graph G is bipartite, then its characteristic polynomial is of the form

$$\phi(G,\lambda) = \sum_{k\geq 0} (-1)^k \, b_k \, \lambda^{n-2k} \,,$$

and $b_k \geq 0$. Then the Coulson integral formula is simplified as

$$E(G) = \frac{2}{\pi} \int_0^{+\infty} \frac{1}{x^2} \ln \left[1 + \sum_{k\geq 1} b_k \, x^{2k} \right] \, dx \,.$$

If G is a tree (or, more generally, a forest), then

$$\phi(G,\lambda) = \sum_{k\geq 0} (-1)^k \, m(G,k) \, \lambda^{n-2k}$$

and

$$E(G) = \frac{2}{\pi} \int_0^{+\infty} \frac{1}{x^2} \ln \left[1 + \sum_{k\geq 1} m(G,k) \, x^{2k} \right] \, dx \,, \tag{7.12}$$

where $m(G,k)$ is the number of matchings of size k of G, i.e., the number of selections of k independent edges in G.

Consider now Equation 7.11 and let G_1 and G_2 be two graphs. If the inequalities

$$\begin{aligned} (-1)^k \, a_{2k}(G_1) &\leq (-1)^k \, a_{2k}(G_2) \\ (-1)^k \, a_{2k+1}(G_1) &\leq (-1)^k \, a_{2k+1}(G_2) \end{aligned} \tag{7.13}$$

are satisfied by all values of k, then from Equation 7.11 it follows that $E(G_1) \leq E(G_2)$. If, in addition, at least one of these inequalities is strict, then $E(G_1) < E(G_2)$.

Bearing this in mind we define a partial order $\prec$ and write $G_1 \preceq G_2$ or $G_2 \succeq G_1$ if the conditions (7.13) are obeyed by all k. If, moreover, at least one of the inequalities in (7.13) is strict, then we write $G_1 \prec G_2$ or $G_2 \succ G_1$. Thus we have:

$$G_1 \preceq G_2 \Rightarrow E(G_1) \leq E(G_2)$$

$$G_1 \prec G_2 \Rightarrow E(G_1) < E(G_2) \,.$$

As a special case of the above, if G_1 and G_2 are bipartite graphs, then [125]

$$G_1 \prec G_2 \Leftrightarrow (\forall k) \; b_k(G_1) \leq , b_k(G_2)$$

whereas if G_1 and G_2 are trees (or, more generally, forests), then

$$G_1 \prec G_2 \Leftrightarrow (\forall k)\ m(G_1, k) \leq m(G_2, k)\,.$$

If for some $k' \neq k''$,

$$(-1)^{k'}\ a_{2k'}(G_1) < (-1)^{k'}\ a_{2k'}(G_2)$$
$$(-1)^{k''}\ a_{2k''}(G_1) > (-1)^{k''}\ a_{2k''}(G_2)$$

or

$$(-1)^{k'}\ a_{2k'+1}(G_1) < (-1)^{k'}\ a_{2k'+1}(G_2)$$
$$(-1)^{k''}\ a_{2k''+1}(G_1) > (-1)^{k''}\ a_{2k''+1}(G_2)\,,$$

then graphs G_1 and G_2 cannot be compared by means of the relation $\prec$ and their energies cannot be compared by using the Coulson integral formula.

Practically all the (above-quoted) results on graphs that are extremal with regard to energy were obtained by establishing the existence of the relation $\prec$ between the elements of some class of graphs.

Theorem 7.31 *[74] If T_n is a tree on n vertices, then*

$$E(S_n) \leq E(G) \leq E(P_n)\,,$$

where S_n and P_n denote, respectively, the star and the path with n vertices. Equality holds only if $G \approx S_n$ or $G \approx P_n$.

Eventually, the first few minimum- and maximum-energy n-vertex trees were determined [88,89]. For instance, let P_n^* be the tree obtained by attaching a P_3 to the third vertex of P_{n-2}. Then P_n^* is a tree with the second-maximum energy [74].

Denote by Φ_n the class of trees on n vertices having a perfect matching and by Ψ_n the subclass of Φ_n consisting of trees whose vertex degrees do not exceed 3. Let F_n be obtained by adding a pendent edge to each vertex of the star $K_{1,(n/2)-1}$, and let B_n be the graph obtained from F_{n-1} by attaching a P_3 to the 2-degree vertex of a pendent edge. Let G_n be obtained by adding a pendent edge to each vertex of the path $P_{n/2}$, D_n be the tree obtained from G_{n+2} by deleting the third and fourth pendent edges.

Theorem 7.32 *[79] (i) F_n and B_n are, respectively, the only trees with minimum and second-minimum energies in Φ_n.*

(ii) G_n and D_n are, respectively, the only trees with minimum and second-minimum energies in Ψ_n.

Eventually, Zhang and Li [80] determined the first four trees with maximum energy in Φ_n.

Let $B_{n,d}$ be obtained from the path P_d with d vertices by attaching $n - d$ pendent edges to an end vertex of P_d.

Theorem 7.33 *[82] Among n-vertex trees with a diameter of at least d, $B_{n,d}$ is the only tree with minimum energy (Figure 7.5).*

Theorem 7.34 *[86, 87] Among n-vertex trees having exactly k pendent vertices, $B_{n,n-k+1}$ is the only tree with minimum energy (Figure 7.5).*

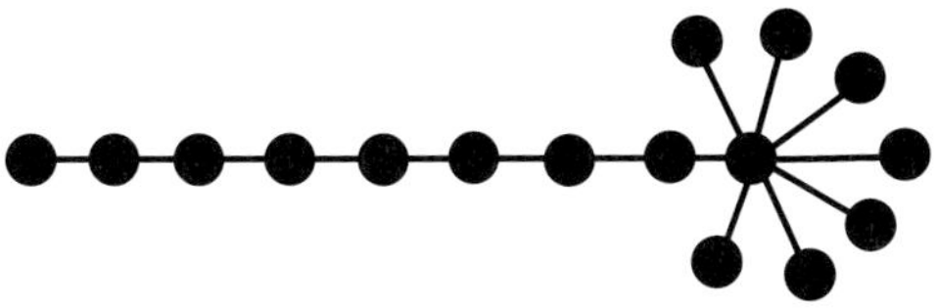

Figure 7.5 A minimum-energy tree with prescribed diameter [82]. This minimal-energy tree also has a prescribed number of pendent vertices [86, 87].

Let $S(n, m, r)$ be obtained by attaching one pendent vertex to each of the m pendent vertices of the star $K_{1,m+r}$. Let $Y(n, m, r)$ be obtained by attaching m P_2s to one end vertex of P_{r+1}. Let $D(n, p, q)$ be obtained from P_2 by adding p and q pendent vertices to the vertices of P_2. Let $T^2_{r,s,t}$ be the tree obtained from P_3 by adding r, s, t pendent vertices to its first, second, and third vertices. Lin, Guo, and Li [84] determined the trees of given maximum degree Δ, having minimum and maximum energies.

Theorem 7.35 *[84] Let T be an n-vertex tree, $n \geq 4$. Let*

$$T_1^*(n,\Delta) \approx \begin{cases} S(n, n-\Delta-1, 2\Delta-n+1) & \text{if } 3 \leq \lfloor \frac{n}{2} \rfloor \leq \Delta(T) \geq n-2 \\ Y(n, \Delta-1, 2\Delta-n+1) & \text{if } 3 \leq \Delta(T) \leq \lfloor \frac{n}{2} \rfloor \\ P_n & \text{if } \Delta(T) = 2 \end{cases}.$$

Then $E(T) \leq E(T_1^*(n,\Delta))$, with equality if and only if $T \approx T_1^*(n,\Delta)$.

Theorem 7.36 *[84] Let T be an n-vertex tree, $n \geq 7$. Let*

$$T_2^*(n,\Delta) \approx \begin{cases} D(n, \Delta-1, n-\Delta-1) & \text{if } \lceil \frac{n}{2} \rceil \leq \Delta(T) \leq n-2 \\ T^2_{\Delta-1,\Delta-1,n-2\Delta-1} & \text{if } \lceil \frac{n}{2} \rceil \leq \Delta(T) \leq \lceil \frac{n}{2} \rceil - 1 \end{cases}.$$

If $\lceil (n+1)/3 \rceil \leq \Delta(T) \leq n-2$, then $E(T) \leq E(T_2^*(n,\Delta))$, with equality if and only if $T \approx T_2^*(n,\Delta)$.

In the above, the trees with a given maximum vertex degree Δ and maximum E happen to be trees with a single vertex of degree Δ. Recently, we [93] offered a simple proof of this result and, in addition, characterized the maximum energy trees having two vertices of maximum degree Δ.

Let $D(p,q)$ be a double star obtained by joining the centers of two stars S_p and S_q by an edge, and let $F(p,q)$ be the tree obtained from $D(p-1,q)$ by attaching a pendent edge to one of the vertices of degree one that joins the vertex of degree q in $D(p-1,q)$.

Theorem 7.37 *[81] Let T be a tree with a (p,q)-bipartition $(p,q \geq 1, p+q \geq 3)$. Then*

$$E(T) \geq \sqrt{2(p+q-1) + 2\sqrt{(p+q-1)^2 - 4(p-1)(q-1)}} + \sqrt{2(p+q-1) - 2\sqrt{(p+q-1)^2 - 4(p-1)(q-1)}},$$

with equality if and only if $T \approx D(p,q)$.

Furthermore, if $q \geq p \geq 2$ and $T \not\approx D(p,q)$, then $E(T) \geq E(F(p,q))$, with equality if and only if $T \approx F(p,q)$.

Let $B(p,q)$ be the graph formed by attaching $p-2$ and $q-2$ vertices to two adjacent vertices of a quadrangle, respectively, and let $H(3,q)$ be the graph formed by attaching $q-2$ vertices to the pendent vertex of $B(2,3)$.

Theorem 7.38 *[97] In the class of bipartite unicyclic graphs with a (p,q)-bipartition, $(q \geq p \geq 2)$, the graph $B(p,q)$ has minimum energy if $p \geq 4$ or $p = 2$, whereas $B(3,q)$ and $H(3,q)$ have minimum energy if $p = 3$.*

Let S_n^3 be the graph obtained from the star graph with n vertices by adding an edge. Hou [94] showed that S_n^3 is the graph with minimum energy among all unicyclic graphs (Figure 7.6).

Let $\mathcal{U}(n,d)$ be the class of connected unicyclic graphs with n vertices and diameter d, where $2 \leq d \leq n-2$. Let $U(n,d)$ be the graph obtained by attaching

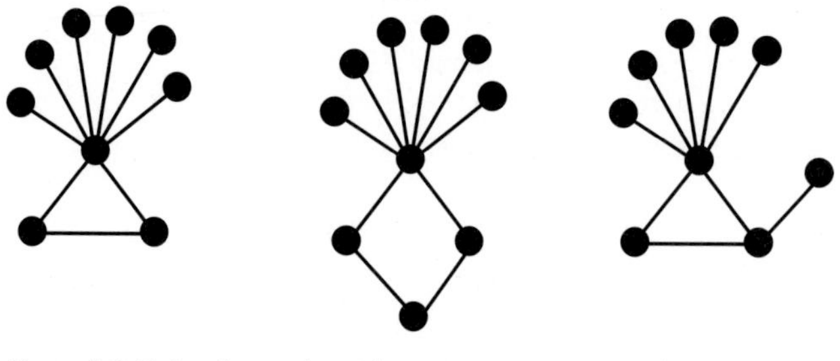

Figure 7.6 Unicyclic graphs with minimum [94], second-minimum, and third-minimum energy [98].

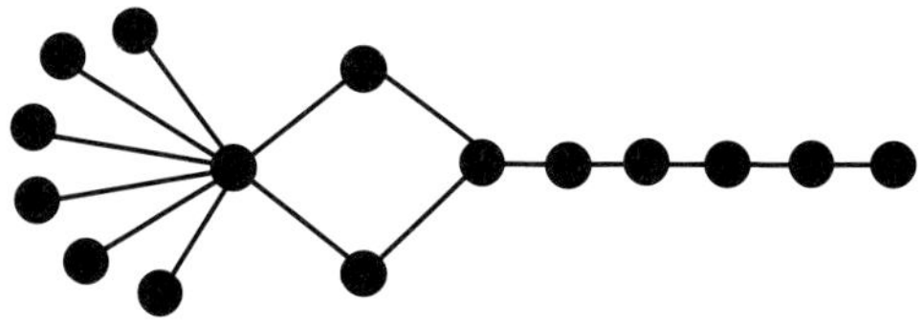

Figure 7.7 A minimum-energy unicyclic graph with prescribed diameter [106].

a path of length $d-3$ at a vertex of C_4 and $n-d-1$ pendent edges at another vertex such that these two vertices are not adjacent (Figure 7.7).

Theorem 7.39 *[106] Let $G \in \mathcal{U}(n,d)$ with $d \geq 3$ and $G \neq U(n,d)$. Then $E(G) > E(U_{n,d})$.*

For the (n,m)-graphs with minimum energy, Caporossi *et al.* [12] put forward the following conjecture.

Conjecture 1 *[12] If $m \leq n+\lfloor(n-7)/2\rfloor$, then the connected (n,m)-graph, $n \geq 6$ and $n-1 \leq m \leq 2(n-2)$, has minimum energy if it is obtained from the star by adding to it $m-n+1$ additional edges all incident to the same vertex. If $m > n+\lfloor(n-7)/2\rfloor$, the minimum-energy graph is the bipartite graph with two vertices in one class, one of which is connected to all vertices on the other class.*

This conjecture is true for $m = n-1, n$ (cf. Theorem 7.31). The conjecture was proved to be true for $m = n-1,\ 2(n-2)$ in [12] by Caporosi *et al.*, and for $m = n$ by Hou [94]. Recently, Li, Zhang, and Wang [78] obtained a positive solution to the second part of the conjecture for bipartite graphs and furthermore determined the graph with the second-minimum energy among connected bipartite (n,m)-graphs, $n \leq m \leq 2n-5$.

Let $S_n^{3,3}$ be the graph formed by joining $n-4$ pendent vertices to a vertex of degree three of $K_4 - e$, and let $P_n^{6,6}$ be the graph obtained from two C_6s by joining them by a path of length $n-10$. Let $G(n)$ be the class of bicyclic graphs G on n vertices containing no disjoint odd cycles of lengths k and ℓ with $k+\ell \equiv 2 \pmod 4$. Then $S_n^{3,3}$ is the graph with minimum energy in $G(n)$ [110].

Let P_n^6 be obtained by connecting a vertex of the cycle C_6 with a terminal vertex of the path P_{n-6}.

Theorem 7.40 *[96] Among n-vertex bipartite unicyclic graphs either P_n^6 or C_n have maximum energy. Thus, if n is odd, then P_n^6 is the maximum-energy unicyclic n-vertex graph.*

Computer-aided calculations show that C_n is the maximum-energy unicyclic graph only for $n = 10$ [95]. However, the proof of the seemingly very simple inequality $E(C_n) < E(P_n^6)$ has not been accomplished so far. The reason for this lies in the fact that graphs C_n and P_n^6 are not comparable by the relation $\prec$.

For bicyclic graphs with maximum energy, the following conjecture was stated, based on computer-aided numerical experiments [75]:

Conjecture 2 *If $n = 14$ and $n \geq 16$, then the maximum-energy bicyclic molecular graph is $P_n^{6,6}$, obtained by attaching six-membered cycles to the end vertices of the path P_{n-12}.*

Recently a partial proof of this conjecture was obtained [111].

Theorem 7.41 *[111] Let $\mathcal{A}(n)$ be the subset consisting of graphs obtained from two cycles C_a and C_b $(a, b \geq 10$ and $a \equiv b \equiv 2 \pmod 4))$ by joining them by an edge. Let $\mathcal{B}_n$ denote the set of all other bipartite bicyclic graphs on n vertices. Then $P_n^{6,6}$ has maximum energy in $\mathcal{B}_n$.*

7.5 Miscellaneous

We state here a few noteworthy results on graph energy that did not fit into the previous sections.

$E(G) \geq 4$ holds for all connected graphs, except for K_1, K_2, $K_{2,1}$, and $K_{3,1}$ [126].

The rank ϱ of a graph is the rank of its adjacency matrix. For a connected bipartite graph G of rank ϱ [126],

$$E(G) \geq \sqrt{(\varrho + 1)^2 - 5}\,.$$

For any graph, $E \geq \varrho$.

Let $\chi(G)$ be the chromatic number of graph G. For any n-vertex graph G, $E(G) \geq 2(n - \chi(\overline{G}))$ [126].

The inequality $E(G) + E(\overline{G}) \geq 2n$ is satisfied by all n-vertex graphs, $n \geq 5$, except by K_n and $K_n - e$ [126].

As an immediate special case of the Koolen–Moulton upper bound (7.2), for an n-vertex regular graph of degree r, we have $E(G) \leq E_0$, where

$$E_0 := r + \sqrt{r(n-1)(n-r)}\,.$$

Balakrishnan [59] showed that for any $\varepsilon > 0$, there exist infinitely many n, for which there are n-vertex regular graphs of degree r, $r < n-1$, such that $E(G)/E_0 < \varepsilon$.

There is no known answer to the question of whether there exist n-vertex regular graphs of degree r for which $E(G)/E_0 > 1-\varepsilon$ [59].

A direct consequence of Equation (7.12) is that by deleting an edge e from a tree (or forest) T, the energy necessarily decreases, $E(T)-E(T-e) > 0$. In the general case the difference $E(G)-E(G-e)$ may be smaller than, greater than, or equal to zero, and the complete solution of this problem is not known. Some partial results along these lines were recently obtained [127, 128].

The energy of a graph is never an odd integer [129] and is never the square root of an odd integer [130].

The way in which the energy depends on various structural features of the underlying (molecular) graph has been much studied in the chemical literature, in most cases empirically [9, 10]. Scores of approximate formulas for E were put forward, in particular formulas that relate the E-value of an (n, m)-graph with n and m [9, 10, 131]. Of these we call the reader's attention to a recent empirical finding that $E(G)$ is an almost perfectly linear (decreasing) function of the number of zeros in the spectrum of G [132, 133].

7.6 Concluding Remarks

At this moment the most significant open problem in the theory of graph energy seems to be the characterization of n-vertex graphs with the greatest energy. Although quite recently much progress has been achieved in this direction (cf. Theorem 7.5), the problem is still far from being completely solved. An additional difficulty that recently emerged [14] is the fact that for some values of n, there exist numerous maximum-energy n-vertex graphs.

There have been several recent attempts to extend the graph-energy concept to eigenvalues of matrices other than the adjacency matrix. Especially much work has been done on the so-called "*Laplacian graph energy*," based on the spectrum of the Laplacian matrix, and on "*distance graph energy*," based on the spectrum of the distance matrix. "Energy" has been redefined so that it could be associated with any matrix, including nonsquare matrices. However, the discussion of such energylike quantities goes beyond the scope of this survey.

References

1 Cvetković, D., Doob, M., Sachs, H., *Spectra of Graphs – Theory and Application*, Academic Press, New York, **1980**.

2 Gutman, I., Trinajstić, N., Graph theory and molecular orbitals, *Topics Curr. Chem.* 42 (**1973**) 49–93.

3 Graovac, A., Gutman, I., Trinajstić, N., *Topological Approach to the Chemistry of Conjugated Molecules*, Springer, Berlin, **1977**.

4 Gutman, I., Polansky, O.E., *Mathematical Concepts in Organic Chemistry*, Springer, Berlin, **1986**.

5 Coulson, C. A., O'Leary, B., Mallion, R.B., *Hückel Theory for Organic Chemists*, Academic Press, London, **1987**.

6 Coulson, C.A., On the calculation of the energy in unsaturated hydrocarbon molecules, *Proc. Cambridge Phil. Soc.* 36 (**1940**), 201–203.

7 Gutman, I., The energy of a graph, *Ber. Math.–Statist. Sekt. Forschungsz. Graz* 103 (**1978**), 1–22.

8 Gutman, I., The energy of a graph: old and new results, in: Betten, A., Kohnert, A., Laue, R., Wassermann, A. (eds.), *Algebraic Combinatorics and Applications*, Springer, Berlin, **2001**, pp. 196–211.

9 Gutman, I., Total π-electron energy of benzenoid hydrocarbons, *Topics Curr. Chem.* 162 (**1992**), 29–63.

10 Gutman, I., Topology and stability of conjugated hydrocarbons. The dependence of total π-electron energy on moleculr topology, *J. Serb. Chem. Soc.* 70 (**2005**), 441–456.

11 McCelland, B.J., Properties of the latent roots of a matrix: The estimation of π-electron energies. *J. Chem. Phys.* 54 (**1971**), 640–643.

12 Caporossi, G., Cvetković, D., Gutman, I., Hansen, P., Variable neighborhood search for extremal graphs. 2. Finding graphs with external energy, *J. Chem. Inf. Comput. Sci.* 39 (**1999**), 984–996.

13 Koolen, J.H., Moulton, V., Maximal energy graphs, *Adv. Appl. Math.* 26 (**2001**), 47–52.

14 Haemers, W.H., Strongly regular graphs with maximal energy, *Lin. Algebra Appl.* 429 (**2008**), 2719–2723.

15 Koolen, J.H., Moulton, V., Gutman, I., Improving the McClelland inequality for total π-electron energy, *Chem. Phys. Lett.* 320 (**2000**), 213–216.

16 Koolen, J.H., Moulton, V., Maximal energy bipartite graphs, *Graphs Combin.* 19 (**2003**), 131–135.

17 Hall, M., *Combinatorial Theory*, Wiley, New York, **1986**.

18 Zhou, B., On spectral radius of nonnegative matrics, *Australas. J. Combin.* 22 (**2000**), 301–306.

19 Zhou, B., Energy of graphs, *MATCH Commun. Math. Comput. Chem.* 51 (**2004**), 111–118.

20 Zhou, B., On the energy of a graph, *Kragujevac J. Sci.* 26 (**2004**), 5–12.

21 Yu, Y., Lu, M., Tian, F., New upper bounds for the energy of graphs, *MATCH Commun. Math. Comput. Chem.* 53 (**2005**), 441–448.

22 Liu, H., Lu, M., Tian, F., Some upper bounds for the energy of graphs, *J. Math. Chem.* 41 (**2007**), 45–57.

23 Hou, Y., Teng, Z., Woo, C., On the spectral radius, k-degree and the upper bound of energy in a graph, *MATCH Commun. Math. Comput. Chem.* 57 (**2007**), 341–350.

24 Liu, H., Lu, M., Sharp bounds on the spectral radius and the energy of graphs, *MATCH Commun. Math. Comput. Chem.* 59 (**2008**), 279–290.

25 Guo, J.M., Sharp upper bounds for total π-electron energy of alternant hydrocarbons, *J. Math. Chem.* 43 (**2008**), 713–718.

26 Rada, J., Tineo, A., Upper and lower bounds for the energy of bipartite graphs, *J. Math. Anal. Appl.* 289 (**2004**), 446–455.

27 Morales, D.A., Bounds for the total π-electron energy, *Int. J. Quantum Chem.* 88 (**2002**), 317–330.

28 Morales, D.A., Systematic search for bounds for total π-electron energy, *Int. J. Quantum Chem.* 93 (**2003**), 20–31.

29 Morales, D.A., The total π-electron energy as a problem of moments: Application of the Backhus–Gilbert method, *J. Math. Chem.* 38 (**2005**), 389–397.

30 Gutman, I., Zare Firoozabadi, S., de la Peña, J.A., Rada, J., On the energy of regular graphs, *MATCH Commun. Math. Comput. Chem.* 57 (**2007**), 435–442.

31 Gutman, I., On graphs whose energy exceeds the number of vertices, *Lin. Algebra Appl.*, in press.

32 Gutman, I, Cyvin, S.J., *Introduction to the Theory of Benzenoid Hydrocarbons*, Springer, Berlin, **1989**.

33 Gutman, I., Bounds for total π-electron energy, *Chem. Phys. Lett.* 24 (**1974**), 283–285.

34 de la Peña, J.A., Mendoza, L., Rada, J., Comparing momenta and π-electron energy of benzenoid molecules, *Discr. Math.* 302 (**2005**), 77–84.

35 Zhou, B., Gutman, I., de la Peña, J.A., Rada, J., Mendoza, L., On the spectral moments and energy of graphs, *MATCH Commun. Math. Comput. Chem.* 57 (**2007**), 183–191.

36 Gutman, I., On the energy of quadrangle-free graphs, *Coll. Sci. Papers Fac. Sci. Kragujevac* 18 (**1996**), 75–82.

37 Gutman, I., McClelland–type lower bound for total π-electron energy, *J. Chem. Soc. Faraday Trans.* 86 (**1990**), 3373–3375.

38 Türker, L., A novel total π-electron energy formula for alternant hydrocarbons – Angle of total π-electron energy, *MATCH Commun. Math. Comput. Chem.* 30 (**1994**) 243–252.

39 Gutman, I., A class of lower bounds for total π-electron energy of alternant conjugated hydrocarbons, *Croat. Chem. Acta* 68 (**1995**), 187–192.

40 Babić, D., Gutman, I., More lower bounds for the total π-electron energy of alternant hydrocarbons, *MATCH Commun. Math. Comput. Chem.* 32 (**1995**), 7–17.

41 Zhou, B., Lower bounds for energy of quadrangle-free graphs, *MATCH Commun. Math. Comput. Chem.* 55 (**2006**), 91–94.

42 Nikiforov, V., The energy of graphs and matrices, *J. Math. Anal. Appl.* 326 (**2007**), 1472–1475.

43 Nikiforov, V., Graphs and matrices with maximal energy, *J. Math. Anal. Appl.* 327 (**2007**), 735–738.

44 Cvetković, D., Gutman, I., The computer system GRAPH: A useful tool in chemical graph theory, *J. Comput. Chem.* 7 (**1986**), 640–644.

45 Walikar, H.B., Ramane, H.S., Hampiholi, P.R., On the energy of a graph, in: Balakrishnan, R., Mulder, H. M., Vijayakumar, A. (eds.), *Graph Connections*, Allied, New Delhi, **1999**, pp. 120–123.

46 Gutman, I., Pavlović, L., The energy of some graphs with large number of edges, *Bull. Acad. Serbe Sci. Arts.* 118 (**1999**), 35–50.

47 Koolen, J.H., Moulton, V., Gutman, I., Vidović, D., More hyperenergetic molecular graphs, *J. Serb. Chem. Soc.* 65 (**2000**), 571–575.

48 Hou, Y., Gutman, I., Hyperenergetic line graphs, *MATCH Commun. Math. Comput. Chem.* 43 (**2001**), 29–39.

49 Stevanović, D., Stanković, I., Remarks on hyperenergetic circulant graphs, *Lin. Algebra Appl.* 400 (**2005**), 345–348.

50 Shparlinski, I., On the energy of some circulant graphs, *Lin. Algebra Appl.* 414 (**2006**), 378–382.

51 Blackburn, S.R., Shparlinski I.E., On the average energy of circulant graphs, *Lin. Algebra Appl.*, 428 (**2008**), 1956–1963.

52 Akbari, S., Moazami, F., Zare, S., Kneser graphs and their complements are hyperenergetic, *MATCH Commun. Math. Comput. Chem.* 61 (**2009**), 361–368.

53 Gutman, I., Hou, Y., Walikar, H.B., Ramane, H. S., Hampiholi, P.R., No Hückel graph is hyperenergetic, *J. Serb. Chem. Soc.* 65 (**2000**), 799–801.

54 Walikar, H.B., Gutman, I., Hampiholi, P.R., Ramane, H.S., Nonhyperenergetic graphs, *Graph Theory Notes New York* 41 (**2001**), 14–16.

55 Gutman, I., Radenković, S., Hypoenergetic molecular graphs, *Indian J. Chem.* 46A (**2007**), 1733–1736.

56 Gutman, I., Li, X., Shi, Y., Zhang, J., Hypoenergetic trees, *MATCH Commun. Math. Comput. Chem.* 60 (**2008**), 415–426.

57 Nikiforov, V., The energy of C_4-free graphs of bounded degree, *Lin. Algebra Appl.* 428 (**2008**), 2569–2573.

58 Brankov, V., Stevanović, D., Gutman, I., Equienergetic chemical trees, *J. Serb. Chem. Soc.* 69 (**2004**), 549–553.

59 Balakrishnan, R., The energy of a graph, *Lin. Algebra Appl.* 387 (**2004**), 287–295.

60 Ramane, H.S., Walikar, H.B., Rao, S., Acharya, B., Hampiholi, P., Jog, S., Gutman, I., Equienergetic graphs, *Kragujevac J. Math.* 26 (**2004**), 5–13.

61 Ramane, H.S., Gutman, I., Walikar, H.B., Halkarni, S.B., Another class of equienergetic graphs, *Kragujevac J. Math.* 26 (**2004**), 15–18.

62 Ramane, H.S., Walikar, H.B., Rao, S., Acharya, B., Hampiholi, P., Jog, S., Gutman, I., Spectra and energies of iterated line graphs of regular graphs, *Appl. Math. Lett.* 18 (**2005**), 679–682.

63 Ramane, H.S., Gutman, I., Walikar, H.B., Halkarni, S.B., Equienergetic complement graphs, *Kragujevac J. Sci.* 27 (**2005**), 67–74.

64 Indulal, G., Vijayakumar, A., On a pair of equienergetic graphs, *MATCH Commun. Math. Comput. Chem.* 55 (**2006**), 83–90.

65 Ramane, H.S., Walikar, H.B., Construction of eqienergetic graphs, *MATCH Commun. Math. Comput. Chem.* 57 (**2007**), 203–210.

66 Xu, L., Hou, Y., Equienergetic bipartite graphs, *MATCH Commun. Math. Comput. Chem.* 57 (**2007**), 363–370.

67 Indulal, G., Vijayakumar, A., Energies of some non-regular graphs, *J. Math. Chem.* 42 (**2007**), 377–386.

68 Liu, J., Liu, B., Note on a pair of equienergetic graphs, *MATCH Commun. Math. Comput. Chem.* 59 (**2008**), 275–278.

69 Alinaghipour, F., Ahmadi, B., On the energy of complement of regular line graph, *MATCH Commun. Math. Comput. Chem.* 60 (**2008**), 427–434.

70 Bonifácio, A.S., Vinagre, C.T. M., de Abreu, N.M. M. Constructing pairs of equienergetic and non-cospectral graphs, *Appl. Math. Lett.* 21 (**2008**), 338–341.

71 Indulal, G., Vijayakumar, A., Equienergetic self–complementary graphs, *Czechoslovak Math. J.*, in press.

72 López, W., Rada, J., Equienergetic digraphs, *Int. J. Pure Appl. Math.* 36 (**2007**), 361–372.

73 Sampathkumar, E., On duplicate graphs, *J. Indian Math. Soc.* 37 (**1973**), 285–293.

74 Gutman, I., Acyclic systems with extremal Hückel π-electron energy, *Theor. Chim. Acta* 45 (**1977**), 79–87.

75 Gutman, I., Vidović, D., Quest for molecular graphs with maximal energy: a computer experiment, *J. Chem. Inf. Comput. Sci.* 41 (**2001**), 1002–1005.

76 Wang, M., Hua, H., Wang, D., Minimal energy on a class of graphs, *J. Math. Chem.*, 44 (2008) 1389–1402.

77 Cvetković, D., Grout, J., Graphs with extremal energy should have a small number of distinct eigenvalues, *Bull. Acad. Serbe Sci. Arts* 134 (**2007**), 43–57.

78 Li, X., Zhang, J., Wang, L., On bipartite graphs with minimal energy, *Discr. Appl. Math.*, in press.

79 Zhang, F., Li, H., On acyclic conjugated molecules with minimal energies, *Discr. Appl. Math.* 92 (**1999**), 71–84.

80 Zhang, F., Li, H., On maximal energy ordering of acyclic conjugated molecules, in: Hansen, P., Fowler, P., Zheng, M. (eds.), *Discrete Mathematical Chemistry*, Amer. Math. Soc., Providence, **2000**, pp. 385–392.

81 Ye, L., Chen, R.S., Ordering of trees with given bipartition by their energies and Hosoya indices, *MATCH Commun. Math. Comput. Chem.* 52 (**2004**), 193–208.

82 Yan, W., Ye, L., On the minimal energy of trees with a given diameter, *Appl. Math. Lett.* 18 (**2005**), 1046–1052.

83 Yan, W., Ye, L., On the maximal energy and the Hosoya index of a type of trees with many pendent vertices, *MATCH Commun. Math. Comput. Chem.* 53 (**2005**), 449–459.

84 Lin, W., Guo, X., Li, H., On the extremal energies of trees with a given maximum degree, *MATCH Commun. Math. Comput. Chem.* 54 (**2005**), 363–378.

85 Zhou, B. Li, F., On minimal energies of trees of a prescribed diameter, *J. Math. Chem.* 39 (**2006**), 465–473.

86 Yu, A., Lv, X., Minimal energy on trees with k pendent vertices, *Lin. Algebra Appl.* 418 (**2006**), 625–633.

87 Ye, L., Yuan, X., On the minimal energy of trees with a given number of pendant vertices, *MATCH Commun. Math. Comput. Chem.* 57 (**2007**), 193–201.

88 Li, N., Li, S., On the extremal energy of trees, *MATCH Commun. Math. Comput. Chem.* 59 (**2008**), 291–314.

89 Gutman, I., Radenković, S., Li, N., Li, S., Extremal energy of trees, *MATCH Commun. Math. Comput. Chem.* 59 (**2008**), 315–320.

90 Ou, J., On acyclic molecular graphs with maximal Hosoya index, energy, and short diameter, *J. Math. Chem.*, in press.

91 Li, S., Li, X., The fourth maximal energy of acyclic graphs, *MATCH Commun. Math. Comput. Chem.*, 61 (**2009**), 383–394.

92 Guo, J.M., On the minimal energy ordering of acyclic conjugated molecules, *Discr. Appl. Math.*, 156 (**2008**), 2598–2605.

93 Li, X., Yao, X., Zhang, J., Gutman, I., Maximum energy trees with two maximum degree vertices, *J. Math. Chem.*, in press.

94 Hou, Y., Unicyclic graphs with minimal energy, *J. Math. Chem.* 29 (**2001**), 163–168.

95 Gutman, I., Hou, Y., Bipartite unicyclic graphs with greatest energy, *MATCH Commun. Math. Comput. Chem.* 43 (**2001**), 17–28.

96 Hou, Y., Gutman, I., Woo, C.W., Unicyclic graphs with maximal energy, *Lin. Algebra Appl.* 356 (**2002**), 27–36.

97 Li, F., Zhou, B., Minimal energy of bipartite unicyclic graphs of a given bipartition, *MATCH Commun. Math. Comput. Chem.* 54 (**2005**), 379–388.

98 Chen, A., Chang, A., Shiu, W.C., Energy ordering of unicyclic graphs, *MATCH Commun. Math. Comput. Chem.* 55 (**2006**), 95–102.

99 Wang, W.H., Chang, A., Zhang, L.Z., Lu, D.Q., Unicyclic Hückel molecular graphs with minimal energy, *J. Math. Chem.* 39 (**2006**), 231–241.

100 Hua, H., On minimal energy of unicyclic graphs with prescribed girth and pendent vertices, *MATCH Commun. Math. Comput. Chem.* 57 (**2007**), 351–361.

101 Hua, H., Bipartite unicyclic graphs with large energy, *MATCH Commun. Math. Comput. Chem.* 58 (**2007**) 57–83.

102 Gutman, I., Furtula, B., Hua, H., Bipartite unicyclic graphs with maximal, second–maximal, and third–maximal energy, *MATCH Commun. Math. Comput. Chem.* 58 (**2007**), 85–92.

103 Hua, H., Wang, M., Unicyclic graphs with given number of pendent vertices and minimal energy, *Lin. Algebra Appl.* 426 (**2007**) 478–489.

104 Wang, W.H., Chang, A., Lu, D.Q., Unicyclic graphs possessing Kekulé structures with minimal energy, *J. Math. Chem.* 42 (**2007**), 311–320.

105 Li, X., Zhang, J., Zhou, B., On unicyclic conjugated molecules with minimal energies, *J. Math. Chem.* 42 (**2007**), 729–740.

106 Li, F., Zhou, B., Minimal energy of unicyclic graphs of a given diameter, *J. Math. Chem.* 43 (**2008**), 476–484.

107 Li, S., Li, X., Zhu, Z., On minimal energy and Hosoya index of unicyclic graphs, *MATCH Commun. Math. Comput. Chem.*, 61 (**2009**), 325–339.

108 Hou, Y., Bicyclic graphs with minimum energy, *Lin. Multilin. Algebra* 49 (**2001**), 347–354.

109 Zhang, J., Zhou, B., Energy of bipartite graphs with exactly two cycles, *Appl. Math. J. Chinese Univ., Ser.A* 20 (**2005**), 233–238 (in Chinese).

110 Zhang, J., Zhou, B., On bicyclic graphs with minimal energies, *J. Math. Chem.* 37 (**2005**), 423–431.

111 Li, X., Zhang, J., On bicyclic graphs with maximal energy, *Lin. Algebra Appl.* 427 (**2007**), 87–98.

112 Furtula, B., Radenković, S., Gutman, I., Bicyclic molecular graphs with greatest energy, *J. Serb. Chem. Soc.* 73 (**2008**), 431–433.

113 Yang, Y., Zhou, B., Minimal energy of bicyclic graphs of a given diameter, *MATCH Commun. Math. Comput. Chem.* 59 (**2008**), 321–342.

114 Liu, Z., Zhou, B., Minimal energies of bipartite bicyclic graphs, *MATCH Commun. Math. Comput. Chem.* 59 (**2008**), 381–396.

115 Li, S., Li, X., Zhu, Z., On tricyclic graphs with minimal energy, *MATCH Commun. Math. Comput. Chem.* 59 (**2008**), 397–419.

116 Zhang, J., On tricyclic graphs with minimal energies, Preprint, **2006**.

117 Li, S., Li, X., On tetracyclic graphs with minimal energy, *MATCH Com-*

mun. Math. Comput. Chem. 60 (**2008**), 395–414.

118 Zhang, F., Li, Z., Wang, L., Hexagonal chain with minimal total π-electron energy, *Chem. Phys. Lett.* 37 (**2001**), 125–130.

119 Zhang, F., Li, Z., Wang, L., Hexagonal chain with maximal total π-electron energy, *Chem. Phys. Lett.* 37 (**2001**), 131–137.

120 Rada, J., Tineo, A., Polygonal chains with minimal energy, *Lin. Algebra Appl.* 372 (**2003**), 333–344.

121 Rada, J., Energy ordering of catacondensed hexagonal systems, *Discr. Appl. Math.* 145 (**2005**), 437–443.

122 Ren, H., Zhang, F., Double hexagonal chains with minimal total π-electron energy, *J. Math. Chem.* 42 (**2007**), 1041–1056.

123 Gutman, I., Mateljević, M., Note on the Coulson integral formula, *J. Math. Chem.* 39 (**2006**), 259–266.

124 Mateljević, M., Gutman, I., Note on the Coulson and Coulson–Jacobs integral formulas, *MATCH Commun. Math. Comput. Chem.* 59 (**2008**), 257–268.

125 Gutman, I., Zhang, F., On the quasiordering of bipartite graphs, *Publ. Inst. Math.* (Belgrade) 40 (**1986**), 11–15.

126 Akbari, S., Ghorbani, E., Zare, S., Some relations between rank, chromatic number and energy of graphs, *Discr. Math.*, in press.

127 Day, J., So, W., Singular value inequality and graph energy change, *El. J. Lin. Algebra* 16 (**2007**), 291–299.

128 Day, J., So, W., Graph energy change due to edge deletion, *Lin. Algebra Appl.* 428 (**2008**), 2070–2078.

129 Bapat, R.B., Pati, S., Energy of a graph is never an odd integer, *Bull. Kerala Math. Assoc.* 1 (**2004**), 129–132.

130 Pirzada, S., Gutman, I., Energy of a graph is never the square root of an odd integer, *Appl. Anal. Discr. Math.* 2 (**2008**), 118–121.

131 Gutman, I., Soldatović, T., (n, m)-Type approximations for total π-electron energy of benzenoid hydrocarbons, *MATCH Commun. Math. Comput. Chem.* 44 (**2001**), 169–182.

132 Gutman, I., Cmiljanović, N., Milosavljević, S., Radenković, S., Dependence of total π-electron energy on the number of non-bonding molecular orbitals, *Monatsh. Chem.* 135 (**2004**), 765–772.

133 Gutman, I., Stevanović, D., Radenković, S., Milosavljević, S., Cmiljanović, N., Dependence of total π-electron energy on large number of non-bonding molecular orbitals, *J. Serb. Chem. Soc.* 69 (**2004**), 777–782.

8
Generalized Shortest Path Trees: A Novel Graph Class by Example of Semiotic Networks

Alexander Mehler

8.1
Introduction

In this chapter we introduce a class of tree-like graphs that combines the efficiency of tree-like structures with the expressiveness of general graphs. Our starting point is the notion of a *generalized tree* (GT)*, that is, a graph with a kernel hierarchical skeleton in conjunction with graph-inducing peripheral edges [17]. We combine this notion with the theory of *network optimization problems* (NOPs) [60] in order to introduce *generalized shortest pathS trees* (GPST) as a subclass of the class of GTs. One advantage of this novel class is that it provides a functional semantics of the different types of edges of GTs. Another is that it naturally gives rise to combining graph modeling with conceptual spaces [28] and, thus, with cognitive or, more generally, semiotic modeling. This chapter provides three examples in support of this combination.

The graph model presented in this chapter focuses on structure formation in semiotic networks. Its background is the rising interest in network models due to the renaissance of, so to speak, functionalist models of networking in a wide range of scientific disciplines starting from the famous work of [45] in social psychology and extending into the area of physics [1], quantitative biology [3, 23], quantitative sociology [6, 66], quantitative linguistics [26, 41], and information science [48], to name only a few. See [21, 40, 47] for surveys of this research in the area of the natural sciences and the humanities.

What all these network models have in common is that they start from a remarkably low-level graph model in terms of simple graphs with at most labeled or typed vertices and edges. That is, for a decade or so networks have been explored almost exclusively in terms of simple graphs [47], in some cases with weighted edges [4, 52], together with a partitioning into

* A list of abbreviations can be found at the end of the chapter.

Analysis of Complex Networks: From Biology to Linguistics. Edited by Matthias Dehmer and Frank Emmert-Streib

ISBN: 978-3-527-32345-6

bipartite models [67]. One exception to this trend is the notion of network motifs [53], which is restricted to the formation of micro-level structures. That is, the main focus of research has been on structures on the macro level (cf., e.g., the bow-tie model of [8]) disregarding intermediate levels of structure formation within complex networks. As a consequence, structure formation is almost exclusively dealt with in terms of network characteristics as a gateway to "universal" laws of network organization [5]. Note that there are many graph cluster algorithms for identifying subgraphs of an above-average cluster-internal homogeneity and cluster-external heterogeneity (see [11] for an overview). This is in the line of supervised or unsupervised learning, which basically decides on the membership of objects to clusters as partitions of the vertex sets of the underlying graph. In this chapter we want to shed light on a graph model in the area of semiotic networks that goes beyond traditional approaches to graph clustering and, at the same time, departs from the predominant model-theoretic abstinence regarding meso-level structures.

Generally speaking, graph models are quite common in semiotics and related disciplines. Whereas in linguistics tree-like models predominate (as, e.g., rhetorical structure trees [37] to name only one example), efforts have been made to build more general graph models in quantitative linguistics [7, 38], partly inspired by category theory [29] and topology [27]. Here we mention only three less cited approaches in this area (see [40] for a survey of network models in linguistics): Firstly, Thiopoulos [61] builds on, amongst others, the categorical notion of product and coproduct in order to model the process of meaning constitution in lexical networks. Secondly, Baas [2] utilizes hierarchical hypergraphs as models of recursive processes of networking. Thirdly, Ehresmann and Vanbremeersch [22] utilize – comparable to [2] – the notion of a colimit in order to give a formal account of emergent structures in complex systems.

In spite of their expressiveness, category theory and topology are hardly found as methodic bases of present-day approaches in quantitative and computational linguistics. One reason is that graph theory seems to be already expressive enough to master a wide range of structure formations in linguistics. In this chapter we follow this methodic conception, however, with a focus on *generalized trees* [16]. In web mining, GTs were introduced in order to grasp the striking gestalt of web documents in-between tree- and graph-like structures [15, 17, 43]. See Figure 8.1 for an example of a GT with a typical kernel hierarchical structure complemented by graph-inducing lateral and vertical links. Recently, Emmert-Streib, Dehmer, and Kilian [25] have shown that this concept is also of interest in modeling biological structures. However, one important question has been left open by this research.

Generally speaking, the search for spanning trees of a given graph that satisfy certain topological constraints is a well-known research topic in graph

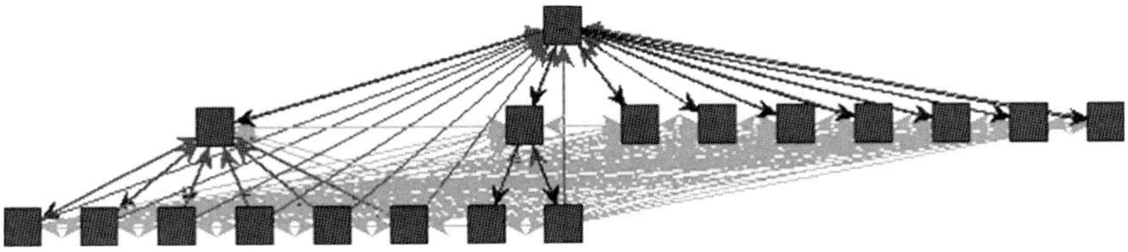

Figure 8.1 A webgraph [11] in the form of a directed GT derived from a conference web site (www.text-technology.de) [15, 43].

theory [60]. Along this line of research we can formulate a central question about GTs as follows: *Given a connected graph G, which GT G′ induced by G satisfies certain desirable topological constraints?* By focusing on this question we do not – unlike related approaches – ask about a similarity model of pairs of predetermined GTs (see [16] for such a model). In contrast to this, we take a step back and ask how to induce GTs from a given connected graph. This problem is at the core of the present chapter. It will be tackled by means of the notion of a *generalized shortest path tree* (GSPT). The basic idea behind this notion is to introduce a functional semantics of edge types of GTs. That is, starting from a graph we justify in functional terms which of its edges would preferably serve as kernel, lateral, or vertical links. In this way, we introduce a functional semantics into the inducement of GTs that goes beyond the approaches mentioned above.

This endeavor is in accordance with Tarjan [60, p. 71], who generally describes the approach of network optimization as follows: Given a weighted graph G – called a *network* – whose edges are weighted by an edge weighting function $\mu : E \rightarrow \mathbb{R}$, the task is to describe a *network optimization problem* (NOP) that consists of finding a subgraph of G that satisfies a set of well-defined properties by optimizing (i.e., minimizing or maximizing) a certain function of μ. It is a basic idea of the present chapter to introduce the notion of context sensitivity into the specification of such NOPs. That is, as distinguished from the notion of a minimum spanning tree, we look for subgraphs in the form of GTs whose generation is nontrivially affected by the choice of some root vertex. This sort of context sensitivity is in accordance with what is known about priming and spreading activation in cognitive networks [46]. As becomes clear by this explanation, the present chapter always strives to provide both a graph-theoretically and empirically well-founded graph model.

What do we gain by such a graph model? Such a model is a first step towards a time- and space-efficient, as well as cognitively plausible, model of information processing in semiotic networks. Although we aim at this model, the present chapter does not cut the Gordian knot. That is, our graph model of structure formation in complex networks does not overcome the disregard mentioned at the beginning. What we provide is a further development of GT as a model of semiotic structures. So the main result of this chapter is a formalization of a promising concept that may help to better understand structure formation in semiotic networks and the processes operating thereon.

The chapter is organized as follows. In Section 8.2, our graph model is developed in detail, including directed and undirected graphs. At the core of this chapter is the notion of a GSPT that enables a detailed semantics of the different edge types provided by GTs. It turns out that this is a step towards combining GTs with the theory of conceptual spaces. This combination is also provided by Section 8.2. Next, Section 8.3 gives an empirical account of our graph model by example of three semiotic systems: social tagging, text networks, and discourse structures.

8.2
A Class of Tree-Like Graphs and Some of Its Derivatives

8.2.1
Preliminary Notions

In this section we briefly define two well-known notions that will be used throughout the chapter to introduce our graph model. This relates to paths in undirected and directed graphs as well as to the notion of geodesic distance in weighted graphs.

Definition 8.1 (Preliminaries) Let $G = (V, E, \mathcal{L}_V, \mathcal{L}_E, \mu)$ be a connected weighted undirected graph whose vertices are uniquely labeled by the function $\mathcal{L}_V : V \to L_V$ for the set of vertex labels L_V and whose edges are uniquely labeled by $\mathcal{L}_E : E \to L_E$ for the set of edge labels L_E. Throughout this chapter we assume that $\mathcal{L}_V \subset \mathbb{N}_0$ and $\mathcal{L}_E \subset \mathbb{N}_0$, that is, vertices and edges are labeled by ordinal numbers. Further, we assume that this numbering is consecutive. Next, let $D = (V, A, \mathcal{L}_V, \mathcal{L}_E, \nu)$ be an orientation of G, that is, a connected weighted digraph such that $\forall a \in A\colon \nu(a) = \mu(e) \Leftrightarrow e = \{\text{in}(a), \text{out}(a)\} \in E$. By $\leq_V \subset L_V^2$ ($\leq_E \subset L_E^2$) we denote the natural order of $L_V \subset \mathbb{N}_0$ ($L_E \subset \mathbb{N}_0$) such that for all $a, b \in L_V$ ($a, b \in L_E$): $a <_V b$ ($a <_E b$) iff $a \leq_V b$ ($a \leq_E b$) and $a \neq b$. This allows us to define the order relation $\leq_a = \leq_V \cup \leq_E$ of vertices *and* edges. Without loss of generality we assume that $\mu\colon E \to \mathbb{R}^+ \setminus \{0\}$ is an edge weighting function that represents the costs of traversing edges in E. Think of μ, for example, as a function of the loss of coherence induced by following hyperlinks. Analogously, we assume that $\nu(a), a \in A$, represents the cost of entering out(a) when coming from in(a). Now let $\mathbb{P}(G)$ be the set of all simple paths in G and $P = (v_{i_0}, e_{j_1}, v_{i_1}, \ldots, v_{i_{m-1}}, e_{j_m}, v_{i_m}) \in \mathbb{P}(G)$ such that $\forall 1 \leq k \leq m$: $e_{j_k} = \{v_{i_{k-1}}, v_{i_k}\} \in E$. Further, let $\mathbb{P}(D)$ be the set of all simple paths in D

and $\boldsymbol{P} = (v_{i_0}, a_{j_1}, v_{i_1}, \ldots, v_{i_{m-1}}, a_{j_m}, v_{i_m}) \in \mathbb{P}(D)$ such that $\forall a_{j_k} \in \{a_{j_1}, \ldots, a_{j_m}\}$: $\text{in}(a_{j_k}) = v_{i_{k-1}} \wedge \text{out}(a_{j_k}) = v_{i_k}$. Then, $V(P) = \{v_{i_0}, v_{i_1}, \ldots, v_{i_{m-1}}, v_{i_m}\} \subseteq V$ is the set of all vertices, $E(P) = \{e_{j_1}, \ldots, e_{j_m}\} \subseteq E$ the set of all edges, and $VE(P) = V(P) \cup E(P)$ the set of all constituents of P. Analogously, we define the sets $V(\boldsymbol{P})$, $E(\boldsymbol{P})$ and $V(\boldsymbol{P})$ for directed paths ($\boldsymbol{P}$). If G (resp. D) is a (directed) tree, then for each $v, w \in V$ the simple path ending at v and w is unique. Such paths will be denoted as P_{vw} (($\boldsymbol{P})_{vw}$), indexed by their end vertices v and w. Next, we define the order relation $\leqslant_a \subseteq \mathbb{P}(G)^2$ over the set of paths $\mathbb{P}(G)$ of G such that for $P = (v_{i_1}, e_{i_2}, \ldots, e_{i_{m_i-1}}, v_{i_{m_i}})$, $P' = (v_{j_1}, e_{j_2}, \ldots, e_{j_{m_j-1}}, v_{j_{m_j}}) \in \mathbb{P}(G)$, $P \neq P'$: $P \leqslant_a P'$ iff $\exists r < \min(m_i, m_j) \forall k \in \{1, \ldots, r\} : VE(P) \ni x_{i_k} = x_{j_k} \in VE(P') \wedge VE(P) \ni x_{i_{r+1}} <_a x_{j_{r+1}} \in VE(P')$. Analogously, we define the order relation $\lessdot_a \subseteq \mathbb{P}(D)^2$ over the set $\mathbb{P}(D)$ of directed paths of D. Further, by $\mathbb{P}_G(v, w)$ we denote the set of all simple paths in G ending at v and w. Finally, for $v_{i_{m_i}} = v_{j_1}$ we define the concatenation $P \circ P' = (v_{i_1}, e_{i_2}, \ldots, e_{i_{m_i-1}}, v_{i_{m_i}}, e_{j_2}, \ldots, e_{j_{m_j-1}}, v_{j_{m_j}})$ of P and P'.

Remark. Throughout this chapter we will always assume the existence of the labeling functions $\mathcal{L}_V$ and $\mathcal{L}_E$ and, thus, of the order relations $\leqslant_a$ and $\lessdot_a$ without explicitly noting this in the subsequent definitions of graphs. The reason for this omission is to keep the formalism simple.

Remark. Why so much effort in defining order relations over paths? The reason is that in semiotic systems, multi- and pseudographs are common (e.g., due to redundancy in the system), while simple graphs seem to be the exception. Think, for example, of graphs as simple as webgraphs in which vertices denote pages while edges stand for hyperlinks. Here, it is not unusual that two pages are linked by different edges distinguished by the location of their anchors within the source page. Using some measure of lexical similarity of interlinked texts [35] such links may be equally weighted. As a consequence, we need a method of distinguishing such edges and the paths built out of them in order to provide uniqueness of the mathematical notions to be defined. This is provided by $\leq_E$, which may explore, for example, the aforementioned positional information.

Based on Definition 8.1 we can now define the notion of a geodesic path.

Definition 8.2 (Geodesic Distance and Geodesic Path) Let $G = (V, E, \mathcal{L}_V, \mathcal{L}_E, \mu)$ be a weighted connected graph according to Definition 8.1. Then, we extend μ as a function of $\mathbb{P}(G)$, that is,

$$\mu \colon \mathbb{P}(G) \to (0, \infty) ,$$

such that for each $P = (v_{i_0}, e_{j_1}, v_{i_1}, \ldots, v_{i_{m-1}}, e_{j_m}, v_{i_m}) \in \mathbb{P}(G)$ we set

$$\mu(P) = \sum_{k=1}^{m} \mu(e_{j_k}) .$$

Based on μ we define the *geodesic path* $GP_\mu(v, w)$ *between* v *and* w *in* G as

$$GP_\mu(v, w) = \inf_{\leqslant_a} \left\{ \arg\min_{P \in \mathbb{P}_G(v,w)} \mu(P) \right\},$$

where $\mathbb{P}_G(v, w)$ is the set of all simple paths in G ending at v and w (Definition 8.1). Further, the *geodesic distance* $\hat{\mu}\colon V \times V \to [0, \infty)$ between $v, w \in V$ is defined as

$$\hat{\mu}(v, w) = \begin{cases} 0 & v = w\,, \\ \mu(GP_\mu(v, w)) & v \neq w\,. \end{cases}$$

Finally, for any weighted graph $G = (V, E, \mu)$ we define

$$\mu(G) = \sum_{e \in E} \mu(e)\,.$$

Note that the definition of geodesic distances and paths makes use of the order relation $\leqslant_a$ of paths. These two notions play a crucial role in defining so-called generalized shortest path trees in Sections 8.2.4 and 8.2.5. Further, they are used to bridge the gap between the graph model introduced here and the cognitive-linguistic notion of a conceptual space [28]. This is done in Sections 8.2.8 and 8.2.9. We are now in a position to introduce the fundamental notion of a GT.

8.2.2
Generalized Trees

What is common to many semiotic networks is their hierarchical skeleton in conjunction with graph-inducing links. Obviously, such networks lie in between tree-like structures on the one hand and more general graphs on the other. This ambivalent nature has been grasped by the notion of a *generalized tree* (GT) [17], which will be developed further in the subsequent sections. Extending the approach presented in [17] and [41] we will distinguish directed from undirected GTs. The reason for doing this is dictated by the nature of semiotic structures: there are semiotic systems that are better described by abstracting from the orientation of arcs used to model relations among their components. This holds, for example, for lexical networks whose nodes are interlinked by multiple arcs or simply by edges. In order to capture the variety of semiotic structures that are spanned over their kernel tree-like skeletons, we utilize and extend the following graph-theoretical apparatus.

Definition 8.3 (Undirected Generalized Tree) Let $T = (V, E', r)$ be an undirected tree rooted in r. Further, let $P_{rv} = (v_{i_0}, e_{j_1}, v_{i_1}, \ldots, v_{i_{n-1}}, e_{j_n}, v_{i_n})$, $v_{i_0} = r$, $v_{i_n} = v$,

$e_{j_k} = \{v_{i_{k-1}}, v_{i_k}\} \in E', 1 \le k \le n$, be the unique path in T from r to $v \in V$ and $V(P_{rv}) = \{v_{i_0}, \ldots, v_{i_n}\}$ the set of all vertices of P_{rv}. An *undirected GT*

$$G = (V, E, \tau, r)$$

induced by T is a pseudograph (i.e., a multigraph that may contain loops) *rooted* in r whose edges are typed by the function $\tau : E \to \mathcal{T} = \{\mathrm{k}, \mathrm{l}, \mathrm{r}, \mathrm{v}\}$ as follows (note that edges $e \in E$ are multisets of exactly two elements that in the case of reflexive edges contain the same element twice):[1)]

$$\forall e \in E: \begin{cases} \tau(e) = \mathrm{k} & \Rightarrow \quad e \in E_{\mathrm{k}} = E' \\ & \quad\ (\textit{kernel edges}) \\ \tau(e) = \mathrm{v} & \Rightarrow \quad e \in E_{\mathrm{v}} = \{\{v, w\} \mid v \in V(P_{rw}) \vee w \in V(P_{rv})\} \\ & \quad\ (\textit{vertical edges}) \\ \tau(e) = \mathrm{r} & \Rightarrow \quad e \in E_{\mathrm{r}} = \{\{v, v\} \mid v \in V\} \\ & \quad\ (\textit{reflexive edges}) \\ \tau(e) = \mathrm{l} & \Rightarrow \quad e \in E_{\mathrm{l}} = [V]^2 \setminus (E_{\mathrm{k}} \cup E_{\mathrm{v}} \cup E_{\mathrm{r}}) \\ & \quad\ (\textit{lateral edges}) \end{cases}$$

such that $E^{\tau}_{[1]} = \{e \in E \mid \tau(e) = \mathrm{k}\}$, $E^{\tau}_{[2]} = \{e \in E \mid \tau(e) = \mathrm{v}\}$, $E^{\tau}_{[3]} = \{e \in E \mid \tau(e) = \mathrm{r}\}$, $E^{\tau}_{[4]} = \{e \in E \mid \tau(e) = \mathrm{l}\}$, where $E = \cup_{i=1}^{4} E^{\tau}_{[i]}$. Because of the interdependence of τ and the sequence of sets $E^{\tau}_{[i]}$, $1 \le i \le 4$, we alternatively denote G by $(V, E^{\tau}_{[1..4]}, r)$, where $e \in E^{\tau}_{[1..4]}$ iff $e \in \cup_{i=1}^{4} E^{\tau}_{[i]}$. In other words, GTs G are interchangeably denoted by (V, E, τ, r) and $(V, E^{\tau}_{[1..4]}, r)$. We say that G is *generalized by its lateral, reflexive, and vertical edges.* Further, r is called the *root (vertex)* of G. The GT $G = (V, E, \tau, r)$ induces the undirected tree $\mathrm{kern}(G) = (V, E^{\tau}_{[1]}, r) = T$, called *kernel (tree)* or *skeleton* of G. Further, the graph $\mathrm{periphery}(G) = (V, \cup_{i=2}^{4} E^{\tau}_{[i]})$ is called *periphery* or *complementary graph* of G. Edges belonging to periphery(G) are called *peripheral edges* (complementing the set of kernel edges). Finally, the GT (V, E, τ, r, μ) with the edge weighting function $\mu : E \to \mathbb{R}$ is called a *weighted undirected GT.*

Remark. The reason why we use the implication instead of the equivalence relation in defining the edge typing function τ is that there may be multiple edges that are typed differently. Note further that since GTs are multigraphs, the sets $E^{\tau}_{[i]}$, $1 \le i \le 4$, do not form a partition of E in the strict sense.

Remark. In order to prevent negative cycles [60, 85], we henceforth assume that μ is a function from E to $\mathbb{R}^+ \setminus \{0\}$. Note that it does not make sense to have zero valued edges, that is, edges e for which $\mu(e) = 0$. Such edges simply do not exist. Further, throughout this chapter we only deal with finite graphs.

1) For a multiset $X = (Y, m)$, $m : Y \to \mathbb{N}_{\ge 1}$, we use the notation $X = \{\underbrace{x, \ldots, x}_{n \text{ times}} \mid x \in Y \wedge m(x) = n\}$. Further, $[X]^k$ denotes the set of all subsets of k elements of X (cf. [19]).

Example *Let graph A in Figure 8.2 be given as a starting point. In this case, we can derive a GT* $C = (V, E^{\tau}_{[1..4]}, 0)$ *from A such that* $E^{\tau}_{[1]} = \{\{0,1\},\{0,4\},\{0,6\},\{1,3\},\{4,5\},\{2,5\}\}$, $E^{\tau}_{[2]} = E^{\tau}_{[3]} = \emptyset$ *and* $E^{\tau}_{[4]} = \{\{1,2\},\{1,6\},\{2,3\}\}$. *Graph D in Figure 8.2 exemplifies another GT of A also rooted in 0 but with a different kernel tree. Finally, graph E in Figure 8.2 shows a GT rooted in vertex 4, thereby exemplifying a vertical edge ending at 0 and 6.*

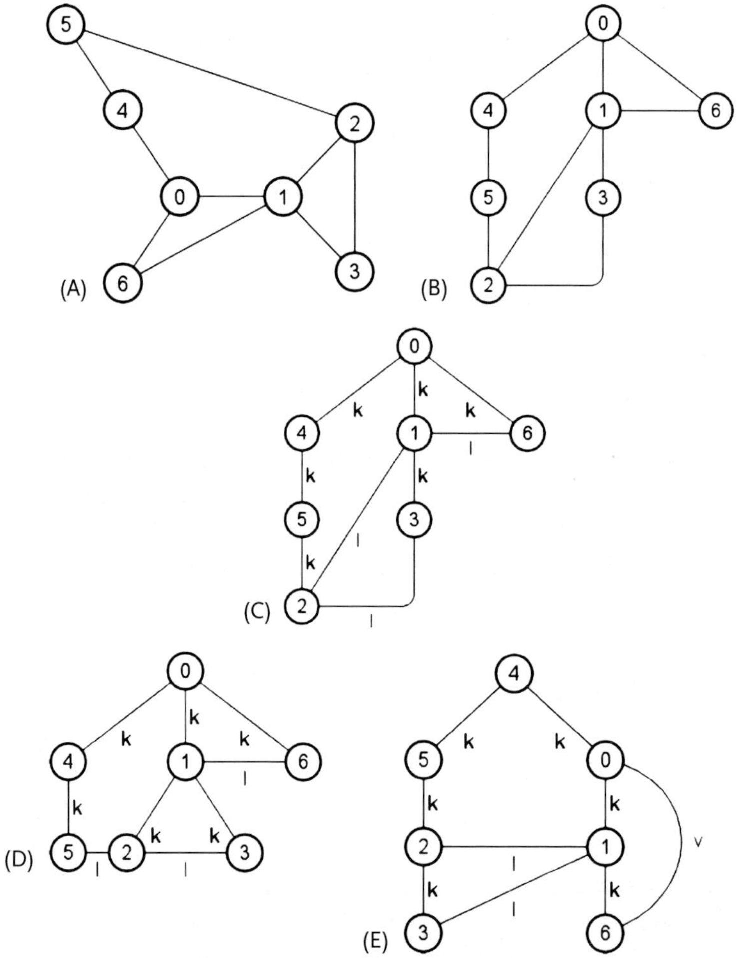

Figure 8.2 A connected graph A. The same graph in a tree-like perspective denoted by graph B. A GT C of A rooted in 0. A GT D of A rooted in 0 with a different kernel than C. Finally, a GT E of A rooted in 4. For reasons of simplification, edge weights are omitted while edge types are noted as edge labels.

Remark. Unlike [24] but in accordance with [17], we do not define GTs by means of a multilevel function. Rather, we focus more generally on *unleveled* graphs. The reason is that in semiotic systems we hardly observe such a mapping of vertices onto distinguished levels of a graph. Look, for example, at the category graph of Wikipedia [64]: despite what its users might claim, the categories in this graph do not span a tree, but a directed cyclic graph. This fact is contrary to mapping vertices to the same level of taxonomic resolution even if they have the same geodesic distance to the root of the category system, which, by the way, does not uniquely exist in this example of social tagging. Further, as we do not focus solely on categorical systems with a hierarchical skeleton but additionally on association networks, the idea of a level function becomes obsolete. Thus, we need a more general definition of GTs as provided by Definition 8.3.

In the sequel of this chapter, the notion of a generalized subtree of a GT will be used. By exploring the type system of GTs such subtrees are defined as follows.

Definition 8.4 (Type-Restricted Generalized Subtree) Let $G = (V, E, \tau, r, \mu)$ be a weighted GT. Further, let $r \in V$ be a vertex in G. Then, for a subset $\mathcal{T}' = \{\mathrm{k}, \ldots\} \subseteq \mathcal{T} \leftarrow E : \tau$ and the restriction τ' of τ to $\mathcal{T}'$, a *type-restricted generalized subtree* of G is a GT $G' = (V, E', \tau', r, \mu')$ of G such that $\forall e \in E : \tau(e) \notin \mathcal{T}' \Rightarrow e \notin E' \subseteq E$. In cases where $\mathcal{T}'$ omits types in descending order of the index of the sequence $E^{\tau}_{[i]}$, $1 < i$, we alternatively denote type-restricted generalized subtrees by $(V, E^{\tau'}_{[1..i]}, r, \mu')$ for $i > 1$. As usually, μ' is the restriction of μ to E'.

Remark. As GTs are connected graphs, we do not consider type-restricted subtrees that exclude kernel edges. Therefore, we always have that $\mathrm{k} \in \mathcal{T}'$.

It seems natural to map the tree-like structure of a semiotic network by means of the kernel edges of a corresponding GT G while its peripheral edges may be used to map the remainder of that network. However, as the same graph induces several GTs (see, e.g., Figure 8.2), we must pose the following question: *Given an undirected connected graph $G = (V, E, \mu)$, how many GTs can we build out of G?* Look, for example, at Figure 8.3: Given graph G with three vertices, we can derive exactly nine GTs from G in which the third edge is either a vertical edge (if ending at the corresponding root) or a lateral edge. More generally, if G is a completely connected graph with n vertices, there exist n^{n-1} GTs of G. The reason is that the cardinality of this set of GTs equals the number of vertices in G times the cardinality of the set of spanning trees of G. And the latter cardinality is determined by n^{n-2} [69]. *But why do both of these sets have the same size?* The reason is that a GT according to Definition 8.3 is determined by its kernel (spanning) tree – remember that

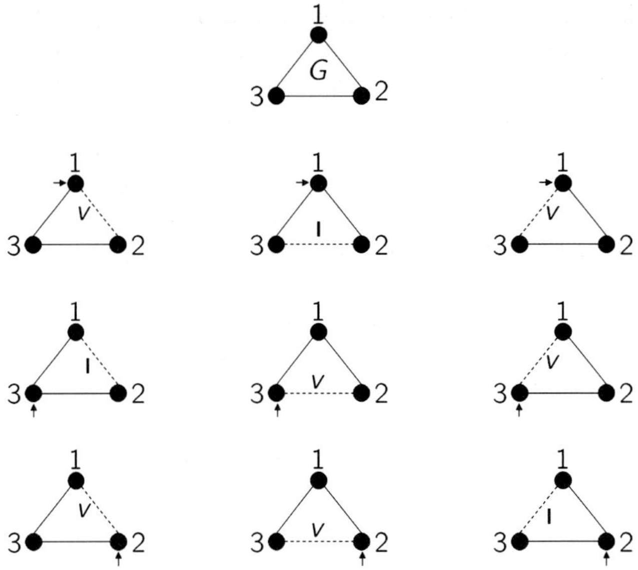

Figure 8.3 A graph G of three vertices together with all GTs derivable from it. Vertical edges are labeled by v, lateral edges by l. Roots are marked by an incoming arrow $\longrightarrow$.

apart from kernel edges all other edge types are defined with respect to the kernel tree.

In order to stress the relationship of a graph $G = (V, E, \mu)$ with its GTs, we add the following definition.

Definition 8.5 (Generalized Spanning Tree) Let $G = (V, E, \mu)$ be a weighted connected undirected graph without negative cycles. Further, let $r \in V$ be a vertex in G. An *undirected generalized spanning tree* (GST) $G' = (V, E^{\tau}_{[1..4]}, r, \mu)$ of G is a GT with the kernel $\text{kern}(G') = (V, E^{\tau}_{[1]}, r, \nu)$ as a spanning tree $T = \text{kern}(G')$ of G such that $E = \cup_{i=1}^{4} E^{\tau}_{[i]}$ and ν is the restriction of μ to $E^{\tau}_{[1]}$. We say that G' *is spanned over* G *by means of* T *starting from* r. G is called the *underlying graph* of G'.

From this perspective, a GT simply denotes a sort of "partitioning" of the set of edges of its underlying graph G: it neither contains more nor less, but exactly the same number of edges as G. The sole but informative exception is that edges in GTs are *typed* according to the structural classes distinguished by Definition 8.3. What these types actually denote depends on the application area in which GTs are empirically observed. Moreover, a GT is determined by the choice of a spanning tree of the underlying graph: once this kernel is specified together with the root of the GT to be built, its peripheral edges are uniquely determined. Therefore, as long as we do not select a subset of edges but retain the complete edge set of an input graph when deriving a GT from it, we have to focus on the choice of the kernel tree in order to specify subclasses of the class of GTs. From a formal point of view, this is the central theme of this chapter: we ask about the impact of this choice on typing peripheral edges and shed light on their semantics as a result of this choice. This semantics is in turn our bridge to empirical systems that, because of their structural constraints, impose different semantics of kernel and peripheral edges. This will be specified in the remainder of the chapter.

Before we proceed introducing subclasses of the class of GTs, we first ask about the time complexity of computing the sort of edge typing induced by them. This is answered by the proof of the following theorem.

Theorem 8.1 *Given a connected graph $G = (V, E, \mu)$ without negative cycles, a vertex $r \in V$ and a spanning tree $T = (V, E', \nu)$ of G, the time complexity of computing the GST $G' = (V, E^{\tau}_{1..4}, r, \mu)$ spanned over G by means of T starting from r is in the order of $\mathcal{O}(|V| + |E|)$.*

Proof. First, we observe that solving this task basically demands distinguishing between vertical and lateral edges. The reason is that while kernel edges are identified by their membership to T, reflexive edges are distinguished by the fact that they contain the same vertex twice. Because of the definition of lateral edges, this further means that we have to decide whether a given edge $e \in (E^{\tau}_{[1..4]} \setminus E^{\tau}_{[1]}) \setminus E^{\tau}_{[3]}$ is a vertical edge. This decision is computed by Algorithm 8.1, whose time complexity can be estimated as follows:

- Line 3 computes a vector of all paths starting from r and ending at some vertex $v \in V$. We suppose that all vertices are indexed consecutively (by the labeling function $\mathcal{L}_V$ – see Definition 8.1) so that any path P_{rv} can be accessed in $\boldsymbol{x}$ by v's index. In this way, generating $\boldsymbol{x}$ can be carried out by a breadth-first search of order $\mathcal{O}(|V| + |E|) = \mathcal{O}(|V| + |V| - 1) = \mathcal{O}(|V|)$.
- Line 8 denotes an index-based access operation that, in the case of a vector, is of constant complexity [59].
- Line 9 denotes a search operation that is also of constant complexity if paths are represented as vectors of length $|V|$. That is, $\boldsymbol{x}[v]$ is a Boolean vector such that for any $w \in V$: $\boldsymbol{x}[v][w] = 1 \Leftrightarrow w \in P_{rv}$; note that we only check for membership of vertices in paths.
- Line 4 requires the repetition of lines 5–14 exactly $|E| - |E'|$ times, which because of the constant complexity of the latter operations is in the order of $\mathcal{O}(|E| - |E'|) = \mathcal{O}(|E|)$.

Thus we get $\mathcal{O}(|V|+|E|)$ as the desired upper bound. Of course, more efficient algorithms can be designed, but they are not of interest in this chapter as Algorithm 8.1 is already sufficiently efficient.

Algorithm 8.1 Spanning Peripheral Edges

Require: A graph $G = (V, E, \mu)$, a spanning tree $T = (V, E', \nu)$ of G, and a vertex $r \in V$ according to Definition 8.5.

Ensure: The set $E^\tau_{[2]}$ of vertical, the set $E^\tau_{[3]}$ of reflexive, and the set $E^\tau_{[4]}$ of lateral edges of the GST G' spanned over G by means of T starting from r.

1: **procedure** SPANNINGPERIPHERALEDGES(G, T, r)
2:     $E^\tau_{[1]} \leftarrow E'$; $E^\tau_{[2]} \leftarrow E^\tau_{[3]} \leftarrow E^\tau_{[4]} \leftarrow \emptyset$
3:     $x \leftarrow$ VECTOROFALLPATHSINTREESTARTINGFROMROOT(T, r)
4:     **for** $e = \{v, w\} \in E \setminus E^\tau_{[1]}$ **do**
5:         **if** $v = w$ **then**
6:             $E^\tau_{[3]} \leftarrow E^\tau_{[3]} \cup \{e\}$
7:         **else**
8:             $v \leftarrow x[v] \wedge w \leftarrow x[w]$
9:             **if** $v[w] \vee w[v]$ **then**
10:                 $E^\tau_{[2]} \leftarrow E^\tau_{[2]} \cup \{e\}$
11:             **else**
12:                 $E^\tau_{[4]} \leftarrow E^\tau_{[4]} \cup \{e\}$
13:             **end if**
14:         **end if**
15:     **end for**
16:     **return** $E^\tau_{[2..4]}$
17: **end procedure**

Remark. Utilizing Algorithm 8.1 the computation of GSTs is divided into two parts: firstly, computing the kernel spanning tree T and, secondly, typing the remainder of lateral, reflexive, and vertical edges. Below we will reuse this greedy approach in order to estimate the time complexity of generating GTs whose kernel trees meet specific constraints.

8.2.3
Minimum Spanning Generalized Trees

Based on the notion of a GST and on the fact that the number of these GTs derivable from a connected graph G is a simple function of the number of its spanning trees, we can pose a more interesting question: *Which GTs among all possible spanning GTs of a connected graph G meet which structural constraints?* This question goes beyond a purely mathematical endeavor as it bridges the area of empirical, that is, semiotic, systems on the one hand and mathematical systems on the other. The reason is that interesting structural constraints are those for which there are relevant semiotic or information-theoretic interpretations. From this point of view not all but only a subset of GTs is worth being considered theoretically thereby stigmatizing the remainder of GTs as semiotically irrelevant. A subclass of GTs of a graph G that is certainly of higher relevance due to its impact on the flow of information within G is the one whose kernel tree is spanned by the *minimum spanning tree* (MST) [60] of G.[2)] This notion relates to the class of GTs G' of a graph $G = (V, E, \mu)$ whose kernel tree minimizes the cost of edge transitions among all candidate spanning trees of G while every peripheral edge ending at some vertices $v, w \in V$ is at least as costly as the kernel edges of G' ending at least at one of these vertices. This notion is captured by the following definition.

Definition 8.6 (Minimum Spanning Generalized Tree) Let $G = (V, E, \mu)$ be a weighted connected graph, $T = (V, E', \nu)$ a MST of G and $r \in V$. The *minimum spanning generalized tree* (MSGT) induced by T is a GT $G' = (V, E^{\tau}_{[1..4]}, r, \mu)$ spanned over G by means of the kernel tree T starting from r.

Corollary 8.1 *Given a graph $G = (V, E, \mu)$ according to Definition 8.6 and a vertex $r \in V$, the MSGT spanned over G starting from r is not necessarily unique.*

This property is a simple consequence of the fact that already the MST of a graph is not necessarily unique, especially if we consider equally weighted multiple edges. In order to secure uniqueness in this case one can proceed as follows:

Definition 8.7 (Minimum Spanning Generalized Tree Revisited)
Let $\mathrm{mst}(G)$ be the set of all MSTs of the weighted connected labeled graph $G = (V, E, \mathcal{L}_V, \mathcal{L}_E, \mu)$ according to Definition 8.1. Then, we define the order relation $\leq_{\mathrm{kern}} \subseteq \mathrm{mst}(G)^2$ such that $\forall T' = (V, E', \nu'), T'' = (V, E'', \nu'') \in \mathrm{mst}(G)$: $T' \leq_{\mathrm{kern}} T'' \Leftrightarrow T' = T'' \vee \sum_{e \in E'} \mathcal{L}_E(e) < \sum_{e \in E''} \mathcal{L}_E(e)$. For a given vertex $r \in V$, the $\leq_{\mathrm{kern}}$*-induced MSGT* of G is a GT $G' = (V, E^{\tau}_{[1..4]}, r, \mu)$ spanned over G by means of $\inf_{\leq_{\mathrm{kern}}} \mathrm{mst}(G)$ starting from r.

This definition shows a way to derive uniquely defined MSGTs wherever needed. As mentioned above, the derivation of a GT from a graph G generally includes two steps: first, generating the spanning tree with its kernel edges

2) Note that the MST of a connected graph G is not necessarily unique. However, utilizing the order relation $\leq_a$ of vertices and edges (Definition 8.1) we can uniquely determine one of these equally weighted spanning trees of minimum weight. This approach can be followed whenever uniqueness is desired with respect to the mathematical constructs introduced later. For an example see Definition 8.7 (below).

and, second, typing the remaining edges of G. Following this approach, the time complexity of generating an MSGT is bound by the sum of the complexity of generating its kernel tree and that of typing its peripheral edges. This idea is reflected by the following corollary. (Note that typing peripheral edges may be done simultaneously with spanning kernel edges so that there are certainly lower-complexity bounds than the one mentioned in the following corollary.)

Corollary 8.2 *Because of Theorem 8.1 the time complexity of generating an MSGT is in the order of* $\mathcal{O}(|V| + |E| + |E| \log |V|)$ *when using a standard algorithm [60] to generate the kernel MST of the MSGT. It reduces to* $\mathcal{O}(|V| + |E| + |E|\alpha(|E|, |V|))$ *when using the algorithm of [12], where* α *is the classical functional inverse of Ackermann's function.*

The following theorem presents a first important statement about the semantics of peripheral edges – in this case as a result of using a MST as the kernel of a GT.

Theorem 8.2 *Let* $G' = (V, E^{\tau}_{[1..4]}, r, \mu)$ *be an MSGT spanned over* $G = (V, E, \mu)$ *by means of the MST* $T = (V, E', \nu)$ *starting from* r*. Then,* $\forall e \in E^{\tau}_{[1..4]} \setminus E'\ \forall f \in E'$: $e \cap f \neq \emptyset \Rightarrow \mu(f) \leq \mu(e)$.

Proof. Let $e = \{v, w\} \in E^{\tau}_{[1..4]} \setminus E'$. Let further $P_{rv} = (r, \ldots, v', f, v)$ and $P_{rw} = (r, \ldots, w', g, w)$ be the unique paths in T from r to $v \in V$ and $w \in V$, respectively. Then, we have to distinguish two cases:

- *Case A (vertical edges):* w is a vertex on P_{rv} or v is a vertex on P_{rw}. In this case we conclude as follows. Without loss of generality we assume that v is a vertex on P_{rw}. Then we can construct a tree $T' = (V, (E' \setminus \{g\}) \cup \{e\})$ such that $\mu(T') < \mu(T)$ – T and T' differ by a single edge. This contradicts the status of T as a MST of G. Note that changing f by e would disconnect T in the present case where we assume that $v \in P_{rw}$.
- *Case B (lateral edges):* w is not a vertex on P_{rv}, nor is v a vertex on P_{rw}. That is, v and w belong to different branches of T. In this case we conclude as follows: If $\mu(e) < \mu(f)$, we can construct a tree $T' = (V, (E' \setminus \{f\}) \cup \{e\})$ such that $\mu(T') < \mu(T)$, once more in contrast to the status of T as a MST of G. Analogously, we conclude when changing g by e.

Remark. Theorem 8.2 separates MSGTs from ordinary graphs G and the remaining GTs derived from G as they distinguish peripheral from kernel edges in terms of μ. In MSGTs, the function of peripheral edges is separated from that of kernel edges: information flow along the former is more costly than along the latter. In spite of this information, MSGTs do not distinguish vertical from lateral links: Irrespective of its status as a vertical link or as a lateral link, an edge may contribute to paths shorter than that on the kernel MST. (Note that MSTs are *not* shortest path trees.) In this sense, MSGTs do not explore the full informational capacity of GTs. Below we will introduce generalized shortest path trees, which additionally provide this latter information value.

Example *Let the graph* $A = (V, E, \mu)$ *be given as shown in Figure 8.4. In this case, we can derive the MSGT* $B = (V, E^{\tau}_{[1..4]}, 0, \mu)$ *with the kernel MST* $T = (V, \{\{0,4\}, \{4,5\}, \{2,4\}, \{1,2\}, \{2,6\}, \{1,3\}\})$ *and the following subsets of edges:* $E^{\tau}_{[2]} = \{\{0,5\}, \{0,3\}\}$, $E^{\tau}_{[3]} = E^{\tau}_{[4]} = \emptyset$.

Now we are in a position to repeat the basic principle of generating GTs as follows. Starting from an underlying graph we select a GT whose kernel and periphery meet certain structural constraints. By specifying these constraints we get more and more informative GTs. This principle is followed in the subsequent sections.

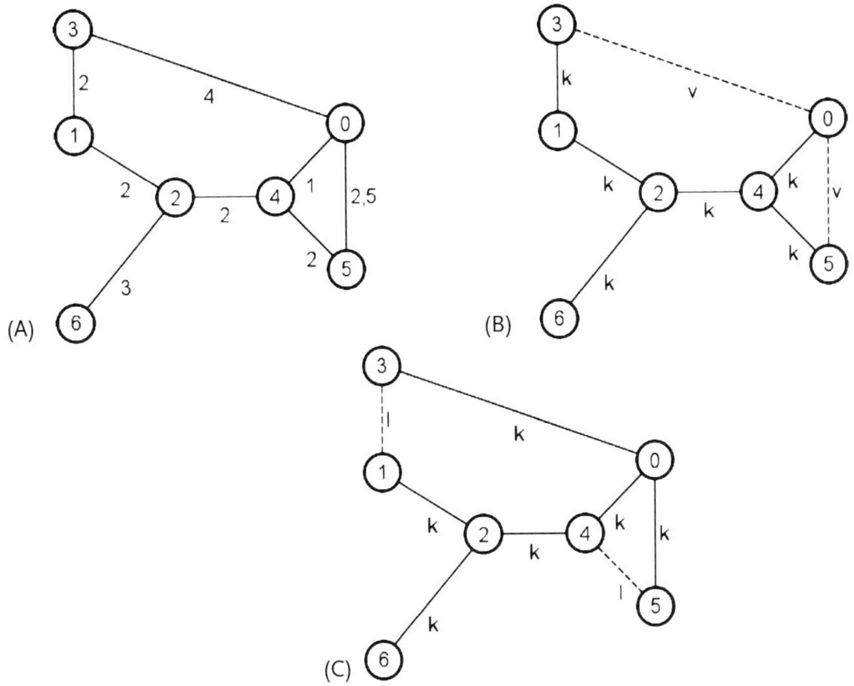

Figure 8.4 A graph A together with an MSGT (cf. graph B) rooted in vertex 0 and a GSPT (cf. graph C) also rooted in 0 derived from A. For reasons of simplification, edge weights are omitted in the graphical representations of graphs B and C.

8.2.4
Generalized Shortest Path Trees

It is Theorem 8.2 that motivates speaking about a *minimum* spanning GT. It divides an underlying graph $G = (V, E, \mu)$ into the set of *low-price* (i.e., kernel) and *high-cost* (i.e., peripheral) edges. In this sense, MSGTs share the restricted view of a graph G with its MSTs but additionally retain complete information about the topology of G. Note that for different vertices $v, w \in V$ used to root the kernel MST of G, E is separated differently into vertical and lateral edges (while the set of kernel edges remains the same if the MST is uniquely defined). Obviously, this is a very low degree of context sensitivity induced by the choice of a root that only affects the typing of edges while leaving the kernel tree untouched. In order to exceed this lower bound of context sensitivity, we can think of GTs whose kernel varies with the choice of the root, that is, GTs whose construction is context sensitive to the root being chosen. A candidate instance of this notion is given by a generalized shortest path trees, which – by analogy to Definition 8.6 – are defined as follows.

Definition 8.8 (Generalized Shortest Path Tree) Let $G = (V, E, \mu)$ be a weighted connected undirected graph without negative cycles, $r \in V$ a vertex, and $T_r = (V, E', r, \nu)$ the *shortest path tree* (SPT) of G rooted in r. The *generalized shortest path tree* (GSPT) induced by T_r is a GT $G' = (V, E^{\tau}_{[1..4]}, r, \mu)$ spanned over G by means of T_r starting from r.

Corollary 8.3 *Given a graph $G = (V, E, \mu)$ according to Definition 8.8 and a vertex $r \in V$, the GSPT induced by T_r is not necessarily unique.*

Once more, this property simply follows from the fact that the underlying graphs of GTs may contain equally weighted multiple edges. As before, we can utilize $\leq_E$ to provide uniqueness. By analogy to Corollary 8.2 we get the following corollary.

Corollary 8.4 *According to Theorem 8.1 the time complexity of generating a GSPT is in the order of $\mathcal{O}(|V| + |E| + |V|^2) = \mathcal{O}(|V|^2)$ when using Dijkstra's algorithm [20] to solve the single-source problem of computing the shortest paths. It reduces to $\mathcal{O}((|V| + |E|) \log |V|) = \mathcal{O}(|E| \log |V|)$ when operating on sparse graphs (for which $|E| \ll |V|^2$).*

Remark. The notion of a GSPT is reminiscent of the notion of a distance-function-based GT as introduced by [24]. The difference is that Emmert-Streib and Dehmer [24] use the geodesic distance of vertices from the root of a GT to map equally distant vertices onto the same level. In contrast to this, we start from a SPT in order to get a kernel tree disregarding any graph levels. It seems that our notion is more general than that of Emmert-Streib and Dehmer [24] as it does not refer to structural constraints hardly observable in empirical systems, that is, the questionable existence of graph levels (see above). Nevertheless, both notions pave the way for more complex kernel trees of GTs whose construction is restricted by observable constraints of natural systems (see Sections 8.2.8 and 8.2.9).

Example *Let graph A in Figure 8.4 be given as a starting point. Then, graph C in Figure 8.4 is a GT induced by G with the kernel SPT* $T = (V, \{\{0, 3\}, \{0, 4\}, \{0, 5\}, \{2, 4\}, \{1, 2\}, \{2, 6\}\}, 0)$ *rooted in* 0 *and the following subsets of edges:* $E^{\tau}_{[2]} = E^{\tau}_{[3]} = \emptyset$, $E^{\tau}_{[4]} = \{\{1, 3\}, \{4, 5\}\}$.

Unlike MSGTs, GSPTs are context sensitive not only with respect to the classification of peripheral edges but also with respect to the kernel tree itself, which may vary with the choice of the root of the GSPT. In this way, we get a one-to-many relation: the same underlying graph $G = (V, E, \mu)$ is nontrivially related to many different kernel SPTs (with different edge sets) and, thus, to as many different GSPTs. In the extreme case we get $|V|$ many different kernel trees of GSPTs starting from the underlying graph $G = (V, E, \mu)$. *Nontrivially* means that we disregard multiple edges in this counting (which may induce different GSPTs rooted in the same vertex). In contrast to this, the kernel trees of all MSGTs induced by the various root vertices $v \in V$ have the same edge set as long as the MST of G is unique. Thus, MSTs lack the kind of context sensitivity of GSPTs and, therefore, do not induce the aforementioned one-to-many relation.

By analogy to Theorem 8.2 we now consider Corollary 8.5 and Theorem 8.3 about GSPTs, which together provide a functional separation of vertical and lateral edges in relation to kernel edges (as absent in MSGTs):

Corollary 8.5 *Let* $G' = (V, E^{\tau}_{[1..4]}, r, \mu)$ *be a GSPT spanned over* $G = (V, E, \mu)$ *by means of the SPT* $T_r = (V, E', r)$ *starting from r. Then, for any* $v \in V$ *each path* $P = (r, e_{i_1}, \ldots, e_{i_m}, v)$ *in* G' *ending at r and v is at least as costly as the unique path* P_{rv} *in* T_r*, that is,* $\mu(P) \geq \mu(P_{rv})$.

This corollary is in a sense obvious so that we can skip its proof (it is a simple consequence of the definition of SPTs). Its meaning is to assign vertical and lateral edges marginal roles in relation to kernel edges: starting from the root of a GSPT it is more costly to traverse lateral or vertical edges than following kernel edges. An obvious implication of Corollary 8.5 runs as follows.

Corollary 8.6 *Any vertical edge in a GSPT is at least as costly as the corresponding subpath of the kernel SPT, which is cut short by this vertical edge.*

So far, we have distinguished lateral and vertical from kernel edges. This does not really make it beyond the notion of an MSGT. Thus, we have to additionally ask: *How can we further separate the function of lateral edges from that of vertical edges in GSPTs?* This question is answered by the proof of the following theorem.

Theorem 8.3 *Let $G' = (V, E^{\tau}_{[1..4]}, r, \mu)$ be a GSPT spanned over $G = (V, E, \mu)$ by means of the SPT $T_r = (V, E', r)$ starting from r. Then, the shortest path $GP_{\mu'}(v, w)$ between any pair of vertices v, w in $T' = (V, E^{\tau}_{[1..3]}, r, \mu')$ is a path in T_r – μ' is the restriction of μ to $E^{\tau}_{[1..3]}$. In other words: apart from lateral edges $e \in E^{\tau}_{[4]}$, the shortest paths in G' contain only kernel edges.*

Proof. For $v, w \in V$, P_{rv}, P_{rw} are the shortest paths in T_r ending at r as well as v and w, respectively. Now we have to consider two cases:

- *Case A*: $v \in V(P_{rw})$: In this case we conclude that the subpath $P = (v, e_{i_1}, \ldots, e_{i_m}, w)$ of the shortest path P_{rw} from r to w is the shortest path between v and w. Otherwise, if there is at least one vertical edge e between two vertices $x, y \in V(P)$ such that $\mu((v, e_{i_1}, \ldots, x, e, y, \ldots, e_{i_m}, w)) < \mu(P)$, then P_{rw}, including the subpath P is not the shortest path between r and w – in contrast to the definition of SPTs. $w \in V(P_{rv})$ is a mirror case.
- *Case B*: $v \notin V(P_{rw}) \wedge w \notin V(P_{rv})$, that is, v and w belong to different branches of T_r rooted in r. In this case, there is a unique *least common predecessor* u of v and w in T_r such that $u \in V(P_{rv})$, $u \in V(P_{rw})$, and for any other vertex $x \neq u \in V$ satisfying the same conditions it holds that $x \in P_{ru}$. Now we conclude that $P_1 = (v, e_{i_1}, \ldots, e_{i_m}, u)$ is the shortest path in T_r from v to u and $P_2 = (u, e_{j_1}, \ldots, e_{j_n}, w)$ the shortest path in T_r from u to w. Further, by *case A* we know that neither P_1 nor P_2 is shortened by including any vertical edge. Thus, $P_1 \circ P_2 = (v, e_{i_1}, \ldots, e_{i_m}, u, e_{i_{j_1}}, \ldots, e_{i_{j_n}}, w) = GP_{\mu'}(v, w)$ in T'.

According to this theorem, vertical edges do not shorten any shortest path in the kernel tree of a GSPT. However, things look different if lateral edges are taken into account that may cut short the paths in T_r. For example, in the GSPT in Figure 8.4 (see graph *C*), the shortest path between vertex 4 and 5 is spanned by a single lateral edge. Together with Theorem 8.3 this observation assigns vertical and lateral edges quite different roles *in GSPTs* so that this class of GTs is more informative about their peripheral edges than MSGTs:

- *Vertical edges* do not shorten any path of the kernel tree of a GSPT. Their role is rather to provide aggregations of such paths at the expense of a more costly transition *of* or less efficient information flow *within* the kernel tree. From the point of view of social taxonomies one can think of vertical edges as condensations in terms of [30]: In a concept hierarchy induced by hypernymy relations, vertical edges provide shortcuts by relating specific to general terms, thereby bypassing (i.e., aggregating or condensing) intermediary hypernyms.[3)]

- Lateral edges do not have this role in GSPTs. Their function changes between genuine *shortcuts* on the one hand – as they enable faster information flow or less costly graph transitions – and *cross references* on the other (which realize more expensive shifts in direction). GSPTs are underspecified with respect to this distinction. Thus, we need a more informative notion of a GT that goes beyond GSPTs; this extension is introduced in Section 8.2.5.

At this point we come back to one of the central themes of the present chapter, that is, the context-sensitive formation of GTs spanned over a certain semiotic network. As seen above, MSGTs are less sensitive to the selection of the root of a GT than GSPTs. From a formal point of view, this difference is manifested by Corollary 8.2 in contrast to Corollary 8.5, 8.6 and Theorem 8.3. From an empirical point of view, the choice between MSGTs, GSPTs, or even more restricted GTs depends on the characteristics of the natural system under consideration to be retained by its formal model. This can be formulated by means of a criterion as follows: *Whenever we observe a sensitivity of having an overview of a given network subject to adopting an initial position in that network, we have to prefer GSPTs to MSGTs. Otherwise, we can rely on MSGTs that provide invariant kernel trees irrespective of our initial (root) position in the network.* Thus, we can formulate our question about the preferred model of a GT pointed by asking: *Are there empirical systems in which rooting is decisive at least in the sense reflected by GSPTs?* This, of course, holds for all cognitive processes based on priming and spreading activation [49, 54, 62]. In this sense, GSPTs are one step toward an empirically well-observable concept of structure formation.

8.2.5
Shortest Paths Generalized Trees

In spite of the latter considerations, we may complain that, unlike MSTs, which realize selections of subsets of edges of the corresponding input graph G, MSGTs and GSPTs always include all edges of G. That is, according to Definitions 8.3 and 8.5, GTs induce classifications of G's edges and, thus,

3) Think, for example, of a situation in which a certain concept, say *small car*, is classified (by a group of interlocutors) nearly as often as a *vehicle* or as a *car*. In this case, the classification by the more general noun *vehicle* bypasses that by *car* without making the latter obsolete.

contain as many edges as G. In order to circumvent this situation, we have at least two alternatives:

- We may start from a kernel MST in order to span the periphery of a GT by the shortest paths between its vertices. This notion preserves information about the cheapest edges (to let the resulting GT be efficiently manageable as a tree) *and* about the shortest paths (as the cheapest graph-like skeleton of the underlying graph) while it disregards the remaining edges.
- Alternatively, we may generalize the notion of a GSPT by SPTs itself. That is, as in the latter case we span the periphery of a GT solely by means of shortest paths, but now around a kernel SPT so that the resulting GT consists solely of SPTs – it declares a single SPT as its kernel while the remaining SPTs span its periphery.

It is the latter notion that is of interest here: it combines the context sensitivity of GSPTs with a finer-grained semantics of peripheral edges, finer than in the case of MSGTs. This combination is grasped by the following definition.

Definition 8.9 (Shortest Paths Generalized Tree) Let $G = (V, E, \mu)$ be a weighted connected undirected graph without negative cycles, $r \in V$ any vertex, and $T_r = (V, E', r, \nu)$ the SPT of G rooted in r (as usual, ν is the restriction of μ to E'). The *shortest paths generalized tree* (SPGT) $G' = (V, E^{\tau}_{[1..4]}, r, \mu')$ derived from G is a GSPT spanned over the simple graph $G'' = (V, E'', \mu')$ by means of T_r starting from r such that

$$E'' = \bigcup_{v \in V, T_v = (V, E_v, v, \mu_v)} E_v \,,$$

where E'' is the set of all edges belonging to any SPT of any vertex of G, T_v is the SPT induced by $v \in V$ in G, and μ' is the restriction of μ to $E^{\tau}_{[1..4]}$.

Remark. We assume that G'' is a simple graph and therefore contains neither loops nor multiple edges. Otherwise, the resulting SPGT of a graph would contain *more* edges than its underlying graph (because of the union that is in use in Definition 8.9).

Based on this definition we can easily prove the following corollary.

Corollary 8.7 *Let $G' = (V, E^{\tau'}_{[1..4]}, r, \mu)$ be a GSPT spanned over $G = (V, E, \mu)$ by means of the SPT $T_r = (V, E', r, \nu)$ starting from $r \in V$. Then, the SPGT $G'' = (V, E^{\tau}_{[1..4]}, r, \mu'')$ derived from G by means of T_r and r satisfies the following equalities and one inequality:*

1. $E^{\tau}_{[2]} = \emptyset$,

2. $E^{\tau}_{[3]} = \emptyset$,
3. $E^{\tau}_{[4]} \subseteq E^{\tau'}_{[4]}$,
4. $\forall e = \{v, w\} \in E^{\tau}_{[4]} : \mu(e) \leq \mu(P_{vw})$.

   Note that in case 4 P_{vw} denotes the unique path in the kernel tree of G' ending at v and w (Definition 8.1).

Proof. Case 1 is a consequence of Theorem 8.3, case 2 is a consequence of the fact that shortest paths are always simple, while cases 3 and 4 are simple consequences of the way SPGTs are defined, that is, for generating a shortest path between two vertices v, w in G'' a lateral edge $\{x, y\}$ is added to $E^{\tau}_{[4]}$ if and only if it shortens the path P_{xy} as a subpath of P_{vw}.

Remark. If we define G'' in Definition 8.9 as a multigraph, then the periphery of the SPGT derived from it is always connected. The reason is that in this case all kernel edges are duplicated as often (in the form of peripheral edges) as there are different SPTs of G to which they belong. This characteristic of "peripheral connectivity" is not necessarily provided by any of the concurrent notions of GTs introduced so far. However, such a highly connected GT would contain more edges than its underlying graph. In the present stage of modeling this characteristic is not desirable. It may be the starting point for future extensions of the notion of a GT.

In Section 8.2.6 we utilize the notion of an SPGT in order to take the next step in specifying a functional semantics of edges in GTs. As will be shown, this is accompanied by an extension of the set of edge types used so far.

8.2.6 Generalized Shortest Paths Trees

Based on the notion of an SPGT we can now define GTs in which shortcuts are provided by a separate, more specific edge type. This is done by means of the notion of a generalized shortest path<u>s</u> tree:

Definition 8.10 (Generalized Shortest Paths Tree) Let $G = (V, E, \mu)$ be a weighted connected undirected graph without negative cycles, $T_r = (V, E', r, \nu)$ the SPT of G rooted in $r \in V$, $G' = (V, E^{\tau'}_{[1..4]}, r, \mu')$ the SPGT, and $G'' = (V, E^{\tau''}_{[1..4]}, r, \mu)$ the GSPT both derived from G by means of T_r and starting from r. The *generalized shortest path<u>s</u> tree* (GPST) $G''' = (V, E^{\tau}_{[1..5]}, r, \mu)$ is derived from G' and G'' by refining the edge typing functions τ' and τ'' in terms of $\tau : E \rightarrow \{\mathrm{c}, \mathrm{k}, \mathrm{r}, \mathrm{s}, \mathrm{v}\} = \mathcal{T}$ such that

$$\forall e \in E: \begin{cases} \tau''(e) \in \{\mathrm{k}, \mathrm{r}, \mathrm{v}\} \Rightarrow \tau(e) = \tau''(e) & \\ \tau(e) = \mathrm{c} \Rightarrow e \in E^{\tau''}_{[4]} \setminus E^{\tau'}_{[4]} & \textit{(cross-reference edges)} \\ \tau(e) = \mathrm{s} \Rightarrow e \in E^{\tau'}_{[4]} & \textit{(shortcut edges)} \end{cases}$$

Further, we set $E^{\tau}_{[1]} = \{e \in E \mid \tau(e) = \mathrm{k}\} = E^{\tau'}_{[1]} = E^{\tau''}_{[1]} = E'$, $E^{\tau}_{[2]} = \{e \in E \mid \tau(e) = \mathrm{v}\} = E^{\tau''}_{[2]}$, $E^{\tau}_{[3]} = \{e \in E \mid \tau(e) = \mathrm{r}\} = E^{\tau''}_{[3]}$, $E^{\tau}_{[4]} = \{e \in E \mid \tau(e) = \mathrm{c}\}$, $E^{\tau}_{[5]} = \{e \in E \mid \tau(e) = \mathrm{s}\} = E^{\tau'}_{[4]}$, where $E^{\tau}_{[4]} \cup E^{\tau}_{[5]} = E^{\tau''}_{[4]}$.

Corollary 8.8 *Apart from shortcut edges $e \in E^{\tau}_{[5]}$, the shortest paths in a GPST G contain only kernel edges.*

This corollary is a simple consequence of Theorem 8.3 and the fact that GPSTs basically induce a partition of lateral edges into the subset of shortcut edges $e \in E^{\tau}_{[5]}$, which contribute to the shortest paths, and the subset of cross-reference edges $e \in E^{\tau}_{[4]}$, which do not. Based on Definition 8.10 and the latter corollary we can now establish a fine-grained semantics of lateral edges in addition to that of vertical edges induced by Theorem 8.3:

- *Short cut edges:* In order to span the shortest paths among vertices subject to the topology of the underlying graph G, an SPGT G' selects a subset of lateral edges. Lateral edges in the GSPT G'' of G with the same kernel as G' that do not shorten paths in the latter sense are excluded from G'. Thus, in SPGTs lateral edges are genuine shortcuts: they are the only means to establish shortest paths apart from the corresponding kernel tree. Think of such edges, for example, as shortcuts in small-world graphs [68] that are used to establish a high cluster value within a network. In contrast to this, vertical edges realize more costly aggregations or, in terms of semiotics, conceptual condensations along chains of hypernymy relations.
- *Cross-reference edges:* Compared with shortcut and vertical edges, cross-reference edges neither aggregate nor shorten any (e.g., conceptual) relation in any other way. They rather serve as *transverse* edges that bridge weakly related regions of a GT. That is, cross-reference edges build shortcuts in a broader sense as they bridge more distant vertices of the underlying graph G. In terms of small-world graphs, cross-reference edges correspond to somehow randomly rewired (and therefore costly) edges. They provide short average geodesic distances and, thus, connectivity even among less related vertices as a precondition of efficient information flow within the network [68] – however, at the price of a loss in coherence among the vertices linked in such a manner.

Now we are in a position to directly interpret the semantics of kernel and peripheral edges in terms of real semiotic systems. Assume, for example, that we model a social (terminological [55]) ontology [56] as spanned by the category system of Wikipedia. In this case, vertical edges can be used to model transitivity among hypernymy relations, while lateral links are a means to map co-classifications or polymorphic categorizations. Obviously, the role of vertical and lateral edges is quite different in this example, so that this distinction should be reflected in the construction of a GT spanning the social ontology. From a *functional* point of view, this distinction can be emphasized as follows:

- *Searching*: In order to efficiently search or walk through a network, kernel edges are the first choice.
- *Changing*: If in contrast to the latter search function agents aim at non-randomly changing the current standpoint (view or topic) in the course of their network transitions, they should select lateral or, more specifically, shortcut edges as long as they are available. A more random walk through the network is instead of this supported by following cross-reference links.
- *Abridging*: Finally, in order to cut short the traversal of the kernel hierarchy, vertical edges are the primary means.

This functional scenario gives a complete and distinguished semantics of the different types of edges in GTs and, therefore, motivates the introduction of this kind of tree-like graph in between complete order – as manifested by trees – and randomness – as manifested by random graphs. That is, unlike the literature about GTs and complex networks introduced so far, we are now in a position where we can assign edges a certain role as a function of their contribution to the topology of the network in which they are spanned. According to the definition of GPSTs, this system of potential roles of edges distinguishes five types. Obviously, this goes much beyond present-day approaches to complex networks in which edges are normally neither labeled nor typed.

Let us now consider an example that exemplifies the notion of a GPST.

Example *Let the graph* $A = (V, E, \mu)$ *in Figure 8.5 be given. Then, the graph* $B = (V, E^{\tau_B}_{[1..4]}, 0, \mu_B)$ *is an SPGT spanned over A by means of the kernel SPT* $T_0 = (V, \{\{0,3\}, \{0,4\}, \{0,5\}, \{2,4\}, \{1,2\}, \{2,6\}\}, 0)$ *and the set of shortest-path-inducing lateral edges* $E^{\tau_B}_{[4]} = \{\{1,3\}, \{4,5\}\}$*. Next, the graph* $C = (V, E^{\tau_C}_{[1..5]}, 0, \mu_C)$ *is a GPST with the following sequence of edge sets:* $E^{\tau_C}_{[1]} = E^{\tau_B}_{[1]}$, $E^{\tau_C}_{[2]} = \{\{4,6\}\}$, $E^{\tau_C}_{[3]} = \emptyset$, $E^{\tau_C}_{[4]} = \{\{2,5\}\}$, $E^{\tau_C}_{[5]} = E^{\tau_B}_{[4]}$.

So far, we have introduced GTs by combining the context-sensitive formation of kernel trees with an increasingly constrained semantics of peripheral edges. This approach goes beyond existing efforts to use GTs as a graph model in between tree-like structures and unconstrained graphs. The reason is that it does not just demarcate generalized from ordinary graphs by means of typed edges. These types are also justified in functional terms that are missing in general graphs. Section 8.2.8 shows that this extension is a prerequisite of a graph model of a certain class of semiotic systems. But before introducing this model we extend our approach by *orientating* GTs.

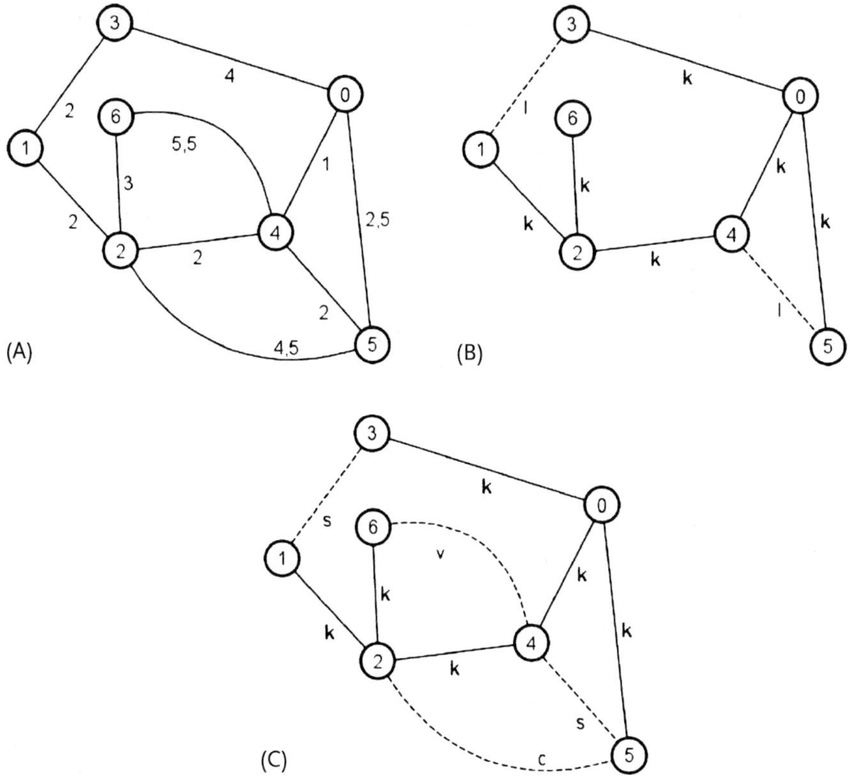

Figure 8.5 A graph A together with its SPGT (graph B) rooted in vertex 0 and the corresponding GPST (graph C). For reasons of simplification, edge weights are omitted in the graphical representations of graph B and C.

8.2.7
Accounting for Orientation: Directed Generalized Trees

So far we have considered only undirected graphs. As is well known from graph theory, things look quite different when dealing with oriented graphs. The MST of a directed graph, for example, cannot be computed in the same way as its undirected counterpart [60]. In the remainder of this section we provide an orientation of GTs. However, we concentrate on complexity statements leaving the proofs of many counterparts of the theorems presented above for future work.

Definition 8.11 (Directed Generalized Tree) Let $T = (V, A', r)$ be a directed tree rooted in $r \in V$. Further, let $P_{rv} = (v_{i_0}, a_{j_1}, v_{i_1}, \ldots, v_{i_{n-1}}, a_{j_n}, v_{i_n}), v_{i_0} = r, v_{i_n} = v, a_{j_k} \in A', \text{in}(a_{j_k}) = v_{i_{k-1}}, \text{out}(a_{j_k}) = v_{i_k}, 1 \leq k \leq n$, be the unique path in T from r to v and $V(P_{rv}) = \{v_{i_0}, \ldots, v_{i_n}\}$ the set of all vertices of P_{rv}. A *directed* *generalized tree* (DGT)

$$G = (V, A, \tau, r)$$

induced by T is a pseudograph (i.e., a multigraph possibly with multiple and parallel arcs or loops) whose arcs are typed by the function $\tau : A \rightarrow \{\mathrm{d}, \mathrm{k}, \mathrm{l}, \mathrm{r}, \mathrm{u}\}$ as follows:

$$\forall a \in A: \begin{cases} \tau(a) = \mathrm{k} & \Rightarrow \quad a \in A_{\mathrm{k}} = A' \\ & \quad \textit{(kernel arcs)} \\ \tau(a) = \mathrm{u} & \Rightarrow \quad a \in A_{\mathrm{u}} = \{a \mid \mathrm{in}(a) = v \in V \\ & \quad \wedge\ \mathrm{out}(a) = w \in V(P_{rv}) \setminus \{v\}\} \\ & \quad \textit{(upward arcs)} \\ \tau(a) = \mathrm{d} & \Rightarrow \quad a \in A_{\mathrm{d}} = \{a \mid \mathrm{in}(a) = w \in V(P_{rv}) \setminus \{v\} \\ & \quad \wedge\ \mathrm{out}(a) = v \in V\} \\ & \quad \textit{(downward arcs)} \\ \tau(a) = \mathrm{r} & \Rightarrow \quad a \in A_{\mathrm{r}} = \{a \mid \mathrm{in}(a) = \mathrm{out}(a) = v \in V\} \\ & \quad \textit{(reflexive arcs)} \\ \tau(a) = \mathrm{l} & \Rightarrow \quad a \in V^2 \setminus (A_{\mathrm{k}} \cup A_{\mathrm{u}} \cup A_{\mathrm{d}} \cup A_{\mathrm{r}}) \\ & \quad \textit{(lateral arcs)} \end{cases}$$

such that $A^{\tau}_{[1]} = \{a \in A \mid \tau(a) = \mathrm{k}\}, A^{\tau}_{[2]} = \{a \in A \mid \tau(a) = \mathrm{u}\}, A^{\tau}_{[3]} = \{a \in A \mid \tau(a) = \mathrm{d}\}, A^{\tau}_{[4]} = \{a \in A \mid \tau(a) = \mathrm{r}\}, A^{\tau}_{[5]} = \{a \in A \mid \tau(a) = \mathrm{l}\}$ is a partition of A such that $A = \cup_{i=1}^{5} A^{\tau}_{[i]}$ and $\forall 1 \leq i < j \leq 5 : A^{\tau}_{[i]} \cap A^{\tau}_{[j]} = \emptyset$. Because of the interdependence of τ and the latter partition, we alternatively denote G by $(V, A^{\tau}_{[1..5]}, r)$, where $a \in A^{\tau}_{[1..5]}$ iff $a \in \cup_{i=1}^{5} A^{\tau}_{[i]}$. In other words, directed GTs G are interchangeably denoted by (V, A, τ, r) and $(V, A^{\tau}_{[1..5]}, r)$. We say that G is *generalized by its lateral, reflexive, upward, and downward arcs.* The directed GT $G = (V, A, \tau, r)$ induces the directed tree $\mathrm{kern}(G) = (V, A^{\tau}_{[1]}, r) = T$ called *kernel (tree)* or *skeleton* of G. Further, the graph $\mathrm{periphery}(G) = (V, \cup_{i=2}^{5} A^{\tau}_{[i]})$ is called *periphery* or *complementary graph* of G. Arcs belonging to periphery(G) are called *peripheral arcs* (complementing the set of kernel arcs). Finally, any GT (V, A, τ, r, μ) with the arc weighting function $\mu : A \rightarrow \mathbb{R}$ is called a *weighted undirected GT.*

Remark. A simple consequence of orientating GTs is the need to distinguish between upward and downward arcs both of which are subsumed under the notion of vertical edges in undirected GTs.

Remark. By analogy to undirected GTs we henceforth assume that μ is a function from A to $\mathbb{R}^{+} \setminus \{0\}$. Once more, the reason is to prevent negative cycles. Further, we only deal with finite graphs.

Definition 8.12 (Directed Generalized Spanning Tree) Let $G = (V, A, \mu)$ be a weighted connected digraph without negative cycles. Further let $r \in V$ be a vertex in G. A *directed generalized spanning tree* (DiGST) $G' = (V, A^{\tau}_{[1..5]}, r, \mu)$ of G is a directed GT whose kernel $\text{kern}(G') = T$ is a directed spanning tree $T = (V, A^{\tau}_{[1]}, r, \nu)$ of G rooted in r such that $A = \cup_{i=1}^{5} A^{\tau}_{[i]}$ and ν is the restriction of μ to $A^{\tau}_{[1]}$. We say that *G' is spanned over G by means of T starting from r*. G is called the *underlying graph* of G'.

Example *Let the digraph A in Figure 8.6 be given. In this case, we have at most six different directed spanning trees (vertex 6 does not root a spanning tree). Starting with graph D in Figure 8.6, we see that it does not induce any DGT since D does not have any directed spanning trees.*

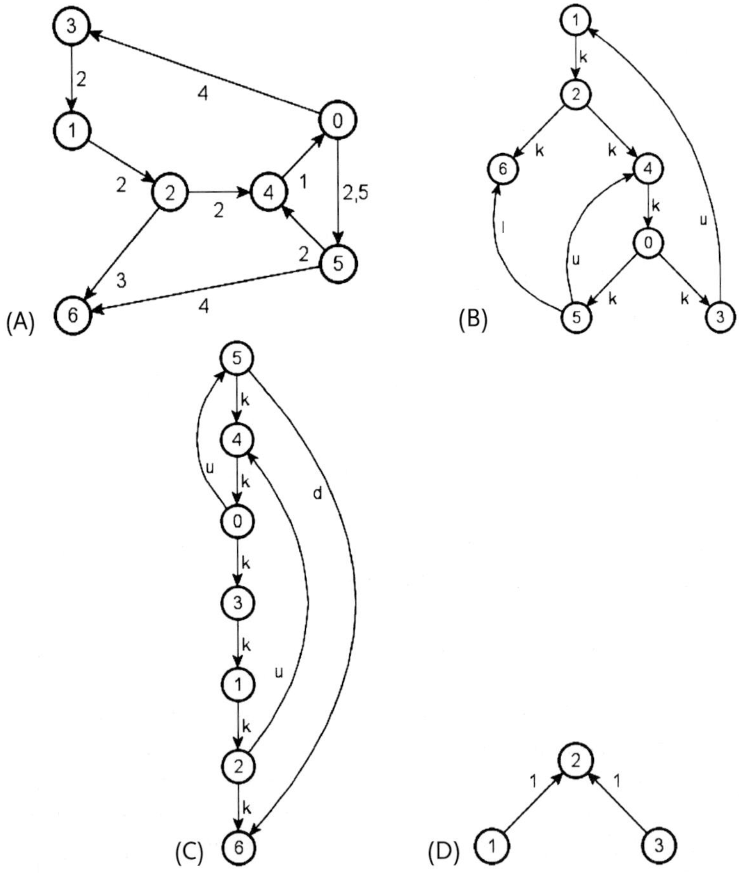

Figure 8.6 A digraph *A* and two directed spanning trees derived from it: rooted in vertex 1 (graph *B*) and alternatively rooted in vertex 5 (graph *C*). For reasons of simplification, edge weights are omitted in graphs *B* and *C*.

8.2 A Class of Tree-Like Graphs and Some of Its Derivatives

Algorithm 8.2 Spanning Peripheral Arcs

Require: A digraph $G = (V, A, \mu)$, a spanning tree $T = (V, A', r, \nu)$ of G, and a vertex $r \in V$ according to Definition 8.12.
Ensure: The set $A^\tau_{[2]}$ of upward, the set $A^\tau_{[3]}$ of downward, the set $A^\tau_{[4]}$ of reflexive, and the set $A^\tau_{[5]}$ of lateral arcs of the DiGST G' spanned over G by means of T starting from r.

1: **procedure** SPANNINGPERIPHERALARCS(G, T, r)
2: $\quad A^\tau_{[1]} \leftarrow A'; A^\tau_{[2]} \leftarrow A^\tau_{[3]} \leftarrow A^\tau_{[4]} \leftarrow A^\tau_{[5]} \leftarrow \emptyset$
3: $\quad x \leftarrow$ VECTOROFALLPATHSINTREESTARTINGFROMROOT(T, r)
4: $\quad$ **for** $a \in A \setminus A^\tau_{[1]}$ **do**
5: $\quad\quad v \leftarrow \text{in}(a), w \leftarrow \text{out}(a)$
6: $\quad\quad$ **if** $v = w$ **then**
7: $\quad\quad\quad A^\tau_{[4]} \leftarrow A^\tau_{[4]} \cup \{a\}$
8: $\quad\quad$ **else**
9: $\quad\quad\quad v \leftarrow x[v] \wedge w \leftarrow x[w]$
10: $\quad\quad\quad$ **if** $v[w]$ **then**
11: $\quad\quad\quad\quad A^\tau_{[2]} \leftarrow A^\tau_{[2]} \cup \{a\}$
12: $\quad\quad\quad$ **else if** $w[v]$ **then**
13: $\quad\quad\quad\quad A^\tau_{[3]} \leftarrow A^\tau_{[3]} \cup \{a\}$
14: $\quad\quad\quad$ **else**
15: $\quad\quad\quad\quad A^\tau_{[5]} \leftarrow A^\tau_{[5]} \cup \{a\}$
16: $\quad\quad\quad$ **end if**
17: $\quad\quad$ **end if**
18: $\quad$ **end for**
19: $\quad$ **return** $A^\tau_{[2..5]}$
20: **end procedure**

Theorem 8.4 *Suppose we have a weighted connected digraph $G = (V, A, \mu)$ without negative cycles, a vertex $r \in V$, and a spanning tree $T = (V, A', r, \nu)$ of G rooted in r. Then, the time complexity of computing the DiGST $G' = (V, A^\tau_{[1..5]}, r, \mu)$ spanned over G by means of T is in the order of $\mathcal{O}(|V| + |A|)$.*

Proof. By analogy with the proof of Theorem 8.1 we observe that solving this task demands differentiating between upward, downward, and lateral arcs. The reason is that while kernel arcs are identified by their membership in T, reflexive arcs are distinguished by the fact that they contain the same vertex twice. Because of the definition of lateral arcs, this further means that we have to decide whether a given arc $a \in (A^\tau_{[1..5]} \setminus A^\tau_{[1]}) \setminus A^\tau_{[4]}$ is an upward or a downward arc. This decision is computed by Algorithm 8.2 by analogy to Algorithm 8.1. The only difference is that we distinguish between upward and downward arcs (using the same format and method). Thus, the time effort of Algorithm 8.2 is basically induced by lines 5–17, which are repeated exactly $|A| - |A'|$ times so that, due to the constant complexity of the latter operations, the order in question is $\mathcal{O}(|A| - |A'|) = \mathcal{O}(|A|)$. Further, the complexity of performing a breadth-first search (line 3) is, as before, of order $\mathcal{O}(|V| + |A|) = \mathcal{O}(|V| + |V| - 1) = \mathcal{O}(|V|)$, so that we get $\mathcal{O}(|V| + |A|)$ as the desired upper bound. Again, more efficient algorithms can be envisioned but are out of the focus of this chapter as Algorithm 8.2 is already sufficiently efficient.

Now we can introduce and exemplify directed MSGTs as follows.

Definition 8.13 (Directed Minimum Spanning Generalized Tree) Let $G = (V, E, \mu)$ be a weighted connected digraph and $T = (V, A', r, \nu)$ an MST of G rooted in some $r \in V$. The *directed minimum spanning generalized tree* (DiMSGT) induced by T is a directed GT $G_{T_r} = (V, A^{\tau}_{[1..5]}, r, \mu)$ spanned over G by means of the kernel tree T starting from r.

Example *Let the digraph $A = (V, E, \mu)$ in Figure 8.7 be given. Then, the digraph $B = (V, A^{\tau_B}_{[1..5]}, 2, \mu_B)$ is a DiMSGT spanned over A by means of the kernel MST $T_2 = (V, A', 0)$, $A' = \{(2,4), (2,6), (4,0), (0,5), (0,3), (3,10), (3,1), (10,12), (1,7), (1,8), (12,11), (8,9)\}$, together with the following partition of the arc set $A^{\tau_B}_{[1..5]}$: $A^{\tau_B}_{[1]} = A'$, $A^{\tau_B}_{[2]} = \{(1,2), (5,4)\}$, $A^{\tau_B}_{[3]} = \{(3,11), (10,11), (3,7)\}$, $A^{\tau_B}_{[4]} = \emptyset$, $A^{\tau_B}_{[5]} = \{(7,9)\}$. Note that the subgraph spanned by vertices 1, 7, 8, and 9 is a typical case that demarcates Prim's algorithm of spanning MSTs from the corresponding algorithms adapted to digraphs. While Prim's algorithm would select the arc $(1,7)$, then the arc $(7,9)$, and finally the arc $(1,8)$, we realize that the choice of $(1,7)$, $(1,8)$, and $(8,9)$ is less costly [70].*

Corollary 8.9 *Because of Theorem 8.4, the time complexity of generating a DiMSGT is on the order of $\mathcal{O}(|V| + |A| + \min\{|A| \log |V|, |V|^2\})$ when using a standard algorithm [60] to generate a directed MST as its kernel.*

This corollary is a simple consequence of separating the generation of spanning trees from spanning peripheral arcs as realized by Algorithm 8.2. As a DiMSGT rooted in a preselected vertex r equals the shortest path tree rooted in the same vertex, it is superfluous to consider *directed* generalized shortest path trees. As an alternative we consider *directed generalized dependency trees* (DiGDTs). These are GTs that are spanned by means of directed dependency trees that, in turn, result from orientating so-called dependency trees. *Dependency trees* (DTs) have been used in computational linguistics to order association data in a tree-like fashion [36, 39, 50]. They are generated as follows. For a distinguished vertex r of a graph $G = (V, A, \mu)$, vertices are inserted into the DT rooted by r in ascending order of their geodesic distance from r where the predecessor of any vertex v to be inserted is chosen to be the vertex w that in terms of μ is closest to v among all vertices already inserted into the DT. Look at Figure 8.7 and vertex 2 as our distinguished root vertex. In this case, we realize first that the vertices of graph A are inserted into a DT rooted by 2 according to their geodesic distance to 2. Thus, we get the following sequence of vertices to be inserted: 4, 6, 0, 5, 3, 1, 10, 11, 12, 7, 8, 9.

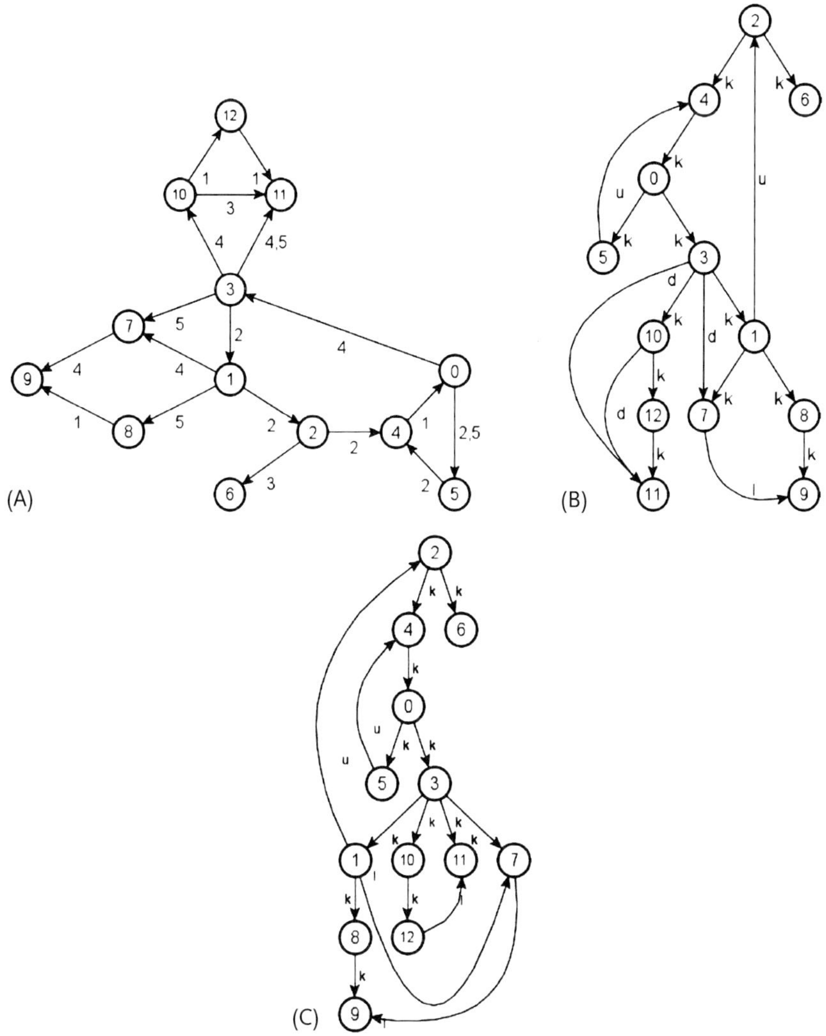

Figure 8.7 A directed graph A together with its DiMSGT (graph B) rooted in vertex 2 and a directed generalized dependency tree (graph C) rooted in the same vertex. For reasons of simplification, arc weights are omitted in the graphical representations of graphs B and C.

Based on this sequence we get a kernel spanning DT as shown in Figure 8.7 by graph C which is a *generalized* DT spanned by means of the latter DT starting from 2. The generation of this DiGDT is more context sensitive than that of the corresponding DiMSGT since the choice of the root uniquely determines the sequence in which vertices are processed, that is, according to their order of being *primed* by the root vertex. As a consequence, DiGDTs realize a sort of construction-integration process [34] in which the root vertex initiates firstly a process of spreading activation or information percolation, which is secondly organized in the form of a GT. In other words: in Figure 8.7, the DiGDT C structures the network given by graph A more from the perspective of vertex 2 than is done by the DiMSGT B. We do not try to formalize this notion here but hint at a publication in which DTs and more general Markov trees are studied as the kernel trees of GTs in detail (cf. [42]).

At this stage, we may envision many theorems about directed GTs by analogy to those proved for their undirected counterparts. However, we resist following this branch of research and go back to elaborating the framework of undirected GTs – this time with a strict view of semiotic modeling.

8.2.8
Generalized Trees, Quality Dimensions, and Conceptual Domains

So far we have introduced GTs as a fairly expressive, though nonetheless well-constrained, model in between the extremal cases of trees and general graphs. In this section, we explore the representational potential of GTs a step further. As was done in previous sections, we do this in graph-theoretical terms. The general story behind this approach is that we seek a model beyond *semantic spaces* as far as they are based on a purely geometric understanding of meaning relations [10, 31, 33, 50, 51]. Although we agree with the conception of usage-based semantics and its quantitative reconstruction by semantic spaces, we quarrel with the space complexity of this model and – as a result of this – with its cognitive implausibility. Without going into the details of this argumentation we simply mention that semantic spaces equal completely connected graphs as their meaning points are always directly relatable in terms of their distance without the need to consider intermediate points. Consequently, they always have a maximal cluster value [68] – far from what is known about real semiotic networks [40, 58] and their small-world-like topology. As an alternative to this undesirable state we seek a less compact model with a realistically sparse topology in conjunction with a tree-like skeleton by analogy to conceptual hierarchies.

In order to approach this model, we refer to Gärdenfors [28], who elaborates *conceptual spaces* as a level of representation in between subsymbolic association networks of lower resolution and symbolic models of higher resolution. Roughly speaking, a conceptual space is based on a system of conceptual domains, which are *integral dimensions* used to map points onto the space. That is, for a set of quality dimensions $\{D_1, \ldots, D_n\}$ one can build a conceptual space in which objects v to be observed are represented as vectors $v' = (v(D_1), \ldots, v(D_n))$, where $v(D_i), 1 \leq i \leq n$, is the value taken by object v on dimension D_i. A central starting point of [28] is to view domains as systems of interrelated quality dimensions. This gives a conceptual space an internal structure as its objects can be characterized by *structured values*, which they take on the corresponding domain.[4)]

Gärdenfors [28] does not completely determine the mathematical notion of a conceptual space but relies on an axiomatic approach by naming necessary conditions of candidate implementations. In this sense, semantic spaces are just one way of implementing conceptual spaces that leave plenty room for developing alternative, topologically more constrained space models. This is exactly our gateway to make a first step in promoting GTs as such an alternative. In this section we show how GTs can be conceived as conceptual domains. In this way, we open the door to less complex representation formats apart from semantic spaces, formats that provide the efficiency of tree-like structures together with the structural freedom of networks. In order to approach this goal we proceed as follows. Firstly, we define a metric space based on GTs. Secondly, we interrelate this definition with the notion of betweenness and equidistance in terms of GTs. Thirdly, we introduce an interpretation of GTs as a sort of conceptual domain by which conceptual spaces are spanned as networks of networks (cf. Section 8.2.9). Once more, it turns out that GSPTs are the valuable starting point of this endeavor.

Corollary 8.10 *Let $G = (V, E, \mathcal{L}_V, \mathcal{L}_E, \mu)$ be a weighted connected graph according to Definition 8.1, $G' = (V, E^{\tau'}_{[1..4]}, r, \mu)$ a GSPT, and $G'' = (V, E^{\tau''}_{[1..5]}, r, \mu)$ a GPST spanned over G by means of the SPT T_r starting from $r \in V$. Then, $\hat{\mu}'$ (see Definition 8.2) is a distance function in $\hat{G}' = (V, E^{\tau'}_{[1..3]}, r, \mu')$ and $\hat{\mu}''$ a distance function in $\hat{G}'' = (V, E^{\tau''}_{[1..4]}, r, \mu'')$ – μ' and μ'' are the restrictions of μ to $E^{\tau'}_{[1..3]}$ and $E^{\tau''}_{[1..4]}$, respectively. That is, $(\hat{G}', \hat{\mu}')$ and $(\hat{G}'', \hat{\mu}'')$ are metric spaces.*

Proof. Because of Theorem 8.3 and Corollary 8.8 we can concentrate on kernel edges when considering shortest paths and geodesic distances. As a trivial consequence of these two theorems, Corollary 8.10 is reduced to a statement about trees since neither vertical nor cross-reference links interfere with the function of kernel links, that is, establishing shortest paths. Thus, the proof simply looks as follows:

4) For more details of this notion see [28].

- *Minimality*: The geodesic distance between two vertices is a nonnegative, real-valued function that, because of the definition of $\hat{\mu}$, is 0 in the case of two vertices v, w if and only if $v = w$ (Definition 8.2).
- *Symmetry*: The symmetry of $\hat{\mu}'$ and $\hat{\mu}''$ simply follows from the fact that we are dealing with undirected graphs.
- *Triangle inequality*: If P is the geodesic path between u and v (which because of Definition 8.2 is uniquely defined) and P' the geodesic path between v and w, then $P \circ P'$ is the geodesic path between u and w so that $\hat{\mu}'(u, v) + \hat{\mu}'(v, w) = \hat{\mu}'(u, w)$. As claimed by Theorem 8.3, this does not interfere with any vertical edge. In the case of $\hat{\mu}''$ we must argue analogously.

Following Gärdenfors [28], we now define the relation of betweenness and the relation of equidistance in terms of GTs where the latter form – according to Corollary 8.10 – a special kind of metric space.

Definition 8.14 (Geodesic Betweenness) Let $G = (V, E, \mathcal{L}_V, \mathcal{L}_E, \mu)$ be a weighted connected graph according to Definition 8.1. Then, we define the relation $B \subseteq V^3$ where

$$\forall u, v, w \in V\colon B(u, v, w) \Leftrightarrow u \neq v \neq w \wedge v \in V(GP_\mu(u, w))\,.$$

Relation B is called a *relation of geodesic betweenness*. This is the relation of all vertices u, v, w for which v is on the geodesic path between u and w. Every vertex v for which $B(u, v, w)$ is called *geodesically between u and w*.

Corollary 8.11 *Let $G = (V, E, \mathcal{L}_V, \mathcal{L}_E, \mu)$ be a weighted connected graph according to Definition 8.1, $G' = (V, E^{\tau'}_{[1..4]}, r, \mu)$ a GSPT, and $G'' = (V, E^{\tau''}_{[1..5]}, r, \mu)$ a GPST spanned over G by means of the SPT T_r starting from $r \in V$. Then, firstly, relation B satisfies Axioms B1–B4 of Betweenness [28] in $\hat{G}' = (V, E^{\tau'}_{[1..3]}, r, \mu')$ and in $\hat{G}'' = (V, E^{\tau''}_{[1..4]}, r, \mu'')$. Secondly, for any $u, v, w \in V : (B(u, v, w) \Leftrightarrow \hat{\mu}'(u, v) + \hat{\mu}'(v, w) = \hat{\mu}'(u, w)) \wedge (B(u, v, w) \Leftrightarrow \hat{\mu}''(u, v) + \hat{\mu}''(v, w) = \hat{\mu}''(u, w))$. μ' and μ'' are the restrictions of μ to $E^{\tau'}_{[1..3]}$ and $E^{\tau''}_{[1..4]}$, respectively.*

Proof. Trees are known to satisfy the axioms of betweenness. By Theorem 8.3 and Corollary 8.8 we know that neither vertical edges in GSPTs nor vertical and cross-reference edges in GPSTs interfere with kernel edges in spanning geodesic paths. Thus, the first part of Corollary 8.11 reduces to a well-known statement about (kernel) trees so that we can claim that B satisfies the following axioms of betweenness:

- $B1$: $\forall u, v, w \in V\colon u \neq v \neq w \Rightarrow (B(u, v, w) \Rightarrow B(w, v, u))$.
- $B2$: $\forall u, v, w \in V\colon u \neq v \neq w \Rightarrow (B(u, v, w) \Rightarrow \neg B(v, u, w))$.

- $B3$: $\forall v, w, x, y \in V$: $v \neq w \neq x \neq y \Rightarrow (B(v, w, x) \wedge B(w, x, y) \Rightarrow B(v, w, y))$.
- $B4$: $\forall v, w, x, y \in V$: $v \neq w \neq x \neq y \Rightarrow (B(v, w, y) \wedge B(w, x, y) \Rightarrow B(v, w, x))$.

The second part of Corollary 8.11 simply interrelates the triangle inequality of Corollary 8.10 with the present corollary.

By analogy with Definition 8.14 and Corollary 8.11 we get the following definition of equidistance in conjunction with its concomitant corollary.

Definition 8.15 (Geodesic Equidistance) Let $G = (V, E, \mathcal{L}_V, \mathcal{L}_E, \mu)$ be a weighted connected graph according to Definition 8.1. Then, we define the relation $E \subseteq V^4$, where

$$\forall v, w, x, y \in V\colon E(v, w, x, y) \quad \Leftrightarrow \quad v \neq w \neq x \neq y \wedge \\ \mu(GP_\mu(v, w)) = \mu(GP_\mu(x, y)).$$

Relation E is called the *relation of geodesic equidistance*. This is the relation of all vertices v, w, x, y for which the geodesic distance between v and w equals the geodesic distance between x and y. Any two pairs of vertices v, w and x, y for which $E(v, w, x, y)$ are called *geodesically equidistant*.

This allows us to formulate the following self-evident corollary.

Corollary 8.12 *Let $G = (V, E, \mathcal{L}_V, \mathcal{L}_E, \mu)$ be a weighted connected graph according to Definition 8.1, $G' = (V, E^{\tau'}_{[1..4]}, r, \mu)$ a GSPT, and $G'' = (V, E^{\tau''}_{[1..5]}, r, \mu)$ a GPST spanned over G by the SPT T_r starting from $r \in V$. Then, firstly, relation E satisfies Axioms E1–E4 of Equidistance [28] in $\hat{G}' = (V, E^{\tau'}_{[1..3]}, r, \mu')$ and in $\hat{G}'' = (V, E^{\tau''}_{[1..4]}, r, \mu'')$. Secondly, for any $v, w, x, y \in V$: $(E(v, w, x, y) \Leftrightarrow \hat{\mu}'(v, w) = \hat{\mu}'(x, y)) \wedge (E(v, w, x, y) \Leftrightarrow \hat{\mu}''(v, w) = \hat{\mu}''(x, y))$. μ' and μ'' are the restrictions of μ to $E^{\tau'}_{[1..3]}$ and $E^{\tau''}_{[1..4]}$, respectively.*

Proof. Once more, the only thing we need to hint at is that the generalized subtrees $\hat{G}'$ and $\hat{G}''$ do not contain any edges that interfere with their kernel edges in spanning geodesic paths. Thus, we can claim that E satisfies the following axioms of equidistance within these generalized subtrees:

- $E1$: $\forall u, v, w \in V$: $E(u, u, v, w) \Rightarrow v = w$.
- $E2$: $\forall v, w \in V$: $E(v, w, w, v)$.
- $E3$: $\forall u, v, w, x, y, z \in V : u \neq v \neq w \neq x \neq y \neq z \Rightarrow (E(u, v, w, x) \wedge E(u, v, y, z) \Rightarrow E(w, x, y, z))$.
- $E4$: $\forall u, v, w, x, y, z \in V$: $u \neq v \neq w \neq x \neq y \neq z \Rightarrow (B(u, v, w) \wedge B(x, y, z) \wedge E(u, v, x, y) \wedge E(v, w, y, z) \Rightarrow E(u, w, x, z))$.

By Corollaries 8.11 and 8.12 we see that well-defined generalized subtrees of GSPTs and GPSTs respect the axioms of betweenness and equidistance. Moreover, by Corollary 8.10 we additionally see that we get a simple metric on instances of these classes of graphs. According to [28], these are basic structural constraints to be satisfied by the dimensions of conceptual spaces. In other words, it seems plausible to build conceptual spaces in terms of a special kind of GT. By their hierarchical skeleton, they respect basic constraints of conceptual spaces but nevertheless share the full expressiveness of graphs. This paves the way for a graph-theoretical model of conceptual structures beyond ordinary trees and below the space complexity of semantic spaces. *But what does it mean to use GTs for spanning conceptual spaces?* More specifically: *How can we think of GTs as quality dimensions?* In order to answer these questions, we utilize the notion of a conceptual domain as a system of interrelated or interdependent quality dimensions. More specifically, we define a conceptual space as a set of quality dimensions with a topological structure as defined by GTs where a single domain equals a GT as a kind of structured dimension. In other words, the vertices of a GSPT or a GPST define basic dimensions that by virtue of their edges form integral dimensions. The integrity of the dimensions is reflected by the hierarchical skeleton of the respective GT. As a consequence, objects o to be mapped onto a conceptual domain $\mathcal{D}$ are interrelated with a subset of vertices of $\mathcal{D}$ such that all edges generated by this measurement operation preserve the GT-like topology of $\mathcal{D}$. In this way, a measurement of o along a basic dimension v of $\mathcal{D}$ equals the geodesic path ending at v and o as a result of adding o as a new vertex to $\mathcal{D}$ subject to preserving the structural constraints of this GT. In Section 8.3 we exemplify this structuralistic notion of measurement in detail. Note that so far we have sketched GT-based conceptual spaces only in terms of undirected graphs leaving the examination of conceptual spaces based on *directed* GTs for future work. The next section shows how this notion of a graph-like representation of conceptual structures is extended in order to cope (in Section 8.3) with distributed knowledge as provided by social ontologies and social encyclopedias.

8.2.9
Generalized Forests as Multidomain Conceptual Spaces

So far we have considered GTs as intermediary units between trees and graphs. We have also related this notion to cognitive modeling in terms of conceptual domains as introduced by Gärdenfors [28]. In this section we take this perspective of spanning conceptual spaces by GTs a step further. This is done by interlinking conceptual domains as separable, inherently structured dimensions whose values – taken by objects to be mapped onto a given conceptual space – are measured separately from each other. From a graph-theoretical point of view, interlinked domains can be modeled by appropriately generalizing the notion of a forest (of trees). Following this approach, we build conceptual spaces in the form of generalized forests of GTs. This is done as follows.

Definition 8.16 (Generalized Forest) A *generalized forest* (GF) is a graph $G = (V, E^{\tau}_{[0..4]}, \mu)$ such that the connected components $D_1, \ldots, D_n$ of the subgraph $G' = (V, E^{\tau}_{[1..4]}, \mu')$ of G are GTs $D_i = (V_i, E^{\tau_i}_{[1..4]}, r_i, \mu_i)$, $1 \leq i \leq n$, that satisfy the following structural constraints.

1. The sequence $V_1, \ldots, V_n$ is a partition of V, that is, $V = \cup_{i=1}^{n} V_i$ and $\forall 1 \leq i < j \leq n \colon V_i \cap V_j = \emptyset$. In order to denote this partition we use the function $\mathrm{V} \colon V \to \{V_1, \ldots, V_n\}$, where $\forall v \in V, \forall 1 \leq i \leq n \colon \mathrm{V}(v) = V_i \Leftrightarrow v \in V_i$.
2. The sequence $E^{\tau_1}_{[1..4]}, \ldots, E^{\tau_n}_{[1..4]}$ is a partition of $E^{\tau}_{[1..4]}$, that is, $E^{\tau}_{[1..4]} = \cup_{i=1}^{n} E^{\tau_i}_{[1..4]}$ and $\forall 1 \leq i < j \leq n \colon E^{\tau_i}_{[1..4]} \cap E^{\tau_j}_{[1..4]} = \emptyset$. Thus, $\forall 1 \leq i \leq n \colon \tau_i \subseteq \tau$.
3. $\tau \colon E^{\tau}_{[0..4]} \to \mathcal{T} = \{\mathrm{e}, \mathrm{k}, \mathrm{l}, \mathrm{r}, \mathrm{v}\}$ is an extended edge typing function such that $\forall e = \{v, w\} \in E^{\tau}_{[0..4]} : \tau(e) = \mathrm{e} \Rightarrow \mathrm{V}(v) \neq \mathrm{V}(w)$. That is, for any $1 \leq i \leq n \colon E^{\tau}_{[0]} \cap E^{\tau_i}_{[1..4]} = \emptyset$. Further, we define $E^{\tau}_{[0]} = \{e \in E^{\tau}_{[0..4]} \mid \tau(e) = \mathrm{e}\}$. Thus, $\forall \{v, w\} \in E^{\tau}_{[0]} \exists 1 \leq i < j \leq n : v \in V_i \wedge w \in V_j$. Edges $e \in E^{\tau}_{[0]}$ are called *external edges*.
4. μ_i, $1 \leq i \leq n$, denotes the restriction of μ to $E^{\tau_i}_{[1..4]}$.

In other words, a GF is a graph that is partitioned into possibly interlinked GTs. We call the GTs D_i the *components* of the GF G linked by external edges as elements of $E^{\tau}_{[0]}$ and denote them by $\mathrm{cmp}(G) = \{D_1, \ldots, D_n\}$.

Remark. A GF is a GT with an extended edge typing function that induces a decomposition of the underlying graph into a sequence of GTs by specifying external edges.

A connected graph with at least two vertices does not uniquely induce a GT. Likewise, such a graph is not uniquely decomposable into the domains of a GF. This simple observation gives plenty room for spanning GFs on a given graph subject to cost functions of external edges or the coherence of single domains. In this sense, we have to think of specific notions of GFs that are specialized by analogy to minimum spanning GTs, GSPTs, and so on. We may think, for example, of the domains D of a GF G as subgraphs that span regions of higher "internal homogeneity" (e.g., in terms of graph clustering [63]) within the underlying graph while its external edges span a sort of MST over these domains. As a sample notion of this kind consider the following definition of conceptual graphs.

Definition 8.17 (Conceptual Graph) Let $G = (V, E, \mu)$ be a weighted connected graph and $\mu^* \colon V^2 \to \mathbb{R}^+_0$ a metric measuring the connectedness of vertices subject to the topology of G spanned by E. Other than μ, μ^* valuates directly as well as indirectly connected vertices. In this sense, it is reminiscent of the notion of a transitive closure. However, we leave it open whether μ^* only explores simple or even only shortest paths (as done by $\hat{\mu}$ – cf. Definition 8.2). Think of μ^*, for example, as a function that measures the degree of unrelatedness or dissimilarity [9] of signs denoted by the vertices of G. Now a *conceptual graph* (CG) is a labeled GF $G = (V, E^{\tau}_{[0..5]}, \zeta, r, \mathcal{L}_V, \mathcal{L}_E, , \mu)$ together with an edge typing function $\zeta : E^{\tau}_{[0]} \rightarrow \mathcal{T} = \{c, k, r, s, v\}$ and the *top-level vertex* $r \in V$, which altogether satisfy the following additional constraints:

1. *Metric basic structure:* Each component $D_i \in \text{cmp}(G)$ – henceforth called a *domain* of G – is a GPST – note that the index of $E^{\tau}_{[0..5]}$ is running from 0 to 5 so that the edge typing functions of the components of G distinguish among cross-reference and shortcut edges.
2. *Micro-level coherence:*
$$\forall e \in E^{\tau}_{[0]} \nexists e' \in E^{\tau}_{[1..5]} : \mu(e) < \mu(e')$$
3. *Meso-level coherence:*
$$\forall v, w \in V \colon \text{V}(v) \neq \text{V}(w) \Rightarrow$$
$$\mu^*(v, w) > \max\{\max_{x,y \in \text{V}(v)}\{\mu^*(x, y)\}, \max_{x,y \in \text{V}(w)}\{\mu^*(x, y)\}\}$$
4. *Macro-level structure:* $G' = (V', E', \zeta', R, \mu')$ is a GPST such that $V' = \{V_i | (V_i, E^{\tau_i}_{[1..5]}, r_i, \mu_i) \in \text{cmp}(G)\}$, $|E'| = |E^{\tau}_{[0]}|$, $\forall e = \{v, w\} \in E^{\tau}_{[0]} : v \in V_i \wedge w \in V_j \wedge \chi_{E^{\tau}_{[0]}}(e) = k \Rightarrow \{V_i, V_j\} \in E' \wedge \chi_{E'}(\{V_i, V_j\}) = k \wedge \zeta'(\{V_i, V_j\}) = \zeta(e) \wedge \mu'(\{V_i, V_j\}) = \mu(e)$. Further, there exists an i such that $1 \le i \le |\text{cmp}(G)|$ and $(V_i, E^{\tau_i}_{[1..5]}, r_i, \mu_i) \in \text{cmp}(G)$ and $r \in R = V_i$. That is, r is the root of the root-building GPST-like component R of G.

We call G a $|\text{cmp}(G)|$*-dimensional* conceptual graph.

Remark. Satisfying Constraint 4 of Definition 8.17 guarantees the persistence of metric characteristics not only within single domains of a CG but also between the different domains of that graph. In this sense, we get the notion of a metric space of metric spaces where each of these spaces is represented by a separate GT that satisfies certain structural constraints guaranteeing a functional semantics of its different edge types.

Remark. The notion of a conceptual graph is not to be confused with that of the same name as introduced by [55]. In contrast to the latter term, Definition 8.17 is reminiscent of the notion of a conceptual space as introduced by Gärdenfors [28] by relying on the notion of a graph-like topology instead of referring to a hyperspace-based geometry. In Section 8.2.8 we have shown how to utilize GTs as graph models of conceptual domains. By Definition 8.17 we get the understanding that these domains, each of which is endowed with a graph-like topology, are interlinked by external edges that connect less coherent and, thus, separable vertices. In other words: Definition 8.17 reconstructs the opposition of separable and integral dimensions by the notion of micro and meso (level) coherence – leaving the definition of macro (level) coherence of conceptual graphs to future work: a domain is an internally structured, externally separable and internally integral dimension of a conceptual graph. We call a conceptual graph G for which $|\mathrm{cmp}(G)| \gg 1$ a *multi-dimensional conceptual space*. In Section 8.3 we exemplify this notion by three different semiotic systems.

Corollary 8.13 *A GF can have micro-level coherence without having meso-level coherence.*

Proof. This corollary can simply be proved by constructing a counterexample as shown in Figure 8.8, which is a GF $G = (V, E^{\tau}_{[0..5]}, \mu)$ with two components $D_1 = (\{1,2,3,4\}, E^{\tau_1}_{[1..5]}, 1, \mu_1)$ and $D_2 = (\{5,6,7,8\}, E^{\tau_2}_{[1..5]}, 5, \mu_2)$ such that $E^{\tau}_{[0]} = \{\{2,8\}\}$, $E^{\tau_1}_{[2]} = E^{\tau_1}_{[3]} = E^{\tau_1}_{[4]} = E^{\tau_2}_{[2]} = E^{\tau_2}_{[3]} = E^{\tau_2}_{[4]} = \emptyset$, $E^{\tau_1}_{[1]} = \{\{1,2\},\{1,4\},\{3,4\}\}$, $E^{\tau_2}_{[1]} = \{\{5,6\},\{5,8\},\{6,7\}\}$, $E^{\tau_1}_{[5]} = \{\{2,3\}\}$ and $E^{\tau_2}_{[5]} = \{\{7,8\}\}$. Further, $\mu(\{1,2\}) = \mu(\{1,4\}) = \mu(\{3,4\}) = \mu(\{5,6\}) = \mu(\{5,8\}) = \mu(\{6,7\}) = 1.5$, $\mu(\{2,3\}) = \mu(\{7,8\}) = 1.55$, and $\mu(\{2,8\}) = 2.0$. Suppose now that $\mu^* = \hat{\mu}$. In this case we see that Constraint 1 of Definition 8.17 is satisfied while $\max\{\max_{x,y \in V_1}\{\mu^*(x,y)\}, \max_{x,y \in V_2}\{\mu^*(x,y)\}\} = 3.0 > 2.0 = \hat{\mu}(v,w)$.

Obviously, Definition 8.17 demands spanning CGs in a sense that vertices of the same domain are "nearer" to each other, more related, or more similar than vertices of different domains. This is the GT-based analog to the distinc-

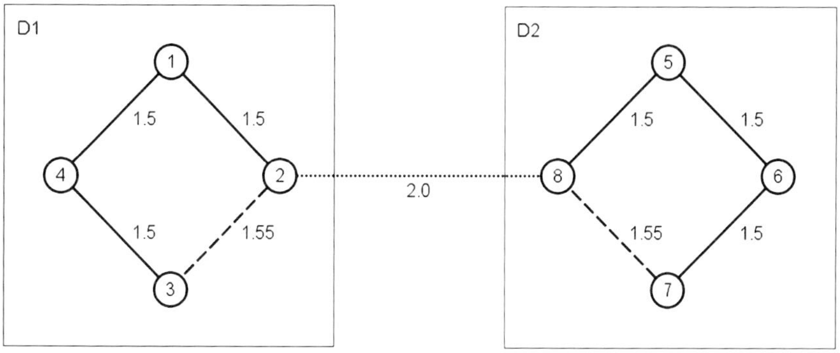

Figure 8.8 A generalized forest with two components D_1 and D_2 (cf. the proof of Corollary 8.13). Kernel edges are denoted by straight lines, lateral edges by dashed lines, and the single external edge by a dotted line. Numeric labels denote the weights of the edges.

tion of integral (domain internal) and separable (domain external) dimensions in conceptual spaces. It reflects a basic idea of explorative data analysis according to which objects of the same cluster shall be more homogeneous than objects of different clusters – irrespective of the operative measure of object similarity. Although we do not specify μ^* in Definition 8.17, good candidates for instantiating μ^* can be derived from algorithms for clustering graphs (cf., e.g., [63]).

At this point we stop extending the graph-theoretical apparatus introduced so far and leave this endeavor for future work. What we finally present in the next section is an overall interpretation of the notion of a GF as introduced so far.

8.3 Semiotic Systems as Conceptual Graphs

So far we have gained several novel subclasses of the class of GTs. It was Dehmer's [15] task – who first and, up to that time, most comprehensively formalized GTs – to define a similarity measure for classifying *given* sets of GTs. That is, for a triple of GTs Dehmer [15] determines the most similar pair of GTs. In this chapter we have taken one step back in order to approach an answer to the following question: *Given a single graph, which of the GTs derivable from it satisfies which topologically and semiotically founded constraints?* Following this line of research we have introduced the notion of a minimum spanning generalized tree (MSGT), of a generalized shortest path tree (GSPT), and of a generalized shortest paths tree (GPST). Especially by the subclass of GPSTs we have gained a detailed semantics of kernel, vertical, reflexive, and lateral edges where the latter have further been divided into the subset of cross-reference and shortcut edges. In Section 8.2.6 we have given a functional semantics of kernel edges as search facilities, of vertical edges as abridging facilities, of shortcut edges as association facilities (in support of large cluster values), and of cross-reference edges as randomization facilities (in support of short average geodesic distances). GTs are a class of graphs that impose functional restrictions on the typing of their edges, which locate this class in between the class of trees and general graphs. In this sense, we have reached an information-added value from our new look on graphs: although each GT is a graph by definition, the latter semantics of edge types provides a detailed classification of edges according to their *function* in processes of information flow in networks. Further, in Sections 8.2.8 and 8.2.9 we have approached the stage of elaborating the notion of a GT in terms of cognitive modeling. It is now that we explain this model by example of three semiotic domains.

In order to do that, let us briefly recapitulate our instantiation of conceptual spaces by means of GFs as presented above: Starting from a graph we denote quality dimensions by its vertices whose networking into GTs defines domains, that is, internally structured dimensions of mutually integral basic dimensions. Starting from this setting, conceptual spaces are spanned by interlinked domains such that the dimensions of different domains are less coherent than those belonging to the same domain (remember our analogue to the notion of separable dimensions). *So what does it mean to map an object onto such a conceptual space?* There are at least two candidate answers to this question – a (neo-)structuralistic one and a conceptualistic one (as we call them).

1. *Structuralistic interpretation – the unipartite model*: According to this view, a conceptual space is a unipartite graph in which all entities – whether dimensions or objects – are mapped onto the same single mode of the graph. In this way, any object is defined in accordance with the general stance of structuralism by its relative position with respect to all other objects of the same space [44] (in terms of direct or indirect links). Following this interpretation – which below is exemplified by text networks – a newly observed object o is mapped onto conceptual space
   a) by finding the domain to which it is best related (in terms of the operative notion of object relatedness or similarity),
   b) by locating o relative to the dimensions of that domain subject to the restrictions of generalized shortest paths trees (GPSTs), and
   c) by establishing external edges that relate o to alternative domains as representations of its additional meanings.

   Under this regime, the root of a GPST representing a certain domain is the prototype of this domain, which by virtue of this structuralist interpretation is an existent sign [46] (and not just a virtual configuration of features). Further, domains (and the CGs spanned by them) necessarily grow as a function of newly made observations, that is, by newly made object measurements (e.g., processes of sign interpretation). This may also affect the rewiring of already established domain-internal or -external links. As a consequence, measurement operations are reconstructed as a sort of object wiring or edge formation. We can represent such measurement operations by means of a vector-like notation as follows: an object o is mapped onto a CG G by a vector $\mathbf{o}^T = (d_1, \ldots, d_n)$, where $d_i = \mu(o, v_i)$, $1 \leq i \leq n$, iff o is linked with vertex v_i, otherwise $d_i = 0$. From this point of view, we naturally gain an interpretation of CGs according to the theory of diachronic structuralism as put forward by [32], [13] and especially by [18].

2. *Conceptualistic interpretation – the bipartite model*: According to this view, a CG spans a bipartite graph with two modes: whereas the top mode is (as above) spanned by the interlinked domains of the graph, objects are now separately mapped onto the graph's bottom mode. There are two alternatives for representing this bottom mode: either objects are represented in

Table 8.1 Overview of the notion of a conceptual graph as a purely graph-theoretical reconstruction of the notion of a *conceptual space* [28] and three of its instances.

CS	Conceptual graph	Text networking	Social tagging	Thematic progression
Dimension	Vertex	Text	Social category	Text segment
Domain	Generalized tree	Text subnetwork	Social category subgraph	Subnetwork of text segments
Space	Generalized forest	Text network	Social category Graph	Network of text segments
Object	Vertex of the same or different mode	Text	Text (segment)	Text (segment)

the usual way as vectors of values along dimensions that span a geometric space or the object space spans itself a GF. The latter variant is preferred here. Following this interpretation, a conceptual space consists of two interrelated CGs, the one representing a system of dimensions, the other a system of objects characterized and interrelated along these dimensions. In order to make sense of this interpretation we are in need of a notion of commutativity by analogy to category theory. That is, links among dimensions restrict the set of links among objects. Under this regime, the root of a GPST representing a certain domain is the prototype of this domain, which by virtue of this conceptualistic interpretation is no longer a real existing sign [46].

Both of these interpretations naturally include hierarchical categorization as a mode of object representation [57]. This is a direct consequence of the topology of GTs by which objects may be mapped onto vertices of different levels of the kernel of the operative domain. Note that both of these interpretations of conceptual spaces are contrary to feature semantics, which (i) establishes *semes* [30] as semantic dimensions that form building blocks of (ii) categories as regions of semantic space onto which objects are (iii) mapped by categorizing them along the latter categories. The reason why we do not follow this approach is the sheer impossibility of finding such reliable semantic dimensions. Thus, in good tradition with neostructuralism we refer to the signs themselves as dimension-building units [18].

Now we can exemplify the latter two interpretations of conceptual spaces in terms of conceptual graphs by means of three semiotic domains:

- *Text networking*: A first example of conceptual graphs can be constructed in the area of social text networking. The most prominent example of this is Wikipedia, where articles and related document units are the vertices that are connected by encyclopedic links [41]. In this area we can build a conceptual graph along a structuralist interpretation by identifying topical domains as subnetworks of Wikipedia's article graph. This can be done by clustering articles according to their content. However, Wikipedia also knows the concept of thematic portals that – as is easily shown [41] – have a GT-like topology. Without elaborating this example in detail let us mention that while in this example kernel and vertical links express relations of thematic hypotaxis, subordination, or containedness among articles, shortcut edges connect articles related by textual entailment, while cross-reference links can be used to represent links among thematically loosely connected articles (e.g., by means of dates, locations, etc.). Following this line of thinking, the Wikipedia article graph gets a semantic space in itself that is subdivided into portals and other clusters of articles where each of these clusters spans a certain thematic domain by means of its thematic homogeneity. By mapping a single text (or a new article) onto this conceptual graph we get, among other things, information about its membership in certain thematic domains. Additionally, for each of these domains representing its ambiguous content we get information about the degree of its thematic resolution (as a function of its geodesic distance to the root of that domain). Finally, by classifying all links starting from that text in terms of kernel, vertical, shortcut, and cross-reference edges we specify its functional position in processes of information flow through that network. This is just a structuralist way of revealing the content of an object by interlinking it with other objects of the same ontological sphere.

- *Social tagging*: Along a conceptualistic reading of conceptual graphs we get a second example. Now instead of directly interrelating a text with the vertices of a given text network we can alternatively map that text onto the category system of Wikipedia [65]. In this way, the category system of Wikipedia [64] is reconstructed as a GF in which different subgraphs span thematically distinguished subject areas (e.g., *culture, science, sports*). That is, we assume that kernel edges model hypernymy relations while vertical edges abbreviate them according to their transitivity. Further, we assume that shortcut edges map socially linked categories that denote coclassifications or polymorphic categorizations within a given thematic domain. Finally, cross-reference edges are seen to denote remote, that is, less obvious coclassifications. Under this regime, external edges combine obviously unrelated domains of categorization – possibly due to an erroneous coclassification of an ambiguous term or so. Obviously, vertical and lateral edges have quite different roles so that this information can be reflected by the buildup of GTs representing single domains of categorizations. Note that this interpretation in terms of interlinked domains relieves us of the burden of deciding on a top-level category. Such a single top-level category is as unrealistic as a purely tree-like skeleton of a category graph in social tagging. As before, when a text is mapped onto the resulting conceptual graph, we perform a hierarchical categorization where ambiguous texts are mapped onto different domains of the graph.

- *Thematic progression*: A third example belongs to the area of discourse analysis. According to the notion of thematic progression [14] we may think of a single discourse as being divided into interlinked thematic domains each of which represents a single topic separated from the other topics of the same discourse. Representing these domains as GTs we decide to map thematic progressions by kernel edges that are supplemented by vertical edges as a means to abridge hypotactic relations among discontinuous text segments. Further, we can think of shortcut edges as links among thematically associated text segments while cross-reference links denote thematically remote connections among randomly linked text parts. In other words, kernel and vertical edges model textual coherence as based on textual entailment, while shortcut and cross-reference edges are used to model coherence relations of text segments based on thematic association. Note that the original model of thematic progression does not account for graph-like discourse structures but unrealistically relies on a tree-like model.

What have we gained by the graph-theoretical apparatus introduced so far? We have invented a graph model that shares the efficiency of SPTs with the expressiveness of graphs. Further, we have elaborated this notion along the notion of a conceptual space. More specifically, we have introduced GFs as a graph model that retains several nontrivial characteristics of semiotic systems. With a view on semantic relations this model accounts for

1. *thematic centralization* according to the choice of prototypes as roots of GTs,
2. *hypotactic unfolding* by means of kernel edges along increasingly specialized nodes starting from the prototypical root of the GT,
3. *thematic condensation* as provided by vertical links abridging taxonomical relations due to transitivity relations among kernel edges,
4. *thematic shortcuts* as a means of representing thematic associations apart from taxonomic or otherwise hierarchical meaning relations,
5. *domain formation* as a result of networking among thematically homogeneous signs,
6. *domain networking* by spanning external edges among different domains in order to gain finally,
7. *conceptual spaces* as reference systems of modeling the content of polysemous signs.

This reference example shows that GFs and their constitutive trees can be seen as a powerful tool for mapping semiotic systems of a wide range of areas (ranging from single texts and social ontologies to whole text networks). This opens the perspective on semiotic measurements beyond semantic spaces and their geometric model of meaning relations.

Acknowledgment

Financial support of the German Federal Ministry of Education (BMBF) through the research project *Linguistic Networks* and of the German Research Foundation (DFG) through Excellence Cluster 277 *Cognitive Interaction Technology* (via the project *Knowledge Enhanced Embodied Cognitive Interaction Technologies* (KnowCIT)), the SFB 673 *Alignment in Communication* (via Project X1 *Multimodal Alignment Corpora: Statistical Modeling and Information Management*), Research Group 437 *Text Technological Information Modeling* (via Project A4 *Induction of Document Grammars for Webgenre Representation*), and the LIS-Project *Entwicklung, Erprobung und Evaluation eines Softwaresystems von inhaltsorientierten P2P-Agenten für die thematische Strukturierung und Suchoptimierung in digitalen Bibliotheken* at Goethe-University Frankfurt am Main and at Bielefeld University is gratefully acknowledged. We also thank Jolanta Bachan, Dafydd Gibbon and the anonymous reviewers for their fruitful suggestions, which helped to reduce the number of errors in this chapter.

List of Abbreviations

Acronym	Meaning	Definition
CG	Conceptual Graph	8.17
DGT	Directed Generalized Tree	8.11
DiGST	Directed Generalized Spanning Tree	8.12
DiMSGT	Directed Minimum Spanning Generalized Tree	8.13
GF	Generalized Forest	8.16
GPST	Generalized shortest PathS Tree	8.10
GSPT	Generalized Shortest Path Tree	8.8
GST	undirected Generalized Spanning Tree	8.5
GT	(undirected) Generalized Tree	8.3
MST	Minimum Spanning Tree	
NOP	Network Optimization Problem	
MSGT	Minimum Spanning Generalized Tree	8.6
SPT	Shortest Path Tree	8.8
SPGT	Shortest Paths Generalized Tree	8.9

References

1 Albert, R. and Barabási, A.-L. (2002). Statistical mechanics of complex networks. *Reviews of Modern Physics*, 74:47.

2 Baas, N.A. (1994). Emergence, hierarchies, and hyperstructures. In Langton, C. G. (ed.), *Artificial Life III, SFI Studies in the Sciences of Complexity*, pp. 515–537. Addison-Wesley, Reading.

3 Barabási, A.-L. and Oltvai, Z.N. (2004). Network biology: Understanding the cell's functional organization. *Nature Reviews. Genetics*, 5(2):101–113.

4 Barrat, A., Barthelemy, M., Pastor-Satorras, R., and Vespignani, A. (2004). The architecture of complex weighted networks. *Proceedings of the National Academy of Sciences USA*, 101(11):3747–3752.

5 Barthelemy, M. (2004). Betweenness centrality in large complex networks. *European Physical Journal B*, 38:163–168.

6 Blanchard, P. and Krüger, T. (2004). The cameo principle and the origin of scale free graphs in social networks. *Journal of statistical physics*, 114(5–6):399–416.

7 Brainerd, B. (1977). Graphs, topology and text. *Poetics*, 1(14):1–14.

8 Broder, A., Kumar, R., Maghoul, F., Raghavan, P., Rajagopalan, S., Stata, R., Tomkins, A., and Wiener, J. (2000). Graph structure in the web. *Computer Networks*, 33:309–320.

9 Budanitsky, A. and Hirst, G. (2006). Evaluating WordNet-based measures of lexical semantic relatedness. *Computational Linguistics*, 32(1):13–47.

10 Burgess, C., Livesay, K., and Lund, K. (1999). Exploration in context space: Words, sentences, discourse. *Discourse Processes*, 25(2&3):211–257.

11 Chakrabarti, S. (2002). *Mining the Web: Discovering Knowledge from Hypertext Data.* Morgan Kaufmann, San Francisco. Das Buch behandelt das Teilgebiet bzw. das Anwendungsgebiet des Web Mining.

12 Chazelle, B. (2000). A minimum spanning tree algorithm with inverse-ackermann type complexity. *Journal of the ACM*, 47(6):1028–1047.

13 Coseriu, E. (1974). *Synchronie, Diachronie und Geschichte. Das Problem des Sprachwandels.* Wilhelm Fink.

14 Daneš, F. (1974). Functional sentence perspective and the organization of the text. In Danes, F. (ed.), *Papers on Functional Sentence Perspective*, pp. 106–128. Mouton, The Hague.

15 Dehmer, M. (2005). *Strukturelle Analyse Web-basierter Dokumente.* Multimedia und Telekooperation. DUV, Berlin.

16 Dehmer, M. and Mehler, A. (2007). A new method of measuring the similarity for a special class of directed graphs. *Tatra Mountains Mathematical Publications*, 36:39–59.

17 Dehmer, M., Mehler, A., and Emmert-Streib, F. (2007). Graph-theoretical characterizations of generalized trees. In *Proceedings of the 2007 International Conference on Machine Learning: Models, Technologies & Applications (MLMTA07), June 25–28, 2007, Las Vegas.*

18 Derrida, J. (1988). *Limited Inc.* Northwestern University Press, Chicago.

19 Diestel, R. (2005). *Graph Theory.* Springer, Heidelberg.

20 Dijkstra, E.W. (1959). A note on two problems in connexion with graphs. *Numerische Mathematik*, 1:269–271.

21 Dorogovtsev, S.N. and Mendes, J.F.F. (2004). The shortest path to complex networks. http://www.citebase.org/abstract?id=oai:arXiv.org:cond-mat/0404593

22 Ehresmann, A.C. and Vanbremeersch, J.-P. (1996). Multiplicity principle and emergence in memory evolutive systems. *SAMS*, 26:81–117.

23 Emmert-Streib, F. and Dehmer, M. (2006). A systems biology approach for the classification of dna microarray data. In *Proceedings of ICANN 2005, Torun, Poland.*

24 Emmert-Streib, F. and Dehmer, M. (2007). Topological mappings between graphs, trees and generalized trees. *Applied Mathematics and Computing*, 186(2):1326–1333.

25 Emmert-Streib, F., Dehmer, M., and Kilian, J. (2005). Classification of large graphs by a local tree decomposition. In Arabnia, H.R. and Scime, A. (eds.), *Proceedings of DMIN 05, International Conference on Data Mining, Las Vegas, Juni 20–23*, pp. 200–207.

26 Ferrer i Cancho, R., Riordan, O., and Bollobás, B. (2005). The consequences

of Zipf's law for syntax and symbolic reference. *Proceedings ofthe Royal Society*, 272:561–565.

27 Fischer, W.L. (1969). Texte als simpliziale Komplexe. *Beiträge zur Linguistik und Informationsverarbeitung*, 17:27–48.

28 Gärdenfors, P. (2000). *Conceptual Spaces*. MIT Press, Cambridge, MA.

29 Goldblatt, R. (1979). *Topoi: the Categorial Analysis of Logic*. Springer, Amsterdam.

30 Greimas, A.J. (2002). *Semantique Structurale*. Presses Universitaires de France, Paris.

31 Gritzmann, P. (2007). On the mathematics of semantic spaces. In Mehler, A. and Köhler, R. (eds.), *Aspects of Automatic Text Analysis*, Vol. 209 of *Studies in Fuzziness and Soft Computing*, pp. 95–115. Springer, Berlin, Heidelberg.

32 Jakobson, R. (1971). *Selected Writings II. Word and Language*. Mouton, The Hague.

33 Jones, W. and Furnas, G. (1987). Pictures of relevance: A geometric analysis of similarity measures. *Journal of the American Society for Information Science*, 38(6): 420–442.

34 Kintsch, W. (1998). *Comprehension. A Paradigm for Cognition*. Cambridge University Press, Cambridge.

35 Landauer, T.K. and Dumais, S.T. (1997). A solution to Plato's problem: The latent semantic analysis theory of acquisition, induction, and representation of knowledge. *Psychological Review*, 104(2):211–240.

36 Lin, D. (1998). Automatic retrieval and clustering of similar words. In *Proceedings of the COLING-ACL '98*, pp. 768–774.

37 Marcu, D. (2000). *The Theory and Practice of Discourse Parsing and Summarization*. MIT Press, Cambridge.

38 Marcus, S. (1980). Textual cohesion and textual coherence. *Revue roumaine de linguistique*, 25(2):101–112.

39 Mehler, A. (2002). Hierarchical orderings of textual units. In *Proceedings of the 19th International Conference on Computational Linguistics (COLING '02), August 24 – September 1, 2002, Taipei, Taiwan*, pp. 646–652, Morgan Kaufmann, San Francisco.

40 Mehler, A. (2008). Large text networks as an object of corpus linguistic studies. In Lüdeling, A. and Kytö, M. (eds.), *Corpus Linguistics. An International Handbook of the Science of Language and Society*. De Gruyter, Berlin, New York, pp. 328–382.

41 Mehler, A. (2008). Structural similarities of complex networks: A computational model by example of wiki graphs. *Applied Artificial Intelligence*, 22(7&8):619–683.

42 Mehler, A. (2009). Minimum spanning Markovian trees: Introducing context-sensitivity into the generation of spanning trees. In Dehmer, M. (ed), *Structural Analysis of Complex Networks*. Birkhäuser Publishing, Basel.

43 Mehler, A. and Gleim, R. (2006). The net for the graphs – towards webgenre representation for corpus linguistic studies. In Baroni, M. and Bernardini, S. (eds.), *WaCky! Working Papers on the Web as Corpus*, pp. 191–224. Gedit, Bologna.

44 Merleau-Ponty, M. (1993). *Die Prosa der Welt*. Fink, München.

45 Milgram, S. (1967). The small-world problem. *Psychology Today*, 2:60–67.

46 Murphy, G.L. (2002). *The Big Book of Concepts*. MIT Press, Cambridge.

47 Newman, M.E.J. (2003). The structure and function of complex networks. *SIAM Review*, 45:167–256.

48 Pastor-Satorras, R. and Vespignani, A. (2004). *Evolution and Structure of the Internet*. Cambridge University Press, Cambridge.

49 Pickering, M.J. and Garrod, S. (2004). Toward a mechanistic psychology of dialogue. *Behavioral and Brain Sciences*, 27:169–226.

50 Rieger, B.B. (1978). Feasible fuzzy semantics. In *7th International Conference on Computational Linguistics (COLING-78)*, pp. 41–43.

51 Schütze, H. (1997). *Ambiguity Resolution in Language Learning: Computational and Cognitive Models*, Vol. 71 of *CSLI Lecture Notes*. CSLI Publications, Stanford.

52 Serrano, M.Á., Boguñá, M., and Pastor-Satorras, R. (2006). Correlations in weighted networks. *Physical Review*, 74(055101(R)):1–4.

53 Shen-Orr, S., Milo, R., Mangan, S., and Alon, U. (2002). Network motifs in the transcriptional regulation network of escherichia coli. *Nature Genetics*, 31(1):64–68.

54 Smolensky, P. (1995). Connectionism, constituency and the language of thought. In Donald, M. and MacDonald, G. (eds.),

Connectionism: Debates on Psychological Explanation, Vol. 2, pp. 164–198. Blackwell, Oxford.

55 Sowa, J.F. (2000). *Knowledge Representation: Logical, Philosophical, and Computational Foundations*. Brooks/Cole, Pacific Grove.

56 Steels, L. (2006). Collaborative tagging as distributed cognition. *Pragmatics & Cognition*, 14(2):287–292.

57 Stein, B. and Meyer zu Eißen, S. (2007). Topic identification. *Künstliche Intelligenz (KI)*, 3:16–22.

58 Steyvers, M. and Tenenbaum, J. (2005). The large-scale structure of semantic networks: Statistical analyses and a model of semantic growth. *Cognitive Science*, 29(1):41–78.

59 Stroustrup, B. (2000). *Die C++-Programmiersprache*. Addison-Wesley, Bonn.

60 Tarjan, R.E. (1983). *Data structures and network algorithms*. Society for Industrial and Applied Mathematics, Philadelphia, Pennsylvania.

61 Thiopoulos, C. (1990). Meaning metamorphosis in the semiotic topos. *Theoretical Linguistics*, 16(2/3):255–274.

62 Tversky, A. and Gati, I. (2004). Studies of similarity. In Shafir, E. (ed.), *Preference, Belief, and Similarity. Selected Writing os Amos Tversky*, pp. 75–95. MIT Press, Cambridge, MA.

63 van Dongen, S. (2000). A cluster algorithm for graphs. Technical Report INS-R0010, National Research Institute for Mathematics and Computer Science in the Netherlands, Amsterdam.

64 Voss, J. (2006). Collaborative thesaurus tagging the wikipedia way. http://www.citebase.org/abstract?id=oai:arXiv.org:cs/0604036

65 Waltinger, U., Mehler, A., and Heyer, G. (2008). Towards automatic content tagging: Enhanced web services in digital libraries using lexical chaining. In *4th Int. Conf. on Web Information Systems and Technologies (WEBIST '08), 4–7 May, Funchal, Portugal*. Barcelona.

66 Wasserman, S. and Faust, K. (1999). *Social Network Analysis. Methods and Applications*. Cambridge University Press, Cambridge.

67 Watts, D.J. (2003). *Six Degrees. The Science of a Connected Age*. W.W. Norton & Company, New York, London.

68 Watts, D.J. and Strogatz, S.H. (1998). Collective dynamics of 'small-world' networks. *Nature*, 393:440–442.

69 Wu, B.Y. and Chao, K.-M. (2004). *Spanning Trees and Optimization Problems*. CRC Press, Boca Raton and London.

70 Yang, S.J. (2000). The directed minimum spanning tree problem. http://www.ce.rit.edu/~sjyeec/dmst.html.

9
Applications of Graph Theory in Chemo- and Bioinformatics

Dimitris Dimitropoulos, Adel Golovin, M. John, and Eugene Krissinel

9.1
Introduction

Chemoinformatics [1–4] and bioinformatics [5–9] may be broadly defined as the use of information technologies in chemistry and biology. This includes the collection and systematization of data with the purpose of converting it into knowledge by the identification of common trends and similarities. It also includes the analysis of data to determine if they support preexisting hypotheses and models. The ultimate goal of chemo- and bioinformatics is to aid discoveries (in particular, drug discoveries [10]) by narrowing the field of search to areas of greater promise.

Chemo- and bioinformatics address a significant number of tasks, which may be classed into several groups. Due to the ever growing amount of chemical and biological data, as well as their specifics and variety, data storage, retrieval, and maintenance become a discipline in itself [11–15]. Statistical analysis of data is important for the identification of common trends, similarities, and relationships [10, 16]. Many problems arise in the field of data treatment and assessment, such as sequence and structure analyses [17–20], analysis of gene and protein expressions [20, 21], measuring biodiversity [22, 23], evolutionary studies [7, 23–25], and many others. A particular type of application where chemo- and bioinformatics merge with computational chemistry and biology relates to modeling of different objects, processes, and phenomena [26–28].

Graph theory is widely used in chemo- and bioinformatics [29, 30]. This is so for a number of reasons, such as:

1. Many practical problems can be conveniently stated in graph terms.
2. Graph properties are well studied.
3. Efficient graph-theoretical algorithms continue to be developed for optimal and almost-optimal solutions.
4. Intractable problems may be easily identified.
5. Graphs are convenient structures for storing, searching, and retrieving data.

Analysis of Complex Networks: From Biology to Linguistics. Edited by Matthias Dehmer and Frank Emmert-Streib

ISBN: 978-3-527-32345-6

Examples of the use of graphs in chemo- and bioinformatics include sequence analysis [18], identification of chemical compounds [29], analysis of metabolic pathways [31–35] and phylogenetic trees [36–39], comparison and analysis of molecular structures [40–48], protein docking [49], identification of macromolecular complexes in crystal packing [50], and the investigation of protein topology [51, 52].

In this chapter, we outline basic graph-theoretical concepts and methods used in structural chemo- and bioinformatics. We also discuss practical issues relating to the Macromolecular Structure Database (MSD) at the European Bioinformatics Institute (EBI) [14], which are associated with graph-theoretical algorithms. MSD is the European project for the collection, management, and distribution of data on macromolecular structures, derived in part from the Protein Data Bank (PDB) [53]. The project involves the construction of a relational database to store structural data on macromolecules (protein and nucleotide chains) and ligands (small molecules, drugs, etc.) found in the PDB and the provision of a number of public web services developed to assist research. The MSD resources are available at http://www.ebi.ac.uk/msd/.

9.2
Molecular Graphs

Chemical diagrams that present molecules have been used in organic chemistry for centuries [54]. The orientation of atoms and the average length of bonds that attach them to each other vary and are influenced by factors such as the number of shared pairs of electrons, the characteristics of the atoms being joined, and the nature of their immediate environment. A molecular graph or chemical graph is an abstract representation of the structural formula of a chemical compound in terms of graph theory, which in this context refers to a collection of vertices (atoms) and a collection of edges that connect pairs of vertices (bonds). Vertices are labeled with the type of atom and bonds with bond types – single, double, or triple. The lengths of bonds and the angles between them are depicted in 2D diagrams and provide a simple representation of the molecule. Bonds have no direction and are described as unidirectional. The treewidth is the maximum number of bonds attached to each atom and is restricted by the valency of atoms, which is 5 or less. This low number makes molecular graphs sparse. Bodlaender *et al.* checked the treewidth of 10,000 chemical structures in a biological database and found the maximum to be 4 [55]. Molecular graphs are typically planar, but nonplanar ones do exist [56].

The details depicted in molecular graphs need to be consistent with known characteristics of the atoms and bonds as a group, which include factors such as the following:

1. The sum of bonds attached to each atom, taking into account bond orders and formal atom charges, must be consistent with the valency of each atom.
2. Bond patterns that do not strictly fit single or double characterization such as those in delocalized aromatic rings and bonds involved in hybridization should be presented accordingly. Often this is not shown explicitly in chemical diagrams, but is implied.
3. The configuration of substructures around chiral centers must be presented so that it is possible to distinguish between enantiomers. This is done by identifying potential chiral atoms (Figure 9.1) and checking to see if a stereochemical descriptor has been assigned to them (and only for them). The Cahn-Ingold-Prelog [57] absolute notation is the preferable notation for graph operations.

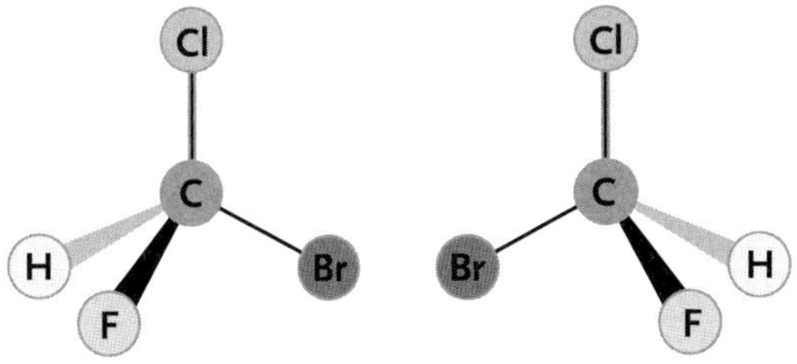

Figure 9.1 Example of a chiral center (the middle carbon atom), with the two enantiomers shown.

Hydrogen atoms can only form single bonds and are often omitted from graphs. Such graphs are referred to as hydrogen-depleted molecular graphs. For molecules that have metal ions and metal complexes, the use of graphs may be problematic as the atomic forces in a molecule are complex and cannot be approximated by the use of standard covalent bonds.

9.3 Common Problems with Molecular Graphs

The first problem that arises when dealing with a collection of molecular graphs such as those that exist in a database is duplication. The best approach to dealing with this task is to find a way to canonicalize [58] the molecular graphs and reorder the nodes to obtain all possible permutations of atoms. By storing the canonicalized forms in a database, the complexity of the task of checking whether an input molecule already exists is reduced significantly. Textual representations of molecules such as absolute SMILES [59] or INChi [60] can also be stored in databases and will help to reduce the cost of searches. It is essential, though, to ensure that the molecular graphs are chemically valid with the correct assignment of aromatic bond types and stereochemical descriptors before canonicalization takes place. The INChi strings in particular are an ideal tool for this task because by definition they have to be produced by the INChi published algorithm proposed by IUPAC and are designed particularly to solve the problem of chemical identity.

Another problem often encountered when dealing with a collection of molecular graphs is that of finding ones that are chemically similar to an input molecule, comparisons with which may involve finding others that:

(a) are supergraphs or subgraphs,
(b) share a large part of their structure (maximum common subgraph),
(c) share many similar subgraphs (common fragments).

A related class of problems is to organize molecules in a collection in classes either by an ad-hoc set of rules regarding their structure or by using a clustering algorithm and defining some method of assigning a similarity distance for two input structures [61]. The majority of elements do not occur frequently in molecules. The same is true for charged ions and triple bonds. This makes it possible to use an algorithm that iterates over the nodes of one graph to examine if a matching equivalent exists in the second graph starting with the rarest properties so that cases that don't match can be discarder earlier.

A similar way to improve the performance of subgraph searching especially over a collection of molecules is to precalculate a bitmap string that indicates the existence of properties or other graph features. The bitmap string could indicate, for example, whether each one from a predefined set of chemical fragments (subgraphs) is contained as a subgraph in the molecule. Using a technique like this may help to discard a large proportion of candidate matches using relatively low-cost bit operations. For cases where a sufficient sample of the molecules is known, the set of fragments can be selected in order to partition the collection more efficiently. An extension of this method is to keep not just a flag but also a counter of the number of occurrences of each feature. The molecular formula, for example, may be used directly to filter out a large proportion of the molecules in a collection during a subgraph search operation.

Some algorithms also take advantage of automorphisms that chemical graphs have by avoiding repeat searches of symmetrical parts, while others reduce graphs using the fact that aromatic structures of rings of five or six atoms are very common in biological molecules. Many of these approaches can be problematic if the input search molecule is not a valid chemical graph. This may be quite common because many of the subgraphs of valid molecules are not valid molecules themselves. For instance, removing a single bond of an aromatic ring will make it invalid unless the other bonds of the ring do not remain aromatic. SMARTS, a notation quite similar to SMILES, is a language designed specifically to specify substructures for subgraph search operations.

9.4 Comparisons and 3D Alignment of Protein Structures

Proteins are natural biopolymers made of a limited number of amino acid residues [62]. There are 20 basic amino acids plus a relatively large number of chemically modified derivatives. In proteins, amino acid residues are connected linearly by peptide links, forming chain structures often referred to as protein primary structure.

In their natural environments, protein chains usually fold into complex 3D structures [63]. Three levels of structural organization of proteins are recognized [64]. Secondary structure refers to local elements of protein fold, such as helices and strands, which normally have a size of 6 to 30 residues. The secondary structure elements (SSEs) are stabilized by hydrophobic interactions and hydrogen bonds [62,63] and are relatively well conserved in protein evolution. Tertiary structure refers to the structure of the whole polypeptide chain, as defined by the atomic coordinates [64], and may be viewed as the way the SSEs assemble. Quaternary structure refers to protein assemblies and represents the arrangement of several folded chains into a complex [64].

The function and chemical activity of proteins depend on their 3D structures [63]. Therefore the identification and measurement of structural similarities are important tasks in protein research, which are used to understand their biological role. Some practical questions that arise are, given two protein structures, how similar are they and which parts of their structures are similar or dissimilar. Also, given a particular 3D structure, how many other proteins share the structure, or, given several structures, which components are common. These questions are difficult to answer due to the overall complexity of protein folds and the expense of existing computational methods. To date, over 52,000 protein structures are in the PDB [53], which makes database screening for similar folds a resource-intensive task.

The problem of identifying residues that occupy geometrically equivalent positions in two or more structures has been investigated by many research groups over the last two decades. Techniques used include comparison of distance matrices [65], analysis of differences in vector distance plots [66], minimization of the soap-bubble surface area between two protein backbones [67], dynamic programming on pairwise distances between the residues [68, 69] and SSEs [70], 3D clustering [71, 72], combinatorial extension of alignment path [73], vector alignment of SSEs [44], depth-first recursive search on SSEs [74], graph theory [40–47], and many others [75–83].

Graph theory is probably the most convenient tool for the problem. If suitable graph representations for proteins can be constructed, graph-matching algorithms may be applied to identify common subgraphs and thus to establish structural similarities. However, a few problems considerably complicate practical implementation of this method. Firstly, protein structures are defined with a finite accuracy with respect to atomic coordinates. Because of that, the representing graphs need to be matched with a level of tolerance, which considerably increases the search field and makes the procedure more computationally demanding. Secondly, protein structure similarity does not necessarily assume chemical identity. In most cases, replacing a few residues in a protein sequence does not result in considerable changes of the overall structure. Therefore, graph-matching techniques need to allow for substructure substitutions. Thirdly, although protein molecules can be represented as graphs (as any other molecule), the size of such graph would be prohibitively large (typically on the order of a few thousand nodes) for any graph-matching algorithm.

A typical simplification, used to reduce the problem size, is to limit protein structures to representative atoms from each residue. Although chemically different, all amino acids have a common structural part, which is used to connect them into a chain in a regular way [62]. Traditionally, the only carbon atom from that part, the alpha carbon, is used for representing residues. This reduces the problem size to the length of protein chains, which typically

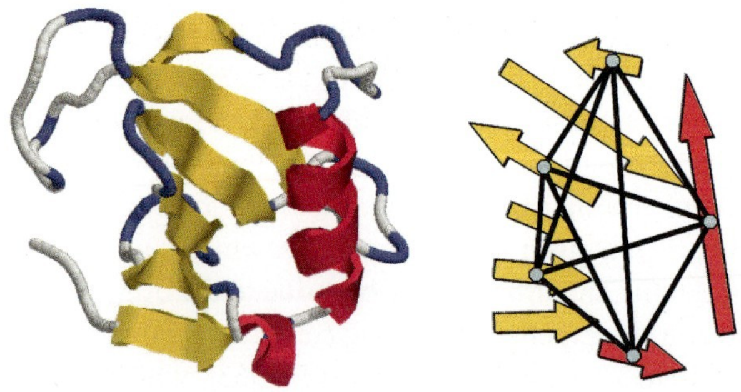

(a) (b)

Figure 9.2 Graph representation of a protein structure. (a) Cartoon image of a protein chain drawn through alpha Carbon atoms, where colors represent secondary structure elements (yellow strands and pink helices). (b) The corresponding graph, where vector representations of SSEs are used as graph vertices. Graph edges are sections connecting mass centers of SSEs (for clarity, not all are shown).

range from 60 to 5000 nodes. This is still too high for graph-theoretical algorithms to be practical. Therefore, combining conserved structural elements into complex graph nodes further reduces the problem. Here the SSEs serve as natural and convenient objects.

Figure 9.2 shows the graph representation of a protein built on the elements of secondary structure [42,45,70]. Each SSE is represented as a vector. These vectors are then used as graph nodes labeled with SSE type and number of residues in the given SSE. Mass centers of all SSEs are connected with sections used as graph edges. The edges are labeled with a number of properties: length, angles between the edge and connected SSEs, and angle between the SSEs and the corresponding torsion angle (Figure 9.3).

Matching of SSE graphs may be performed using any available algorithm, modified to compare complex labels of vertices and edges. A system of tolerances should be employed when comparing particular properties (lengths and labels) [45]. Special care should be taken when matched graphs correspond to different structural motifs due to the different connectivity of SSEs along the protein chain. An example of such a case is given in Figure 9.4, where motifs A and B have a different biological context and belong to chains with distinctly different topologies, but the spatial arrangements of their SSEs are identical. In such a case, the decision of whether A and B are similar depends on the context of the biological problem under consideration. This problem may be tackled by assigning serial numbers P_i to graph vertices and adding them to the list of edge properties. In this way, SSE connectivity is preserved if, during graph matching, the edge e_{ij}^A of graph A may be matched to the edge e_{kl}^B of graph B only if $sign\left(P_i^A - P_j^A\right) = sign\left(P_k^B - P_l^B\right)$ [45].

For most proteins, SSE graphs include 20 to 50 vertices, which represents a moderately hard problem for graph-matching algorithms. Note that because SSE graphs are fully connected, the subgraph search tree may be

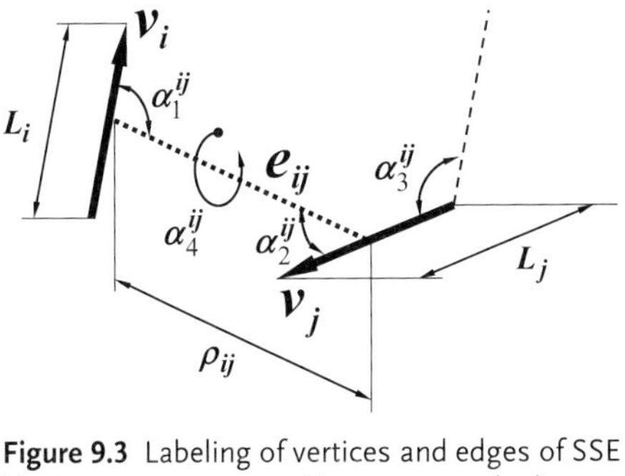

Figure 9.3 Labeling of vertices and edges of SSE graphs. Vertices are represented by vectors and edges connect their mass centers. Edge properties length and angles define mutual positions and orientations of all vertices in the graph.

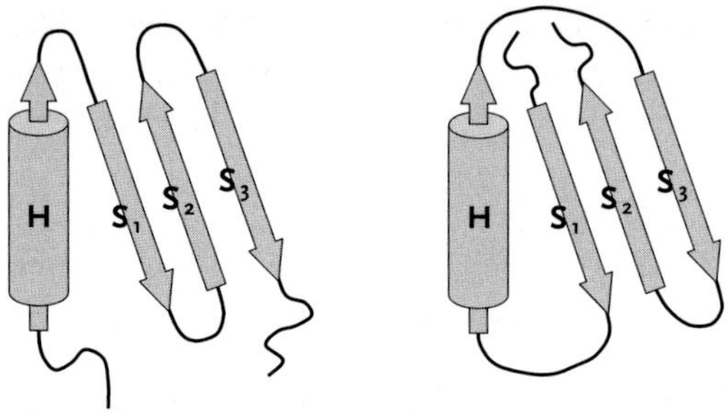

Figure 9.4 Example of SSE motifs of helix H and 3 strands S1, S2, and S3 each having different SSE connectivity. Motifs A and B form 3D SSE graphs that are geometrically identical but relate to different biological functions.

quite wide. However, this effect is compensated by the complex nature of SSE graph labels. Because of this, the number of potential vertex and edge matches is considerably decreased, which narrows the search field substantially.

The performance of a subgraph search in SSE graphs appears to be insufficient for whole-scale database screening, in which a target is matched to each and every one of more than 52,000 PDB structures. Graph-based database screening may be helped significantly by decision-tree algorithms [84], which include a preliminary decomposition and fragment classification of graphs. While this approach seems to be very attractive, it is impractical due to a high demand of computer memory [84]. Thus another approach, based on the assumption that only sizeable matches, with common subgraphs including 30% and more vertices, are of practical interest, was developed [45]. An algorithm for the detection of minimum-size graph isomorphisms has been developed [85]. This algorithm is able to identify, without additional computational expense, whether a given branch of a subgraph search tree is able to yield a subgraph isomorphism equal to or larger than a given size, and terminates the branch early if that is not the case. Thus, a large number of insignificant small-size matches are filtered out without computation resulting in substantial performance gains.

An implementation of the described approach is publicly available at http://www.ebi.ac.uk/msd-srv/ssm/ssmstart.html. The server performs pairwise (target-to-query), serial (target-to-database) protein structure comparisons, as well as identification of common structural motifs in sets of more than two structures. After general structural similarity is established through SSE graph matching, the server performs further refinement on the level of alpha C atoms, providing information on the correspondence of amino acid residues in structures being compared and giving their best possible superposition. Running on a variable (up to 16) number of parallel CPUs, the server provides an average response time of 1 min for queries that include exhaustive structural search of the PDB [45].

9.5
Identification of Macromolecular Assemblies in Crystal Packing

The biochemical functions of many proteins are dependent on their ability to form complexes. The way in which protein chains assemble in a complex represents the protein quaternary structure (PQS). Experimental identification of PQS is a complicated procedure that normally involves use of several complementing techniques, such as mobility and mass measurements [86, 87], light scattering [86], neutron and X-ray scattering [88, 89], nuclear magnetic resonance [90], electron microscopy [91], and others. However, most protein structure data (80% of the PDB) come from experiments on X-ray diffraction on macromolecular crystals [92]. It is reasonable to expect that stable protein complexes do not change during crystallization, and therefore they should be identifiable in crystal packing.

The inference of PQS from crystallographic data is not a straightforward procedure. A complex may be represented as a graph, where vertices are protein chains and edges are interchain interfaces that bind chains together. The binding force of interfaces comes from hydrophobic interactions [93, 94] and weak bonds, such as hydrogen bonds, salt bridges, and disulphide bonds [94–96]. There is no obvious way, in general, to discriminate between inter- and intracomplex interfaces. Numerous attempts to find a set of discriminating parameters have had limited success [97–99].

Chemically, a complex represents a stable structure if its free Gibbs energy of dissociation is positive [100]:

$$\Delta G_0 = -\Delta G_i - T\Delta S > 0\,, \tag{9.1}$$

whereΔG_i stands for the binding energy of dissociated subunits and ΔS is the entropy change. These quantities depend on the choice of dissociating subunits. For example, a homohexameric complex A_6 may dissociate into trimers $A_3 + A_3$, dimers $A_2 + A_2 + A_2$, or six monomeric units $6A$. While all possible dissociation scenarios take place, thermodynamically, the most probable one is that corresponding to minimal ΔG_0.

Applying Equation 9.1 to all complexes that may be formed in a given crystal, one could identify the largest stable complex, which will be the most probable solution to the problem [50]. Considering a crystal as an infinite periodic graph [50], this method may be described as looking for a subgraph with specific properties. The outline of an algorithm to do this is as follows:

- Define a "crystal" graph, where vertices are protein chains and edges are interchain interfaces. Vertex labels are the protein chain type, as defined by chain sequence and conformation. Edge labels refer to the type of connected vertices and their relative orientation.
- Mark edges as "engaged" if the corresponding interface connects chains from the same complex, and "disengaged" otherwise. Initially, set all edges disengaged.
- Find all possible complexes in the crystal by enumeration of all combinations of engaged and disengaged graph edges. Due to crystal symmetry, if an edge is engaged, all other edges with the same label must be engaged. An edge cannot be engaged if engagement results in a complex containing identical protein chains in parallel orientations. In crystal, all identical parallel chains are found in geometrically equivalent positions with respect to each other; therefore, having two parallel chains in a complex would automatically mean that the complex is infinite [50].
- Leave only stable complexes in the list, and select the largest one on the top as the most probable solution (see detail score in [50]).

Enumeration of all edge engagements may be conveniently addressed by a recursive backtracking scheme [50] with complexity $O(2^n)$, where n is the number of different edge labels. In some cases, n may reach 50 to 100, which makes a straightforward approach computationally intractable, despite considerable trimming of the recursion tree due to restrictions on edge engagements mentioned above. Two further enhancements allow one to overcome the difficulty. Firstly, after engaging or disengaging an edge, the algorithm should check whether the current configuration of engaged edges means automatic engagement of other edges. For example, engagement of two edges in a three-edge subgraph A_3 means that all three vertices belong to the same subgraph and therefore the remaining edge may be only in an engaged state. As practice shows, this trick drastically reduces the recursion tree.

Secondly, the algorithm may estimate whether the current branch of the recursion tree may result in subgraphs representing stable complexes on higher recursion levels, and terminate the branch if the answer is negative. This technique is based on the observation that the entropy change ΔS in Equation 9.1 do not decrease as more edges get engaged [50]. In the most favorable situation, engagement of additional edges will decrease the binding term in Equation 9.1, ΔG_i, without changing ΔS. Therefore, if

$$\Delta G_0^* - \sum_{k>*} \Delta G_i^k < 0\,, \tag{9.2}$$

where the asterisk denotes the current recursion level and ΔG_i^k is the binding energy of the interface that may be engaged further along the recursion tree, no stable complexes may be found and the algorithm should retreat to lower recursion levels.

As mentioned above, the assessment of complex stability (Equation 9.1) includes the calculation of a dissociation scenario. This may also be done using the recursive scheme of engaging and disengaging edges. First represent the protein complex as a graph and mark all edges as engaged. Then calculate all possible dissociation scenarios by the enumeration of all combinations of engaged and disengaged edges subject to the same restrictions as above. If a scenario with negative ΔG_0 is found, the complex is classed as unstable and the algorithm quits.

The outlined approach to the identification of protein complexes in crystal packing has been implemented in a web service called PISA (Protein Interfaces, Surfaces and Assemblies), publicly available at http://www.ebi.ac.uk/msd-srv/prot{_}int/pistart.html. The service provides a searchable database of protein complexes and interfaces calculated for all PDB structures. The service also allows the submission of structures for analysis, for which calculations are distributed over a variable number of CPUs depending on task complexity. Typical response times are under 1 min. However, difficult cases may take up to 20 min. The service also provides detailed descriptions of assemblies and interfaces, their visualization, and database search tools.

9.6 Chemical Graph Formats

A large number of file formats have been used to encode chemical graphs in various software packages. CACTVS [101], a popular chemical software package, supports many file formats and incorporates an input/output module manager. BABEL [102] is one of the first chemoinformatics tools evolved from a format conversion utility. One of the most popular file formats, MDL molfile [103], provides a simple way to represent molecules, is easy to use, and includes support for all chemical properties of atoms and bonds including atomic coordinates.

The Chemical Markup Language [104] (CML) uses SGML and XML and provides a rich and flexible format for providing chemical information. A more compact popular format that includes all the necessary information in chemical diagrams, including stereodescriptors and aromaticity, is the SMILES string notation. It is easy to understand and naturally extends the notion of the molecular formula, but it was not designed as a way to provide a molecule's identity since it does not include a clear, documented method for canonicalizing. This contrasts with IUPAC INChi, which is by definition unique for each molecule for various layers of isomerization.

9.7 Chemical Software Packages

Popular chemical software packages useful for processing molecular graphs are as follows:

1. CACTVS: a chemical package in C++ with a TCL-based scripting interface that is free for academic use.
2. Daylight: commercial software package that introduced smiles. It is ideal for chemical databases with very efficient subgraph and similarity capabilities and specialized indexing.
3. BABEL/OpenBabel: open-source package for converting chemical files but also incorporates a lot of chemical algorithms such as aromaticity detection.
4. OpenEyes: commercial package with Java interface and low-level support for most of the operations required by chemical databases and processing applications.
5. Chemistry Development Kit [105] (CDK): open-source package implemented in Java, useful also as a reference point for understanding chemical algorithms.

9.8 Chemical Databases and Resources

While it is always possible to generate sets of molecules computationally, it is often a lot more useful to work with molecules that have been experimentally observed or at least are available from some source. Databases that provide chemical data are:

1. PubChem [106]: 18.4 million entries, contains pure and characterized chemical compounds.
2. PDB ligands [107]: ligands and small molecules interacting with biopolymers.
3. ChEBI [108]: chemical entities of biological interest.
4. CCDC [109]: depositions of crystal structure data from X-ray and neutron diffraction studies, organic compounds, and metal–organic compounds.
5. KEGG [110]: information on molecular interaction networks.

9.9 Subgraph Isomorphism Solution in SQL

The concept of organizing data into relations was first proposed by Codd in 1970 [111] together with rules for integrity constraints and operators for the manipulation of data. These relational operators form the basis of SQL, which is standard across different database platforms today. The relational model is versatile and many problems can be tackled using it in preference to custom-written applications or algorithms. One such problem is subgraph isomorphism for which data can be visualized as a hypergraph [112] defined by a pair (*X*, *E*), where *X* is a set of vertices and *E* is a set of hyperedges, each of which is a nonempty subset of *X*. Thus one hyperedge can connect with more than one pair of vertices. In relational database terms, *X* is a set of attributes describing the application fields and *E* is a set of relations (tables without ordered rows). Chemical graphs where edges and vertices have attributes, like bond type and element type, can be represented as hypergraphs and therefore can be stored in a relational database using two tables: atoms and bonds. The atoms table contains vertices and their attributes, while the bonds table contains data on pairs of atoms and the attributes associated with their bond type. These tables are shown below:

Atoms table	
atom_id	Unique identified
molecule	Name of molecule
atom	Name of atom
symbol	Symbol for atom

Bonds table	
bond_id	Unique identified
atom1	Identifier for the first atom
atom2	Identifier for the second atom
bond-type	Type of covalent bond

As the graph is nondirectional, the bonds table must have two records for each pair of atoms. An example of a SQL query that can be used for searching an O=C–N fragment is presented below; its implementation is fairly straightforward.

```
SELECT
        a1.molecule, a1.atom, a2.atom, a3.atom
FROM
        atoms a1, bonds b1, atoms a2, bonds b2, atoms a3
WHERE
        a1.symbol = 'O' AND
        b1.atom1 = a1.atom_id AND
        a2.symbol = 'C' AND
        a2.atom_id = b1.atom2 AND
        b2.atom1 = a2.atom_id AND
        a3.symbol = 'N' AND
        a3.atom_id = b2.atom2;
```

Although modern relational database management systems (RDBMS) can execute complex searches very quickly, they cannot alter the complexity of the problem. The drawback of the approach described above is the issue of NP-complete, which is best illustrated using an example. Consider a fragment consisting of five carbon atoms in a star architecture such that the one carbon atom in the center is covalently bonded to the other four carbons via single bonds. How can a RDBMS be used to search for this fragment? The first approach is to use four nested loops of C–C bonds that can be presented as:

```
FOR EACH C–C bond
    FOR EACH C–C bond
        FOR EACH C–C bond
            FOR EACH C–C bond
```

The number of operations needed by a RDBMS in this case is $N \times N \times N \times N$ where N is the number of C–C bonds in the database. How can the search speed be improved? We can make use of the property that chemical graphs are sparse. If the bonds table is accessed by the first atom identifier, the number of hits is considerably reduced. This can be effected by building an index on the atom1 attribute. The number of operations now becomes $N \times M \times M \times M$, where M is the average number of bonds to C atoms, which could be between 2 and 4 and considerably less than N. Thus by introducing an index we reduce the number of operations by ~ $(N/3)$ power of bonds minus one. The conclusion is that simply by indexing the tables we can largely improve the performance; however, the complexity still remains NP-complete. The big challenge is reducing the complexity. Looking again at the nested loops figure and thinking about other ways of resolving these loops one can see that the number of valid hits is the number of combinations, that is, factorial rather than exponential. The reason for this is that we must exclude the hits where the same bond appears more than once. So, to reduce the exponential problem to a combinatorial one we add nonequal constraints to the SQL query. Now the question is, if we have a total number N of C–C bonds, then why after joining two bonds do we have N*4 (4 because a carbon atom can form four single bonds). The N*4 figure is with repetitions where the first and second bonds can swap positions; they produce a symmetry equivalent solution. The overall example has a fourfold symmetry and therefore 4 factorial (24) possible combinations of the bonds that give the same graph. This illustrates the point that it is symmetry that causes the NP-complex problem. The theory of the constraint satisfaction problem (CSP) [113] discusses methods of breaking symmetry [114]. CSP on a finite domain (FCSP) is a hypergraph, and therefore a relational database can be used to study it.

The constraints added to SQL queries to break the symmetry impose order on the vertices in the search subgraph, which can be found by solving the graph isomorphism problem on the graph. The number of constraints will then be the factorial of the number of graph vertices. Software that can be used for this purpose are SAUCY and NAUTY. The cost of solving this problem is not a critical factor as it is only executed once after which the resulting template can be applied to the whole database of graphs (or chemical compounds in our case). Examples of services that use the above technology are available at:
http://www.ebi.ac.uk/msd-srv/chemsearch (chemical compound searches)
http://www.ebi.ac.uk/msd-srv/msdmotif/chem (ligand searches of PDB)

9.10 Cycles in Graphs

The identification of ring structures constitutes a large part of the structural topology in the study and characterization of molecular structures. Chemical compounds with rings can be represented as graphs. If a graph represents a compound without cycles, then the number of edges it has will be equal to the number of atoms minus 1. Graphs of compounds without cycles are called acyclic and can be presented as trees. To draw a graph in 2D, a common practice is to start with the largest ring from the set of smallest rings, which highlights the importance of algorithms for determining the size of rings. Rings are characterized as essential or nonessential. Nonessential rings are those that are tied, multitied, or dependent rings. A tied ring is defined as a ring with one transannular bond that links directly two non-adjacent nodes of rings. Essential rings are rings other than nonessential rings.

Many different approaches for the extraction of cycles from molecular graphs such as the smallest set of smallest rings (SSSR), essential set of essential rings (ESER), extended set of smallest rings (ESSR), and the set of smallest cycles at edges (SSCE) have been used. The popular graph theory problem of finding the SSSR is ambiguous and, unlike the ERER, is not used in organic chemistry. The definition of what is essential varies, and in practice the problem is solved by finding an ESSR. To solve it a tree is constructed from each vertex of the graph, which terminates at any branch where an intersection occurs (where a loop is found). Only branches that grow from the selected atom and which do not encounter loops are allowed to grow further. This set of rings is used to determine the aromatic properties of individual rings as well as of ring subsets. It is also used to draw chemical structures in 2D and to annotate and search databases of chemical compounds. Both rings and atoms can take part in bonds such as hydrogen bonds and salt

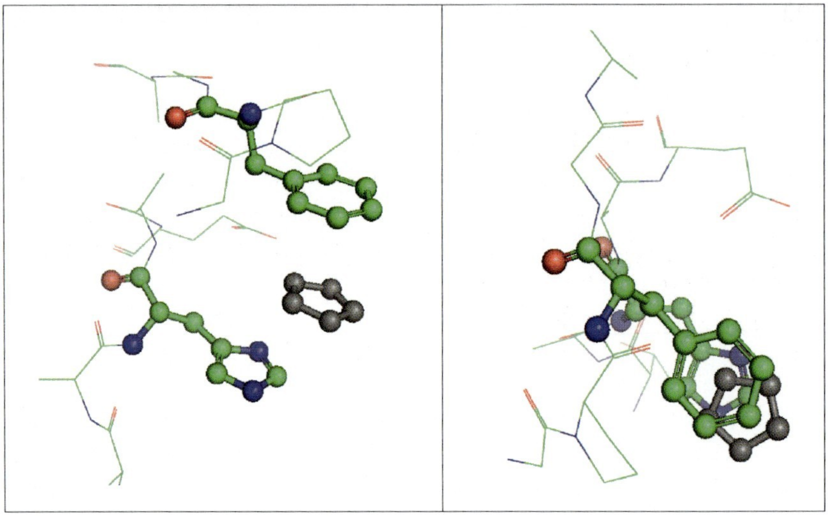

Figure 9.5 Ligand sandwich example found in PDB entries 1xmi, 1vc8, 2fl5, 1ewj, 1kll, 1l5q, 154d, and 4dcq.

bridges that occur in biological molecules. There are two special types of molecular interactions, ring-ring and ring-atom. The first one occurs when two aromatic rings are parallel and their pi electron clouds intersect. If this type of interaction happens on both sides of a ring, a sandwich is formed. In cases where the molecule in the middle is a ligand, the term ligand sandwich is used. There are a number of such sandwiches in the Protein Data Bank (PDB) (Figure 9.5).

For ring-atom interaction the favorable 3D conformation is when the atom approaches the ring orthogonally and close to the ring side rather than the center.

9.11
Aromatic Properties

Aromatic rings are made of a conjugated system of unsaturated bonds with delocalized pi electron clouds [115,116]. Such systems are planar and exhibit stronger stabilization than would be expected by conjugation alone. To assign aromatic properties to rings we select the smallest rings first and then use the breadth-first search (BFS) algorithm to identify conjugated bonds. This is not enough to assign aromaticity to individual rings; the whole rings system has to be taken into account.

9.12
Planar Subgraphs

There are two different contexts in which the term planarity can be used. Here we distinguish between chemical planarity and graph planarity. The latter is used for an abstract presentation of chemical compounds in 2D. In accordance with Kuratowski's theorem, graphs that do not have K3,3 and K5 graphs as subgraphs are planar, that is, they can be drawn in 2D without edge intersections (Figure 9.6).

In practice testing a graph against this theorem is considered to be a last choice due to its complexity. The simple test that says that a graph is nonplanar is as follows: given graph G(X,Y), where X is the vertices, Y is the edges, N is the number of vertices, and M is the number of edges, the graph with at least three vertices where $M > 3N - 6$ is nonplanar. Another test is to see if there are no three member rings and $M > 2N - 4$. Although all chemical compounds are probably planar in terms of graph theory, they are sometimes classified as 3D to underline essential properties influenced by 3D conformation. As well as aromatic ring systems conjugated bonds in other compounds exhibit chemical planarity. An example is the amide group shown in Figure 9.7.

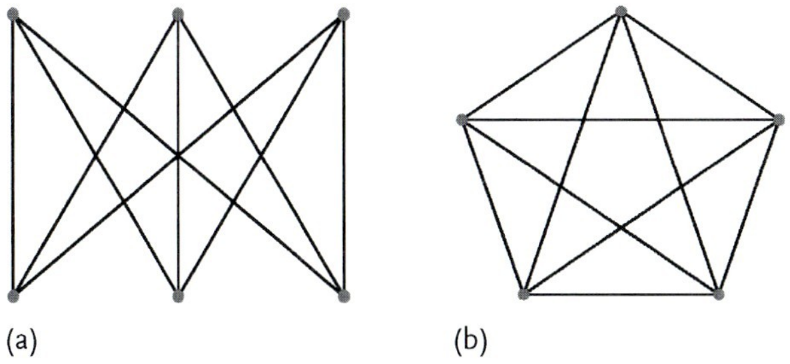

Figure 9.6 (a) K3,3 complete bipartite graph of six vertices three of which connect to each of the other three; (b) K5 complete graph of five vertices.

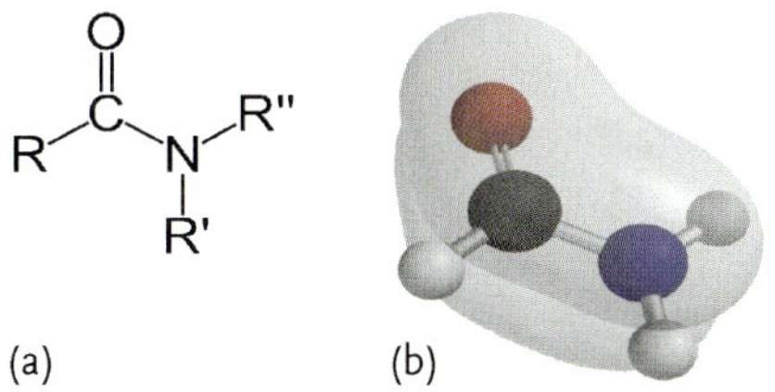

Figure 9.7 (a) Amide schematic 2D representation; (b) amide 3D representation with the lone pair electrons of the oxygen and nitrogen atoms delocalized.

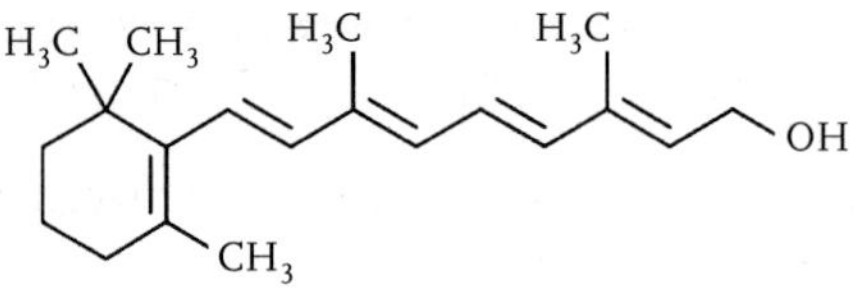

Figure 9.8 Carbon atoms donate pi electrons in vitamin A.

Amides are the most stable of all carbonyl groups. This stability is gained from delocalization of the lone pair of electrons of the oxygen and nitrogen atoms. Similar electron distributions are observed in most conjugated systems.

An example of how carbon atoms donate pi electrons in vitamin A is shown in Figure 9.8. However, the sequence of single and double bonds is not enough to gain additional stability and other checks must be applied.

9.13 Conclusion

Graph theory is widely used in chemo- and bioinformatics, where it addresses a range of practical problems, ranging from identification of molecular similarities to the analysis of macromolecular interactions and crystal packing. Being a well-developed mathematical discipline, it saves researchers from searching for technical implementations of their computational problems, allowing them to focus on the formulation of their problems in graph-theoretical terms. This, however, is not always straightforward and often represents a real challenge. Here, two main difficulties may be highlighted.

Firstly, objects in chemo- and bioinformatics are complex by nature. Graph representations of chemical molecules and, even more so, biological macromolecules should reflect both chemical and topological (3D) properties. A proper understanding of these properties is a necessary prerequisite for the development of practical algorithms within the field. A simplified description of chemical molecules as planar graphs is suitable for many purposes, however there needs to be special labeling when their chemical properties depend on a particular 3D conformation. In the case of biological macromolecules, this is yet more entangled, given that their functionality is most often associated with conformational changes.

Secondly, graph theory has size limits for the objects it works with. Many useful graph-theoretical problems, such as graph matching, are known to be NP-complete, which implies factorial complexity on the size of graphs. In the recent past, this was to a certain degree compensated for by an exponential growth of computing power; however, this progress has slowed to a linear trend more recently. The state of things today allows us to work comfortably in a graph-theoretical framework with molecules up to medium size (100 atoms), while larger ones require special treatment. Therefore, much effort is expended in finding suitable representations of chemical and biological objects in order to reduce the size of their graphs. Inevitably, this results in approximations, which may sometimes hinder the quality of results obtained from graph-theoretical approaches.

It appears that the future of graph-theoretical techniques in chemo- and bioinformatics will be most closely associated with the choice of particular descriptions in graph terms, which would allow the achievement of the most accurate solutions at controllable simplifications. Most of the practical research in molecular biology today is based on similarity studies, which requires powerful tools for comparison of different type of objects. It is obvious that molecular biology is moving towards a systematic approach and towards more complex objects: from small molecules to macromolecules, macromolecular assemblies and real biological objects like viruses. All this creates demand for more and more robust and efficient pattern recognition techniques. While various methods of pattern recognition are in use today, due to its flexibility and sound mathematical foundation, graph theory has become one of the main tools.

References

1 F.K. Brown, Chemoinformatics: what is it and how does it impact drug discovery? *Annual Reports in Medicinal Chemistry*, **1998**, 33, 375.

2 F. Brown, *Current Opinion in Drug Discovery & Development*, **2005**, 8(3), 296–302.

3 J. Gasteiger, T. Engel (eds), *Chemoinformatics: A Textbook*, John Wiley & Sons, Ltd, **2004**.

4 A.R. Leach, V.J. Gillet, *An Introduction to Chemoinformatics*, Springer, **2003**.

5 A. S. Nair. *Computational Biology & Bioinformatics – A Gentle Overview*, Communications of Computer Society of India, **2007**.

6 P. Baldi, S. Brunak, *Bioinformatics: The Machine Learning Approach*, 2nd edn, MIT Press, **2001**.

7 M.R. Barnes, I.C. Gray (eds), *Bioinformatics for Geneticists*, 1st edn, John Wiley & Sons, Ltd, **2003**.

8 J.M. Claverie, C. Notredame, *Bioinformatics for Dummies*, John Wiley & Sons, Ltd, **2003**.

9 D. Gilbert, *Briefings in Bioinformatics*, **2004**, 5(3), 300–304.

10 A. Patani, E.J. LaVoie, *Chemical Reviews*, **1996**, 96, 3147–3176.

11 D.L. Wheeler, T. Barrett, D.A. Benson, S.H. Bryant, K. Canese, V. Chetvernin, D.M. Church, M. DiCuccio, R. Edgar, *Nucleic Acids Research*, **2007**, 35, D5–D12.

12 C. Brooksbank, G. Cameron, J. Thornton, *Nucleic Acids Research*, **2005**, 33, D46–D53.

13 D.S. Wishart, C. Knox, A.C. Guo, S. Shrivastava, M. Hassanali, P. Stothard, Z. Chang, J. Woolsey, *Nucleic Acids Research*, **2006**, 34, D668–D672.

14 A. Golovin, T.J. Oldfield, J.G. Tate, S. Velankar, G.J. Barton, H. Boutselakis, D. Dimitropoulos, J. Fillon, A. Hussain, J.M.C. Ionides, M. John, P.A. Keller, E. Krissinel, P. McNeil, A. Naim, R. Newman, A. Pajon, J. Pineda, A. Rachedi, J. Copeland, A. Sitnov, S. Sobhany, A. Suarez-Uruena, J. Swaminathan, M. Tagari, S. Tromm,

W. Vranken, K. Henrick, *Nucleic Acids Research*, **2004**, 32, D211–D216.
15 Y.M. Galperin, *Nucleic Acids Research*, **2008**, 36, D2–D4.
16 L. Pachter, B. Sturmfels, *Algebraic Statistics for Computational Biology*, Cambridge University Press, **2005**.
17 R. Durbin, S. Eddy, A. Krogh, G. Mitchison, *Biological Sequence Analysis*, Cambridge University Press, **1998**.
18 D. Gusfield, *Algorithms on Strings, Trees, and Sequences: Computer Science and Computational Biology*, Cambridge University Press, **1997**.
19 D.W. Mount, *Bioinformatics: Sequence and Genome Analysis*, Spring Harbor Press, **2002**.
20 A.D. Baxevanis, B.F.F. Ouellette (eds), *Bioinformatics: A Practical Guide to the Analysis of Genes and Proteins*, 3rd edn, John Wiley & Sons, Ltd, **2005**.
21 I.A. Kohane, A. Kho, A.J. Butte, *Microarrays for an Integrative Genomics*. MIT Press, **2002**.
22 E.O. Wilson, F.M. Peter (eds), *Biodiversity*. National Academy Press, **1988**.
23 C.B. Charlesworth, D. Charlesworth, *Evolution*. Oxfordshire: Oxford University Press, **2003**.
24 R.H. Whittaker, *Taxon*, **1972**, 21, 213–251.
25 S.R. Freeman, J.C. Herron, *Evolutionary Analysis*, Prentice-Hall, **2003**.
26 J.L. Snoep, H.V. Westerhoff, From isolation to integration, a systems biology approach for building the silicon cell. In: L. Alberghina, H. V Westerhoff (eds), *Systems Biology: Definitions and Perspectives*, Springer, p 7, **2005**.
27 M.D. Mesarovic, *Systems Theory and Biology*, Springer, **1968**.
28 A. Nayeem, D. Sitkoff, S. Krystek, Jr, *Protein Science*, **2006**, 15, 808–824.
29 D. Bonchev, D.H. Rouvray, *Chemical Graph Theory: Introduction and Fundamentals*, Gordon and Breach Science Publishers, **1990**.
30 D.H. Rouvray, A.T. Balaban, Chemical applications of graph theory. In: R.J. Wilson, L.W. Beineke (eds), *Applications of Graph Theory*, Academic Press, **1979**, 177–221.
31 C. Francke, R.J. Siezen, B. Teusink, *Trends in Microbiology*, **2005**, 13(11), 550–558.
32 J.A. Papin, N.D. Price, B.O. Palsson, *Genome Research*, **2002**, 12, 1889–1900.
33 N.D. Price, J.L. Reed, J.A. Papin, S.J. Wiback, B.O. Palsson, *Journal of Theoretical Biology*, **2003**, 225, 185–194.
34 J.A. Papin, J. Stelling, N.D. Price, S. Klamt, S. Schuster, B.O. Palsson, *Trends in Biotechnology*, **2004**, 22(8), 400–405.
35 S. Schuster, D.A. Fell, T. Dandekar, *Nature Biotechnology*, **2000**, 18, 326–332.
36 D. Penny, M.D. Hendy, M.A. Steel, *Trends in Ecology and Evolution*, **1992**, 7, 73–79.
37 A. Dress, K.T. Huber, V. Moulton, *Documenta Mathematica LSU*, **2001**, 121–139.
38 F.D. Ciccarelli, *Science*, **2006**, 311(5765), 1283–1287.
39 I. Letunic, *Bioinformatics*, **2007**, 23(1), 127–128.
40 Y. Ye, A. Godzik, *Bioinformatics*, **2005**, 21(10), 2362–2369.
41 N.N. Alexandrov, *Protein Engineering*, **1996**, 9, 727–732.
42 E.M. Mitchell, P.J. Artymiuk, D.W. Rice, P. Willett, *Journal of Molecular Biology*, **1990**, 212, 151–166.
43 H.M. Grindley, P.J. Artymiuk, D.W. Rice, P. Willett, *Journal of Molecular Biology*, **1993**, 229, 707–721.
44 J.F. Gibrat, T. Madej, S.H. Bryant, *Current Opinion in Structural Biology*, **1996**, 6, 377–385.
45 E. Krissinel, K. Henrick, *Acta Crystallographica*, **2004**, D60, 2256–2268.
46 E. Krissinel, K. Henrick, Multiple alignment of protein structures in three dimensions. In: M.R. Berthold *et al.* (eds), *CompLife*, Springer, **2005**, 67–78.
47 M. Randic, C.L. Wilkins, *Journal of Chemical Information and Computer Sciences*, **1979**, 19, 31.
48 E.H. Sussenguth, *Journal of Chemical Documentation*, **1965**, 5, 36–43.
49 E.J. Gardiner, P. Willett, P.J. Artymiuk, *Journal of Chemical Information and Computer Sciences*, **2000**, 40, 273–279.
50 E. Krissinel, K. Henrick, *Journal of Molecular Biology*, **2007**, 372, 774–797.
51 I. Koch, F. Kaden, J. Selbig, *Proteins: Structure, Function, Genetics*, **1992**, 12, 314–323.
52 E.C. Ujah, *A study of beta sheet motifs at different levels of structural abstraction using graph theoretic and dynamic programming*

techniques, PhD Thesis, University of Sheffield, **1992**.

53 H.M. Berman, J. Westbrook, Z. Feng, G. Gilliland, T.N. Bhat, H. Weissig, I.N. Shindyalov, P.E. Bourne, *Nucleic Acids Research*, **2000**, 28,235–242.

54 J.J. Sylvester, On the application of the new atomic theory to the graphical representation of the invariants and covariants of binary quantics, with three appendices, *American Journal of Mathematics*, **1878**, 1:64–128.

55 H.L. Bodlaender, J.R. Gilbert, H. Hafsteinsson, T. Kloks, Approximating treewidth, pathwidth, and shortest elimination tree height, *Journal of Algorithms*, **1995**, 18(2), 238–255.

56 C. Rucker, M. Meringer, How many organic compounds are graph-theoretically nonplanar?, *Communications in Mathematical and in Computer Chemistry*, **2002**, 45,153–172, ISSN:0340-6253.

57 R.S. Cahn, C.K. Ingold, V. Prelog, Specification of molecular chirality, *Angewandte Chemie – International Edition 5*, **1996**, 4, 385–415, ISSN:1433-7851, DOI:10.1002/anie.196603851.

58 J. Braun, R. Gugish, A. Kerber, R. Laue, M. Meringer, C. Rucker, MOLGEN-CID – a canonizer for molecules and graphs accessible through the Internet, *Journal of Chemical Information and Computer Sciences*, **2004**, 44(2), 542–548, ISSN:0095-2338.

59 C.A. James *et al.*, *Daylight Theory Manual Daylight 4.9, Daylight Chemical Information Systems, Inc.*, **2008**, URL:http://www.daylight.com/dayhtml/doc/theory/.

60 S.E. Stein, S.R. Heller, D. Tchekhovski, *An Open Standard for Chemical Structure Representation – The IUPAC Chemical Identifier*. Proceedings of the 2003 Nimes International Chemical Information Conference, **2003**, 131–143, ISBN:1-873699-93-X.

61 G.M. Downs, J.M. Barnard, Clustering methods and their uses in computational chemistry, *Reviews in Computational Chemistry*, **2003**, 18, ISBN:9780471215769 DOI:10.1002/0471433519.ch1.

62 J.M. Berg, J.L. Tymoczko, L. Stryer, *Biochemistry*. W.H. Freeman and Co., **2002**.

63 C.M. Dobson, The nature and significance of protein folding. In: *Mechanisms of Protein Folding*, 2nd edn, R.H. Pain (ed), Frontiers in Molecular Biology Series. Oxford University Press, New York, **2000**.

64 C. Branden, J. Tooze, *Introduction to Protein Structure*. 2nd edn, Garland Publishing, **1999**.

65 L. Holm, C. Sander, *Journal of Molecular Biology*, **1993**, 233, 123–138.

66 C.A. Orengo, W.R. Taylor, *Methods in Enzymology*, **1996**, 266, 617–635.

67 A. Falicov, F.E. Cohen, *Journal of Molecular Biology*, **1996**, 258, 871–892.

68 S. Subbiah, D.V. Laurents, M. Levitt, *Current Biology*, **1993**, 3(3), 141–148.

69 M. Gerstein, M. Levitt, *Using Iterative Dynamic Programming to ObtainAccurate Pairwise and Multiple Alignments of Protein Structures*. Proceedings of the 4th International Conference on Intelligent Systems for Molecular Biology, Menlo Park, CA, AAAI Press, **1996**, 59–67.

70 A.P. Singh, D.L. Brutlag, *Hierarchical Protein Structure Superposition Using Both Secondary Structure and Atomic Representations*. Proceedings of the 4th International Conference on Intelligent Systems for Molecular Biology, ISMB-97, AAAI Press, **1997**, 284–293.

71 G. Vriend, C. Sander, *Proteins*, **1991**, 11, 52–58.

72 K. Mizuguchi, N. Go, *Protein Engineering*, **1995**, 8(4), 353–362.

73 I.N. Shindyalov, P.E. Bourne, *Protein Engineering*, **1998**, 11(9), 739–747.

74 G.J. Kleywegt , T.A. Jones, *Methods in Enzymology*, **1997**, 277, 525–545.

75 M. Zuker, R.L. Somorjai, *Bulletin of Mathematical Biology*, **1989**, 51(1), 55–78.

76 W. Taylor, C. Orengo, *Journal of Molecular Biology*, **1989**, 208, 1–22.

77 A. Godzik, J. Skolnick, *CABIOS* **1994**, 10, 587–596.

78 R.B. Russell, G.B. Barton, *Proteins: Structure, Function, and Genetics*, **1992**, 14, 309–323.

79 A. Sali, T. Blundell, *Journal of Molecular Biology*, **1990**, 212, 403–428.

80 D.W. Barakat, P.M. Dean, *Journal of Computer-Aided Molecular Design*, **1991**, 5, 107–117.

81 J. Leluk, L. Konieczny, I. Roterman, *Bioinformatics*, **2003**, 19(1), 117–124.

82 J. Jung, B. Lee, *Protein Engineering*, **2000**, 13(8), 535–543.

83 H. Kato, Y. Takahashi, *Journal of Chemical Software*, **2001**, 7(4), 161–170.

84 K. Shearer, H. Bunke, S. Venkatesh, *Pattern Recognition*, **2001**, 34, 1075–1091.

85 E. Krissinel, K. Henrick, *Software – Practice and Experience*, **2004**, 34(6), 591–607.

86 T. Liu, B. Chu, Light scattering by proteins. In: *Encyclopedia of Surface and Colloid Science*, A. Hubbard (ed), Marcel Dekker, **2002**, pp. 3023–3043.

87 C. Dass, *Principles and Practice of Biological Mass Spectrometry*. John Wiley & Sons, Ltd, **2001**.

88 L.A. Feigin, D.I. Svergun, *Structure Analysis by Small Angle X-ray and Neutron Scattering*, Plenum Press, **1987**.

89 D.I. Svergun, M.H.J. Koch, *Current Opinion in Structural Biology*, **2002**, 12, 654–660.

90 J. Cavanagh, W.J. Fairbrother, A.G. Palmer III, N.J. Skelton, *Protein NMR Spectroscopy*, Academic Press, **1996**.

91 J. Frank, *Three-Dimensional Electron Microscopy of Macromolecular Assemblies*, Oxford University Press, **2006**.

92 T.L. Blundell, L.N. Johnson, *Protein Crystallography*, Academic Press, **1976**.

93 N. Horton, M. Lewis, *Protein Science*, **1992**, 1, 169–181.

94 J. Janin and F. Rodier, *Proteins: Structure, Function, and Genetics*, **1995**, 23, 580–587.

95 E.N. Baker, R.E. Hubbard, *Progress in Biophysics and Molecular Biology*, **1984**, 44, 97–179.

96 J. Janin, S. Miller, C. Chothia, *Journal of Molecular Biology*, **1988**, 204, 155–164.

97 S. Jones, J.M. Thornton, *Proceedings of the National Academy of Science of the USA*, **1996**, 93, 13–20.

98 K. Henrick, J. Thornton, *Trends in Biochemical Sciences*, **1998**, 23, 358–361.

99 H. Ponstingl, T. Kabir, J. Thornton, *Journal of Applied Crystallography*, **2003**, 36, 1116–1122.

100 W.J. Moore, *Physical Chemistry*. Prentice-Hall, **1972**.

101 W.D. Ihlenfeldt, Y. Takahashi, H. Abe, S. Sasaki, Computation and management of chemical properties in CACTVS: an extensible networked approach toward modularity and flexibility, *Journal of Chemical Information and Computer Sciences*, **1994**, 34, 109–116, ISSN:0095-2338.

102 R. Guha, M.T. Howard, G.R. Hutchison, P. Murray-Rust, H. Rzepa, C. Steinbeck, J.K. Wegner, E.L. Willighagen, The blue obelisk-interoperability in chemical informatics, *Journal of Chemical Information and Modeling*, **2006**, ISSN:1549-9596, DOI:10.1021/ci050400b.

103 A. Dalby, J.G. Nourse, W.D. Hounshell, A.K.I. Gushurst, D.L. Grier *et al.*, Description of several chemical structure file formats used by computer programs developed at Molecular Design Limited, *Journal of Chemical Information and Computer Sciences*, **1992**, 32, 244–255, ISSN:0095-2338.

104 P. Murray-Rust, H.S. Rzepa, Chemical Markup, XML, and the Worldwide Web. 1. Basic Principles, *Journal of Chemical Information and Computer Sciences*, **1999**, 39,928–39,942, ISSN:0095-2338, DOI:10.1021/ci990052b.

105 C. Steinbeck, Y.Q. Han, S. Kuhn, O. Horlacher, E. Luttmann, E.L. Willighagen, The Chemistry Development Kit (CDK): an open-source Java library for chemo- and bioinformatics, *Journal of Chemical Information and Computer Sciences*, **2003**, 43,493–500. ISSN:0095-2338, DOI:10.1021/ci025584y.

106 E. Sayers, PubChem: an Entrez database of small molecules. *NLM Technical Bulletin*, **2005**, Jan–Feb, 342:e2.

107 D. Dimitropoulos, J. Ionides, K. Henrick, UNIT 14.3: Using MSDchem to search the PDB ligand dictionary, *Current Protocols in Bioinformatics*, **2006**, 14.3.1–14.3.3, Hoboken, NJ, ISBN: 978-0-471-25093-7.

108 K. Degtyarenko, P. de Matos, M. Ennis, J. Hastings, M. Zbinden, A. McNaught, R. Alcantara, M. Darsow, M. Guedj, M. Ashburner, ChEBI: a database and ontology for chemical entities of biological interest, *Nucleic Acids Research*, **2008**, 36, D344–D350, ISSN 0305-1048.

109 F.H. Allen, The Cambridge Structural Database: a quarter of a million crystal structures and rising, *Acta Crystallographica*, **2002**, B58, 380–388, ISSN:0108-7681, DOI:10.1107/S0108768102003890.

110 M. Kanehisa, S. Goto, M. Hattori, K.F. Aoki-Kinoshita, M. Itoh, S. Kawashima, T. Katayama, M. Araki, M. Hirakawa, From genomics to chemical genomics: new developments in KEGG, *Nucleic Acids Research*, **2006**, 34, D354–357, ISSN 0305-1048.

111 E.F. Codd A relational model of data for large shared data banks. *Communications of the ACM* **1970**, **13**, 377–387.

112 R. Fagin, Degrees of acyclicity for hypergraphs and relational database schemes. *Journal of the ACM*, **1983**, 30(3), 514–550.

113 M. Rudolf, Utilizing constraint satisfaction techniques for efficient graph pattern matching. *Lecture Notes in Computer Science*. **1998**, 1764, 238–251.

114 J. Crawford, M. Ginsberg, E. Luks, A. Roy, *Symmetry-Breaking Predicates for Search Problems*. Proceedings of the 5th International Conference on Principles of Knowledge Representation and Reasoning; KR'96, **1996**.

115 P.v.R. Schleyer, Aromaticity (editorial). *Chemical Reviews*, **2001**, 101, 1115–1118.

116 T. Balaban, P.v.R. Schleyer, H.S. Rzepa, Crocker, not Armit and Robinson, begat the six aromatic electrons. *Chemical Reviews*, **2005**, 105, 3436–3447.

10
Structural and Functional Dynamics in Cortical and Neuronal Networks

Marcus Kaiser and Jennifer Simonotto

10.1
Introduction

Nervous systems are complex networks par excellence, capable of generating and integrating information from multiple external and internal sources. Within the neuroanatomical network (structural connectivity), the nonlinear dynamics of neurons and neuronal populations result in patterns of statistical dependencies (functional connectivity) and causal interactions (effective connectivity), defining three major modalities of complex brain networks [61]. How does the structure of the network relate to its function and what effect do changes of edge or node properties have [37]? Since 1992 [1,76] tools from graph theory and network analysis [18] have been applied to study these questions in neural systems (cf. http://www.biological-networks.org).

After describing the properties of neural systems – fiber tracts between brain areas of the mammalian cortex and axons between individual neurons in the nematode *Caenorhabditis elegans* – we describe dynamics in structure and function. Structural dynamics concern changes in network topology by deleting or adding edges or nodes. We describe the deletion of components in terms of the removal of tissue during strokes or head injuries and the addition of components during the development and growth of neural systems. The network topology is robust to random attacks but reacts critically to targeted attacks – similar to a scale-free network. The simulations on network evolution show that spatial growth and time windows during development are sufficient for generating small-world and multicluster networks.

Functional dynamics concern changes in the activity level of individual neurons or of cortical regions. We observe the spreading of activation, first describing spreading in excitable media and then in cardiac tissue before moving to spreading of epileptic seizures in neural networks. In neural networks, the hierarchical and modular topology provides novel mechanisms to limit spreading in addition to the known influence of inhibitory nodes. Finally, we discuss principles of neural organization. Whereas we applied concepts of network analysis to neuroscience, we reverse this process by

Analysis of Complex Networks: From Biology to Linguistics. Edited by Matthias Dehmer and Frank Emmert-Streib

ISBN: 978-3-527-32345-6

suggesting theoretical and computational challenges of network science that appeared during the analysis of neural systems.

10.1.1
Properties of Cortical and Neuronal Networks

Cortical areas are brain modules that are defined by structural (microscopic) architecture. Observing the thickness and cell types of the cortical layers, several cortical areas can be distinguished [10]. Furthermore, areas also show a functional specialization. Within one area further subunits (cortical columns) exist. Using neuroanatomical techniques, it can be tested which areas are connected, that means that projections in one or both directions between the areas do exist. If a fiber projection between two areas is found, the value '1' is entered in the adjacency matrix; the value '0' defines absent connections or cases where the existence of connections was not tested (Figure 10.1a). For a brain with N defined areas there would be N injection studies and, in total, $N * (N - 1)$ potential connections to be tested (loops are not tested). For the neural systems described here, except for the neuronal network of *C. elegans*, most but not all connections were tested. Ideas how to predict missing connections by deciding which nodes are more likely to be linked have included the combination of topological and spatial features ([15], see Section 10.1.2) as well as the combination of topological information and response latencies [11].

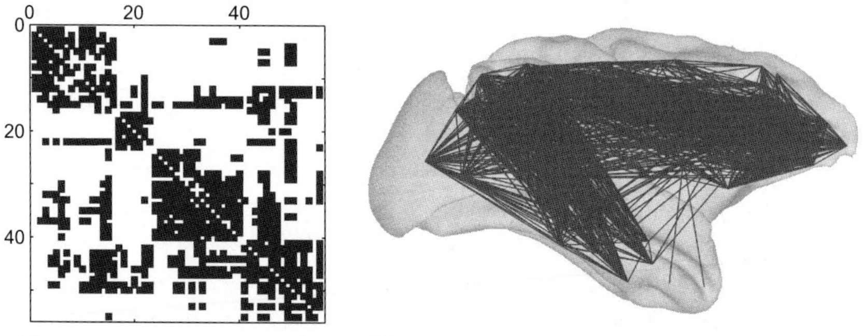

Figure 10.1 (a) Adjacency matrix of the cat connectivity network (55 nodes; 891 directed edges). Dots represent "ones" and white spaces the "zero" entries of the adjacency matrix. (b) Macaque cortex (95 nodes; 2402 directed edges).

10.1.1.1 Modularity

Contrary to popular belief, cortical networks are not completely connected, i.e., *not* "everything is connected to everything else": only about 30% of all possible connections (arcs) between areas do exist. Instead, highly connected sets of nodes (*clusters*) are found that correspond to functional differentiation of areas. For example, clusters corresponding to visual, auditory, somatosensory, and frontolimbic processing were found in the cat cortical connectivity network [34]. Furthermore, about 20% of the connections are unidirectional [30], i.e., a direct projection from area A to area B, but not vice versa, exists. Although some of these connections might be bidirectional as the reverse direction was not tested, there were several cases where it was confirmed that projections were unidirectional. Therefore, measures that worked for directed graphs were used.

Until now, there has not been enough information about connectivity in the human brain that would allow network analysis [17]. However, several new noninvasive methods, including diffusion tensor imaging [66] and resting state networks [2], are under development and might help to define human connectivity in the future. At the moment, however, we are bound to analyze known connectivity in the cat and the macaque (rhesus monkey, Figure 10.1b cortical networks [52, 61]. Both networks exhibit clusters, i.e., areas belonging to a cluster have many existing connections between them but there are few connections to areas of different clusters [54, 77]. These clusters are also functional and spatial units. Two connected areas tend to be spatially adjacent on the cortical surface and tend to have a similar function (e.g., both take part in visual processing). Whereas there is a preference for short-length connections to spatially neighboring areas for the macaque, about 10% of the connections cover a long distance (≥ 40 mm) – sometimes close to the maximum possible distance (69 mm) between two areas of one hemisphere [41].

10.1.1.2 Small-World Features

Many complex networks exhibit properties of small-world networks [71]. In these networks neighbors are better connected than in comparable Erdös–Rényi random networks [27] (called random networks throughout the text), whereas the average path length remains as low as in random networks. Formally, the average shortest path (ASP, similar, though not identical, to characteristic path length ℓ [70]) of a network with N nodes is the average number of edges that has to be crossed on the shortest path from any one node to another:

$$ASP = \frac{1}{N(N-1)} \sum_{i,j} d(i,j) \quad \text{with} \quad i \neq j, \tag{10.1}$$

where $d(i,j)$ is the length of the shortest path between nodes i and j.

The neighborhood connectivity is usually measured by the clustering coefficient. The clustering coefficient of one node v with k_v neighbors is

$$C_v = \frac{|E(\Gamma_v)|}{\binom{k_v}{2}}, \tag{10.2}$$

where $|E(\Gamma_v)|$ is the number of edges in the neighborhood of v and $\binom{k_v}{2}$ is the number of possible edges [70]. In the following analysis, we use the term clustering coefficient as the average clustering coefficient for all nodes of a network.

Small-world properties were found on different organizational levels of neural networks: from the tiny nematode *C. elegans* with about 300 neurons [71] to cortical networks of the cat and the macaque [33, 34]. Whereas the clustering coefficient for the macaque is 49% (16% in random networks), the ASP is comparatively low with 2.2 (2.0 in random networks). That is, on average only one or two intermediate areas are on the shortest path between two areas. Note that a high clustering coefficient does not necessarily correlate with the existence of multiple clusters. Indeed, the standard model for generating small-world networks by rewiring regular networks [71] does not lead to multiple clusters.

10.1.1.3 Scale-Free Features

Features of scale-free networks, such as power-law degree distributions, were found for *functional* brain networks. Dodel [22] developed a deterministic clustering method that combines cross-correlations between fMRI (functional magnetic resonance imaging) signal time courses and elements of graph theory to reveal brain functional connectivity. Three-dimensional image voxels (volume elements) form nodes of a graph, and their temporal correlation matrix forms the weight matrix of the edges between the nodes. Thus a network can be implemented based entirely on fMRI data, defining as "connected" those voxels that are functionally linked, that is, that are correlated beyond a certain threshold r_c. A set of experiments examined the resulting functional brain networks [25] obtained from human visual and motor cortex during a finger-tapping task. Over a wide range of threshold values r_c the functional correlation matrix resulted in clearly defined networks with characteristic and robust properties. Their degree distribution and the probability of finding a link versus metric distance both decay as a power law. Their characteristic path length is short (similar to that of equivalent random networks), while the clustering coefficient is several orders of magnitude larger. Scaling and small-world properties persisted across different tasks and within different locations of the brain. The power-law degree distribution indicates a scale-free network, but one has to keep in mind that the use of connections between individual voxels is not very meaningful in terms of a physiological functional unit.

The question about a power-law degree distribution for *structural* brain networks is even more difficult [45]. Such analysis is hindered by two problems: the potentially incomplete network data and the low number of network nodes (usually less than 100 brain areas). Whereas highly connected nodes exist, the network is too small to test the degree distribution for a power-law behavior. For the cortical network of the cat with 56 and the macaque with 66 nodes, there are some highly connected nodes; the top 5 nodes for each network are shown in Table 10.1. The neuronal network of *C. elegans* with about 300 neurons – enough for an analysis of the degree distribution – was found to be small-world but not scale-free [5].

Whereas a direct analysis of the degree distribution of cortical networks was impossible, we used an indirect approach in testing the response of cortical networks toward structural damage (Section 10.2.1.2). We found that effects of damage on the modeled cat and macaque brain connectivity networks are largely similar to those observed in scale-free networks. Furthermore, the similarity of scale-free and original cortical networks, as measured by graph similarity, was higher than for other benchmark networks [45].

We note that this issue remains controversial. A study of the human resting state network between cortical areas [2] concluded that the resting state network is not a scale-free network as (a) it is more resilient toward targeted attack compared to a scale-free benchmark network, (b) the degree distribution is not a power law, and (c) late developing areas such as the dorsolateral prefrontal cortex are among the hubs of the network. The structural network

Table 10.1 Overview of the most highly connected regions in the cat and macaque network. The table shows the total number of connections of the region (degree) as well as the number of incoming/afferent (in-degree) and outgoing/efferent (out-degree) connections. The maximal possible number of connections would have been 110 for the cat and 130 for the macaque.

Rank	Area	Total	Incoming	Outgoing
Cat				
1	AES	59	30	29
2	Ia	55	29	26
3	7	54	28	26
4	Ig	52	22	30
5	5al	49	30	19
Macaque				
1	A7B	43	23	20
2	LIP	42	19	23
3	A46	42	23	19
4	FEF	38	19	19
5	TPT	37	18	19

that we analyzed, however, differed from the resting state functional network. First, the resilience toward targeted attack was comparable with that of a scale-free network. Second, though the degree does not follow a power-law distribution, this might be due to the small size of the network and incomplete sampling of connections between regions (cf. [65] for the effect of sampling on the classification of networks).

10.1.1.4 **Spatial Layout**

Whereas neural components such as individual neurons or axons may undergo changes, their positions remain the same for mature neural systems after the process of early migration. Observing the pattern of neural connectivity, one might ask whether there are underlying principles for the spatial organization of neural systems. An early concept, posed in 1994 [12], was inspired by the layout of artificial information processing systems. For microchips, increasing the length of electric wires increases the energy loss through heat dissipation. In a similar way, neural systems were thought to be optimized to reduce wiring costs as well. In the brain, energy is consumed for establishing fiber tracts between areas and for propagating action potentials over these fibers. Thus, the total length of all wires should be kept as short as possible. This has led to the idea of optimal component placement according to which modules are arranged in a way so that every rearrangement of modules would lead to an increase in total wiring length [12–14].

It has been proposed for several neural systems – including the *C. elegans* neural network and subsets of cortical networks – that components are indeed optimally placed [12]. This means that all node position permutations of the network – while connections are unchanged – result in higher total connection length. Therefore, the placement of nodes is optimized to minimize the *total* wiring length. However, using larger data sets than were used in the original study, we found that a reduction in wiring length by swapping the position of network nodes was possible [43].

For the macaque, we analyzed wiring length using the spatial three-dimensional positions of 95 areas and their connectivity. We also looked at the wiring of individual neurons in *C. elegans*. The total wiring length was between the case of only establishing the shortest possible connections and establishing connections randomly regardless of distance. For the original networks, a further reduction of total wiring length by 32% for the macaque and by almost 50% for *C. elegans* was possible (Figure 10.2). Reducing the total wiring length was possible due to the number of long-distance connections in the original networks [41], some of them even spanning almost the largest possible distance between areas. Why would these metabolically expensive connections exist in such large numbers? We tested the effect of removing all long-distance connections and replacing them by short-distance connections. Whereas several network measures improved, the value for the

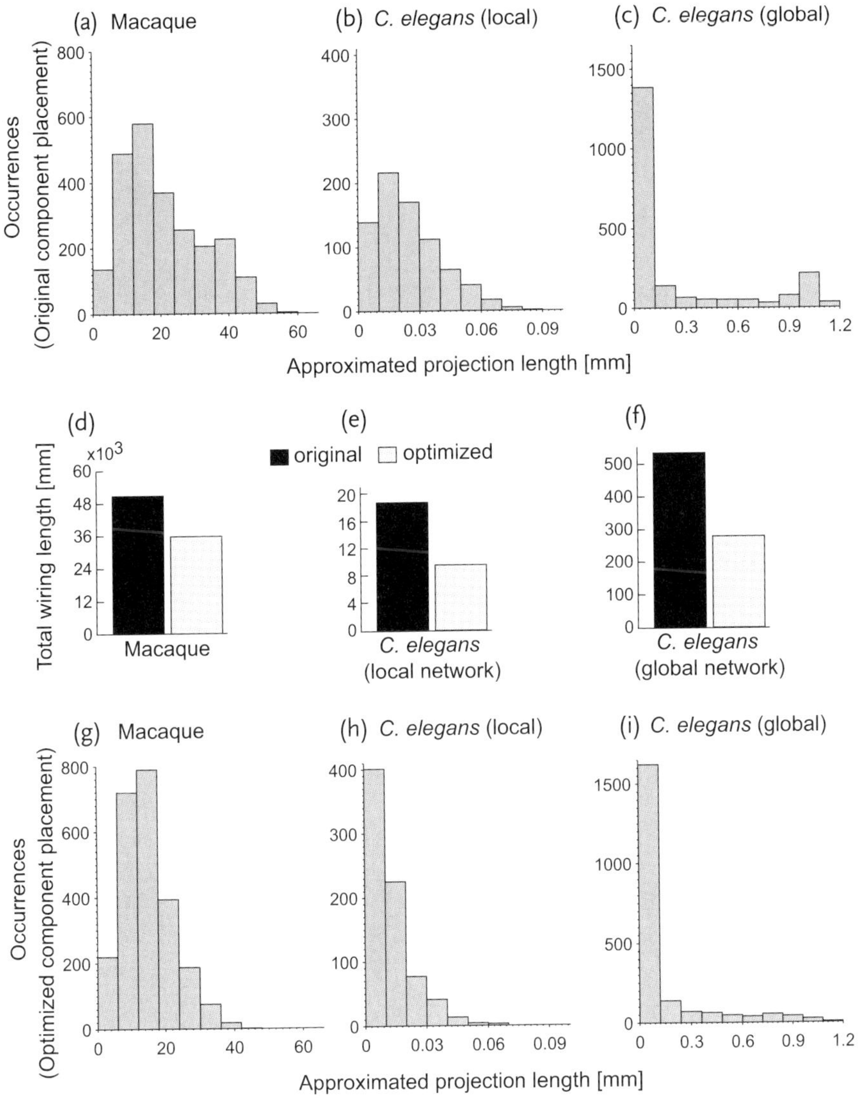

Figure 10.2 Projection length distribution and total wiring length for original and rearranged neural networks [43]. (a–c) Approximated projection length distribution in neural networks. Macaque monkey cortical connectivity network with 95 areas and 2402 projections (a). Local distribution of connections within rostral ganglia of *C. elegans* with 131 neurons and 764 projections (b). Global *C. elegans* neural network with 277 neurons and 2105 connections (c). (d–f) Reduction in total wiring length by rearranged layouts yielded by simulated annealing for macaque cortical network (d), *C. elegans* local network (neurons within rostral ganglia) (e), and global *C. elegans* network (f). (g–i) Approximated projection length distribution in neural networks with optimized component placement. Macaque monkey cortical connectivity network (g). Local distribution of connections within rostral ganglia of *C. elegans* (h). Global *C. elegans* neural network (i). For all optimized networks, the number of long-distance connections is reduced compared to the original length distribution in (a–c).

ASP increased when long-distance connections were unavailable. Retaining a lower ASP has two benefits. First, there are fewer intermediate areas that might distort the signal. Second, as fewer areas are part of shortest paths, the transmission delay along a pathway is reduced. The propagation of signals over long distances, without any delay imposed by intermediate nodes, has an effect on synchronization as well: both nearby (directly connected) areas and faraway areas are able to receive a signal at about the same time and could have synchronous processing [43]. A low ASP might also be necessary because of the properties of neurons: John von Neumann, taking into account the low processing speed and accuracy of individual neurons, suggested that neural computation needed to be highly parallel with using a low number of subsequent processing steps [68].

10.1.2
Prediction of Neural Connectivity

As for other biological systems, incomplete data sets are a problem for structural brain connectivity. Because tract-tracing studies require an enormous effort and take a long time, is there any way to predict missing connections? We tried to predict whether two nodes of a neural network are connected based on topological as well as spatial features of the nodes [15]. Starting with all areas being disconnected, pairs of areas with similar sets of features are linked together, in an attempt to recover the original network structure. Topological features included node degree, clustering coefficient, average shortest path, and matching index (also called Jaccard clustering coefficient). Spatial or geometrical features included local density of nodes, coefficient of variation of the nearest distances, area size of each cortical region, and cartesian coordinates of the cortical areas' center of mass.

Inferring network connectivity from the properties of the nodes already resulted in remarkably good reconstructions of the global network organization, with the topological features allowing slightly superior accuracy to the geometrical ones. Analogous reconstruction attempts for the *C. elegans* neuronal network resulted in substantially poorer recovery of known connections, indicating that cortical area interconnections are relatively stronger with respect to the considered topological and geometrical properties than neuronal projections in the nematode [15].

We tested the performance of the prediction for the network of the macaque visual cortex reviewing reconstructed networks in light of whether they were able to predict previously unknown connections. For the combination of the best two topological and two spatial measures, 111 currently unknown projections were predicted to exist and 174 connections were predicted to be absent, yielding a realistic ratio for predicted existing connections of 39%, out of all unknown connections. The predicted projections are shown as yellow fields in the reconstructed subgraph matrix in Figure 10.3. The figure also indicates mismatches (red fields) between the original and reconstructed matrices, either existing connections that were left out of the reconstructed matrix (90 cases) or absent connections filled in the reconstructed matrix (106 cases). Most entries (in green fields), however, were confirmed to exist (207 cases) or to be absent (212 cases).

The close relationship between area-based features and global connectivity may hint at developmental rules and constraints for cortical networks. Particularly, differences between the predictions from topological and geometrical properties, together with the poorer recovery resulting from geometric properties, indicate that the organization of cortical networks is not entirely determined by spatial constraints. This is also in line with the results in the

	V1	V2	V3	PIP	V3A	V4	PO	MT	V4t	DP	VOT	LIP	VIP	MIP	MDP	MSTd	MSTl	PITd	PITv	7a	STPp	CITd	CITv	STPa	AITv	FEF	TF	46	FST	TH	AITd
V1		0	0	0	0	1	0	0	0	0	0	0	0	0	0	0	0	1	1	1	1	1	1	1	1	0	1	1	0	1	1
V2	0		1	0	1	1	1	1	1	1	0	1	1	0	0	1	1	0	1	1	1	0	1	0	0	1	1	1	1	0	0
V3	0	1		0	1	1	1	1	1	1	0	1	1	0	0	1	1	0	1	1	0	0	0	0	0	1	1	1	1	0	0
PIP	0	0	0		0	0	0	0	0	0	0	0	0	0	0	0	0	0	0	0	0	0	0	0	0	0	0	0	0	0	0
V3A	0	1	1	0		1	0	1	1	1	0	1	1	0	0	1	1	0	1	1	0	0	0	0	0	1	1	1	1	0	0
V4	1	1	1	0	1		0	1	1	1	0	1	1	0	0	1	1	1	1	1	1	1	1	0	0	1	1	1	0	1	1
PO	0	1	1	0	0	0		0	1	1	0	0	1	0	0	0	1	0	0	0	0	0	0	0	0	0	0	0	1	0	0
MT	0	1	1	0	1	1	0		1	1	0	1	1	0	0	1	1	0	0	0	0	0	0	0	0	1	0	1	1	0	0
V4t	0	1	1	0	1	1	1	1		1	0	1	1	0	0	1	1	0	1	1	1	0	1	0	0	1	1	1	1	0	0
DP	0	1	1	0	1	1	1	1	1		0	1	1	0	0	1	1	0	1	1	1	0	1	0	0	1	1	1	1	0	0
VOT	0	0	0	0	0	0	0	0	0	0		0	0	0	0	0	0	0	0	0	0	0	0	0	0	0	0	0	0	0	0
LIP	0	1	1	0	1	1	0	1	1	1	0		1	0	0	1	1	0	0	0	0	0	0	0	0	1	0	1	1	0	0
VIP	0	1	1	0	1	1	1	1	1	1	0	1		0	0	1	1	0	1	1	1	0	1	0	0	1	1	1	1	0	0
MIP	0	0	0	0	0	0	0	0	0	0	0	0	0		1	0	0	0	0	0	0	0	0	0	0	0	0	0	0	0	0
MDP	0	0	0	0	0	0	0	0	0	0	0	0	0	1		0	0	0	0	0	0	0	0	0	0	0	0	0	0	0	0
MSTd	0	1	1	0	1	1	0	1	1	1	0	1	1	0	0		1	0	1	1	0	0	0	0	0	1	1	1	1	0	0
MSTl	0	1	1	0	1	1	1	1	1	1	0	1	1	0	0	1		0	1	1	1	0	1	0	0	1	1	1	1	0	1
PITd	1	0	0	0	0	1	0	0	0	0	0	0	0	0	0	0	0		1	1	1	1	1	1	1	0	1	1	0	1	1
PITv	1	1	1	0	1	1	0	0	1	1	0	0	1	0	0	1	1	1		1	1	1	1	0	0	0	1	1	0	1	1
7a	1	1	1	0	1	1	0	0	1	1	0	0	1	0	0	1	1	1	1		1	1	1	0	0	0	1	1	0	1	1
STPp	1	1	0	0	0	1	0	0	1	1	0	0	1	0	0	0	1	1	1	1		1	1	1	1	0	1	1	0	1	1
CITd	1	0	0	0	0	1	0	0	0	0	0	0	0	0	0	0	0	1	1	1	1		1	1	1	0	1	1	0	1	1
CITv	1	1	0	0	0	1	0	0	1	1	0	0	1	0	0	0	1	1	1	1	1	1		1	1	0	1	1	0	1	1
STPa	1	0	0	0	0	0	0	0	0	0	0	0	0	0	0	0	0	1	0	0	1	1	1		1	0	0	0	0	1	1
AITv	1	0	0	0	0	0	0	0	0	0	0	0	0	0	0	0	0	1	0	0	1	1	1	1		0	0	0	0	1	1
FEF	0	1	1	0	1	1	0	1	1	1	0	1	1	0	0	1	1	0	0	0	0	0	0	0	0		1	1	0	0	0
TF	1	1	1	0	1	1	0	0	1	1	0	0	1	0	0	1	1	1	1	1	1	1	1	0	0	1		1	0	1	1
46	1	1	1	0	1	1	0	1	1	1	0	1	1	0	0	1	1	1	1	1	1	1	1	0	0	1	1		0	1	1
FST	0	1	1	0	1	0	1	1	1	1	0	1	1	0	0	1	1	0	0	0	0	0	0	0	0	0	0	0		0	0
TH	1	0	0	0	0	1	0	0	0	0	0	0	0	0	0	0	0	1	1	1	1	1	1	1	1	0	1	1	0		1
AITd	1	0	0	0	0	1	0	0	0	0	0	0	0	0	0	0	1	1	1	1	1	1	1	1	1	0	1	1	0	1	

Figure 10.3 Confirmation or mismatch of connections, and prediction of unknown connections in a reconstructed submatrix of the visual cortex of the macaque monkey. Green fields denote confirmed existing (1) and absent (0) connections, respectively, whereas red fields indicate a mismatch between the original and the shown reconstructed connectivity (either by inserting connections into the matrix or removing them from the original). Yellow fields highlight connections that were predicted to exist (1) or to be absent (0) by the reconstruction approach and whose status was previously not known.

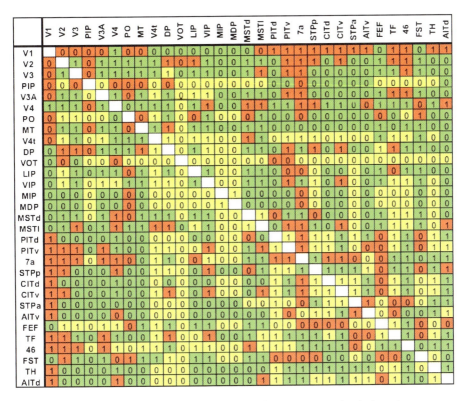

previous section in that there are more long-distance connections than expected for optimal wiring [43] and therefore prediction methods based on spatial proximity are of limited success.

10.1.3
Activity Spreading

One example of neural dynamics is the origin and spreading of epileptic seizures. Whereas we will investigate the spreading of activity later (Section 10.3.2), we will give a brief overview of epileptic seizures here.

Epilepsy affects 3 to 5% of the population worldwide, affecting persons without regard to age, sex, or race. Seizures are the clinical manifestation of an abnormal and excessive excitation and synchronization of a population of cortical neurons. These seizures can spread along network connections to other parts of the brain (depending on the type and severity of the seizure) and can be quite debilitating in terms of quality of life, cognitive function, and development. In the vast majority of cases, seizures arise from medial temporal structures that have been damaged (due to injury or illness) months to years before the onset of seizures [26]. Over this "latent period," cellular and network changes are thought to occur that precipitate the onset of seizures. Loss of inhibitory neurons, excitatory axonal sprouting, or loss of excitatory neurons "driving" inhibitory neurons are all thought to contribute to epileptogenesis [26].

It is not understood exactly how these seizures come about, but it is thought to be due to structural changes in the brain, as in the loss of inhibitory neurons, the strengthening of excitatory networks, or the suppression of GABA (Gamma-aminobutyric acid) receptors [46]. Cranstoun *et al.* [16] reported on the detection of self-organized criticality[1)] in EEG recordings from human epileptic hippocampus; thus the study and understanding using network analysis in this system may reveal useful information about the development (and possible prevention) of seizures. As these networks that support the spread of seizure activity are the very same networks that also support normal cognitive activity, it is important to understand how this type of activity arises in networks in general [32]. The question of how seizures are initiated (ictogenesis) is also of great interest, as further elucidation to either epileptogenesis or ictogenesis has great impact in the treatment and (possible cure) of epilepsy [36].

The phase transition to the epileptic ("ictal") state is abrupt from a behavioral point of view (seizures start suddenly), but from an electrical/network point of view, there are subtle changes in network activity that can indicate

1) Self-organized criticality [6] has also led to other research in neuroscience concerning neural avalanches [8].

that a seizure will occur soon (with a prediction window ranging from minutes to hours). This so-called "preictal" period, in which one is not "interictal" (between seizure states) or currently having a seizure, has been the subject of intense debate in the literature, but more and more evidence points to its existence [36]. Epileptogenesis typically has a longer timescale of development (months to years) than ictogenesis (weeks to days), but understanding the changes of epileptogenesis and how seizures become more easily generated is also of intense interest, as characterization of network changes may allow the treatment of epilepsy in a more precise manner.

10.2 Structural Dynamics

The previous section described the topology and spatial organization of neural systems. The structure of neural systems, however, can change over time. First, edges or nodes of the network could be removed. This can occur either due to learning or cell death (apoptosis) or to head injuries or strokes affecting fiber tracts or cortical tissue. Second, the network structure changes during neural development. Whereas development consists of both the formation of network connections and the loss of connections, especially in the pruning phase of early development, we will focus on the establishment of neural networks during development. In this section, we describe simulations of robustness against structural changes and of the spatial development of neural networks.

10.2.1 Robustness Toward Structural Damage

Compared to technical networks (power grids or communication networks), the brain is remarkably robust toward damage. On the local level, Parkinson's disease in humans only becomes apparent after more than half of the cells in the responsible brain region have been eliminated [20]. On the global level, the loss of the whole primary visual cortex (areas 17, 18, and 19) in kittens can be compensated by another region, the posteromedial suprasylvian area (PMLS) [60]. On the other hand, the removal of a small number of nodes or edges of the network can lead to a breakdown of functional processing. As functional deficits are not related to the number or size of removed connections or brain tissue, it might be the role within the network that makes some elements more critical than others. Identifying these critical components has applications in neurosurgery where important parts of the brain should remain intact even after the removal of a brain tumor and its surrounding tissue. The following sections describe simulations for the effect of removing edges or nodes from cortical networks. These simulations show that neural systems are robust toward random removal of components (average case) but show a rapid breakdown after the removal of few critical components (worst-case scenario).

10.2.1.1 Removal of Edges

We found that the robustness toward edge removal is linked to the high neighborhood connectivity and the existence of multiple clusters [40]. For connections within clusters, many alternative pathways of comparable length do exist once one edge is removed from the cluster (Figure 10.4a). For edges between clusters, however, alternative pathways of comparable length are unavailable and removal of such edges should have a larger effect on the network. The damage to the macaque network was measured as the increase in the average shortest path (ASP) after single edge removal. Among several measures, edge frequency (approximate measure of edge betweenness) of an edge was the best predictor of the damage after edge elimination (linear correlation $r = 0.8$ for macaque). The edge frequency of an edge counts the number of shortest paths in which the edge is included.

Furthermore, examining comparable benchmark networks with three clusters, edges with high edge frequency are those between clusters. In addition, removal of these edges causes the largest damage as measured by the increase in ASP (Figure 10.4b). Therefore, intercluster connections are critical for the network. Concerning random loss of fiber connections, however, in most cases one of the many connections within a cluster will be damaged with little effect on the network. The chances of eliminating the fewer inter-

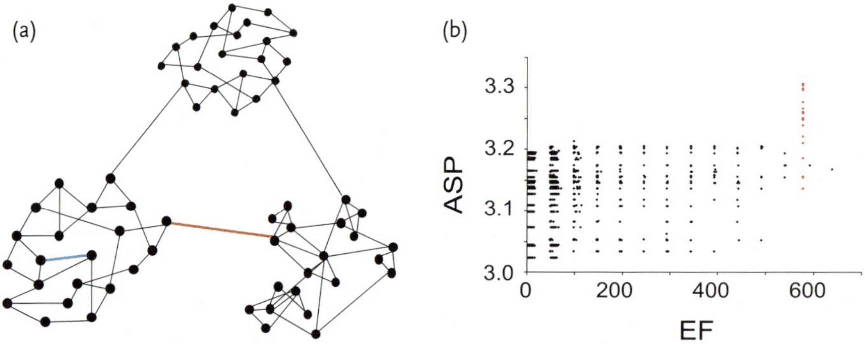

Figure 10.4 (a) Schematic drawing of a network with three clusters showing examples for an intra- (blue line) and intercluster (red line) connection. (b) Edge frequency of the eliminated edge vs. ASP after edge removal (20 generated networks with three clusters, defined intercluster connections, and random connectivity within clusters; intercluster connections: red; intracluster connections: black).

cluster connections are lower. Therefore, the network is robust to random removal of an edge [40].

10.2.1.2 **Removal of Nodes**

In addition to high neighborhood clustering, many real-world networks have properties of scale-free networks [7]. In such networks, the probability for a node possessing k edges is $P(k) \propto k^{-\gamma}$. Therefore, the degree distribution – where the degree of a node is the number of its connections – follows a power law. This often results in highly connected nodes that would be unlikely to occur in random networks. Technical networks such as the worldwide web of links between web pages [35] and the Internet [28] at the level of connections between domains/autonomous systems. Do cortical networks, as natural communication networks, share similar features?

In cortical networks, some structures (e.g., evolutionary older structures like the Amygdala) are highly connected. Unfortunately, the degree distribution cannot be tested directly as less than 100 nodes are available in the cat and macaque cortical networks. However, using the node elimination pattern as an indirect measure, cortical networks were found to be similar to scale-free benchmark networks [45].

In that approach, we tested the effect on the ASP of the macaque cortical network after subsequently eliminating nodes from the network until all nodes were removed [3]. For random elimination, the increase in ASP was slow and reached a peak for a high fraction of deleted nodes before shrinking due to network fragmentation (Figure 10.5a). When taking out nodes in a targeted way ranked by their connectivity (deleting the most highly

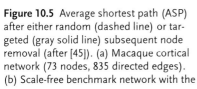

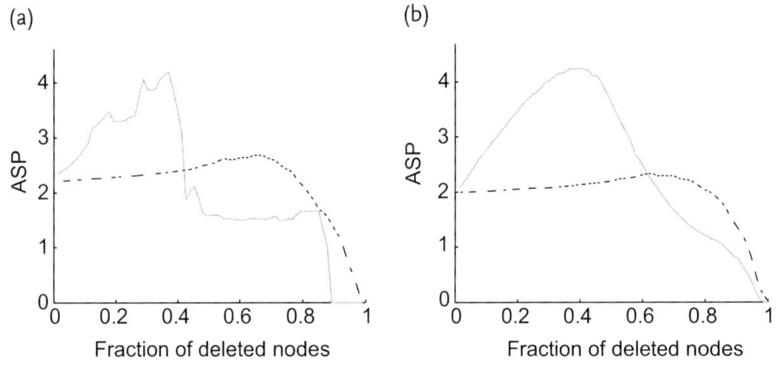

Figure 10.5 Average shortest path (ASP) after either random (dashed line) or targeted (gray solid line) subsequent node removal (after [45]). (a) Macaque cortical network (73 nodes, 835 directed edges). (b) Scale-free benchmark network with the same number of nodes and edges (lines represent the average values over 50 generated networks and 50 runs each in the case of random node removal). Similar results were obtained for the cat cortical network (not shown).

connected nodes first), however, the increase in ASP was steep and a peak was reached at a fraction of about 35%. The curves for random and targeted node removal were similar for the benchmark scale-free networks (Figure 10.5b) but not for generated random or small-world [71] networks [45]. Therefore, cortical as well as scale-free benchmark systems are robust to random node elimination but show a larger increase in ASP after removing highly connected nodes. Again, as for the edges, only few nodes are highly connected and therefore critical so that the probability to select them randomly is low.

10.2.2
Network Changes During Development

Neural systems evolved over millions of years. Starting from diffuse homogeneous networks such as nerve nets, ganglia or network clusters evolved when different tasks had to be implemented. During individual brain development, the neural architecture is formed by a combination of genetic blueprint and self-organization [64]. Here, we focus on the role of random processes and self-organization on the development of neural systems.

10.2.2.1 Spatial Growth Can Generate Small-World Networks

What are the mechanisms of self-organization during network development? A possible algorithm for developing spatial networks with long-distance connections and small-world connectivity is spatial growth [42]. In this approach, the probability to establish a connection decays with the spatial (Euclidean) distance, thereby establishing a preference for short-distance connections. This assumption is reasonable for neural networks as the concentration of growth factors decays with the distance to the source so that faraway neurons have a lower probability to detect the signal and sent a projection toward the source region of the growth factor. In addition, anatomical studies have shown that the probability of establishing a connection decreases with the distance between neurons.

In contrast to previous approaches that generated spatial graphs, the node positions were not determined before the start of connection establishment. Instead, starting with one node, a new node was added at each step at a randomly chosen spatial position. For all existing nodes, a connection between the new node u and an existing node v was established with probability

$$P(u, v) = \beta \, e^{-\alpha \, d(u,v)} , \qquad (10.3)$$

where $d(u, v)$ is the spatial distance between the node positions and α and β are scaling coefficients shaping the connection probability. A new node that did not manage to establish connections was removed from the network. Node generation was repeated until the desired number of nodes was established. Parameter β ("density") served to adjust the general probability of edge formation. The nonnegative coefficient α ("spatial range") regulated the dependence of edge formation on the distance to existing nodes. Depending on the parameters α and β, spatial growth could yield networks similar to small-world cortical networks and scale-free highway-transportation networks, as well as networks in non-Euclidean spaces such as metabolic networks [42]. Specifically, it was possible to generate networks with wiring organization similar to that of the macaque cortical network [41].

10.2.2.2 Time Windows Generate Multiple Clusters

Whereas spatial growth can generate high neighborhood clustering, such as in small-world networks, it does not generate multiple clusters. However, the use of different time domains for connection development, where several spatial regions of the network establish connections in partly overlapping time windows, allows the generation of multiple clusters or communities [44, 51].

In this extended algorithm, the establishment of an edge depends on the distance between the nodes [42, 47, 72] and the current likelihood of establishing a connection given by the time windows of both nodes. The distance-dependent probability is

$$P_{\text{dist}} = \beta\, e^{-\gamma\, d}\,, \tag{10.4}$$

where d is the spatial Euclidean distance between two nodes, $\gamma = 6$, and $\beta = 6$. The effect of varying γ and β has been described previously [42].

The time-dependent probability P_{time} of a node is influenced by its distance to pioneer nodes ($N_1 \ldots N_k \in \mathbb{R}^3$, where $k \in \mathbb{N}$ is the desired number of time windows). The reasoning is that nodes originate from other nodes in the region, thereby inheriting their time domains from previous nodes. These regions are the basis for network clusters. Each node has a preferred time for connection establishment, and the probability decays with the temporal distance to that time.

The investigated approach was able to generate small-world multiple-cluster networks. We found that network topology is mainly influenced by the number of time windows and the spatial position of pioneer nodes and thus time domains.

The case of three time windows behaves remarkably different from the other numbers of time windows. A potential explanation could be that the low overlap of the time windows allows only for a very slow network growth, in the sense that most nodes are discarded as they were not able to link to the existing network.

Another critical parameter is the spatial position of pioneer nodes. For the connectivity within and between clusters, "artifacts" can be seen independently of the number of time windows. This suggests that these values depend on the actual placement of the pioneer nodes. Therefore, the interconnectivity and the size of network clusters could be adjusted by merely changing the placement of pioneer nodes while preserving the number of time windows.

10.3 Functional Dynamics

The previous section looked at changes in network topology but as an information processing system the brain also shows functional changes regarding the activity levels of nodes and to information transfer and processing. We now look how activity spreads in neural systems starting with simple systems that exhibit local neighborhood but no global long-distance connections.

10.3.1 Spreading in Excitable Media

In some cases, a network may be more simply connected than described for neural systems in that only direct spatial neighbors are connected as in lattice or regular networks. A simple model for activation spreading is that of excitable media where waves can propagate but cannot pass another wave until a certain amount of time (the refractory time) has passed. Such a model can be implemented using either partial differential equations or cellular automata.

One example of excitable media studies in neuroscience is the phenomenon of spreading depression over the cortical surface. Whereas most models only study a flat two-dimensional surface [67], some models start to take into account the three-dimensional folding structure of the cortex and the mapping of area function [19].

An earlier field, and in fact the first application of excitable media [73], was the study of spreading activation in cardiac tissue. For cardiac tissue, muscle cells are linked by electric synapses (gap junctions) so that changes in the internal cellular potential can quickly spread to neighboring cells leading to a rapid contraction of the heart. The tissue thereby forms a so-called *syncytium* where all cells are thus excited in turn by a wave of excitation for a normal sinus rhythm (the sinus or sinoatrial node is a pacemaker for the heart that initiates action potential and following muscle contractions). This rapid spreading allows for coordinated contraction of atria and ventricles to pump blood efficiently and effectively through the body. This connectivity has been simplified in some models to two-dimensional sheets of nearest-neighbor connected nodes, allowing one to study spreading more easily.

10.3.1.1 Cardiac Defibrillation as a Case Study

Ventricular fibrillation (VF) is a serious medical problem; in the Western world, VF claims more lives than any other heart disease. But little is understood about how normal, ordered behavior (sinus rhythm) can suddenly change into the complex activity that characterizes VF. It has been established [74, 75] that when there are wavebreaks on the surface of the heart (a literal breakup of the smoothly propagating electrical wavefront), spiral waves can result. Further breakdown can lead to VF, but it is unclear exactly how changes in the system can lead to such a breakdown.

Currently the only effective therapy for VF is the delivery of a large electrical shock (cardioversion) across the myocardium with the goal of terminating VF and allowing the sinus node to reestablish sinus rhythm. However, external defibrillation shocks (DS), which for humans can be 200 to 360 Joules in magnitude, are not always successful. It is important to understand why defibrillation failure occurs in order to make DS more effective.

Experimentally recorded data from porcine ventricles (taken from a whole heart preparation) undergoing repeated defibrillation attempts was analyzed to (1) understand the mechanism of defibrillation failure, (2) determine the earliest point in time after the DS that defibrillation had failed, and (3) determine when in time it would be most effective to intervene again.

High spatial and temporal resolution optical mapping techniques and voltage-sensitive dyes were used to visualize the surface excitations of porcine (pig) heart. The data were taken at 1000 frames per second by a CCD camera with an 80×80 pixel resolution covering a 5×5 cm area [57].

The data sets used in analysis consisted of 10 s of high-speed CCD video recording, which was sufficient to cover the transition from fibrillation to DS to defibrillation failure [56,58]. The data were processed with a custom image processing program that corrected for offset and gain differences across the image plane, and removed noise and anatomical information from the data by means of bandpass filtering the signal across two bands and subtracting one signal from another.

10.3.1.2 Critical Timing for Changing the State of the Cardiac System

Simonotto *et al.* [56, 58] observed the evolution of failed defibrillation and spontaneous focal beating through synchronization and recurrence analysis. The synchronization analysis categorized the tissue behavior into categories: similar to focal behavior or similar to reentry behavior. One can clearly see periods in which the entire tissue behaved similarly to the focal area and the gradual breakdown of this type of behavior, with reentry gradually becoming the more dominant type of behavior toward the end of the data set. This switchover is captured in RQA (recurrence quantification analysis), showing a changeover in orderedness, which can be interpreted as a time of maximal orderedness that can be determined with analysis of very short noisy data sets (a strength of RQA). It can be extrapolated that this maximally ordered time could be a window in which a minimal corrective shock could be applied to a failed cardioversion and restore the entire tissue to order. What we found was a robust marker to demonstrate when a corrective shock could be applied with the energy landscape at its lowest point, in agreement with several points of defibrillation failure.

The implications of this work are potentially profound, with minimally invasive smart cardioversion being the end goal of many researchers all over the world. Current external defibrillators simply deliver enough energy to rapidly excite and de-excite cardiac tissue blindly, allowing normal sinus rhythm to reinitialize a beat. Current internal defibrillators attempt to pace the heart of an arrhythmia, but in a blind rate-driven manner. Both types of defibrillators work in a blind manner, without looking at characteristics of signal order before defibrillation; indeed, the voltage settings are left up to the operator. With a consistent measure for determining relative order in the complex spatiotemporal dynamics, perhaps a minimal but properly timed cardioversion shock would be more effective. This would mean less damage to tissue due to excess voltage applied or multiple shocks applied, less pain, and an equal, if not greater, success rate in cardioversion.

10.3.2
Topological Inhibition Limits Spreading

Excitable media models of spreading in lattice or small-world networks [49] have a long tradition in statistical mechanics but are not realistic for large-scale neural systems as these models lack long-distance connections and multiple clusters, respectively. As seen earlier (Section 10.1.1.4), long-distance connections lead to a reduction in path length and therefore to faster information transmission and processing. However, short path lengths in cortical networks also pose a potential problem as local activation patterns could potentially spread through the whole brain. Such large-scale activations in the form of increased activity can be observed in the human brain during epileptic seizures [38].

An essential requirement for the representation of functional patterns in complex neural networks, such as the mammalian cerebral cortex, is the existence of stable network activations within a limited critical range. In this range, the activity of neural populations in the network persists between the extremes of quickly dying out, or activating a large part of the network as during epileptic seizures. The standard model would achieve such a balance by having interacting excitatory and local inhibitory neurons. Whereas such models are of great value on the local level of neural systems, they are less meaningful when trying to understand the global level of connections between columns, areas, or area clusters.

Global corticocortical connectivity (connections between brain areas) in mammals possesses an intricate, nonrandom organization. Projections are arranged in clusters of cortical areas, which are closely linked among each other but less frequently with areas in other clusters. Such structural clusters broadly agree with functional cortical subdivisions. This cluster organization is found at several levels: neurons within a column, area, or area cluster (e.g., visual cortex) are more frequently linked with each other than with neurons in the rest of the network [34].

Using a basic spreading model without inhibition, we investigated how functional activations of nodes propagate through such a hierarchically clus-

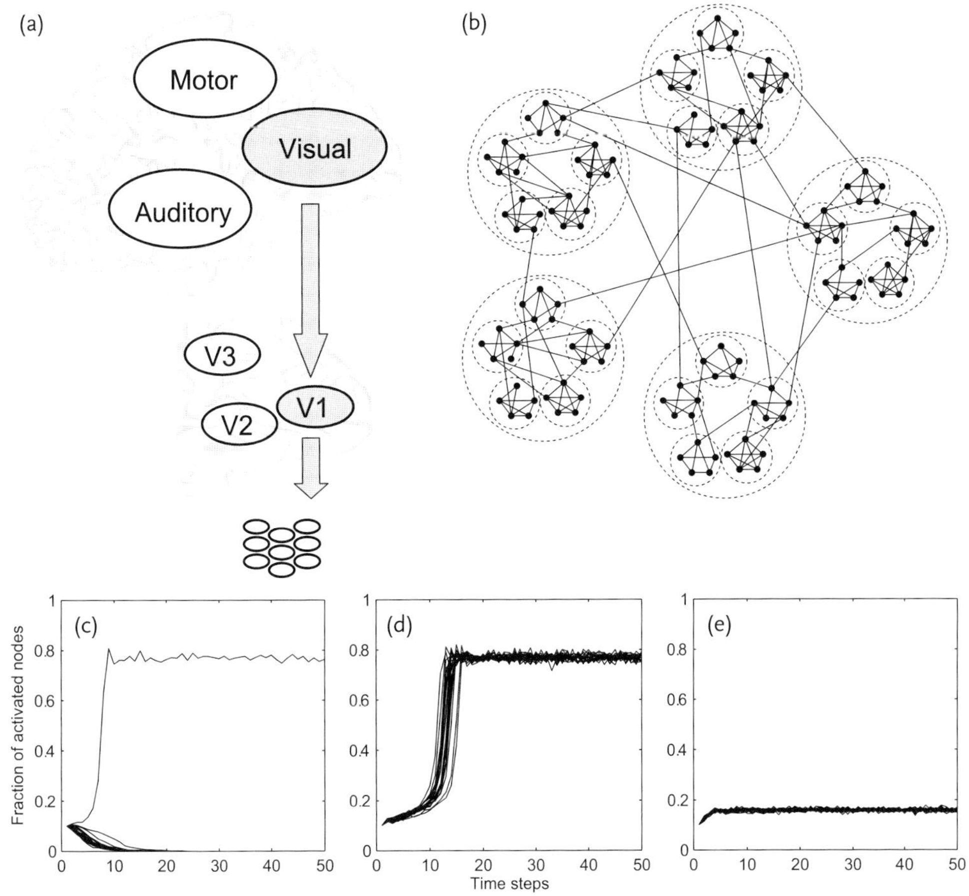

Figure 10.6 Spreading in hierarchical modular networks (after [39]). (a) The hierarchical network organization ranges from cluster such as the visual cortex to subcluster such as V1 to individual nodes being cortical columns; (b) schematic view of a hierarchical cluster network with five clusters containing five subclusters each. Examples of spread of activity in (c) random, (d) small-world, and (e) hierarchical cluster networks ($i = 100, i_0 = 150$), based on 20 simulations for each network.

tered network [39]. The hierarchical network consisted of 1000 nodes made of 10 clusters with 100 nodes each. In addition, each cluster consisted of 10 subclusters with 10 nodes each (Figure 10.6a,b). Connections were arranged so that there were more links within (sub-)clusters than between (sub-)clusters. Starting with activating 10% of randomly chosen nodes, nodes became activated if at least six directly connected nodes were active. Furthermore, at each time step, activated nodes could become inactive with a probability of 30% (for a more complete overview of the parameter space, see [39]).

The simulations demonstrated that persistent and scalable activation could be produced in clustered networks, but not in random or small-world networks of the same size (Figure 10.6c–e). Robust sustained activity also occurred when the number of consecutive activated states of a node was limited due to exhaustion. These findings were consistent for threshold models as well as integrate-and-fire models of nodes, indicating that the topology rather than the activity model was responsible for balanced activity. In conclusion, hierarchical cluster architecture may provide the structural basis for the stable and diverse functional patterns observed in cortical networks.

10.4 Summary

Using methods and concepts of network analysis [4], we discussed the topological organization and structural and functional dynamics of neural networks. The topology of neural systems is influenced by several constraints: the specialization into different subsystems such as modules for visual, auditory, or sensorimotor processing leads to topological clusters in cortical networks [34] or multiple ganglia for the neuronal network of *C. elegans*. However, not all connections are limited to the local neighborhood of individual modules: several spatially long-distance connections exist for both integrating different processing streams as well as enabling rapid processing due to reduction in the characteristic path length (analogously to shortcuts in small-world networks) [43]. Cortical networks show maximal structural and dynamic complexity, which is thought to be necessary for encoding a maximum number of functional states and might arise as a response to rich sensory environments [63]. Sporns and Kötter [62] have looked at how the structure links to the degrees of freedom for network function. Motif analysis of cortical (macaque) and neuronal (*C. elegans*) systems shows that these systems are optimized for a maximal number of possible functional motifs over the total network where the number of functional motifs of one structural motif is the set of distinct configurations of activated edges. More recent results concern frequency patterns [53] and synchronization [23] in neural systems.

Neural systems can be remarkably robust toward structural dynamics where brain structures change due to lesions or head injuries, a result that is both visible in clinical studies as well as in computer models [45]. Whereas removing the few highly connected nodes has a large effect on network structure, a more likely random removal of nodes or edges has a small effect in most cases. For neural development, spatial network growth is able to generate small-world networks, and the inclusion of time windows for connection establishment results in multiple clusters as observed for real-world neural systems.

The hierarchical and modular structure of neural systems also influences the functional dynamics of activity spreading. First, persistent but contained network activation can occur in the absence of inhibitory nodes. This might explain why cortical activity does not normally spread through the whole brain, even though top-level links between cortical areas are exclusively formed by excitatory fibers [48]. Second, in hierarchical clustered networks, activity can be sustained without the need for random input or noise as an external driving force. Third, multiple clusters in a network influence activity spreading in two ways: bottleneck connections between clusters limit global spreading, whereas a higher connection density within clusters sustains recurrent local activity.

Challenges for Network Science

The application domain of neural systems presents several challenges for the field of network science. Neural systems differ from Erdös–Rényi and traditional small-world networks in that they are modular and hierarchical. Network properties can be studied at different levels ranging from connectivity between brain areas, connectivity within areas, connectivity within columns [9], or connectivity of groups and ensembles. Another difference with respect to standard network models is that nodes, although treated as uniform at the global level of analysis, differ at the neuronal level in their response modality (excitatory or inhibitory), their functional pattern due to the morphology of the dendritic tree and properties of individual synapses, and their current threshold due to the history of previous excitation or inhibition. Such inhomogeneous node properties can also be expected at the global level in terms of the size and layer architecture of cortical and subcortical regions. Another theoretical challenge is the comparison of network topologies and dynamics, e.g., between experiments and *in silico* studies. Recent studies on network topologies have used the spectrum of a network given by the eigenvalues of its adjacency matrix [29, 31] or information-theoretic measures such as graph entropy [21].

In addition to theoretical challenges, the simulation of the dynamics in these networks also poses computational problems. The analysis of experimental network data, such as of correlation network between electrodes for multi-electrode recordings, can take a considerable amount of time. Whereas detecting all network motifs [50] in a 100-node correlation network is computationally feasible, recordings over 20 h can generate hundreds of such correlation networks. Moreover, multielectrode units with 1000 electrodes are now becoming available. One recent e-science project addressing the storing, comparing, and analyzing of large electrophysiological data sets is the CARMEN Neuroinformatics project (http://www.carmen.org.uk; [24,59,69]). On the other hand, high-performance computing is also needed for large-scale simulations of neural circuits, such as the Blue Brain project for simulating activity within a single cortical column [55].

References

1 T.B. Achacoso and W.S. Yamamoto. *AY's Neuroanatomy of C. elegans for Computation.* CRC Press, Boca Raton, FL, 1992.

2 S. Achard, R. Salvador, B. Whitcher, J. Suckling, and E. Bullmore. A resilient, low-frequency, small-world human brain functional network with highly connected association cortical hubs. *Journal of Neuroscience*, 26:63–72, 2006.

3 R. Albert, H. Jeong, and A.-L. Barabási. Error and attack tolerance of complex networks. *Nature*, 406:378–382, 2000.

4 R. Albert and A.-L. Barabási. Statistical mechanics of complex networks. *Reviews of Modern Physics*, 74(1):47–97, 2002.

5 L.A.N. Amaral, A. Scala, M. Barthélémy, and H.E. Stanley. Classes of small-world networks. *Proceedings of the National Academy of Sciences of the USA*, 97(21):11149–11152, 2000.

6 P. Bak, C. Tang, and K. Wiesenfeld. Self-organized criticality. *Physical Review A*, 38:364–374, 1988.

7 A.-L. Barabási and R. Albert. Emergence of scaling in random networks. *Science*, 286:509–512, 1999.

8 J.M. Beggs and D. Plenz. Neuronal avalanches in neocortical circuits. *Journal of Neuroscience*, 23:11167–11177, 2003.

9 T. Binzegger, R.J. Douglas, and K.A.C. Martin. A quantitative map of the circuit of cat primary visual cortex. *Journal of Neuroscience*, 24:8441–8453, 2004.

10 K. Brodmann. *Vergleichende Lokalisationslehre der Grosshirnrinde in ihren Prinzipien dargestellt auf Grund des Zellenbaues.* Barth, Leipzig, 1909.

11 M. Capalbo, E. Postma, and R. Goebel. Combining structural connectivity and response latencies to model the structure of the visual system. *PLoS Computational Biology*, 4:e1000159, 2008.

12 C. Cherniak. Component placement optimization in the brain. *Journal of Neuroscience*, 14(4):2418–2427, 1994.

13 D.B. Chklovskii, T. Schikorski, and C.F. Stevens. Wiring optimization in cortical circuits. *Neuron*, 34:341–347, 2002.

14 D.B. Chklovskii and A.A. Koulakov. Maps in the brain: what can we learn from them? *Annual Review of Neuroscience*, 27:369–392, 2004.

15 L. da Fontoura Costa, M. Kaiser, and C.C. Hilgetag. Predicting the connectivity of primate cortical networks from topological and spatial node properties. *BMC Systems Biology*, 1:16, 2007.

16 S. Cranstoun, G. Worrell, J. Echauz, and B. Litt. Self-organized criticality in the epileptic brain. *(Engineering in Medicine and Biology, 2002. 24th Annual Conference and the Annual Fall Meeting of the Biomedical Engineering Society) EMBS/BMES Conference, 2002. Proceedings of the Second Joint*, 1:232–233, Vol. 1, 2002.

17 F. Crick and E. Jones. Backwardness of human neuroanatomy. *Nature*, 361(6408):109–110, 1993.

18 L. da Fontura Costa, F.A. Rodrigues, G. Travieso, and P.R. Villas Boas. Characterization of complex networks: a survey of measurements. *Advances in Physics*, 56:167–242, 2007.

19 M.A. Dahlem, R. Engelmann, S. Löwel, and S.C. Müller. Does the migraine aura reflect cortical organization? *European Journal of Neuroscience*, 12:767–770, 2000.

20 P. Damier, E.C. Hirsch, Y. Agid, and A.M. Graybiel. The substantia nigra of the human brain. II. Patterns of loss of dopamine-containing neurons in Parkinson's disease. *Brain*, 122:1437–1448, 1999.

21 M. Dehmer and F. Emmert-Streib. Structural information content of networks: Graph entropy based on local vertex functionals. *Computational Biology and Chemistry*, 32:131–138, 2008.

22 S. Dodel, J.M. Herrmann, and T. Geisel. Functional connectivity by cross-correlation clustering. *Neurocomputing*, 44:1065–1070, 2002.

23 J. Dyhrfjeld-Johnsen, V. Santhakumar, R.J. Morgan, R. Huerta, L. Tsimring, and I. Soltesz. Topological determinants of epileptogenesis in large-scale structural and functional models of the dentate gyrus derived from experimental data. *Journal of Neurophysiology*, 97:1566–1587, 2007.

24 S.J. Eglen, C. Adams, C. Echtermeyer, M. Kaiser, J. Simonotto, and E. Sernagor. The carmen project: large-scale analysis of spontaneous retinal activity during development. In *European Retina Meeting*, 2007.

25 V.M. Eguíluz, D.R. Chialvo, G. Cecchi, M. Baliki, and A.V. Apkarian. Scale-free brain functional networks. *Physical Review Letters*, 94:018102, 2005.

26 J. Engel. *Surgical Treatment of the Epilepsies*. Lippincott Williams & Wilkins, 1993.

27 P. Erdös and A. Rényi. On the evolution of random graphs. *Publications of the Mathematical Institute of the Hungarian Academy of Sciences*, 5:17–61, 1960.

28 M. Faloutsos, P. Faloutsos, and C. Faloutsos. On power-law relationships of the internet topology. *ACM SIGCOMM Computer Communication Review*, 29:251–262, 1999.

29 I.J. Farkas, I. Derényi, A.-L. Barabási, and T. Vicsek. Spectra of "real-world" graphs: beyond the semi-circle law. *Physical Review E*, 64:026704, 2001.

30 D.J. Felleman and D.C. van Essen. Distributed hierarchical processing in the primate cerebral cortex. *Cerebral Cortex*, 1:1–47, 1991.

31 K.I. Goh, B. Kahng, and D. Kim. Spectra and eigenvectors of scale-free networks. *Physical Review E*, 64:051903, 2001.

32 J. Gómez-Gardeñes, Y. Moreno, and A. Arenas. Synchronizability determined by coupling strengths and topology on complex networks. *Physical Review E*, 75:066106, 2007.

33 C.C. Hilgetag, G.A.P.C. Burns, M.A. O'Neill, J.W. Scannell, and M.P. Young. Anatomical connectivity defines the organization of clusters of cortical areas in the macaque monkey and the cat. *Philosophical Transactions of the Royal Society of London B*, 355:91–110, 2000.

34 C.C. Hilgetag and M. Kaiser. Clustered organisation of cortical connectivity. *Neuroinformatics*, 2:353–360, 2004.

35 B.A. Huberman and L.A. Adamic. Growth dynamics of the world-wide web. *Nature*, 401:131, 1999.

36 P. Jung and J. Milton. *Epilepsy as a Dynamic Disease*. Biological and Medical Physics Series. Springer, 2003.

37 M. Kaiser. Brain architecture: a design for natural computation. *Philosophical Transactions of the Royal Society A*, 365:3033–3045, 2007.

38 M. Kaiser. *Multiple-scale hierarchical connectivity of cortical networks limits the spread of activity*, Chapter 9. Academic Press, New York, 2008.

39 M. Kaiser, M. Goerner, and C.C. Hilgetag. Criticality of spreading dynamics in hierarchical cluster networks without inhibition. *New Journal of Physics*, 9:110, 2007.

40 M. Kaiser and C.C. Hilgetag. Edge vulnerability in neural and metabolic networks. *Biological Cybernetics*, 90:311–317, 2004.

41 M. Kaiser and C.C. Hilgetag. Modelling the development of cortical networks. *Neurocomputing*, 58–60:297–302, 2004.

42 M. Kaiser and C.C. Hilgetag. Spatial growth of real-world networks. *Physical Review E*, 69:036103, 2004.

43 M. Kaiser and C.C. Hilgetag. Nonoptimal component placement, but short processing paths, due to long-distance projections in neural systems. *PLoS Computational Biology*, page e95, 2006.

44 M. Kaiser and C.C. Hilgetag. Development of multi-cluster cortical networks by

time windows for spatial growth. *Neurocomputing*, 70(10–12):1829–1832, 2007.
45 M. Kaiser, R. Martin, P. Andras, and M.P. Young. Simulation of robustness against lesions of cortical networks. *European Journal of Neuroscience*, 25:3185–3192, 2007.
46 I. Khalilov, M. Le Van Quyen, H. Gozlan, and Y. Ben-Ari. Epileptogenic actions of gaba and fast oscillations in the developing hippocampus. *Neuron*, 48:787–796, 2005.
47 J. Kleinberg. *The Small-World Phenomenon: An Algorithmic Perspective.* Proceedings of the 32nd ACM Symposium on Theory of Computing, pp. 163–170, 2000.
48 P.E. Latham and S. Nirenberg. Computing and stability in cortical networks. *Neural Computation*, 16:1385–1412, 2004.
49 N. Masuda and K. Aihara. Global and local synchrony of coupled neurons in small-world networks. *Biological Cybernetics*, 90:302–309, 2004.
50 R. Milo, S. Shen-Orr, S. Itzkovitz, N. Kashtan, D. Chklovskii, and U. Alon. Network motifs: simple building blocks of complex networks. *Science*, 298:824–827, 2002.
51 F. Nisbach and M. Kaiser. Developmental time windows for spatial growth generate multiple-cluster small-world networks. *European Physical Journal B*, 58:185–191, 2007.
52 R.E. Passingham, K.E. Stephan, and R. Kötter. The anatomical basis of functional localization in the cortex. *Nature Reviews Neuroscience*, 3:606–616, 2002.
53 A.K. Roopun, M.A. Kramer, L.M. Carracedo, M. Kaiser, C.H. Davies, R.D. Traub, N.J. Kopell, and M.A. Whittington. Period concatenation underlies interactions between gamma and beta rhythms in neocortex. *Frontiers in Cellular Neuroscience*, 2:1, 2008.
54 J.W. Scannell, C. Blakemore, and M.P. Young. Analysis of connectivity in the cat cerebral cortex. *Journal of Neuroscience*, 15(2):1463–1483, 1995.
55 G. Silberberg, S. Grillner, F.E.N. LeBeau, R. Maex, and H. Markram. Synaptic pathways in neural microcircuits. *Trends in Neuroscience*, 28(10):541–551, 2005.
56 J. Simonotto, M. Furman, W. Ditto, M. Spano, G. Liu, and K. Kavanagh. Nonlinear analysis of failed ventricular defibrillation. In *Engineering in Medicine and Biology Society, 2003.* Proceedings of the 25th Annual International Conference of the IEEE, Vol. 1, pp. 196–199, 2003.
57 J.D. Simonotto, M.D. Furman, W.L. Ditto, A. Miliotis, M.L. Spano, and T.M. Beaver. Dynamic transmurality: cardiac optical mapping reveals waves travel across transmural ablation line. *International Journal of Bifurcation and Chaos*, 17(9):3229–3234, 2007.
58 J.D. Simonotto, M.D. Furman, W.L. Ditto, M.L. Spano, G. Liu, and K. M. Kavanaugh. Nonlinear analysis of cardiac optical mapping data reveals ordered period in defibrillation failure. In *Computers in Cardiology*, pp. 551–554, 2005.
59 L.S. Smith, J. Austin, S. Baker, R. Borisyuk, S. Eglen, J. Feng, K. Gurney, T. Jackson, M. Kaiser, P. Overton, S. Panzeri, R. Quian Quiroga, S.R. Schultz, E. Sernagor, V.A. Smith, T.V. Smulders, L. Stuart, M. Whittington, and C. Ingram. The CARMEN e-Science pilot project: neuroinformatics work packages. Proceedings of the UK e-Science All Hands Meeting, 2007.
60 P.D. Spear, L. Tong, and M.A. McCall. Functional influence of areas 17, 18 and 19 on lateral suprasylvian cortex in kittens and adult cats: implications for compensation following early visual cortex damage. *Brain Research*, 447(1):79–91, 1988.
61 O. Sporns, D.R. Chialvo, M. Kaiser, and C.C. Hilgetag. Organization, development and function of complex brain networks. *Trends in Cognitive Science*, 8:418–425, 2004.
62 O. Sporns and R. Kötter. Motifs in brain networks. *PLoS Biology*, 2:1910–1918, 2004.
63 O. Sporns, G. Tononi, and G.M. Edelman. Theoretical neuroanatomy: relating anatomical and functional connectivity in graphs and cortical connection matrices. *Cerebral Cortex*, 10:127–141, 2000.
64 G.F. Striedter. *Principles of Brain Evolution.* Sinauer Associates, October 2005.
65 M.P.H. Stumpf, C. Wiuf, and R.M. May. Subnets of scale-free networks are not scale-free: sampling properties of networks. *Proceedings of the National Academy of Sciences of the USA*, 102:4221–4224, 2005.
66 D.S. Tuch, J.J. Wisco, M.H. Khachaturian, L.B. Ekstrom, R. Kötter, and W. Vanduffel.

Q-ball imaging of macaque white matter architecture. *Philosophical Transactions of the Royal Society B*, 360:869–879, 2005.

67 H.C. Tuckwell and R.M. Miura. A mathematical model for spreading cortical depression. *Biophysical Journal*, 23:257–276, 1978.

68 J. von Neumann. *The Computer and the Brain*. Yale University Press, 1958.

69 P. Watson, T. Jackson, G. Pitsilis, F. Gibson, J. Austin, M. Fletcher, B. Liang, and P. Lord. The CARMEN Neuroscience Server. Proceedings of the UK e-Science All Hands Meeting, 2007.

70 D.J. Watts. *Small Worlds*. Princeton University Press, 1999.

71 D.J. Watts and S.H. Strogatz. Collective dynamics of 'small-world' networks. *Nature*, 393:440–442, 1998.

72 B.M. Waxman. Routing of multipoint connections. *IEEE Journal of Selected Areas in Communication*, 6(9):1617–1622, 1988.

73 N. Wiener and N. Rosenbluth. The mathematical formulation of the problem of conduction of impulses in a network of connected excitable elements, specifically in cardiac muscle. *Archivas de Instituta Cardiologica Mexicana*, 16:205–265, 1946.

74 A. T. Winfree. Persistent tangled vortex rings in generic excitable media. *Nature*, 371(6494):233–236, 1994.

75 F.X. Witkowski, L.J. Leon, P.A. Penkoske, W.R. Giles, M.L. Spano, W.L. Ditto, and A.T. Winfree. Spatiotemporal evolution of ventricular fibrillation. *Nature*, 392(6671):78–82, 1998.

76 M.P. Young. Objective analysis of the topological organization of the primate cortical visual system. *Nature*, 358(6382):152–155, 1992.

77 M.P. Young. The organization of neural systems in the primate cerebral cortex. *Philosophical Transactions of the Royal Society*, 252:13–18, 1993.

11
Network Mapping of Metabolic Pathways

Qiong Cheng and Alexander Zelikovsky

11.1
Introduction

The explosive growth of cellular network databases requires novel analytical methods constituting a new interdisciplinary area of computational systems biology. The main problems in this area are finding conserved subnetworks, integrating interacting gene networks, protein networks, and biochemical reactions, discovering critical elements or modules, and finding homologous pathways. With the immense increase in good-quality data from high-throughput genomic and proteomic technologies, studies of these questions are becoming more and more challenging from analytical and computational perspectives.

This chapter deals with network mappings, a central tool for comparing and exploring biological networks. When mapping metabolic pathways by matching similar enzymes and chemical reaction chains, one can match homologous pathways. Network mapping can be used for predicting unknown pathways, fast and meaningful searching of databases, and potentially establishing evolutionary relations. This tool integrated with protein database search can be used for filling pathway holes.

Let the *pattern* be a pathway for which one is searching for homologous pathways in the *text* such as the known metabolic network of a different species. Existing mapping tools on this problem are mostly based on isomorphic and homeomorphic embeddings (see [31–43] and [3–14, 16]), effectively solving a problem that is NP-complete [30] even when searching a match for a tree in acyclic networks.

Given a linear length-ℓ pathway as the pattern and a graph as the text, Kelley *et al.* [7–9] find the image of the pattern in the text such that no consecutive mismatches or gaps in the pattern and the text are allowed. The path-to-path mapping algorithm builds a global alignment graph and decomposes it into linear pathway mapping. A single enzyme in one pathway may replace a few sequential enzymes in a homologous pathway and vice versa.

Analysis of Complex Networks: From Biology to Linguistics. Edited by Matthias Dehmer and Frank Emmert-Streib

ISBN: 978-3-527-32345-6

Pinter *et al.* [5,6] find the optimal homeomorphic tree-to-tree mapping allowing an arbitrary number of gaps.

In contrast to these approaches, this chapter considers network alignments that allow mapping of different enzymes from a pattern into the same enzyme from the text while keeping the freedom to map a single edge from the pattern to a path in the text. For such mappings (homomorphisms), the undesirable node collapse can be prevented by appropriately setting the cost on the matching of different nodes; furthermore, trees can be optimally mapped into arbitrary networks in polynomial time.

A metabolic pathway hole may be the result of ambiguity in identifying a gene and its product in an organism or when the gene encoding an enzyme is not identified in an organism's genome [18]. Due to gaps in sequence motif research, several sequences may not get specific annotations. The specific function of a protein may not be known during annotation. Reactions catalyzed by those proteins result in metabolic pathway holes. An error in reading an open reading frame (ORF) may also lead to a pathway hole. With further research, some of those proteins will get specific annotations, and pathway descriptions should be updated in pathway/genome databases. As a consequence, filling pathway holes can facilitate the improvement of both the completeness and accuracy of the pathway database and the annotation of its associated genome. In this chapter, a framework is proposed for identifying and filling metabolic pathway holes based on protein function homology with matching prosites and significant amino acid sequence alignment.

This chapter describes:

- an efficient dynamic programming-based algorithm that can be used to find the minimum-cost homomorphism from a directed graph with restricted cyclic structure to arbitrary networks;
- a generalization of this algorithm to allow different types of pattern vertex deletion;
- efficient implementations of this algorithms;
- an experimental study comparing the pathways of four unrelated organisms using enzyme matching cost based on EC notation;
- the application of network mapping to detecting and filling pathway holes and finding conserved pathways.

The remainder of the chapter is organized as follows. The next section describes previous work. Section 11.3 presents the proposed models, necessary definitions, and graph-theoretical problem formulation. Section 11.4 presents an effective dynamic programming algorithm that handles cycles in patterns and allows pattern vertex deletion and runtime analysis. Section 11.5 describes the computational study of metabolic pathways of four organisms based on the mapping algorithm. The analysis and validation of experimental studies is given in Section 11.6. The section also describes two pathway hole types and proposes the framework for finding and filling these holes based on pathway mapping and database search. Finally, conclusions are drawn in Section 11.7.

11.2
Brief Overview of Network Mapping Methods

This section first gives a brief review of network mapping and then introduces previous work on identifying and filling pathway holes.

The first class in network mapping is comprised of graph and subgraph isomorphisms. An intuitive enumeration algorithm to obtain isomorphisms of pattern graph P to text graph T is to generate a state-space representation tree that represents all possible mappings between the nodes of the two graphs and to check whether each generated mapping in the tree is a good alignment. In a tree with a vertex size of $\frac{|V_T|^{|V_P|+1}-1}{(|V_T|-1)}$, a vertex represents a pair of matched nodes; a path from a root down to a leaf represents a mapping between the two graphs. Any path across k levels in the tree represents a possible subgraph isomorphism between P and T.

For circumventing the hardness, part of the computation is filtered by using more selective feasibility rules to cut the state search space [35]. Another part is performed in an intensive preprocessing step so that the alignment process based on subgraph isomorphisms runs in polynomial time when one ignores the exponential preprocessing time. The first examples of this approach were presented by Shasha, Wang, and Giugno [37], Yan, Yuz, and Hany [40], Yan and Han [36], Giugno and Shasha [38], Messmer [33], and Bunke [32], which convert the database of graphs individually into DFS a code tree, label path, and decisive tree.

For the second class-subgraph homeomorphism, earlier works restrict their topology to a linear path [1–4]. They focus only on similarities between vertices such as proteins or genes composing pathways rather than their connectivity structure.

Pinter *et al.* [6] model metabolic pathways as outgoing trees; they reduce the problem to the approximately labeled tree homeomorphism problem. They solve the problem using a bottom-up dynamic programming algorithm with a runtime of $O(\frac{|V_P|^2|V_T|}{\log|V_P|} + |V_P||V_T|\log|V_T|)$, where $|V_P|$ and $|V_T|$ are the number of vertices in the pattern and text, respectively.

Koyuturk, Grama, and Szpankowski [11] introduce and employ a duplication/divergence model for the evolution of PPI networks. The authors' solution allows one to delete pattern and text vertices, and different vertices in patterns and text can be mapped to one vertex respectively in the text and patterns. The authors rebuild an alignment graph with $|V'| = O(|V_P||V_T|)$

vertices and $|E'| = O((|V_P||V_T|)^2)$ edges and propose a greedy heuristic error estimation algorithm. SAGA [16] converted graphs to fragment indices first and calculated the differences between paths in patterns and text. Its greedy algorithm may obtain the minimum subgraph distance. The heuristic greedy algorithms cannot guarantee that an optimal solution will be found.

Kelly *et al.* [7–9] have taken into account the nonlinearity of protein network topology and formulated the mapping problem as follows. Given a linear length-ℓ pathway *pattern* $T = (V_P, E_P)$ and *text* graph $G = (V_G, E_G)$, find an image of the pattern in the text without consecutive gaps and minimizing mismatches between proteins. A global alignment graph in [7] was built in which each vertex represents a pair of proteins and each edge represents a conserved interaction, gap, or mismatches; their objective is to find the k-highest-scoring path with limited length ℓ and no consecutive gaps or mismatches based on the built global graph. The approach takes $O(|V_T|^{\ell+2}|V_G|^2))$ in runtime.

PathBlast with the same problem formulation as [7] was presented in [8]. However, PathBlast's solution is to randomly decompose the text graph into linear pathways, which are then aligned against the pattern, and then to obtain optimal mapping based on standard sequence alignment algorithms. The algorithm requires $O(\ell!)$ random decompositions to ensure that no significant alignment is missed, effectively limiting the size of the query to about six vertices.

Li *et al.* [10] formulate the problem as an integer quadratic problem for alignment of networks based on both the protein sequence similarity and the network architecture similarity. An exhaustive searching approach was employed in [13] to find the vertex-to-vertex and path-to-path mappings with the maximal mapping score under the condition of limited length of gaps or mismatch. The algorithm has a worst-case time complexity of $O(2^m \times m^2)$. Wernicke [12] proposes a label-diversity backtrack algorithm to align two networks with cycles based on the mapping of as many path-to-path similarities as possible. Finally, Cheng, Harrison, and Zelikovsky [14] give an efficient dynamic programming-based algorithm for the case of the mapping of tree pattern to arbitrary text.

Yang and Sze [13] focused on two problems: path matching and graph matching. The authors reduced the path matching problem to finding a longest weighted path in a directed acyclic graph and showed that the problem of finding the top k suboptimal paths could be solved in polynomial time. Their graph matching reduced the graph matching problem to finding the highest score subgraphs in a graph. They allow one to delete dissociated vertices or induced subnetworks in a query network and then align what's left of it to target the network by a subgraph isomorphism. Their exact algorithm solved the problem when the query graph is of moderate size.

Implicitely, their solutions allow several different vertices of the pattern graph to be associated with a single vertex of the text graph and a single vertex of the pattern graph to be associated with several vertices of the text graph. This is equivalent to allowing that different vertices in a query path can be mapped to a vertex in text in an optimal graph matching.

The existing frameworks identifying pathway holes are based on a DNA homology. Enzyme annotations in pathway/genome databases are used to predict metabolic pathways present in an organism. Green and Karp apply a Bayesian method [18] on the assumption of conditional independence among the evidence nodes in their model and the same probability distributions calculated from known reactions as those for missing reactions. Methods based on the similarity of nucleotide sequences to known enzyme coding genes [23] and on the similarity in pathway expression in related organisms [24] have been used to fill pathway holes.

Following [25] this chapter shows a fresh approach to predicting the function of proteins by studying the annotations of similar sequences, because similar sequences usually have common descendents and, therefore, a similar structure and function. A framework has been designed that will find potential enzymes for filling pathway holes by searching functionally similar proteins using online protein databases.

11.3
Modeling Metabolic Pathway Mappings

A metabolic pathway is a series of chemical reactions catalyzed by enzymes that occur within a cell. Metabolic pathways are represented by directed networks in which vertices correspond to enzymes and there is a directed edge from one enzyme to another if the product of the reaction catalyzed by the first enzyme is a substrate of the reaction catalyzed by the second.

Mapping metabolic pathways should capture the similarities of enzymes represented by proteins as well as topological properties that cannot always be reduced to sequential reactions represented by paths. The commonly used measure of enzyme dissimilarity is the logarithm of the number of different enzymes in the lowest common upper class (see, e.g., [2, 6]).

A different approach (Cheng, Harrison, and Zelikovsky [14]) makes full use of the Enzyme Commission number (EC number) and the tight reaction property classified by EC. The EC number is expressed with a four-level hierarchical scheme. The 4-digit EC number, $d_1.d_2.d_3.d_4$, represents a subsubsubclass indication of a biochemical reaction. If $d_1.d_2$ of two enzymes are different, and their similarity score is infinite; if d_3 of two enzymes are different, their similarity score is 10; if d_4 of two enzymes are different, their similarity score is 1; or else the similarity score is 0. The corresponding penalty score for a gap is 0.5. Cheng, Harrison, and Zelikovsky's [14] experimental study indicates that the proposed similarity score scheme results in biochemically more relevant pathway matches.

The topology of most metabolic pathways is a simple path, but frequently pathways may branch or have several incoming arcs – all such topologies are instances of a *multisource tree* (for example, a directed graph that becomes an undirected tree when edge directions are disregarded). The query pathways are usually simple and can be represented as a multisource tree, but in some cases they can have a cycle or alternative ways to reach the same vertex. Then, by removing edges, such cycles or paths are broken and then the standard practice of tree-to-tree mapping follows.

The obvious way to preserve the pathway topology is to use isomorphic embedding – one-to-one correspondence between vertices and edges of a pattern and its image in the text. The requirement on edges can be relaxed – an edge in the pattern can be mapped to a path in the text [5,6] and the corresponding mapping is called a *homeomorphism*. The computational drawback of isomorphic embedding and homeomorphism is that the problem of finding optimal mapping is NP-complete and, therefore, requires severe constraints on the topology of the text to become efficient. In [5, 6], the text is supposed to be a tree, the pattern should be a directed tree, while allowing multisource-tree patterns complicates the algorithm. Their algorithm is complex and slow because it repeatedly finds minimum weight perfect matchings.

The authors propose to additionally relax one-to-one correspondence between vertices – instead, one allows different pattern vertices to be mapped to a single text vertex. The corresponding mapping is called a *homomorphism*. Such relaxation may sometimes cause confusion – a path can be mapped to a cycle. For instance, if two enzymes with similar functions belong to the same path in a pattern and a cycle with a similar enzyme belongs to the text, then the path can be mapped into a cycle (Figure 11.1). However, if the text graph is acyclic, this cannot happen. Even if there are cycles in the text, still one can expect that functionally similar enzymes are very rare in the same path.

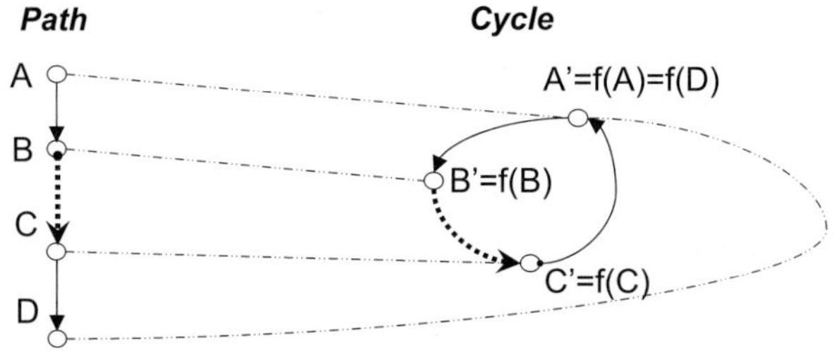

Figure 11.1 Homomorphism of a path $(A, B, \ldots, C, D)$ in a pattern onto a cycle $(A', B', \ldots, C', D' = A')$ in text.

Computing minimum-cost homomorphisms is much simpler and faster than do so for homeomorphisms. Wat follows will show that a fast dynamic programming algorithm can find the best homomorphism from one pathway to another.

11.3.1 Problem Formulation

Let $P = (V_P, E_P)$ and $T = (V_T, E_T)$ be directed graphs with vertex sets V_P and V_T, respectively, and edge sets E_P and E_T, respectively. One can further refer to P as the *pattern* and to T as the *text*. A mapping $f: P \to T$ is called a *homomorphism* if

1. every vertex in V_P is mapped to a vertex in V_T;
2. every edge $e = (u, v) \in E_P$ is mapped to a directed path in G.

The cost of a homomorphism consists of vertex and edge matching parts. Following [6], if the edge is matched with the path, the homomorphism cost should increase proportionally to the number of extra hops in the edge images, i.e.,

$$\sum_{e \in E_P} (|f(e)| - 1) ,$$

where $|f(e)| = k$ is the number of hops in the path $f(e) = (u_0 = f(u), u_1, u_2, \ldots, u_k = f(v))$.

Let $\Delta(u, v)$, $u \in V_P$, $v \in V_T$, be the cost of mapping an enzyme corresponding to the pattern vertex u into an enzyme corresponding to the text vertex v. Thus the total cost of a homomorphism $f: P \to T$ is

$$\cos t(f) = \sum_{v \in V_P} \Delta(v, f(v)) + \lambda \sum_{e \in E_P} (|f(e)| - 1) ,$$

where λ is the cost of a single extra hop in an edge-to-path mapping. Finally, the graph-theoretical problem formulation is as follows.

Minimum Cost Homomorphism Problem. Given a pattern graph P and a text graph T, find the minimum cost homomorphism $f: P \to T$.

11.4 Computing Minimum Cost Homomorphisms

Computing minimum cost homomorphisms is much simpler and faster than computing homeomorphisms. This chapter will describe a fast dynamic programming algorithm that finds the minimum cost homomorphism from a tree pattern to an arbitrary text graph and then show how to generalize this algorithm to patterns with the bounded cyclic structure.

Formally, a *multisource tree* is a directed graph whose underlying undirected graph is a tree. If the pattern is a multisource tree, then the MCH problem can be solved efficiently with the dynamic programming algorithm. The next subsection describes and analyzes the runtime of such algorithm. When the pattern graph is allowed to have cycles in the underlying undirected graph, then the MCH problem becomes NP-complete. An efficient algorithm still exists when the cyclic structure of the pattern is bounded. The complexity of cyclic structure can be measured by the size of the *vertex feedback set* $F(P) \subseteq V_P$ of the pattern $P = (V_P, E_P)$. $F(P)$ covers all cycles of P, so that the subgraph $P_{V \setminus F}$ of P induced by $V \setminus F$ is acyclic. Section 11.4.2 shows how to efficiently handle patterns with small vertex feedback sets and prove the following theorem.

Theorem (Cheng, Harrison, and Zelikovsky [14]) The Minimum cost homomorphism problem with the pattern $P = (V_P, E_P)$ and the text $T = (V_T, E_T)$ can be solved in time

$$O\left(|V_T|^{1+|F(P)|}\left(|E_T| + |V_T||V_P|\right)\right)$$

Section 11.4.3 further generalizes this result to the case when vertices are allowed to be deleted from the pattern.

11.4.1
The Dynamic Programming Algorithm for Multi-source Tree Patterns

This section will first describe preprocessing of the text graph T and ordering of vertices of the pattern graph P; and then it will define the dynamic programming table and show how to fill that table in a bottom-up manner. The runtime of the entire algorithm will be analyzed .

Text Graph Preprocessing. To compute the cost of a homomorphism, it is necessary to know the number of hops for any shortest path in the text graph T. Although finding single-source shortest paths in general graphs is slow, in this chapter's case it is sufficient to run a breadth-first search with runtime $O(|E_T| + |V_T|)$. Assuming that G is connected ($|E_T| \geq |V_T|$), one can conclude that the total runtime of finding all shortest paths is $O(|V_T||E_T|)$. In the resulting transitive closure $T' = (V_T, E'_T)$ of graph T, each edge $e \in E'_T$ is supplied with the number of hops $h(e)$ in the shortest path connecting its ends.

Pattern Graph Ordering. One will further need a certain fixed order of vertices in V_P as follows. Let $P' = (V_P, E'_P)$ be the undirected tree obtained from P by disregarding edge directions. Let us choose an arbitrary vertex $r \in V_P$ as a root and run depth-first search (DFS) in P' from r. Let $\{r = v_1, \ldots, v_{|V_P|}\}$ be the order of the DFS traversal of V_P and let $e'_i = (v_i, v) \in E'_P$ (corresponding to the directed edge $e_i \in E_P$) be the unique edge connecting v_i to the set $\{v_1, \ldots, v_{i-1}\}$. The vertex $v \in \{v_1, \ldots, v_{i-1}\}$ is called a *parent* of v_i and v_i is called a *child* of v.

DP Table. Now the dynamic programming table $DT[1, \ldots, |V_P|][1, \ldots, |V_T|]$ will be described. Each row and column of this table corresponds to a vertex of P and T, respectively. The columns $u_1, \ldots, u_{|V_T|}$ of DT are in no particular order. The rows $\{r = v_1, \ldots, v_{|V_P|}\}$ of DT are sorted in the order of the depth-first search traversal of the undirected tree underlying P from an arbitrary vertex $r \in V_P$. The unique vertex $v \in \{v_1, \ldots, v_{i-1}\}$ connected to v is called a *parent* of v_i and v_i is called a *child* of v.

Filling the DP Table. Each element $DT[i, j]$ is equal to the best cost of a homomorphism from the subgraph of P induced by vertices $\{v_{|V_P|}, v_{|V_P|-1}, \ldots, v_i\}$ into P', which maps v_i into u_j. The table DT is filled bottom-up for $i = |V_P|, |V_P| - 1, \ldots, 1$ as follows. If v_i is not a parent for any vertex in T, then v_i is a leaf and $DT[i, j] = \Delta(v_i, u_j)$. In general, one should find the cheapest mapping of each of v_i's children $v_{i_1}, \ldots, v_{i_k}$ subject to v_i being mapped to u_j. The mappings of the children do not depend on each other since the only connection between them in the tree T is through v_i. Therefore, each child v_{i_l}, $l = 1, \ldots, k$ should be mapped onto u_{j_l} minimizing the contribution of v_{i_l} to the total cost

$$C[i_l, j_l] = DT[i_l, j_l] + \lambda\left(h(j, j_l) - 1\right) ,$$

where $h(j, j_l)$ depends on the direction of e_{i_l}, i.e., $h(j, j_l) = h(u_j, u_{j_l})$ if $e_{i_l} = (v_i, v_{i_l})$ and $h(j, j_l) = h(u_{j_l}, u_j)$ if $e_{i_l} = (v_{i_l}, v_i)$. Finally,

$$DT[i, j] = \Delta(v_i, u_j) + \sum_{l=1}^{k} \min_{j'=1,\ldots,|V_T|} C[i_l, j'] .$$

Runtime Analysis. The runtime for constructing the transitive closure $T' = (V_T, E'_T)$ is $O(|V_T||E_T|)$. The runtime to fill a cell $DT[i, j]$ is proportional to

$$t_{ij} = \deg_P(v_i)\,\deg_{T'}(u_j) ,$$

where $\deg_P(v_i)$ and $\deg_{T'}(u_j)$ are degrees of v_i and u_j in graphs P and T', respectively. Indeed, the number of children of v_i is $\deg_P(v_i) - 1$, and for each child v_{i_l} of v_i there are at most $\deg_{T'}(u_j)$ feasible positions in T' since $f(v_i)$ and

$f(v_{i_l})$ should be adjacent. The runtime to fill the entire table DT is proportional to

$$\sum_{j=1}^{|V_T|}\sum_{i=1}^{|V_P|} t_{ij} = \sum_{j=1}^{|V_T|} \deg_{T'}(u_j) \sum_{i=1}^{|V_P|} \deg_T(v_i) = 2|E'_T||E_P|$$

Thus the total runtime is $O(|V_T||E_T| + |E'_T||E_P|)$. Even though G is sparse, $|E'_T|$ may be as large as $O(|V_T|^2)$, and therefore the runtime is $O(|V_T|(|E_T| + |V_T||E_P|))$.

11.4.2
Handling Cycles in Patterns

The dynamic programming algorithm for multisource patterns relies heavily on the existence of sorting of P such that for any vertex v no children can communicate with each other except through v. In order to have the same property for patterns with cycles, one can "fix" the images of the vertices from $F(P)$ in the text T, called the feedback vertex set or cycle cut vertex set. In other words, it is assumed that for each $v \in F(P)$ one knows its image $f(v) \in V_T$.

Searching for the minimum cost homomorphism $f: P \to T$ among only mappings that preserve mapping pairs $(v, f(v)), v \in F(P)$ can be done efficiently. Indeed, let K be a connected component of $P \setminus F(P)$ and let K' be the connected component of $K \cup F(P)$ containing K. The vertices of K' are sorted in such DFS order that feedback vertices are leaves. One can then run the algorithm discussed in the previous section on the assumption that the text images of feedback vertices are fixed.

In order to find the overall optimal homomorphism one should repeat the above procedure with all possible fixed mappings of the feedback vertices. The total number of such mappings is $O(V_T^{|F(P)|})$ and can be very large if $|F(P)|$ is large.

One can further improve the runtime of the algorithm by reduction to the minimum weighted feedback set problem. The text T usually contains very few vertices corresponding to enzymes that have EC annotation similar to the EC annotation of v. Let $t(v)$ be the number of possible text images of a given pattern vertex v. Then the total number of all possible feedback set mappings is $O(\prod_{v \in F(P)} t(v))$ rather than $O(V_T^{|F(P)|})$. To minimize that amount, one can minimize its logarithm $\sum_{v \in F(P)} \log t(v)$. Finally, one needs to find the minimum weight feedback set of the pattern P where the weight of each vertex v is $\log t(v)$. This problem is NP-complete and Bafna, Berman, and Fujito suggest that a 2-approximation algorithm can be implemented [26].

11.4.3
Allowing Pattern Vertex Deletion

Metabolic networks, like other biological networks such as protein–protein interaction networks and gene regulatory networks, are experimentally derived with substantial false-positive and false-negative errors [27]. Network alignment, which is used to identify regions of similarity and dissimilarity and also to detect conserved subnetworks among species, can be employed to identify true functional modules. As a consequence, false functional submodules can further be detected and deleted through the process of network alignment. Additionally, it is well known that a single enzyme in one pathway may replace a few sequential enzymes in a homologous pathway. So allowing pattern vertex deletion, especially the deletion of degree 1 or 2 vertices or submodules, becomes as reasonable as allowing text vertex deletion in subgraph homomorphism mapping. Of course, without loss of generality, one also needs to consider pattern vertex deletion with degree larger than 3.

For simplicity, one can first consider pattern vertex deletion in mapping multisource trees (Figure 11.2). Two types of pattern vertex deletion can be considered. *Strong deletion* corresponds to the operation of deleting a subtree of the pattern – all edges incident to the deleted vertices are deleted. The *bypass deletion* of a vertex of degree 2 corresponding to replacing a path across the vertex with a single edge and deleting the degree 2 vertex. If the pattern is a directed graph, then a vertex v can be bypassed only if it belongs to a directed path $a \to v \to b$; as a result of deletion the incoming and outgoing edges are replaced by a single edge $a \to b$. One vertex v with degree greater than 2 can be handled in the combination of a strong deletion of all its child vertices except one and the following bypass deletion. As for the vertex, there are different bypass deletion scores due to the possible

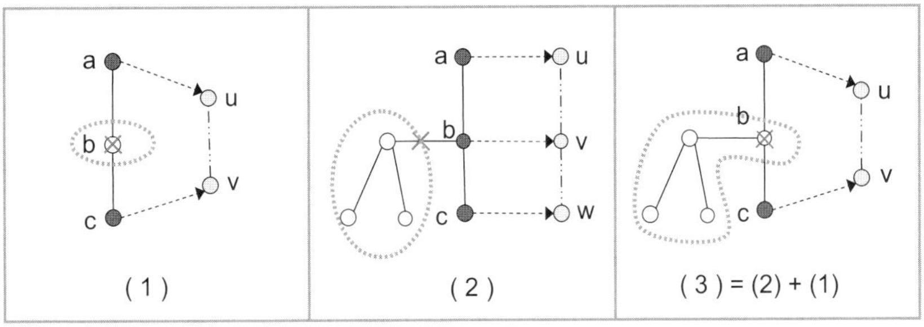

Figure 11.2 Examples of pattern vertex deletion. Solid lines represent pattern edges; dashed lines represent text paths; dashed arrows connect pattern vertices with their images in the text graph. (1) Bypass deletion of a patten vertex of degree 2; (2) branch deletion of three pattern vertices; (3) = (2) + (1) composition of strong and bypass deletions: after strong deletions a pattern vertex becomes eligible to be deleted by bypass deletion.

bypass paths. Every deleted pattern vertex has a NULL vertex as its image in text.

Further, as a result of allowing pattern vertex deletion, the mapping problem becomes a problem of finding a pair of subgraphs in the pattern graph and the text graph that can be properly aligned with each other. Maximizing the similarity between pattern and text is equivalent to minimizing the dissimilarity between them. To quantify the dissimilarity, one needs to take into account all penalties for vertex substitution and text vertex deletion as well as pattern vertex deletion. By preprocessing the patterns, one can calculate and obtain the corresponding penalty score of strong deletions and bypass deletions for every vertex. The dynamic programming algorithm from Section 11.4.1 can still be applied by adding two pseudovertices to the text graph: one vertex s corresponding to the strong deletion and the other vertex b corresponding to the bypass deletion. If a pattern vertex u is mapped to s (resp. b) then u is a strong (resp. bypass) deletion and the cost of a strong (resp. bypass) deletion of u is added. By looping through every vertex in the pattern as a root and iteratively mapping the generated pattern to the text, one can search through all possibilities and obtain an optimal solution.

For arbitrary graphs with cycles, by "fixing" the feedback vertex set of pattern to text vertices, one can still decompose the problem into a set of multisource tree mappings. Unfortunately, this decomposition is not as straightforward as in the case where vertex deletion is not allowed. Nonetheless, strong and bypass deletions do not drastically increase the runtime when the size of the feedback vertex set is small.

11.5 Mapping Metabolic Pathways

This section first describes the metabolic pathway data, then explains how to measure the statistical significance of homomorphisms and reports the results of pairwise mappings between four species.

Data. The genomescale metabolic network data in the studies were drawn from BioCyc [19–21], a collection of 260 Pathway/Genome Databases, each of which describes metabolic pathways and enzymes of a single organism. In this chapter, authors have chosen the metabolic networks of *Escherichia coli*, the yeast *Saccharomyces cerevisiae*, the eubacterium *Bacillus subtilis*, and the archeabacterium *Thermus thermophilus* so that they cover the major lineages Archaea, Eukaryotes, and Eubacteria. The bacterium *E. coli*, with 256 pathways, is the most extensively studied prokaryotic organism. *T. thermophilus*, with 178 pathways, belongs to Archaea. *B. subtilis*, with 174 pathways, is one of the best understood Eubacteria in terms of molecular biology and cell biology. *S. cerevisiae*, with 156 pathways, is the most thoroughly researched eukaryotic microorganism.

Statistical Significance of Mapping. Following a standard randomization procedure, one can randomly permute pairs of edges (u, v) and (u', v') if no other edges exist between these four vertices u, u', v, v' in the text graph by reconnecting them as (u, v') and (u', v). This allows one to keep the incoming and outgoing degrees of each vertex intact. One finds the minimum cost homomorphism from the pattern graph in the full randomization of the text graph and checks if its cost is at least as great as the minimum cost before randomization of the text graph. It is said that a homomorphism is statistically significant with $p < 0.001$ if one finds at most 9 greater costs in 10000 randomizations of the text graph.

Experiments. Two different extraction tools are used in [14] to retrieve the enzyme-oriented models of metabolic pathways for four species. One is from Pinter *et al.* [6], and the other is from Wernicke [12]. For different extraction tools, different mapping algorithms are applied and experiments are conducted. In their experiments, Cheng *et al.* [14] found the minimum cost homomorphism from each pathway of *E. coli* to each pathway of the four

Table 11.1 Pairwise statistical mapping of *E. coli* to *T. thermophilus*, *B. subtilis*, *E. coli* and *S. cerevisiae*. The numbers of pathways in the first column are from the BioCyc web site.

Text network (Parse)	Statistically significant	Pattern network (number of pathways)	
	$p < 0.01$	*Tree pathways*	*Graph pathways*
T. thermophilus (178)	# of mapped pairs	864	1050
	# of mapped pattern pathways	11	124
	# of mapped text pathways	97	118
B. subtilis (174)	# of mapped pairs	842	1240
	# of mapped pattern pathways	12	129
	# of mapped text pathways	94	143
E. coli (256)	# of mapped pairs	1031	1657
	# of mapped pattern pathways	12	122
	# of mapped text pathways	114	201
S. cerevisiae (156)	# of mapped pairs	264	1611
	# of mapped pattern pathways	11	117
	# of mapped text pathways	30	156

species (*B. subtilis, E. coli, T. thermophilus,* and *S. cerevisiae*) using algorithms described in Section 11.4 (see examples in Figures 11.3 and 11.4); furthermore, they checked to see if this homomorphism was statistically significant; then they compared the obtained homomorphism sets from when only tree

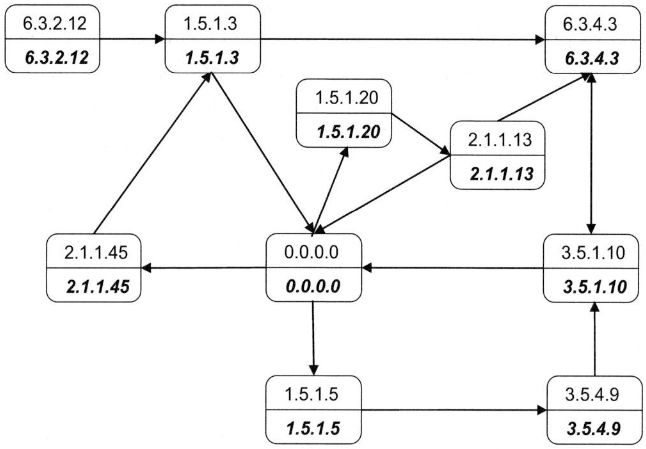

Figure 11.3 Mapping of pentose phosphate pathway onto *E. coli* to a superpathway of oxidative and nonoxidative branches of a pentose phosphate pathway in *S. cerevisiae* ($p < 0.01$). The node with the upper part and lower part represents a vertex-to-vertex mapping. The upper part represents the pattern enzyme and the lower part the text enzyme. The node with the dashed box represents a gap. (The same representation of a homomorphism will be employed in the following figures).

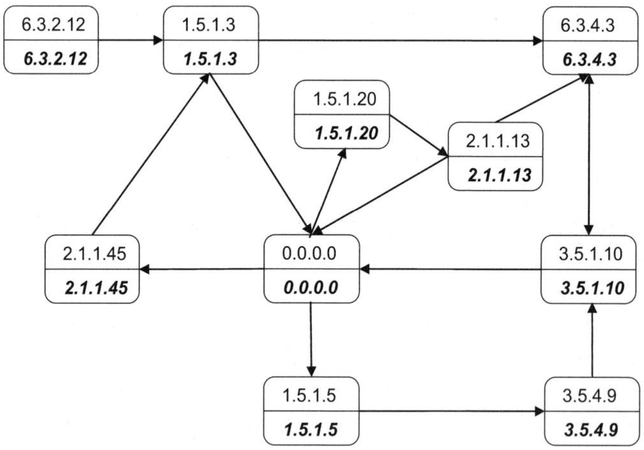

Figure 11.4 Mapping of formylTHF biosynthesis I pathway in *E. coli* to formylTHF biosynthesis I pathway in *S. cerevisiae*.

pathways were available as patterns in the pairwise mapping and when both nontree and tree pathways were available as patterns.

Results. The results of the experiments are reported in Table 11.1. The first column contains the name of the species from *E. coli* for whose metabolic network the text pathways were chosen. The last two columns demonstrate the results of applying two different algorithms in one of which only tree pattern pathways are mapped and in the other of which all pathways, including tree and nontree pathways, as patterns are mapped. For every species-to-species mapping, the authors computed the number of mapped pairs with $p < 0.01$, the number of pattern pathways that have at least one statistically significant homomorphic image and the number of text pathways that have at least one statistically significant homomorphic preimage.

11.6 Implications of Pathway Mappings

This section identifies pathways conserved across multiple species, shows how one can resolve enzyme ambiguity, identifies potential holes in pathways, and phylogenetically validates the pathway mappings.

Identifying Conserved Pathways. The authors first identified the pathways that were conserved across all 4 species under consideration. Table 11.2 contains a list of all 20 pathways in *B. subtilis* that had statistically significant homomorphic images simultaneously in all species. The lower part of Table 11.2 contains 4 more pathways with different names in *E. coli, T. thermophilus*, and *S. cerevisiae*, which have simultaneous statistically significant images in all species.

Besides the 24 pathways conserved across all 4 species, this chapter shows that 18 pathways have been found that are only common for triples of these species. Table 11.3 gives the pathway names for each possible triple of species (the triple *E. coli, T. thermophilus*, and *S. cerevisiae* does not have extra conserved pathways).

Phylogenetic Validation. One can measure the similarity between species based on the number of conserved pathways. The largest amount of conserved pathways is found between *B. subtilis* and *T. thermophilus* – two species-to-species mappings have in total 183 statistically significant pairs of pathways. The next closest two species are *E. coli* and *B. subtilis*, which have 126 statistically significant pairs of pathways. This agrees with the fact that *B. subtilis, T. thermophilus*, and *E. coli* are prokaryotic and *S. cerevisiae* is eukaryotic.

Table 11.2 List of all 20 pathways in *B. subtilis* that have statistically significant homomorphic images simultaneously in all 3 other species *E. coli*, *T. thermophilus*, and *S. cerevisiae*. The lower part contains 4 more different pathways with statistically significant images in all 4 species.

Pathway name
alanine biosynthesis I
biotin biosynthesis I
coenzyme A biosynthesis
fatty acid beta
fatty acid elongation saturated
formaldehyde oxidation V (tetrahydrofolate pathway)
glyceraldehyde 3 phosphate degradation
histidine biosynthesis I
homoserine biosynthesis
lysine biosynthesis I
ornithine biosynthesis
phenylalanine biosynthesis I
phenylalanine biosynthesis II
polyisoprenoid biosynthesis
proline biosynthesis I
quinate degradation
serine biosynthesis
superpathway of gluconate degradation
tyrosine biosynthesis I
UDP galactose biosynthesis
alanine biosynthesis
biotin biosynthesis
fatty acid oxidation pathway
fructoselysine and psicoselysine degradation

Filling Holes in Metabolic Pathways. This section describes two pathway hole types and proposes the framework for finding and filling these holes based on pathway mappings and database searches. It concludes with the analysis of two examples of pathway holes.

One can distinguish two types of pathway holes:

1. **Visible pathway holes:** an enzyme with partially or completely unknown EC notation (e.g., 1.2.4.- or -.-.-.-) in the currently available pathway description. This type of hole is caused by ambiguity in identifying a gene and its product in an organism.
2. **Hidden pathway holes:** an enzyme that is completely missing from the currently available pathway description. This type of hole occurs when the gene encoding an enzyme is not identified in an organism's genome.

Mapping of an incomplete metabolic network of a pattern organism into a better known metabolic network can identify possible hidden pathway

Table 11.3 small List of 14 pathways conserved across *B. subtilis*, *E. coli*, and *T. thermophilus*; 2 more pathways conserved across *B. subtilis*, *E. coli*, and *S. cerevisiae*; 2 more pathways conserved across *B. subtilis*, *T. thermophilus*, and *S. cerevisiae*.

Pathway name
Triple: *B. subtilis*, *E. coli*, and *T. thermophilus*
4 aminobutyrate degradation I
de novo biosynthesis of pyrimidine deoxyribonucleotides
de novo biosynthesis of pyrimidine ribonucleotides
enterobacterial common antigen biosynthesis
phospholipid biosynthesis I
PRPP biosynthesis II
salvage pathways of pyrimidine deoxyribonucleotides
ubiquinone biosynthesis
flavin biosynthesis
glycogen biosynthesis I (from ADP D Glucose)
L idonate degradation
lipoate biosynthesis and incorporation I
menaquinone biosynthesis
NAD biosynthesis I (from aspartate)
Triple: *B. subtilis*, *E. coli*, and *S. cerevisiae*
oxidative branch of the pentose phosphate pathway
S adenosylmethionine biosynthesis
Triple: *B. subtilis*, *T. thermophilus*, and *S. cerevisiae*
tyrosine biosynthesis I
fatty acid elongation unsaturated I

holes in the pattern as well as suggest possible candidates for filling visible or hidden pathway holes.

The candidates for filling pathway holes may have been previously identified in the pattern organism. Then the pathway description with visible holes should be simply updated. In the case of hidden holes, the adjoining enzymes may have been incorrectly annotated. Otherwise, if candidates for filling pathway holes have not been identified in the pattern organism, then the adjoining enzymes can be searched for matching the corresponding text enzyme.

The proposed framework for filling pathway holes is based on an amino acid sequence homology and, therefore, should be superior to existing frameworks based on a DNA homology [23, 24], since amino acids may be coded by multiple codons. Below, the framework is applied to two examples of such holes.

The authors have analyzed an example of how one can fill a visible pathway hole. The homomorphism from glutamate degradation VII pathway in *B. subtilis* to glutamate degradation VII pathway in *T. thermophilus* is shown in Figure 11.7. The pattern contains two visible pathway holes (the corre-

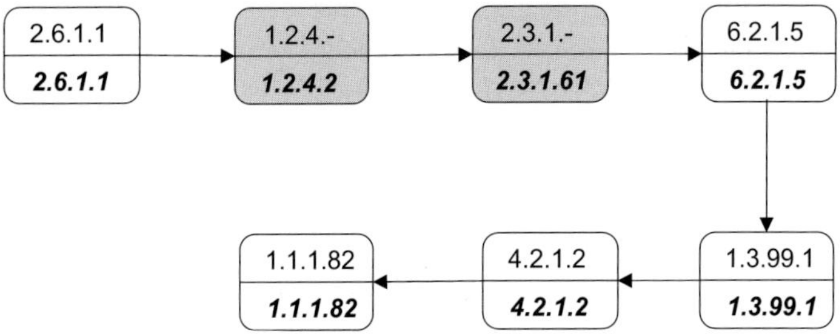

Figure 11.5 Mapping of glutamate degradation VII pathways from *B. subtilis* to *T. thermophilus* ($p < 0.01$). The node with the upper part and lower part represents a vertex-to-vertex mapping. The upper part represents the query enzyme and the lower part the text enzyme. The shaded node represents an enzyme homology.

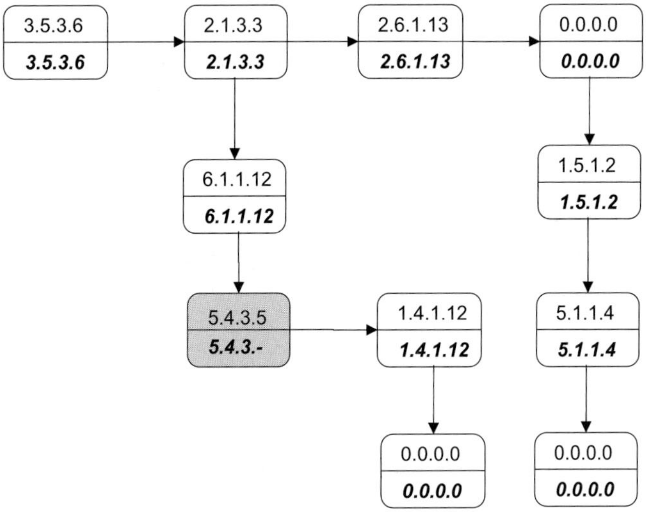

Figure 11.6 Mapping of interconversion of arginine, ornithine, and proline pathway from *T. thermophilus* to *B. subtilis* ($p < 0.01$). The node with the upper part and lower part represents a vertex-to-vertex mapping. The upper part represents the query enzyme and the lower part the text enzyme. The shaded node represents an enzyme homology.

sponding enzymes are shaded). The mapping results indicate that similar corresponding enzymes 2.3.1.61 and 1.2.4.2 with similar functions can be found in *T. thermophilus*. Their tool queries the Swiss-Prot and TrEMBL databases to see if enzymes 2.3.1.61 and 1.2.4.2 have been reported for *B. subtilis*. The authors found that these two enzymes have been reported in the Swiss-Prot database for *B. subtilis* as P16263 and P23129, respectively. There-

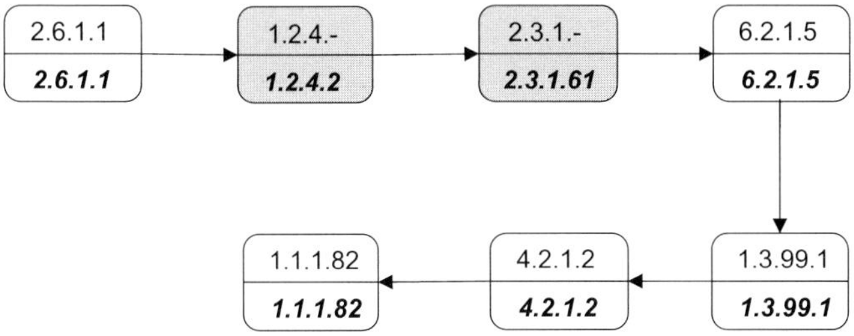

Figure 11.7 Mapping of glutamate degradation VII pathways from *B. subtilis* to *T. thermophilus* ($p < 0.01$). The shaded node represents an enzyme homology.

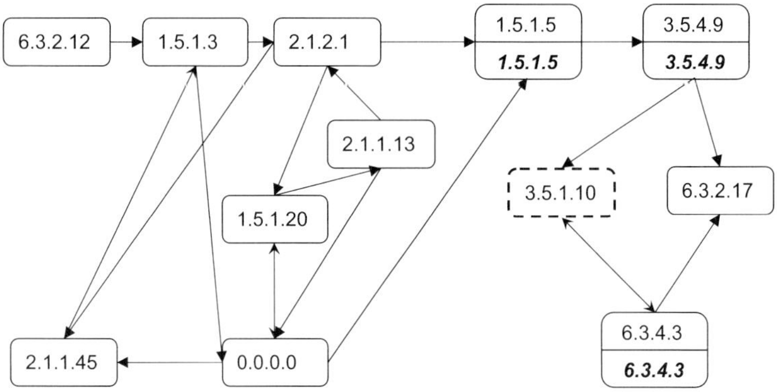

Figure 11.8 Mapping of formaldehyde oxidation V pathway in *B. subtilis* to the formy1THF biosynthesis pathway in *E. coli* ($p < 0.01$) (only vertices in the image of the pattern in the text are shown).

fore, they recommend filling these pathway holes with enzymes 2.3.1.61 and 1.2.4.2.

Now this section will proceed with an example of a hidden pathway hole. Mapping of formaldehyde oxidation V pathway in *B. subtilis* to the formy1THF biosynthesis pathway in *E. coli* is shown in Figure 11.8. In this case enzyme 3.5.1.10 is present between 3.5.4.9 and 6.3.4.3 in *E. coli* but absent in the pathway description for *B. subtilis*. The Swiss-Prot database search shows that this enzyme is completely missing from *B. subtilis* and therefore this hole does not allow an easy fix. Still it is possible that this enzyme has not yet been included in the database but has already been identified either in the literature (this can be detected through keyword search) or in closely related organisms. The Swiss-Prot database search shows that 3.5.1.10 has been reported for *B. clausii*, which is very close to *B. subtilis*. Therefore, it is recommended that this pathway hole be filled with enzymes 3.5.1.10. If such a search would not return hits in close relatives, then it would be investigated if the function of this enzyme has been taken up by one of the adjoining enzymes (in this case 3.5.4.9 or 6.3.4.3) or if there is an alternative pathway existing for this function.

Allowing Pattern Vertex Deletion. The authors have implemented their algorithm to support pattern vertex deletion (see Figure 11.9 as an example). Their experimental results have shown more advantages. The first advantage of the new network alignment with respect to the homomorphisms from [14] is in the symmetry of network alignment. The homomorphism is inherently asymmetrical since it can delete vertices only from the text but aligns all pattern vertices. Further, it could be used to discover conserved motifs. The second advantage that has been observed is significant deletion. Statistically significant deletions in the network alignment of metabolic pathways correspond to either the existence of an alternative pathway producing the same nutrient or the addition of this nutrient to the minimal media required for the growth of the text organism. Their experimental result is consistent with this statement. The third advantage that has been observed is that the average number of mismatches and gaps in the solution has outperformed the solution without the support pattern vertex deletion.

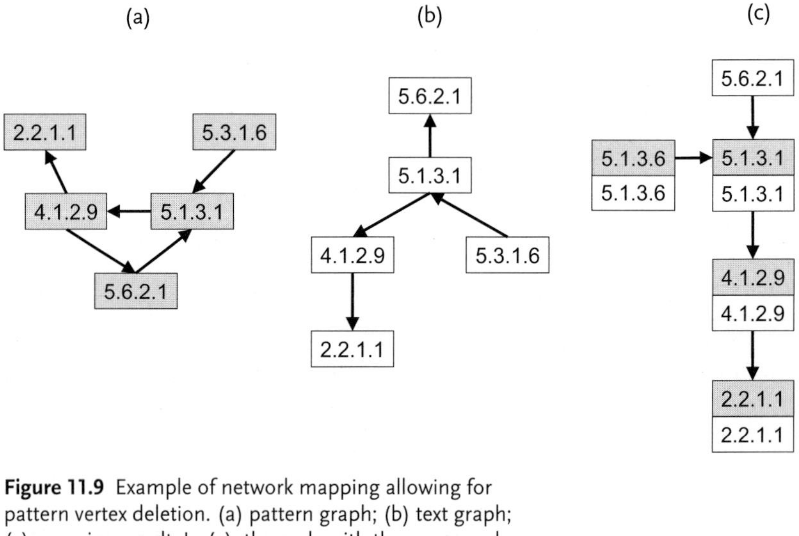

Figure 11.9 Example of network mapping allowing for pattern vertex deletion. (a) pattern graph; (b) text graph; (c) mapping result. In (c), the node with the upper and lower part represents a vertex-to-vertex mapping. Labels in the light gray background represent pattern vertices and those in the white background represent text vertices.

11.7 Conclusion

This chapter has introduced a new method of efficiently finding optimal homomorphisms from a directed graph with a restricted cyclic structure to an arbitrary network using enzyme matching cost based on EC notation. The proposed approach allows one to map different enzymes of the pattern pathway into a single enzyme of a text network. It also efficiently handles cycles in patterns. The authors have applied their mapping tool in pairwise mapping of all pathways for four organisms (*E. coli, S. cerevisiae, B. subtilis,* and *T. thermophilus* species) and found a reasonably large set of statistically significant pathway similarities. Furthermore, they have compared the obtained set for when only tree pathways can be available as a pattern and that for when both nontree and tree pathways are available as a pattern. The chapter also shows that the authors' mapping tool can be used for identification of pathway holes and have proposed a framework for finding and filling these holes based on pathway mapping and database search.

References

1 Forst, C.V., Schulten, K. (1999) Evolution of metabolism: a new method for the comparison of metabolic pathways using genomics information. *Journal of Computational Biology* **6**, 343–360.

2 Tohsato, Y., Matsuda, H., Hashimoto, A. (2000) A Multiple Alignment Algorithm for Metabolic Pathway Analysis using Enzyme Hierarchy. Proceedings of ISMB 2000, pp. 376–383.

3 Chen, M., Hofestaedt, R. (2004) PathAligner: metabolic pathway retrieval and alignment. *Applied Bioinformatics*, **3**(4), 241–252.

4 Chen, M., Hofest, R. (2005) An algorithm for linear metabolic pathway alignment. *Silico Biology*. **5**, 111–128. ISSN 1386-6338.

5 Pinter, R.Y., Rokhlenko, O., Tsur, D., Ziv-Ukelson, M. (2004) Approximate labeled subtree homeomorphism. Proceedings of the 15th Annual Symposium on Combinatorial Pattern Matching, Lecture Notes in Computer Science 3109. pp. 59–73. Springer, Berlin.

6 Pinter, R.Y., Rokhlenko, O., Yeger-Lotem, E., Ziv-Ukelson, M. (2005) Alignment of metabolic pathways. Bioinformatics. LNCS 3109. Springer, Berlin, **21** (16), 3401–3408.

7 Kelly, B.P., Sharan, R., Karp, R.M., Sittler, T. *et al.* (2003) Conserved pathways within bacteria and yeast as revealed by global protein network alignment. *Proceedings of the National Academy of Sciences of the USA*, **100** (20), 11394–11399.

8 Kelly, B.P., Sharan, R., Karp, R.M., Sittler, T. *et al.* (2004) PathBLAST: a tool for alignment of protein interaction networks. *Nucleic Acids Research*, **32**, W83–W88.

9 Sharan, R., Suthram, S., Kelley, R.M., Kuhn, T., McCuine, S. *et al.* (2005) Conserved patterns of protein interaction in multiple species. *Proceedings of the National Academy of Sciences of the USA*, **102**, 1974–1979.

10 Li, Z., Wang, Y., Zhang, S., Zhang, X. *et al.* (2006) Alignment of Protein Interaction Networks by Integer Quadratic Programming. Proceedings of EMBS '06, pp. 5527–5530.

11 Koyuturk, M., Grama, A., Szpankowski, W. (2006) Pairwise local alignment of protein interaction networks guided by model evaluation. *Journal of Computational Biology*, **13**, 182–199.

12 Wernicke, S. (2006) Combinatorial Algorithms to Cope with the Complexity of

Biological Networks. Ph.D. dissertation, University of Jena, Germany, 2006.
13 Yang, Q., Sze, S. (2007) Path matching and graph matching in biological networks. *Journal of Computational Biology*, **14** (1), 56–67, 5527–5530.
14 Cheng, Q., Harrison, R., Zelikovsky, A. (2007) Homomorphisms of Multisource Trees into Networks with Applications to Metabolic Pathways. Proceedings of BIBE'07.
15 Cheng, Q., Harrison, R., Zelikovsky, A. (2007) Homomorphisms of Multisource Trees into Networks with Applications to Metabolic Pathways. RECOMB Satellite Conference on Systems Biology.
16 Tian, Y., McEachin, R.C., Santos, C., States, D.J. *et al.* (2007) SAGA: a subgraph matching tool for biological graphs. *Bioinformatics Journal*, **23** (2), 232–239.
17 Dandekar, T., Schuster, S., Snel, B., Huynen, M., Bork, P. (1999) Pathway alignment: application to the comparative analysis of glycolytic enzymes. *Biochemical Journal*, **1**, 115–124.
18 Green, M.L., Karp, P.D. (2004) A Bayesian method for identifying missing enzymes in predicted metabolic pathway databases. BMC Bioinformatics, Sept. 2004.
19 http://www.biocyc.org/.
20 Keseler, I.M., Collado-Vides, J., Gama-Castro, S., Ingraham, J. *et al.* (2005) EcoCyc: a comprehensive database resource for Escherichia coli. *Nucleic Acids Research*, **33** (1), D334–D337.
21 Krieger, C.J., Zhang, P., Mueller, L.A., Wang, A. *et al.* (2006) MetaCyc: a microorganism database of metabolic pathways and enzymes. *Nucleic Acids Research*, **32** (1), D438–D442.
22 Green, M.L., Karp, P.D. (2007) Using genome-context data to identify specific types of functional associations in pathway/genome databases. *Bioinformatics*, **23** (13), i205–i211.
23 Kharchenko, P., Chen, L., Freund, Y., Vitkup, D. *et al.* (2006) Identifying metabolic enzymes with multiple types of association evidence, BMC Bioinformatics, March 2006.
24 Kharchenko, P., Vitkup, D., Church, G.M. (2004) Filling gaps in a metabolic network using expression information. *Bioinformatics*. August, Suppl 1, i178–i185.
25 Abascal, F., Valencia, A. (2003) Automatic annotation of protein function based on family identification. *Proteins*, **53** (3), 683–692.
26 Bafna, V., Berman, P., Fujito, T. (1999) A 2-approximation algorithm for the undirected feedback vertex set problem. *SIAM Journal of Discrete Mathematics*, **12** (3), 289–297.
27 von Mering, C., Krause, R., Snel, B., Cornell, M. *et al.* (2002) Comparative assessment of large-scale data sets of protein–protein interactions. *Nature* **417**, 399–403.
28 Valiente, G., Martinez, C. (1997) An Algorithm for Graph Pattern-Matching. Fourth South American Workshop on String Processing, 1997.
29 Hefferon, J. (2008) Linear Algebra, excellent textbook with complete solutions manual. Downloadable at http://joshua.smcvt.edu/linearalgebra/.
30 Garey, M., Johnson, D. (1979) *Computers and Intractability: A Guide to the Theory of NP-Completeness*. Freeman and Company.
31 Tsai, W.H., Fu, K.S. (1979) Error-correcting isomorphisms of attributed relational graphsfor pattern recognition. *IEEE Transactions on Systems, Man, and Cybernetics*, **9**, 757–768.
32 Bunke, H. (2000) Graph Matching: Theoretical Foundations, Algorithms, and Applications. International Conference on Vision Interface, Montreal, Quebec, Canada, May, pp. 82–88.
33 Messmer, B.T. (1996) Efficient graph matching algorithm for preprocessing model graphs. Ph.D. thesis, University of Bern, Switzerland.
34 Bunke, H. (1998) Error-tolerant Graph Matching: A Formal Framework and Algorithms. Proceedings of the Joint IAPR International Workshops on Advances in Pattern Recognition. Lecture Notes in Computer Science, Vol. 1451, Springer, Berlin.
35 Foggia, P., Sansone, C., Vento, M. (2001) A performance comparison of five algorithms for graph isomorphism. Proceedings of the 3rd IAPR TC-15 Workshop on Graph-based Representations in Pattern Recognition, pp. 188–199.
36 Yan, X., Han, J. (2002) gspan: Graph-based substructure pattern mining. Proceedings of ICDM, pp. 721–724.

37 Shasha, D., Wang, J.T.-L., Giugno, R. (2002) Algorithmics and Applications of Tree and Graph Searching. Proceedings of PODS 2002, pp. 39–52.

38 Giugno, R., Shasha, D. (2002) Graphgrep: A Fast and Universal Method for Querying Graphs. Proceedings of ICPR 2002.

39 Ambauen, R., Fischer, S., Bunke, H. (2003) Graph Edit Distance with Node Splitting and Merging and Its Application to Diatom Identification. IAPR-TC15 Workshop on Graph-based Representation in Pattern Recognition. Lecture Notes in Computer Science, Vol. 2726, Springer, Berlin, pp. 95–106.

40 Yan, X., Yuz, P.S., Hany, J. (2004) Graph Indexing: A Frequent Structure-based Approach. Proceedings of SIGMOD 2004, pp. 335–346.

41 Cordella, L., Foggia, P., Sansone, C., Vento, M. (2004) A (sub)graph isomorphism algorithm for matching large graphs. *IEEE Transactions on Pattern Analysis and Machine Intelligence*, **26** (10), 1367–1372.

42 Ketkar, N., Holder, L., Cook, D., Shah, R. *et al.* (2005) Subdue: Compression-based Frequent Pattern Discovery in Graph Data. Proceedings of ACM KDD, 2005.

43 Borgwardt, K.M. (2007) GRAPH KERNELS. Ph.D. thesis in computer science, Ludwig-Maximilians-University, Munich.

44 Sharan, R., Ideker, T. (2006) Modeling cellular machinery through biological network comparison. *Nature Biotechnology*, **24** (4), 427–433.

45 Vazirani, V.V. (2001) *Approximation Algorithms*. Springer, Berlin.

12
Graph Structure Analysis and Computational Tractability of Scheduling Problems

Sergey Sevastyanov and Alexander Kononov

12.1
Introduction

Scheduling theory is one of those mathematical disciplines that are focused on real-life applications [1,2]. Its main feature is consideration and optimization of various processes running in time. Basically we speak about *discrete mathematical models* of those processes and *discrete optimization problems* targeted to optimize some characteristics of those processes. Each such optimization problem can be solved by different algorithms [3], but for real-life applications it is essential that the algorithm be efficient in running time and required memory [4]. Designing such algorithms is one of the main objectives of discrete optimization in general and scheduling theory in particular [5,6].

It is clear that nowadays the variety of processes organized to serve various human demands is so huge that practically, to model those processes, one has to involve the whole pool of mathematical tools and models. That is why the main feature of scheduling theory is a huge variety of different models. It would not be exaggerating to say that, practically, scheduling theory contains the whole Discrete Mathematics inside. And one of the most popular tools being explored in scheduling theory is the model of *graph* [7,8].

Different types of graphs are used for convenient visualization of various constraints imposed on feasible solutions. Normally, the more general the type of graph that is used, the more time is required to verify those constraints specified by the graph. As a rule, the type of graph is characterized by various numerical parameters. Specifying constraints on those parameters, we can define different types of graphs, and the complexity of the corresponding optimization problem significantly depends on the way in which those constraints are imposed. In this chapter we consider the *Connected List Coloring* (CLC) problem – an interesting and relatively new graph theoretical problem closely related to various practical problems arising in scheduling and other areas of discrete optimization. Due to this "relationship", many questions formulated for scheduling problems can be reformulated in terms

Analysis of Complex Networks: From Biology to Linguistics. Edited by Matthias Dehmer and Frank Emmert-Streib

ISBN: 978-3-527-32345-6

of graphs. And one of the most important questions concerns the complexity status of the problem under investigation.

Our main target here is the complexity analysis of the CLC problem with respect to four *key parameters* of a given input graph $\widehat{G}$ (for a definition we refer the reader to Section 12.4): the type of graph G (representing the "left part" of graph $\widehat{G}$), Δ_V, $\Delta_{V,C}$, and $\Delta_{C,V}$. We introduce an infinite family CLC($\mathcal{X}$) of CLC(x) problems defined for various 4-dimensional vectors $x \in \mathcal{X}$ bounding from above the characteristic vector consisting of four key parameters. Next we investigate how different combinations of constraints on the key parameters affect the problem complexity. And despite the fact that the family of problems under investigation is of infinite cardinality, we have managed to obtain the whole picture of complexities of all its items, due to a basis notion of the *multiparametric complexity analysis* – the notion of a *basis system of problems* (Section 12.6). As we learn from Section 12.6, such a basis system exists also for our infinite family CLC($\mathcal{X}$), and it consists of exactly eight problems considered in Section 12.5.

12.2
The Connected List Coloring Problem

The main objective of our paper is the following *Connected List Coloring problem* introduced by Vising in 1999 [9].

Suppose we are given a graph $G = (V, E)$ and a finite set C of *colors*. A given function $A : V \to 2^C$ specifies for each vertex $v \in V$ a subset $A(v) \subseteq C$ of admissible colors. Function A will be referred to as a *prescription*. Function $\varrho : V \to C$ is called a *connected list coloring* (CLC) of graph G, if $\varrho(v) \in A(v)$, $\forall v \in V$, and for each color $i \in C$ the subset of all identically colored vertices $\varrho^{-1}(i)$ specifies a connected subgraph of G. (The subgraph defined on the empty set of vertices is supposed to be connected.) Given an input I of the CLC problem, we wish either to find a CLC or to prove that it does not exist.

This problem is closely related to the problem of determining *sufficient conditions* for the existence of a CLC for a given input $I = (G, C, A)$. The main question is: for which graphs and under which constraints on the lists of admissible colors does the desired coloring exist? In [9] this question was investigated from the viewpoint of constraints on the cardinality of lists. Let $A_{\min} = \min_{v \in V} |A(v)|$ and $A_{\max} = \max_{v \in V} |A(v)|$ be, respectively, the minimum and maximum list cardinalities. Let $\alpha(G)$ be the least positive integer k such that for any lists of admissible colors with $A_{\min} \geq k$ there exists a CLC.

Parameter $\alpha(G)$ is an important characteristic of graph G when G is investigated subject to the feasibility of its connected list coloring. A few simple cases where this characteristic can be easily calculated are presented below.

If G is a complete graph, then $\alpha(G) = 1$. Indeed, even if the list of each vertex consists of a single color, then the assigning of those unique colors generates a feasible coloring, because any subset of vertices in the complete graph is connected.

In the case that G is an n-vertex chain, $\alpha(G) = \lceil \frac{n}{2} \rceil$. Finally, the maximum value of $\alpha(G)$ among all connected n-vertex graphs G is attained on the star and is equal to n–1. These two facts are corollaries of the equality $\alpha(G) = \varepsilon(G)$, proved in [9] for an arbitrary bipartite graph G, where $\varepsilon(G)$ is the *independence number* of graph G, that is, the cardinality of the maximal pairwise disjoint subset of vertices in G. It is also clear that the maximum of $\alpha(G)$ over all n-vertex graphs is equal to n and is attained on graph G consisting of n isolated vertices. For some other classes of graphs the question of calculating the function $\alpha(G)$ (or estimating it when its exact calculating is not tractable) may also be of interest. Some properties and bounds on the parameter $\alpha(G)$ in terms of other characteristics of graph G were obtained in [9]. It would be natural to assume that calculating the parameter $\alpha(G)$ for an arbitrary graph G must be NP-hard, but this question remains open.

The main target of the current paper is another algorithmic problem resolving the question: given a graph G and a prescription A, does there exist a CLC for vertices of graph G? It is not difficult to invent an infeasible instance of this problem by means of vertices with unique colors in their prescriptions (i.e., vertices $\{v\}$ having $A(v) = 1$), thereby enforcing a disconnected coloring. A less trivial instance with $A(v) > 1, \forall\, v \in V$, is presented in Figure 12.1. Clearly, there exist different proofs that no CLC exists for this particular instance. One such proof is presented below.

Suppose that a feasible CLC exists. First observe that the prescription of vertex v_1 does not contain the stripy color. This means that the coloring of either the left or the right part of graph G cannot use the stripy color. Let it be the right part (due to the symmetry of graph G). This implies that vertices v_3 and v_4 must be black and, thus, must be connected in a black-colored

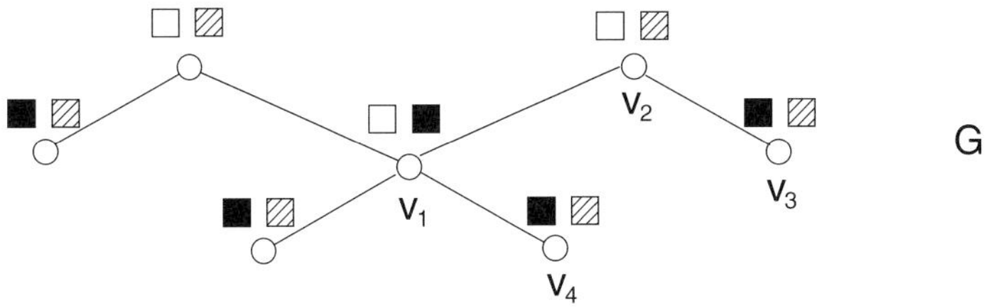

Figure 12.1 An infeasible instance of the CLC problem.

subgraph G_{bl}. Due to the structure of graph G, vertex v_2 must belong to G_{bl}; but v_2 has no black in its prescription – a contradiction.

It becomes rather clear from this example that the problem of checking the existence of a feasible CLC for a given input (G, C, A) is nontrivial and, probably, computationally hard (in terms of complexity theory, it must be NP-complete). Vising made a stronger supposition that it is NP-complete even if G is a simple chain. Indeed, as can be seen from the results of our paper, this supposition proved to be true. Moreover, in Section 12.5 a detailed complexity analysis of this problem with respect to various combinations of constraints on four main parameters of this problem is presented.

12.3 Some Practical Problems Reducible to the CLC Problem

In fact, the CLC problem, while having an independent theoretical interest, is nothing but a convenient mathematical model for presenting various practical problems arising in real-life situations. A few such situations are presented below as discrete optimization problems originating from different areas of human activity.

12.3.1 The Problem of Connected Service Areas

The CLC problem may have the following interpretation. The set of vertices of graph G is the set of clients consuming some resource or service. (Some vertices may correspond to dummy clients.) To each client $v \in V$ we need to assign a unique provider $\varrho(v) \in C$ from a prespecified list $A(v)$ of admissible providers. At that, the connectivity of the service area for each provider is often required.

This problem is closely related to the optimization *Plant Location Problem* (PLP, or *PL problem*) in which, instead of a prespecified set of providers $A(v)$, we define a cost $g_{i,v}$ of serving client v by provider i. The objective is to find the assignment $\varrho : V \to C$ that minimizes the function

$$\Phi(\varrho) = \sum_{i \in C} g_i(\varrho) + \sum_{v \in V} g_{\varrho(v),v} , \tag{12.1}$$

where $g_i(\varrho)$ is the cost of using client i in the assignment ϱ; it is equal to g_i^0 if $\varrho^{-1}(i) \neq \emptyset$, and to zero otherwise.

It can be seen that for each input $I = (G, C, A)$ of the CLC problem we can define an input of the PL problem such that $\Phi(\varrho)$ is finite if and only if ϱ is feasible with respect to prespecified lists $A(v)$. To that end, it is sufficient to define the matrix $(g_{i,v})$ as

$$g_{i,v} = \begin{cases} 0, & i \in A(v) \\ \infty, & i \notin A(v) . \end{cases} \tag{12.2}$$

Although the connectivity property is not among the direct requirements of the problem, sometimes it is possible to show that for matrices satisfying certain conditions there always exists an optimal solution of the PLP in which the serving area of each provider is connected (we say that such a solution is *connected*). One such condition was found by Gimadi [10]. In view of these results, Ageev [11] defined (for any given graph G and a set C of providers) the class $\mathrm{Con}(G, C)$ of all matrices $(g_{i,v})$ such that for any vector $(g_1^0, \ldots, g_{|V|}^0)$ there exists an optimal solution of the PLP that is connected. Next he introduced the function

$$\Omega(\varrho) = \sum_{i \in C} g_i^0 h\left(\varrho^{-1}(i)\right) + \sum_{v \in V} g_{\varrho(v),v} , \tag{12.3}$$

where $h(X)$ is the number of connected components of the subgraph based on a subset of vertices $X \subseteq V$. Let PLP' denote the problem of minimization of the function $\Omega(\varrho)$. Ageev proved the following lemma.

Lemma 12.1 (Ageev, 1992) *If matrix $(g_{i,v})$ belongs to class $\mathrm{Con}(G, C)$, then, given a vector $(g_1^0, \ldots, g_{|V|}^0)$, the solution of PLP' provides a connected solution to the PL problem.*

Lemma 12.1 enables one to obtain connected solutions for the PL problem without requiring directly the connectivity property for feasible solutions. Unfortunately, in the case of the CLC problem matrix $(g_{i,v})$ (defined by Equation 12.2) does not belong to the class $\mathrm{Con}(G, C)$. (To be exact, we cannot say that it belongs to $\mathrm{Con}(G, C)$ for **every** graph G.) Indeed, let graph G be a chain consisting of three vertices: $v_1 \to v_2 \to v_3$; the set of providers consists of two elements: $C = \{1, 2\}$; matrix $(g_{i,v})$ is defined according to

$$(g_{i,v}) = \begin{pmatrix} 0 & \infty & 0 \\ \infty & 0 & \infty \end{pmatrix} .$$

Let us consider the vector $(g_1^0, \ldots, g_{|V|}^0) = (1, \ldots, 1)$. It can be seen that the solution ϱ that minimizes $\Phi(\varrho)$ determines a disconnected solution in which vertices v_1 and v_3 receive provider 1, while v_2 receives provider 2.

Thus, for solving the CLC problem we cannot take advantage of Lemma 12.1 and the solution obtained by minimization of the function $\Omega(\varrho)$. Instead, we introduce another, but similar, function:

$$\Omega'(\varrho) = \max_{i \in C} h(\varrho^{-1}(i)) + \sum_{v \in V} g_{\varrho(v),v} . \tag{12.4}$$

We can establish the following:

Lemma 12.2 *Given a triple* (G, C, A) *defining an input of the CLC problem, a connected solution for the CLC problem exists, iff for the assignment* ϱ^* *minimizing the function* $\Omega'(\varrho)$ *defined by (12.4) we have* $\Omega'(\varrho^*) = 1$.

Lemma 12.2 enables one to apply the complexity results obtained in Section 12.5 to the problem of minimization of function $\Omega'(\varrho)$.

12.3.2
No-Idle Scheduling on Parallel Machines

For the CLC problem on a chain (which is NP-complete, as will be shown in Section 12.5.2) there exist some other practical interpretations belonging to the area of scheduling theory. One such practical problem arising on parallel machines is presented below.

PM Problem. There are m parallel machines and one job consisting of n successive unit-length operations. Each operation can be processed on any machine. For each machine a family of time intervals is specified when the machine is available. The question is: does there exist a feasible schedule for processing the job on those machines such that each machine works without idle time and the job is processed without waiting time?

First we show that the CLC problem defined on a family of chains can be reduced to a CLC problem on a single chain. The latter will then be polynomially reduced to the PM problem.

Suppose that we are given a graph $G = (V, E)$ that is a family of l chains. We connect them to a single chain $G' = (V', E')$ by adding $l-1$ intermediate vertices $\{w_1, \ldots, w_{l-1}\}$ to set V and by connecting each vertex $w_j (j = 1, \ldots, l-1)$ with the end-vertex of the jth chain and the begin-vertex of the $(j+1)$th chain. The prescription of vertex w_j is defined to consist of a single color u_j that does not belong to any other prescription. (The "extended" set of colors is denoted by C'.) It can be seen that a CLC exists for graph G' if and only if it exists for graph G.

Let us consecutively number the vertices of chain G' from left to right by numbers from 1 to $n' = n + l - 1$. Note that a subset of vertices $\widetilde{V} \subseteq V'$ constitutes a connected area in graph G' *iff* their numbers constitute a connected interval of integers. To each color $i \in C'$ we assign a machine M_i. No machine is available outside the time interval $[0, n']$; in addition, machine M_i, $i \in C'$, is available in the kth unit length interval $[k-1, k]$ *iff* $i \in A(v_k)$. There is also a job consisting of n' unit-length operations. (Thus, we have to process this job continuously within the time interval $[0, n']$.) At that, each machine M_i should work without idle time.

Evidently, a feasible schedule for the PM problem exists *iff* there exists a connected list coloring for the given instance of the CLC problem.

Interval Edge Coloring Problem. It should be noted that the requirement for a mutual "idleness" of jobs and machines in a feasible schedule already occurred in some scheduling settings known in the literature (at least in *open shop* problems). For example, such an open shop problem with a simultaneous *no-idle* requirement for machines and a *no-wait* requirement for jobs and with unit execution times of operations was considered in [12]. Afterwards, a real-life interpretation of this problem was presented in [13], where a mutually convenient schedule for consultations of parents with college teachers was the main target of the problem.

In [12] it was shown that the problem could be formulated in terms of a *proper interval* edge coloring of a bipartite graph.[1)] Clearly, such an interval edge coloring problem can be formulated for arbitrary graphs. In this general form the problem represents a special case of the well-known "2-DIMENSIONAL CONSECUTIVE SETS" problem (see Garey and Johnson [4] [SR 19], p. 230), which is nothing but an *interval edge coloring problem for hypergraphs*. The latter was proved to be NP-complete in [14], while the NP-completeness of its special case (interval edge coloring of ordinary graphs) was proved in [12]. As for the problem on bipartite graphs, Asratyan and Kamalyan [12] conjectured that the desired coloring exists for every bipartite graph. The conjecture was partly confirmed by Kamalyan [15], who proved that the desired interval coloring always exists for complete bipartite graphs and trees (representing somewhat "opposite" cases of an arbitrary graph – the "maximal" and the "minimal" connected graphs defined on a given set of vertices). However, the question concerning arbitrary bipartite graphs remained open until a counterexample to Asratyan and Kamalyan's conjecture was presented in [16]. In the same paper, the NP-completeness of the problem for arbitrary bipartite graphs was proved.

12.3.3
Scheduling of Unit Jobs on a p-Batch Machine

Let us formulate the following *batch machine scheduling* problem (*BMS problem*). A machine is called a *batch machine* if it can process several jobs at a time, assuming them to be a single job J'. The processing time p' of this job is defined by a certain formula via processing times $\{p_j\}$ of its compo-

1) We remind the reader that a coloring of vertices (edges) is called a *proper coloring* if adjacent vertices (edges) receive different colors. In this sense the above-defined *connected coloring* of vertices is definitely "improper." Edge coloring is called an *interval* one if for any given vertex the set of colors used for coloring its incident edges represents an interval of integers.

nents. If $p' = \max p_j$ (as if supposing that the components are processed on that machine in parallel), then the machine is called a *p-batch machine.*

A time interval $B_t = [t, t+1]$ defined for an integral $t \geq 0$ will be referred to as a *basic interval*; $\mathcal{B} = \{B_t \mid t = 0, 1, \ldots\}$.

Given a set $\mathcal{J}$ of unit-length jobs, an *integral schedule* is an assignment of jobs from $\mathcal{J}$ to basic intervals from $\mathcal{B}$.

BMS Problem

Instance: There is a finite set $\mathcal{J} = \{J_1, \ldots, J_n\}$ of n unit-length jobs that have to be processed on a single p-batch machine. Each job $J_j \in \mathcal{J}$ is available for processing in one of the basic intervals from a prespecified list $A_j \subseteq \mathcal{B}$.

On the set of jobs a relationship is specified by a graph $G = (\mathcal{J}, E)$ with the set of vertices $\mathcal{J}$. Two jobs $J_i, J_j \in \mathcal{J}$ are said to be *relatives iff* $(J_i, J_j) \in E$. A subset of jobs $\mathcal{J}' \subseteq \mathcal{J}$ is *G-connected* if the subgraph $G' = G|_{\mathcal{J}'}$ defined on the subset of vertices $\mathcal{J}'$ is connected. A partition of jobs into batches is *G-connected* if each batch is *G-connected.*

A schedule for jobs in $\mathcal{J}$ is *feasible* if it is feasible with respect to lists $\{A_j\}$ of basic intervals and if it generates a partition of jobs into batches that is *G-connected* for a given graph G.

Question: Does there exist a feasible integral schedule for jobs in $\mathcal{J}$?

It can be easily seen that the BMS problem is reducible to the CLC problem in which jobs remain vertices of graph G, while basic intervals become colors. Clearly, a desired schedule exists for a given instance of the BMS problem *iff* there exists a connected list coloring for the corresponding instance of the CLC problem.

12.4 A Parameterized Class of Subproblems of the CLC Problem

To perform the complexity analysis of the CLC problem, it is convenient to represent its input in the form of graph $\widehat{G} = (V, C; E, E_A)$ depicted in Figure 12.2. The set of vertices of this graph consists of two parts: V and C; set V will be referred to as the *left part* of graph $\widehat{G}$, while C will be called the *right part*. The set of edges will also consist of two parts: E and E_A; the edges from E connect vertices in the left part, while the edges from E_A connect parts V and C (i.e., constitute a bipartite subgraph of $\widehat{G}$); an edge (v, i) belongs to E_A *iff* $i \in A(v)$ (i.e., color i is admissible for vertex v).

The complexity of this problem will depend significantly on four parameters of graph $\widehat{G}$: T, Δ_V, $\Delta_{V,C}$, $\Delta_{C,V}$.

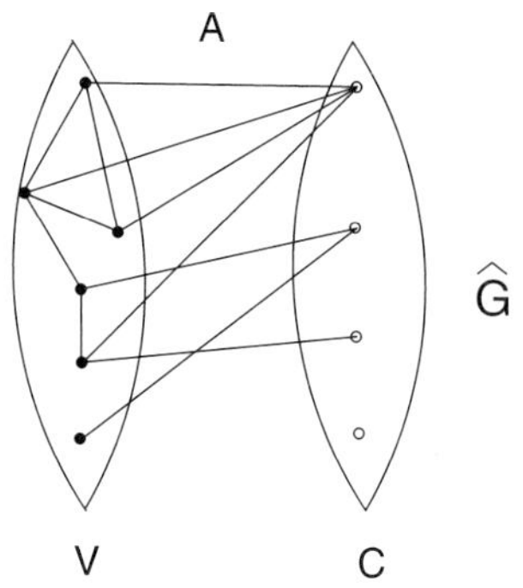

Figure 12.2 Graph $\widehat{G}$ defining an example of the CLC problem.

The first parameter T defines the *type* of graph G. We will distinguish between two types of graphs, depending on whether G is acyclic or not. Respectively, T will take only two values: 0 and ∞; $T = 0$ denotes the case of an acyclic graph G; otherwise $T = \infty$.

The second parameter Δ_V denotes the maximum vertex degree of graph G (where we take into account only the edges from V to V).

The third parameter $\Delta_{V,C}$ denotes the maximum vertex degree in V with respect to edges from E_A. It corresponds to the maximum cardinality of lists $A(v)$.

Finally, the fourth parameter $\Delta_{C,V}$ denotes the maximum vertex degree in C with respect to edges from E_A. It shows the number of vertices from V for which a given color $i \in C$ is admissible (at maximum).

We will impose upper bounds on these parameters and analyze how the complexity of the resulting subproblem changes under various combinations of those constraints. Each such combination will be specified by a 4-dimensional vector $x = (x^1, x^2, x^3, x^4)$, where $x^1 \in \{0, \infty\}$ and $x^2, x^3, x^4 \in \{0, 1, \ldots, \infty\}$. The value $x^i = \infty$ will mean that the ith parameter is not restricted. The set of all possible values of vector x will be denoted by $\mathcal{X}$.

Clearly, each vector $x = (x^1, x^2, x^3, x^4) \in \mathcal{X}$ uniquely defines a CLC(x) problem based on the set of inputs

$$\mathcal{I}(x) = \{\widehat{G} \mid T \le x^1,\ \Delta_V \le x^2,\ \Delta_{V,C} \le x^3,\ \Delta_{C,V} \le x^4\}.$$

Thus, we defined an infinite class of problems CLC($\mathcal{X}$) = $\{\text{CLC}(x) \mid x \in \mathcal{X}\}$, each being a subproblem of the original CLC problem. (The latter is also contained in CLC($\mathcal{X}$). It corresponds to the subproblem CLC($\infty, \infty, \infty, \infty$).)

In the next section we will consider eight problems from CLC($\mathcal{X}$) and perform their complexity analysis.

12.5
Complexities of Eight Representatives of Class CLC($\mathcal{X}$)

In this section eight specific subproblems of the CLC problem belonging to class CLC($\mathcal{X}$) will be analyzed in light of their complexity. As will be shown, three of them are NP-complete, thereby confirming the NP-completeness of the original CLC problem. For the other five subproblems polynomial-time algorithms will be presented solving those special cases to the optimum. At first glance, the selection of these eight representatives of class CLC($\mathcal{X}$) seems random. But this is not the case, as we will see in Section 12.6.

12.5.1
Three NP-Complete Subproblems

In this section it will be shown that three subproblems CLC(x_6), CLC(x_7), and CLC(x_8) defined for $x_6 = (0, 1, 3, 3)$, $x_7 = (0, 4, 2, 3)$, $x_8 = (0, 3, 2, 4)$ are NP-complete. Sample instances of these three subproblems are shown in Figures 12.3 to 12.5.

Figure 12.3 A sample instance of the CLC(0, 1, 3, 3) problem (NP-complete).

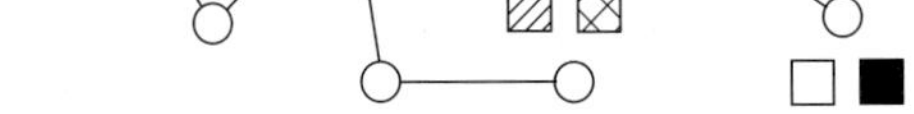

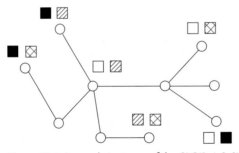

Figure 12.4 A sample instance of the CLC(0, 4, 2, 3) problem (NP-complete).

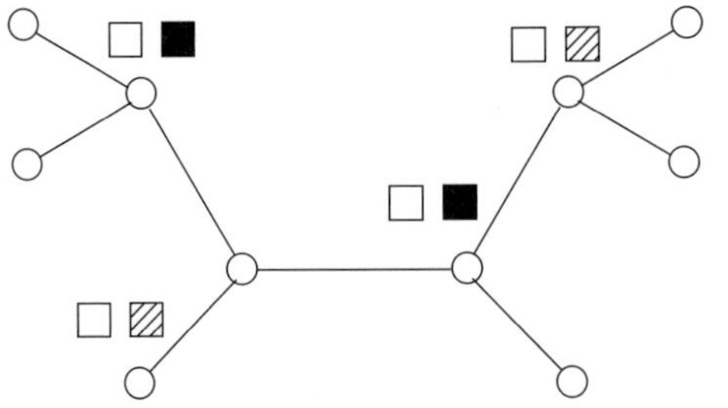

Figure 12.5 A sample instance of the CLC(0, 3, 2, 4) problem (NP-complete).

In the proof of NP-completeness of each subproblem the following NP-complete problem is used.

3-SAT Problem

Instance: A finite family of Boolean variables $z_1, \ldots, z_n$ and a family of clauses $C_1, \ldots, C_m$ each consisting of at most tree literals, that is, either Boolean variables z_i or their negations.

Question: Do there exist values of variables z_i such that each clause contains at least one true variable?

Theorem 12.1 *The CLC(0, 1, 3, 3) problem is NP-complete.*

Theorem 12.2 *The CLC(0, 4, 2, 3) problem is NP-complete.*

Theorem 12.3 *The CLC(0, 3, 2, 4) problem is NP-complete.*

(Detailed proofs of these theorems can be found in [17].)

12.5.2
Five Polynomial-Time Solvable Subproblems

Once some subproblems of the CLC problem turned out to be NP-complete, this immediately implied NP-completeness of the original CLC problem. Moreover, the problem remains NP-complete even if graph G is a single chain, as directly follows from Theorem 12.1 and the reduction (from the case with a family of chains to the case with a single chain) presented in Section 12.3.2. Thus, we have proved that Vising's conjecture on the complexity of the chain-based CLC problem is true.

NP-completeness of this problem in its general form makes relevant the search for its most representative polynomial-time solvable subproblems. In this section we will get acquainted with five such subproblems, as well as with polynomial-time algorithms for their solution. Thus there are five CLC(x_i) problems ($i = 1, \ldots, 5$) from CLC($\mathcal{X}$) that are polynomial-time solvable for the following vectors $x_i \in \mathcal{X}$:

$$x_1 = (\infty, 0, \infty, \infty)\,,$$
$$x_2 = (\infty, \infty, \infty, 2)\,,$$
$$x_3 = (\infty, \infty, 1, \infty)\,,$$
$$x_4 = (\infty, 2, 2, \infty)\,,$$
$$x_5 = (\infty, 3, 2, 3)\,.$$

Sample instances of these subproblems (except the CLC(x_3) problem, which is trivial) are shown in Figures 12.6–12.9.

Figure 12.6 A sample instance of the CLC($\infty, 0, \infty, \infty$) problem (polynomially solvable).

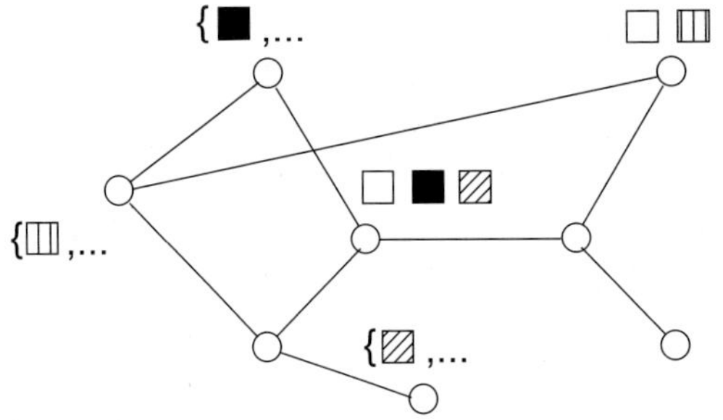

Figure 12.7 A sample instance of the CLC($\infty, \infty, \infty, 2$) problem (polynomially solvable).

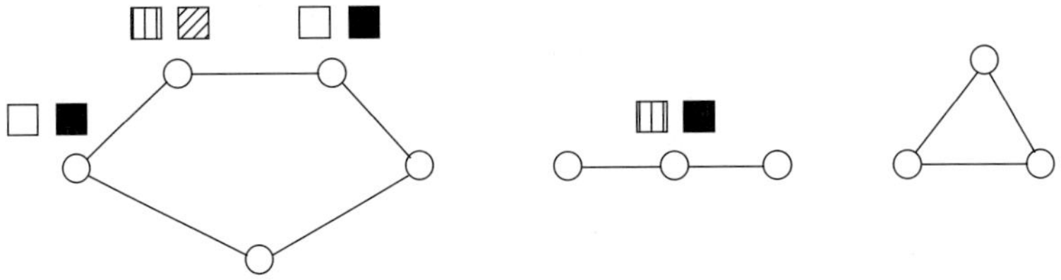

Figure 12.8 A sample instance of the CLC($\infty, 2, 2, \infty$) problem (polynomially solvable).

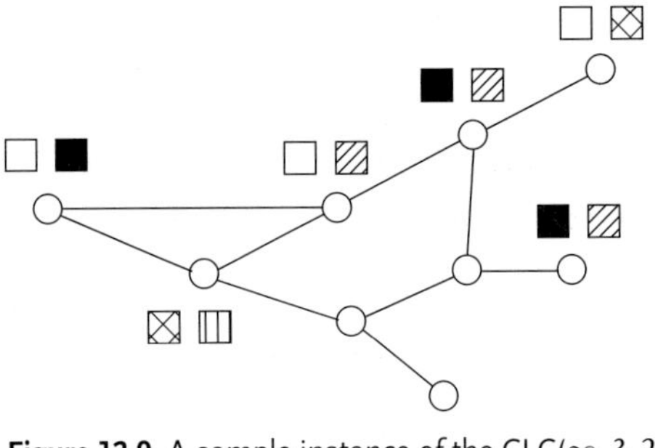

Figure 12.9 A sample instance of the CLC($\infty, 3, 2, 3$) problem (polynomially solvable).

Theorem 12.4 *The CLC($\infty, 0, \infty, \infty$) problem is solvable in polynomial time.*

Proof. For any instance of the CLC($\infty, 0, \infty, \infty$) problem all vertices of graph G belong to different connected components. Therefore, they must be colored in pairwise different colors. Thus, the CLC problem reduces to maximum matching problem in the bipartite graph $\overline{G} = (V, C; E_A)$. If the maximum matching found in that problem covers all vertices in V, this provides a solution for the CLC problem. Otherwise, the desired solution does not exist. Once the maximum matching problem in bipartite graphs is polynomial-time solvable ([18]), the CLC($\infty, 0, \infty, \infty$) problem is also solvable in the same running time.

Theorem 12.5 *The CLC$(\infty, \infty, \infty, 2)$ problem is solvable in polynomial time.*

Proof. Let us describe algorithm $\mathcal{A}_2$ for solving this problem. To that end, a variable graph $\widehat{G}' = (V', C'; E', E'_A)$ will be used initially coinciding with graph $\widehat{G} = (V, C; E, E_A)$ defined in Section 12.4. Graph $\widehat{G}'$ will undergo changes while running algorithm $\mathcal{A}_2$.

Algorithm $\mathcal{A}_2$ is divided into two stages. The first stage consists of $m = |C|$ steps, and each step is dedicated to its own color $i \in C$. Without loss of generality we may assume that step $i = 1, \ldots, m$ deals with color i.

Step i ($i = 1, \ldots, m$). If the area $\mathcal{O}'_i = \{v \in V' \mid i \in A(v)\}$ is connected, then all vertices $v \in \mathcal{O}'_i$ receive color i and are removed from graph $\widehat{G}'$, while color i is removed from the right part C' of graph $\widehat{G}'$.

At the second stage the remaining vertices $v \in V'$ are colored in pairwise different (remaining) colors $i \in C'$, as we did in the proof of Theorem 12.4 (i.e., applying the maximum matching algorithm).

Clearly, the algorithm just described is polynomial in time. Let us show that it finds a feasible coloring for the CLC problem, if such coloring exists.

Let $\varrho : V \to C$ be a feasible coloring of graph G, and let $\widehat{G}^i = (V^i, C^i; E^i, E^i_A)$ denote the state of graph $\widehat{G}'$ at the completion of step i, where $\widehat{G}^0 = \widehat{G}$. The desired property of algorithm $\mathcal{A}_2$ will follow from several propositions.

Proposition 12.1 *The feasible coloring of vertices of the set $\mathcal{O}'_i = \{v \in V' \mid i \in A(v)\}$ in color i attained at one of the steps of the algorithm $\mathcal{A}_2$ cannot be violated in the subsequent steps.*

Proof. Indeed, the connectivity of the area $\varrho^{-1}(i)$ attained at one of the steps of the algorithm can, theoretically, be violated in the subsequent steps in two cases:

(a) as a result of eliminating some vertices from the area $\varrho^{-1}(i)$;
(b) as a result of coloring new vertices in color i.

Yet case (a) is impossible because just colored vertices are immediately eliminated from graph $\widehat{G}'$ (thereby preventing their further recoloring), while case (b) is impossible due to the deletion of color i from set C' right after the first time we used it.

Proposition 12.2 *Let $\widehat{G} = (V, C; E, E_A)$ be an instance of a CLC(x) problem with $x^4 \leq 2$, and let an instance $\widehat{G}' = (V', C'; E', E'_A)$ be obtained from $\widehat{G}$ by deleting a subset $C'' \subseteq C$ of colors together with the subset of vertices $\cup_{i \in C''} \mathcal{O}_i$ (where*

$\mathcal{O}_i = \{v \in V \,|\, i \in A(v)\}$) and together with all incident edges. Let a feasible coloring $\varrho : V \to C$ exist for the instance $\widehat{G}$. Then the restriction $\varrho|_{V'}$ of coloring ϱ to the set of vertices V' is a feasible coloring for the instance $\widehat{G}'$.

Proof. First let us make sure that $\varrho(v) \in C', \forall\, v \in V'$. If we suppose the contrary (that $\varrho(v) \notin C'$ for some vertex $v \in V'$), this would mean that color $i = \varrho(v)$ was deleted from C. But in this case, together with color i, the whole area $\mathcal{O}_i$ was deleted from set V, including vertex v, which would mean that $v \notin V'$ – a contradiction.

Next it should be ascertained that the area $\varrho^{-1}(i)$ remains connected in V' for each color $i \in C'$. Again, as in the proof of Proposition 12.1, case (b) is impossible. Case (a), in principle, is possible, but not for instances of the CLC(x) problem with $x^4 \leq 2$. Indeed, since the area $\varrho^{-1}(i)$ in such instances contains at most two vertices, a deletion of any number of vertices from it remains the area connected.

Thus, coloring ϱ on the remaining set of vertices V' remains feasible for the instance $\widehat{G}'$.

As a direct corollary, we have the following proposition.

Proposition 12.3 *Let a feasible coloring $\varrho : V \to C$ exist for the instance $\widehat{G}$. Then there exists a feasible coloring for $\widehat{G}^m = (V^m, C^m; E^m, E^m_A)$ – the instance obtained from $\widehat{G}$ by the completion of the first stage of algorithm $\mathcal{A}_2$.*

Still we need a proof for the following proposition.

Proposition 12.4 *No feasible coloring for $\widehat{G}^m = (V^m, C^m; E^m, E^m_A)$ may have two identically colored vertices.*

Proof. Suppose the contrary, that is, two vertices $v_1, v_2 \in V^m$ received the same color i in a feasible coloring $\varrho : V^m \to C^m$. This means that both vertices belong to $\varrho^{-1}(i)$. Let $\mathcal{O}_i = \{v \in V \,|\, i \in A(v)\}$ and $\mathcal{O}'_i = \{v \in V^m \,|\, i \in A(v)\}$. Clearly, $\mathcal{O}'_i \subseteq \mathcal{O}_i$, and due to the relations

$$2 \leq |\varrho^{-1}(i)| \leq |\mathcal{O}'_i| \leq |\mathcal{O}_i| \leq x^4 = 2\,,$$

we may conclude that $\varrho^{-1}(i) = \mathcal{O}'_i = \mathcal{O}_i = \{v_1, v_2\}$.

Since the area $\varrho^{-1}(i)$ must be connected, this means that vertices v_1 and v_2 are adjacent. Therefore, the area $\mathcal{O}_i$ is also connected while considering color i at the corresponding step of the first stage, and by the description of algorithm $\mathcal{A}_2$ both vertices v_1, v_2 must be deleted from the set of vertices immediately after their coloring. Thus, they cannot be presented in the set V^m. The contradiction completes the proof of the proposition.

As a corollary, we have the following proposition.

Proposition 12.5 *Let a feasible coloring* $\varrho : V \rightarrow C$ *exist for the instance* $\widehat{G}$. *Then for the instance* $\widehat{G}^m = (V^m, C^m; E^m, E_A^m)$ *obtained from* $\widehat{G}$ *by the completion of the first stage of algorithm* $\mathcal{A}_2$ *there exists a feasible coloring in pairwise different colors.*

As we know, in the case where such a coloring exists, it can be found at the second stage of algorithm $\mathcal{A}_2$ by applying the maximum matching algorithm. Thus, in view of Proposition 12.1, we may conclude that a feasible coloring is found for the whole set of vertices V. Theorem 12.5 is proved.

Theorem 12.6 *The* CLC$(\infty, \infty, 1, \infty)$ *problem is solvable in polynomial time.*

Proof. Since the list $A(v)$ for each vertex v contains a single color, the coloring ϱ feasible with respect to the prescription A is uniquely defined. To check the connectivity of this coloring, it is sufficient to ascertain for each color $i \in C$ that the area $\mathcal{O}_i = \{v \in V \mid i \in A(v)\}$ is connected. Clearly, this can be done in polynomial time.

To prove the polynomial solvability of the next two subproblems, we need to introduce new notions.

Definition 12.1 A function $\mu : V \rightarrow 2^C$ assigning to each vertex $v \in G$ a subset of colors $\mu(v) \subseteq C$ will be referred to as a *labeling*, while $\mu(v)$ is the *label* of vertex v. The set of vertices $\mu^{-1}(i) = \{v \in V \mid i \in \mu(v)\}$ will be called an *area of labeling with color i*. A labeling μ is called a *connected list labeling* (CLL) of vertices of graph $G = (V, E)$ if it meets the following conditions.

- $\mu(v) \neq \emptyset, \; v \in V$;
- $\mu(v) \subseteq A(v), \; v \in V$;
- for each color $i \in C$ the area $\mu^{-1}(i)$ is connected in G.

A labeling μ that meets the above requirements will also be called *feasible* with respect to a given input $I = (G, C, A)$.

It is clear that any connected list coloring is a special case of CLL when we assign to each vertex a one-element subset of colors. Thus, the existence of CLC for a given input I implies the existence of CLL for that input. The converse is not true, in general.

In the **CLL problem** we need to design an algorithm that for any given input I either finds its CLL or establishes the nonexistence of such labeling.

Theorem 12.7 *Any* CLL(x) *problem with* $x^3 \leq 2$ *(which is thus defined on the set of inputs* $I = (G, C, A)$ *with* $A_{\max} \leq 2$*) is solvable in time* $O(nu)$ *if graph* G *has* n *vertices and* u *edges.*

Proof. To solve the CLL problem, we designed an algorithm $\mathcal{A}_\mu(I)$ consisting of the initialization stage and a loop of the form **repeat**... **until false**. The algorithm will use Boolean variables TENLAB and SUCCESS. TENLAB = **true** will mean that a *tentative labeling* is applied that allows its cancellation in the future; SUCCESS=**true** will mean that the desired CLL is found.

Let $\mathcal{W}$ denote the set of pairs (V', A') such that $V' \subseteq V$; $A' : V' \to 2^C$; $A'(v) \subseteq A(v)$, $v \in V'$. We will say that $(V', A') \in \mathcal{W}$ is less than the pair $(V'', A'') \in \mathcal{W}$ if $V' \subseteq V''$ and $A'(v) \subseteq A''(v)$ for every vertex $v \in V'$. If, when that happens, at least one inclusion is strict, we say that pair (V', A') is strictly less than (V'', A'').

The labeling procedure will deal with two variables: a current set of non-labeled vertices V_T and a current prescription A_T; at each moment the pair (V_T, A_T) will take a value from $\mathcal{W}$.

Algorithm $\mathcal{A}_\mu(I)$

Initialization

Set $(V_T, A_T) := (V, A)$; TENLAB := **false**; SUCCESS := **false**.

It follows from $A_{max} \leq 2$ that the list $A(v)$ of vertices $v \in V$ consists of at most two colors. We scan the set V_T and form two lists V_0 and V_1, where V_i is the list of vertices $v \in V_T$ with $|A_T(v)| = i$. Then we define the initial labeling: $\mu_T(v) = \emptyset$ $(v \in V)$.

Loop

At each iteration of the loop the following five *conditions* are verified, and corresponding *actions* are performed.

Condition 1. If $V_T = \emptyset$, then {SUCCESS := **true**; **stop**} (CLL is found).

Condition 2. If $V_0 = \emptyset$ and $V_1 = \emptyset$, then do

Action 2 (starting on a tentative labeling)

TENLAB:=**true**; $(\widetilde{V}, \widetilde{A}) := (V_T, A_T)$; $\tilde{\mu} := \mu_T$ (the original value of (V_T, A_T) and the labeling μ_T are stored prior to commencing the tentative labeling); a vertex $\tilde{v} \in V_T$ and a color $\tilde{i} \in A_T(\tilde{v})$ are chosen and the routine ExpandColor $(\tilde{v}, \tilde{i})$ is applied.

Condition 3. If $V_0 = \emptyset$ and $V_1 \neq \emptyset$, then do

Action 3

Take an arbitrary vertex $v' \in V_1$ and apply routine ExpandColor (v', i') with the uniquely defined color $i' \in A_T(v')$.

Condition 4. If $V_0 \neq \emptyset$ and TENLAB=**true** , then do

Action 4 (cancellation of the tentative labeling)

TENLAB:=**false** ; $(V_T, A_T) := (\widetilde{V}, \tilde{A})$; $\mu_T := \tilde{\mu}$; $V_0 := \emptyset$; $V_1 := \emptyset$. Choose a color $i'' \in A_T(\tilde{v})$ different from $\tilde{i}$ and apply the routine ExpandColor $(\tilde{v}, i'')$.

Condition 5. If $V_0 \neq \emptyset$ and TENLAB=**false** , then { Print: "*No feasible labeling exists*"; **stop**}.
END of the Loop

Routine ExpandColor (v', i')
Find the maximum connected subarea $\mathcal{O}'$ of the area $\mathcal{O}_{i'}$ containing the vertex v'; all vertices in $\mathcal{O}'$ are labeled with color i' and are deleted from V_T and V_1. The remaining vertices $v \in V_T$ are looked through: if $i' \in A_T(v)$, then
{color i' is deleted from $A_T(v)$, and vertex v is added to the list $V_{|A_T(v)|}$}.
END of the routine ExpandColor.

The description of algorithm $\mathcal{A}_\mu(I)$ is completed.

Let us show that algorithm $\mathcal{A}_\mu$ runs in polynomial time. Besides, if there exists a CLL in graph G, then algorithm $\mathcal{A}_\mu$ terminates with the value SUCCESS=**true**, and the resulting labeling $\mu(v)$ (defined for all vertices $v \in V$, since the condition $V' = \emptyset$ is valid) is feasible.

The totality of actions of algorithm $\mathcal{A}_\mu$ prior to the first implementation of Action 2, the one between two successive implementations of Action 2 (from the beginning of the first one to the beginning of the next one), and the one after starting the last implementation will be called *the beginning round, the intermediate round*, and *the ending round*, respectively. The vertex $\tilde{v}$ and the color $\tilde{i}$ chosen at Action 2 of the current round will be referred to as *the initial vertex* and *the initial color* of the round.

To prove the finiteness of the algorithm, we first observe that at each iteration of the loop at least one of those five conditions is satisfied. The validity of Conditions 1 and 5 implies termination of the algorithm. The validity of the remaining three conditions implies performance of the actions, at which point the current pair (V_T, A_T) strictly decreases. The only exception is Action 4, at the completion of which we return to the initial value (V_T, A_T) of the current round. Yet it is clear that between two implementations of Action 4 Boolean variable TENLAB should change its value from **false** to **true**. To that end, Action 2 must be performed. As follows from the definition of the round, Action 4 is performed in each round at most once.

Since right after Action 2 or 4 we have to perform Action 3 (in which the current set V_T decreases by at least one vertex), the assignment of the value $(\widetilde{V}, \widetilde{A}) := (V_T, A_T)$ at each implementation of Action 2 generates the set $\widetilde{V}$ with a smaller number of vertices. This implies that the algorithm consists of at most n rounds.

Action 3 may be performed in a round many times. Yet since at each implementation of this action at least one vertex is being labeled and deleted from V_T, we may conclude that in each round Action 3 is being performed at most $2n$ times. By means of the lists of incident edges specified for each vertex, and using the lists of vertices $\mathcal{O}_i$ $(i \in C)$, we can implement the algorithm $\mathcal{A}_\mu$ so that the running time of each round will not exceed $O(u)$ and the overall bound on the running time of the algorithm $\mathcal{A}_\mu$ will be $O(nu)$, where u is the number of edges in graph G.

Now suppose that for a given instance I there exists a feasible labeling μ. Let us prove that the labeling $\mu_{\mathcal{A}}$ found by algorithm $\mathcal{A}_\mu$ is feasible.

The feasibility of $\mu_{\mathcal{A}}$ with respect to connectivity and prescriptions is evident. It remains to prove that the final labeling $\mu_{\mathcal{A}}$ is complete. To that end, it is sufficient to make certain that the algorithm terminates at Condition 1 (when $V_T = \emptyset$, that is, when all vertices are labeled).

Suppose the contrary, that is, that in some round (called *final*) the algorithm terminates at Condition 5. We will show that this supposition contradicts the supposition made above on the existence of CLL for input I.

Input I is called *CLL-positive* if it admits a CLL. Otherwise, it is *CLL-negative.*

Next we will prove three auxiliary statements.

Proposition 12.6 *Let A' and A'' be the values of the prescription A_T prior to and after the implementation of the routine* ExpandColor (v', i'). *Let there exist a feasible labeling μ for the input $I' = (G, C, A')$, and color i' is used in the label $\mu(v')$. Then the labeling μ is feasible for the input $I'' = (G, C, A'')$.*

Proof. Since while processing this routine we expand label i' to the maximal connected area $\mathcal{O}' \subseteq \mathcal{O}_{i'}$ containing vertex v', we may assert that none of the vertices $v \notin \mathcal{O}'$ is labeled with color i' in labeling μ (because there is no connected area $\mathcal{O}'' \subseteq \mathcal{O}_{i'}$ that would include both vertex v' and vertex v at the same time). Hence, the removal of color i' from the list $A(v)$ of each such vertex $v \notin \mathcal{O}'$ does not violate the feasibility of the labeling μ, which implies that μ is feasible with respect to input I''.

Proposition 12.7 *Let (V', A') and (V'', A'') be the values of the pair (V_T, A_T) at the beginning and at the end of some intermediate round, respectively, and let the input $I' = (G, C, A')$ be* CLL*-positive. Then the input $I'' = (G, C, A'')$ is also* CLL*-positive.*

Proof. Let μ be a CLL for input I', and let $\mu_{\mathcal{A}}$ be the labeling obtained by the algorithm by the completion of the intermediate round. Then we can acceptably extend the labeling $\mu_{\mathcal{A}}$ to the vertices in V''. Indeed, since at the completion of the intermediate round we have $V_0 = \emptyset$, $V_1 = \emptyset$, we may conclude that none of the colors $i \in A(v)$ prescribed to vertices $v \in V''$ was removed while processing the algorithm $\mathcal{A}_\mu$. This means that those colors have not been used in the labeling $\mu_\mathcal{A}$ yet, and therefore no color $i \in A(v)$ ($v \in V''$) has been removed from the lists of the remaining vertices $v \in V$. Thus, by assigning the colors $i \in A(v)$ ($v \in V''$) to the same sets of vertices as in the feasible labeling μ, we thereby acceptably extend the labeling $\mu_\mathcal{A}$ to the set of vertices V''. The proposition is proved.

Proposition 12.8 *Let Action 4 be implemented in the current round of the algorithm $\mathcal{A}_\mu$, and let A', A'' be the values of A_T at the beginning and at the end of the current round, respectively. Let the input $I' = (G, C, A')$ be* CLL*-positive. Then the input $I'' = (G, C, A'')$ is also* CLL*-positive.*

Proof. Let μ be a feasible labeling of input I', and let $\tilde{v}$ and $\tilde{i}$ be, respectively, the vertex and the color chosen to be *initial* in the current round of the algorithm $\mathcal{A}_\mu$. Suppose that color $\tilde{i}$ is used in the label $\mu(\tilde{v})$. Then, by Proposition 12.6, labeling μ is also feasible for the input $I_1 = (G, C, A_1)$ obtained by the time of the completion of Action 2. If for that input I_1 we have $V_1 \neq \emptyset$, and at the next Action 3 vertex $v' \in V_1$ is chosen, then, clearly, the only color $i' \in A_1(v')$ is used in the label $\mu(v')$. Hence, by Proposition 12.6, labeling μ also remains feasible for the input obtained by the completion of Action 3. The subsequent series of performances of Action 3 also results in a CLL-positive input and cannot be completed with the condition $V_0 = \emptyset$ (i.e., with a CLL-negative input). But this contradicts the fact that Action 4 is implemented in the current round. Therefore, our supposition that color $\tilde{i}$ is used in the label $\mu(\tilde{v})$ of a feasible labeling μ was wrong. But this immediately implies that the second color from $A(\tilde{v})$ is used for sure in the label $\mu(\tilde{v})$, and hence Action 4 and the subsequent series of performances of Action 3 of the current round terminates with a CLL-positive input. The proposition is proved.

Let us proceed with the proof of Theorem 12.7. Since by the above assumption input I is CLL-positive, this implies that, due to Proposition 12.6, the series of Action 3 performed in the initial round preserves the CLL-positiveness of the current input. CLL-positiveness is also preserved during the subsequent intermediate rounds, due to Proposition 12.7. Finally, since we assumed that the algorithm terminates its work with Condition 5, this implies that the final round performs Action 4. Therefore, by Proposition 12.8, the final round also generates a CLL-positive input. On the other hand, an input produced by the algorithm with termination by Condition 5 contains a vertex with an empty list and, thus, cannot possess a CLL. The contradiction shows that if the initial input I admits CLL, then the algorithm $\mathcal{A}_\mu$ cannot terminate its work by Condition 5, and hence it has to find a feasible labeling for input I. Theorem 12.7 is proved.

The next lemma establishes a strong connection between CLLs and connected list colorings for inputs $I \in \mathcal{I}(x_4)$.

Lemma 12.3 *For a given input $I \in \mathcal{I}(x_4)$ (where $x_4 = (\infty, 2, 2, \infty)$) there exists a CLL iff there exists a CLC. There is an algorithm with running time $O(n)$ which, given an input $I \in \mathcal{I}(x_4)$ and his CLL, finds its CLC.*

Proof. To transform a CLL μ to a CLC ϱ, it is sufficient to eliminate the double labeling of some vertices $v \in V$, while retaining in the label $\mu(v)$ a single color for each vertex. For the inputs $I \in \mathcal{I}(x_4)$ such a transformation of a CLL μ to a CLC ϱ can be performed by the algorithm $\mathcal{A}^4_{\mu,\varrho}(I)$ described below.

Let us note that for each input $I = (G, C, A) \in \mathcal{I}(x_4)$ the degree of each vertex in graph G is at most 2; therefore, each connected component of graph G represents either a simple chain or a cycle. Thus, we can successively visit all vertices of a connected component in one of two possible directions. Since the area $\mu^{-1}(i)$ of vertices labeled by color $i \in C$ is connected, we may conclude that for the class of inputs under consideration this area also represents a chain or a cycle. The intersection of two such areas (for two different colors i' and i'') is also a cycle, a chain, or a couple of chains. (The last case occurs when each of the areas $\mu^{-1}(i')$, $\mu^{-1}(i'')$ is a chain, they are contained in the same cyclic component, and the chains $\mu^{-1}(i')$, $\mu^{-1}(i'')$ cover the whole connected component, while overlapping at their end fragments.)

Algorithm $\mathcal{A}^4_{\mu,\varrho}(I)$

In the algorithm we will use three different procedures of scanning the vertices that will be referred to as *scanning 1, scanning 2, and scanning 3*. Scanning 1 is an outer loop on graph vertices, inside of which scanning 2 and scanning 3 are called.

Scanning 1. The vertices of graph G are scanned in some order (for instance, in increasing order of their numbers). If a vertex $v' \in V$ has a labeling by two colors (i' and i''), then we start

Scanning 2. Scan the vertices of that connected component, starting from vertex v' in one of two possible directions, until one of the following two events happens:

- *Event 1:* a vertex v'' was encountered whose label does not contain at least one of the colors $\{i', i''\}$;
- *Event 2:* vertex v' was encountered.

Suppose that event 1 happened (let us assume for certainty that $i' \notin \mu(v'')$), then we start

Scanning 3. Scan the vertices of that connected component, starting from vertex v'' and moving in the direction opposite to that used in scanning 2.

Eliminate color i' from the labels of all scanned vertices until one of two events happens:

- *Event* $1'$: a vertex v was encountered that does not contain at least one of the colors $\{i', i''\}$;
- *Event* $2'$: the end of the chain was reached (when the component is a chain).

Clearly, one of the two events $1', 2'$ will be encountered for sure, since in the case where the component is a cycle (not a chain), event $1'$ has to be met either in vertex $v = v''$ or earlier. If one of the events $1', 2'$ happens, stop scanning 3 and proceed with scanning 1.

If scanning 2 terminates at event 2, the connected component is a cycle (C) such that all its vertices are labeled in both colors i', i''. To eliminate the double labeling of vertices $v \in C$, it is sufficient to remove one (arbitrary, but the same) color from all vertices $v \in C$. After that we return to scanning 1.

Algorithm $\mathcal{A}^4_{\mu,\varrho}(I)$ is completely described.

Suppose that a feasible labeling μ is known for a given input $I \in \mathcal{I}(x_4)$. First we note that the running time of algorithm $\mathcal{A}^4_{\mu,\varrho}(I)$ is linear in n. It is also clear that while removing color i' from labels of vertices $v \in V$, the area $\mu^{-1}(i')$ remains connected. (If that area was a cycle, then we remove it wholly; otherwise, if it was a chain, then it shrinks at one of its ends.) Thus, the coloring obtained after the double labeling is eliminated meets the color prescription and is connected. It is thereby proved that the existence of a CLL implies the existence of a CLC for a given input I. If no CLL exists, then no CLC (as a special case of CLL) exists either. Lemma 12.3 is proved.

Next we consider the subproblem specified by vector $x_5 = (\infty, 3, 2, 3)$. The latter means that graphs G with the maximum vertex degree 3 are only allowed in any input $I \in \mathcal{I}(x_5)$, that the list of prescribed colors for each vertex $v \in G$ contains at most two colors, and that each color is included in the prescribed lists of at most three vertices.

Lemma 12.4 *For a given input $I \in \mathcal{I}(x_5)$ (where $x_5 = (\infty, 3, 2, 3)$) there exists a CLL iff there exists a CLC. There is an algorithm with running time $O(n)$ that, given an input $I \in \mathcal{I}(x_5)$ and its CLL, finds its CLC.*

Proof. The desired transformation of a given CLL μ to a CLC ϱ for the inputs $I \in \mathcal{I}(x_5)$ can be performed by the following algorithm.

Algorithm $\mathcal{A}^5_{\mu,\varrho}(I)$

For each vertex $v' \in V$ we do the following verification. If $\mu(v')$ consists of two colors (i' and i''), then we inspect the labels of the vertices adjacent to v' (let V' be the set of such vertices; since $I \in \mathcal{I}(x_5)$, we have $|V'| \leq 3$). By the end of this verification we come to one of the following two cases:

- There exists a color $i \in \mu(v')$ included in the label of at most one vertex from V'; in this case we just remove color i from $\mu(v')$.
- Each of the colors $i', i'' \in \mu(v')$ is included in labels of two vertices from V'; in this case there exists a vertex $v'' \in V'$ labeled by both colors i' and i''; remove color i' from $\mu(v')$ and $\mu(v'')$.

Algorithm $\mathcal{A}^5_{\mu,\varrho}(I)$ is completely described.

To prove that for any input $I = (G, C, A) \in \mathcal{I}(x_5)$ algorithm $\mathcal{A}^5_{\mu,\varrho}(I)$ transforms a feasible labeling μ of the vertices of graph G into a feasible coloring ϱ, it is enough to make certain that deleting a color from the label of some vertex does not violate the connectivity of labeling with that color. Indeed, for any color $i \in C$ the area $\mu^{-1}(i)$ consists of at most three vertices. Deleting the vertex v from the area $\mu^{-1}(i)$ (which happens while deleting color i from the label of vertex v) can violate the connectivity of that area only in the case where:

- $|\mu^{-1}(i)| = 3$;
- vertex v is adjacent to the remaining two vertices from $\mu^{-1}(i)$.

(A vertex with the above properties will be referred to as a *middle vertex* for color i.)

If vertex v is the middle vertex for only one of two colors of label $\mu(v)$, deleting the other color (i) from $\mu(v)$ does not violate the connectivity of labeling with color i. Thus, a violation of connectivity may occur only when vertex v is the middle one for both colors from $\mu(v)$. But in this case some vertex v' adjacent to v must be labeled by the same two colors (i.e., $\mu(v') = \mu(v)$). Removing the color $i \in \mu(v)$ from the labels of both vertex v and vertex v' does not violate the connectivity of the area $\mu^{-1}(i)$, simply because after the removal of vertices v and v' from that area it will contain at most one vertex.

Therefore, the coloring ϱ that the algorithm $\mathcal{A}^5_{\mu,\varrho}(I)$ outputs at its completion is a feasible CLC for input I. Clearly, the running time of the algorithm can be estimated as $O(n)$. Finally, if for a given input I there is no CLL, then there is no CLC either. Lemma 12.4 is proved.

Now we are able to derive ultimate conclusions on the computational complexity of CLC problems defined for classes of inputs $\mathcal{I}(x_4)$ and $\mathcal{I}(x_5)$.

Theorem 12.8 *The CLC(x_4) problem (where $x_4 = (\infty, 2, 2, \infty)$) is solvable in time $O(n^2)$, where n is the number of vertices of graph G.*

Proof. To prove the CLC problem on the class of inputs $I \in \mathcal{I}(x_4)$, it is sufficient to apply algorithm $\mathcal{A}^4$ consisting of two stages.

- Suppose we are given an input $I \in \mathcal{I}(x_4)$. At stage 1 by means of algorithm $\mathcal{A}_\mu(I)$ we find a solution to the CLL problem.
- If the problem has a positive solution (labeling μ), at stage 2 we transform this labeling μ (by means of the algorithm $\mathcal{A}^4_{\mu,\varrho}(I)$) to a feasible CLC ϱ.

Clearly, if there exists a solution to the CLC problem, then it will be found by means of algorithm $\mathcal{A}^4$. Alternatively, if for a given input I there is no CLC, then, as follows from Lemma 12.3, there is no CLL for input I either. In this case algorithm $\mathcal{A}_\mu(I)$ outputs: "CLL does not exist for input I." It is clear that the total running time of algorithms $\mathcal{A}_\mu(I)$ and $\mathcal{A}^4_{\mu,\varrho}(I)$ coincides with the upper bound declared in the theorem, since $u = O(n)$ for $I \in \mathcal{I}(x_4)$. Theorem 12.8 is proved.

Theorem 12.9 *The CLC(x_5) problem (where $x_4 = (\infty, 3, 2, 3)$) is solvable in time $O(n^2)$, where n is the number of vertices of graph G.*

The proof of Theorem 12.9 is similar to that of Theorem 12.8 (one should only change algorithm $\mathcal{A}^4_{\mu,\varrho}(I)$ to algorithm $\mathcal{A}^5_{\mu,\varrho}(I)$ and Lemma 12.3 to Lemma 12.4).

12.6 A Basis System of Problems

First let us introduce a few necessary notations.

An inequality of the form $x' \leq x''$ between two 4-dimensional vectors $x', x'' \in \mathcal{X}$ will mean the validity of four inequalities $x'_i \leq x''_i, i = 1, \ldots, 4$. We say that vectors $x', x'' \in \mathcal{X}$ are incomparable if neither $x' \leq x''$ nor $x'' \leq x'$ holds.

We denote $D^-(x') \doteq \{x \in \mathcal{X} \mid x \leq x'\}$, $D^+(x') \doteq \{x \in \mathcal{X} \mid x \geq x'\}$, $D^-(\mathcal{X}') \doteq \cup_{x \in \mathcal{X}'} D^-(x)$, $D^+(\mathcal{X}') \doteq \cup_{x \in \mathcal{X}'} D^+(x)$; $\mathcal{X}(I) \doteq (T, \Delta_V, \Delta_{V,C}, \Delta_{C,V})$.

Letter $\mathcal{I}$ will denote the set of all inputs of the CLC problem.

In the previous section eight CLC(x) problems for eight different values of the constraining vector x were analyzed. Yet we recall that in Section 12.4 an infinite family CLC($\mathcal{X}$) of CLC(x) problems over all possible vectors $x \in \mathcal{X}$ was defined. Does this mean that, once we have committed ourselves to obtaining the whole picture of complexity over all problems in CLC($\mathcal{X}$), we have to perform a similar complexity analysis for **every** CLC(x) problem from CLC($\mathcal{X}$)?

If this were so, it would be very unfortunate, because the family CLC($\mathcal{X}$) is infinite, and thus, the efforts of the whole scheduling community would be insufficient to cope with this single problem. Fortunately, the situation is not that bad: it is saved due to a simple fact formulated below in Proposition 12.9.

Let $x', x'' \in \mathcal{X}$ and $x' \le x''$. Then, clearly, the set of inputs $\mathcal{I}(x') \doteq \{I \in \mathcal{I} \mid \chi(I) \le x'\}$ is contained in the set of inputs $\mathcal{I}(x'')$, and thereby, the CLC(x') problem is a subproblem of the CLC(x'') problem. This implies that any algorithm that copes efficiently with the CLC(x'') problem is able to cope (not less efficiently) with the CLC(x') problem. And vice versa: if the CLC(x') problem cannot be solved efficiently, then the CLC(x'') problem cannot be solved either.

In terms of the complexity theory this can be formulated as follows.

Proposition 12.9 *Suppose that a family of problems $\mathcal{P} = \{P(x) \mid x \in \mathcal{X}'\}$ is defined, and we consider $x' \in \mathcal{X}'$. Then*

- *if the $P(x')$ problem turns out to be polynomially solvable, then every problem $P(x)$ ($x \in D^-(x')$) is polynomially solvable as well;*
- *if the $P(x')$ problem turns out to be NP-complete, then every problem $P(x)$ ($x \in D^+(x')$) is NP-complete as well.*

Since we have already established that the CLC(x_i) problems ($i = 1, \ldots, 5$) are polynomially solvable, then all CLC(x) problems ($x \in \cup_{i=1}^{5} D^-(x_i)$) are polynomially solvable. In addition, since it has been established that the CLC(x_i) problems ($i = 6, 7, 8$) are NP-complete, then all CLC(x) problems ($x \in \cup_{i=6}^{8} D^+(x_i)$) are NP-complete as well.

Yet what can we say about the remaining CLC(x) problems ($x \in \mathcal{X}$)? We can say nothing, because *there are no other problems* in CLC($\mathcal{X}$). (For the proof of this fact, we refer the reader to [17].) Thus, we are able to perform the complete complexity analysis of the whole infinite family of problems CLC($\mathcal{X}$).

This lucky situation was possible due to the fact that the family of problems CLC($\mathcal{X}$) has a so-called *complete basis system* $\mathcal{B}_{\mathrm{CLC}}(\mathcal{X})$ consisting of the eight above-mentioned *basis problems* CLC(x_i), $i = 1, \ldots, 8$. Let us formally define this important notion.

Definition 12.2 Suppose we consider a problem P and let a parameterized family $\mathcal{P} = \{P(x) \mid x \in \mathcal{X}\}$ of its subproblems be defined. Its subset $\widehat{\mathcal{P}}_{\mathrm{P}} = \{P(x) \mid x \in \widehat{\mathcal{X}}_{\mathrm{P}}\}$ defined on some subset of vectors $\widehat{\mathcal{X}}_{\mathrm{P}} \subseteq \mathcal{X}$ is called a *basis system of polynomially solvable problems* of family $\mathcal{P}$ if it meets the following three conditions.

- *Polynomial solvability:* any problem $P(x) \in \widehat{\mathcal{P}}_{\mathrm{P}}$ is solvable in polynomial time.
- *Independence:* any two vectors $x', x'' \in \widehat{\mathcal{X}}_{\mathrm{P}}$ ($x' \ne x''$) are incomparable.
- *PS-completeness:* for any polynomially solvable problem $P(x) \in \mathcal{P}$ there exists a *basis problem* $P(x') \in \widehat{\mathcal{P}}_{\mathrm{P}}$ such that $x \le x'$ (and so, due to Proposition 12.9, the polynomial solvability of $P(x)$ directly follows from the polynomial solvability of $P(x')$).

Definition 12.3 A set of problems $\widehat{\mathcal{P}}_{\mathrm{NP}} = \{P(x) \mid x \in \widehat{\mathcal{X}}_{\mathrm{NP}}\}$ defined on some subset of vectors $\widehat{\mathcal{X}}_{\mathrm{NP}} \subseteq \mathcal{X}$ is called a *basis system of NP-complete problems* of family $\mathcal{P} = \{P(x) \mid x \in \mathcal{X}\}$ if it meets the following three conditions.

- *NP-completeness:* any problem $P(x) \in \widehat{\mathcal{P}}_{\mathrm{NP}}$ is NP-complete.
- *Independence:* any two vectors $x', x'' \in \widehat{\mathcal{X}}_{\mathrm{NP}}$ ($x' \neq x''$) are incomparable.
- *NPC-completeness:* for any NP-complete problem $P(x) \in \mathcal{P}$ there exists a *basis problem* $P(x') \in \widehat{\mathcal{P}}_{\mathrm{NP}}$ such that $x \geq x'$ (and so, due to Proposition 12.9, the NP-completeness of $P(x)$ directly follows from the NP-completeness of $P(x')$).

Definition 12.4 A set of problems $\widehat{\mathcal{P}} = \widehat{\mathcal{P}}_{\mathrm{P}} \cup \widehat{\mathcal{P}}_{\mathrm{NP}}$ (compound of all polynomially solvable and all NP-complete basis problems) is called a *complete basis system* for the family of problems $\mathcal{P}$ if

$$D^{-}(\widehat{\mathcal{X}}_{\mathrm{P}}) \cup D^{+}(\widehat{\mathcal{X}}_{\mathrm{NP}}) = \mathcal{X}. \tag{12.5}$$

We would like to make two remarks regarding the above definitions.

Remark 12.1 Let $\mathcal{X}_{\mathrm{P}}$ and $\mathcal{X}_{\mathrm{NP}}$ be, respectively, the sets of vectors defining polynomially solvable and all NP-complete problems in $\mathcal{P}$. As shown in [19], if there exists a set of vectors $\widehat{\mathcal{X}}_{\mathrm{P}}$ defining a basis system of polynomially solvable problems from $\mathcal{P}$, then

- each vector in $\widehat{\mathcal{X}}_{\mathrm{P}}$ is a maximum vector in $\mathcal{X}_{\mathrm{P}}$;
- each maximum vector in $\mathcal{X}_{\mathrm{P}}$ must belong to $\widehat{\mathcal{X}}_{\mathrm{P}}$;
- therefore, $\widehat{\mathcal{X}}_{\mathrm{P}}$ has to coincide with the set $\mathcal{X}_{\mathrm{P,max}}$ of all maximum elements from $\mathcal{X}_{\mathrm{P}}$.

It could be assumed that the set $\widehat{\mathcal{X}}_{\mathrm{P}}$ is identical to $\mathcal{X}_{\mathrm{P,max}}$. But in fact, this is not true! The reader should pay attention to the phrase "if there exists the set of vectors $\widehat{\mathcal{X}}_{\mathrm{P}}$...." The fact of the matter is that for some families of problems $\mathcal{P}$ there is no basis system $\widehat{\mathcal{X}}_{\mathrm{P}}$, while the set $\mathcal{X}_{\mathrm{P,max}}$ always exists (possibly empty).

A similar remark is valid with respect to the set $\mathcal{X}_{\mathrm{NP}}$: if there exists a set of vectors $\widehat{\mathcal{X}}_{\mathrm{NP}}$ defining a basis system of NP-complete problems from $\mathcal{P}$, then it coincides with the set $\mathcal{X}_{\mathrm{NP,min}}$ of all minimum elements from $\mathcal{X}_{\mathrm{NP}}$. (However, if $\widehat{\mathcal{X}}_{\mathrm{NP}}$ does not exist, the set $\mathcal{X}_{\mathrm{NP,min}}$ exists anyway.)

Remark 12.2 It is well known from complexity theory [4] that if the conjecture about P≠NP turns out to be true (and most people from our optimization community believe that it should be so), then it implies that there must exist problems in NP that are neither NP-complete nor polynomially solvable (let us call them *middle complexity problems*). Condition 12.5 in Definition 12.4 means that we exclude the existence of middle complexity problems in our parameterized family $\mathcal{P}$ in cases where $\mathcal{P}$ has a (complete) basis system. In other words, the existence of such a basis system in $\mathcal{P}$ implies the *dichotomy property* [20] for the family $\mathcal{P}$.

Knowledge on the complete basis system $\widehat{\mathcal{P}}$ of the family $\mathcal{P}$ lightens considerably the complexity analysis of any particular problem $P(x) \in \mathcal{P}$. Indeed, it is sufficient to compare vector x with every vector $x' \in \widehat{\mathcal{X}} \doteq \widehat{\mathcal{X}}_{\mathrm{P}} \cup \widehat{\mathcal{X}}_{\mathrm{NP}}$. If a vector $x' \in \widehat{\mathcal{X}}_{\mathrm{P}}$ occurs such that $x \leq x'$, this immediately implies the polynomial solvability of problem $P(x)$. Alternatively, if a vector $x'' \in \widehat{\mathcal{X}}_{\mathrm{NP}}$ occurs such that $x \geq x''$, this implies that $P(x)$ is NP-complete. (Evidently, both vectors x' and x'' with the above properties cannot exist simultaneously.)

Admittedly, to guarantee the efficiency of the above procedure (namely, that of enumerating all vectors from $\widehat{\mathcal{X}}$), set $\widehat{\mathcal{X}}$ must be no more than finite. And luckily, as was shown in [19], such a property can be guaranteed for most cases.

Theorem 12.10 (Sevastianov, 2005) *If all key parameters of a problem $\mathcal{P}$ are bounded from below, then for any set of vectors $Y \subseteq \overline{\mathbb{Z}}^n$, the basis system of hard problems for a parameterized family of problems $\mathcal{P}(Y)$ exists and is finite. The basis system of polynomially solvable problems for $\mathcal{P}(Y)$ is also finite, provided that it exists.*

It is also natural to pose the question: how many subsets of problems in $\mathcal{P}(Y)$ possess the properties of a basis system? The answer to this question is contained in the following theorem.

Theorem 12.11 (Sevastianov, 2005) *If for a given parameterized family of problems $\mathcal{P}(Y)$ there exists a complete basis system (basis system of polynomially solvable problems, basis system of NP-complete problems), then it is unique.*

It can be easily verified that three NP-complete subproblems and five polynomially solvable subproblems of the CLC problem considered in Section 12.5 meet all properties formulated in Definitions 12.2 and 12.3. Thus, the eight problems constitute as a whole the uniquely determined **complete basis system** for the parameterized family of problems CLC($\mathcal{X}$).

12.7 Conclusion

In this paper, an interesting graph coloring problem was introduced. It was shown that the problem has useful applications to scheduling and location problems. Next, a parameterized family CLC($\mathcal{X}$) of subproblems of the CLC problem was defined. In Section 12.5, a complexity analysis was performed for eight problems from that infinite family. For five of them, efficient algorithms for their exact solution were presented and analyzed, while the remaining three problems turned out to be NP-complete. And as the reader could observe, graph structure is a significant factor affecting the complexity of a CLC problem.

Finally, from Section 12.6 we learned that those eight problems constitute a uniquely determined complete basis system for the family of problems CLC($\mathcal{X}$). Knowing such a basis system enables one to easily determine the complexity of any other problem from this family.

The question is: what further research could be done for this particularly interesting problem?

To begin with, it is worth mentioning that, although the complexity classification presented in this paper for the family of problems CLC($\mathcal{X}$) is complete, this does not exclude the possibility of performing a similar complexity classification of some other families of problems determined for other collections of key parameters, or even for other types of constraints imposed on the same key parameters. For instance, in this paper parameters Δ_G, $\Delta_{V,C}$, and $\Delta_{C,V}$ were bounded from above, and the NP-completeness of those problems was attained at relatively small values of those bounds. On the other hand, when graph G is close to the complete graph (and thus vertex degrees and the parameter Δ_V become large), the connectivity of the colored area for each color i is no longer a limited factor. As a result, checking the existence of a CLC becomes an easy problem. Thus, an alternative parameterization for the CLC problem is possible based on imposing upper bounds on the parameter $\Delta(\bar{G})$, where $\bar{G}$ is the graph complement to G. The complexity classification of the resulting "alternative" family of problems is of definite interest.

Another open question is the complexity status of the problem of calculating the parameter $\alpha(G)$ (defined in Section 12.2) for an arbitrary graph G. Clarifying the dependence of the problem complexity on the maximum vertex degree of graph G and on its other parameters would be of interest.

The third (but not last) possible research direction is related to the notion of *connected list labeling* introduced in Section 12.5.2. Performing its multiparametric complexity analysis (similar to the one presented in this paper) is a very interesting challenge. To begin with, it would be nice to prove its NP-completeness.

Acknowledgment

This research was partly supported by the Russian Foundation for Basic Research (grant no. 08-0100370) and the joint Russian–Taiwan research grant 05-06-90606-HHC.

References

1 Pinedo, M. *Scheduling: Theory, Algorithms, and Systems* (**1995**), Prentice-Hall, Englewood Cliffs.

2 Pinedo, M., Chao, X. *Operations Scheduling with Applications in Manufacturing and Services* (**1999**), Irwin/McGraw-Hill, New York.

3 Brucker, P. *Scheduling Algorithms*, 2nd edn. (**1998**), Springer, Berlin.

4 Garey, M.R., Johnson, D.S. *Computers and Intractability: A Guide to the Theory of NP-Completeness* (**1979**), Freeman, San Francisco.

5 Tanaev, V.S., Gordon, V.S., Shafransky, Y.M. *Scheduling Theory. Single-Stage Systems* (**1994**), Kluwer, Dordrecht.

6 Tanaev, V.S., Sotskov, Yu.N., Strusevich, V.A. *Scheduling Theory. Multi-Stage Systems* (**1994**), Kluwer, Dordrecht.

7 Harary, F. *Graph Theory* (**1969**), Addison-Wesley, Menlo Park, CA and London.

8 Christofides, N. *Graph Theory: An Algorithmic Approach* (**1975**), Academic Press, New York.

9 Vising, V.G. On the connected list coloring of graphs, *Diskret. Analiz i Issled. Oper., Ser. 1* (**1999**), **6** (4), 36–43 (in Russian).

10 Gimadi, E.Kh. On the location problem on a network with centre-connected areas of cervice, *Upravlyaemye sistemy* (**1984**), **25**, 38–47 (in Russian).

11 Ageev, A.A. A criterion of polynomial-time solvability for the network location problem, in: *Integer Programming and Optimization*. Carnegie Mellon University, Campus Printing (**1992**), pp. 237–245.

12 Asratyan, A.S., Kamalyan, R.R. Interval edge coloring of a multigraph, *Prikladnaya matematika* (**1987**), **5**, 25–34 (in Russian).

13 Hanson, D., Loten, C.O.M., Toft, B. On interval colourings of bi-regular graphs, *ARS Combinatoria* (**1998**), **50**, 23–32.

14 Lipsky, J. Jr. One more polynomial complete consecutive retrieval problem, *Information Processing Letters* (**1977**), **6**, 91–93.

15 Kamalyan, R.R. *Interval colorings of complete bipartite graphs and trees*, preprint (**1989**), Computer Center of the Armenian Academy of Sciences, Yerevan (in Russian).

16 Sevastianov, S.V. Interval edge colorability of a bipartite graph, *Metody Diskret. Analiz.* (**1990**), **50**, 61–72 (in Russian).

17 Kononov, A., Sevastyanov, S. On the complexity of the connected list vertex-coloring problem, *Diskret. Analiz i Issled. Oper., Ser. 1* (**2000**), **7** (2), 21–46 (in Russian).

18 Hopcroft, J.E., Karp R.M. An $n^{5/2}$ algorithm for maximum matching in bipartite graphs, *SIAM Journal of Computing* (**1973**), **2** (2), 225–231.

19 Sevastianov, S. An introduction to multi-parameter complexity analysis of discrete problems, *European Journal of Operational Research* (**2005**), **165** (2), 387–397.

20 Schaefer, T. The complexity of satisfiability problems, in: *Proceedings of the 10th Annual ACM Symposium on Theory of Computing, San Diego, CA, May 1–3, 1978*, pp. 216–226.

13
Complexity of Phylogenetic Networks: Counting Cubes in Median Graphs and Related Problems

Matjaž Kovše

We survey various results on counting hypercubes and related problems. Since median graphs are built in a very special way from hypercubes, the number of hypercubes of different dimensions can also be considered as a measure of complexity for this class of graphs. Applications to phylogenetics are also mentioned.

13.1
Introduction

Probably most mathematicians know the following formulas:

$$n - m = 1 \tag{13.1}$$

and

$$n - m + f = 2\,. \tag{13.2}$$

The first formula is a characteristic property of trees, where n and m denote the number of vertices and number of edges, respectively. The second formula is Euler's formula for planar graphs, where, in addition to the same meaning of n and m, f denotes the number of faces of a planar graph. In this chapter we present different equalities and inequalities of a similar nature for a special class of bipartite graphs, with rich structural properties, that allow one to count their special subgraphs. These graphs are called median graphs, and hypercubes can be considered in some special way as their building blocks. Therefore, the number of different hypercubes of given dimensions can be regarded also as a measure of complexity for median graphs. To count induced (maximal) hypercubes in median graphs is as hard a problem as counting complete graphs in arbitrary graphs. However, many nice theoretical results have been obtained and applications found. The most interesting is the application of counting hypercubes in phylogenetic analysis;

Analysis of Complex Networks: From Biology to Linguistics. Edited by Matthias Dehmer and Frank Emmert-Streib

ISBN: 978-3-527-32345-6

in particular it has been used to detect phantom mutations in sequenced mitochondrial DNA data.

The chapter is organized as follows. In Section 13.2 some basic notions from metric graph theory are defined. Median graphs, which play a central role in the chapter, are defined, and we also present their basic characterizations. In particular, we present an expansion procedure that provides an important tool when considering median graphs and related classes of graphs. In Section 13.3 we present treelike equalities and some Euler-type inequalities for median graphs, quasi-median graphs, partial cubes, and cage-amalgamation graphs. We also discuss the complexity of counting all hypercubes in median graphs. Section 13.4 is devoted to cube polynomials. We give characteristic properties of cube polynomials. Locations of real and rational zeros are presented. Graphs of acyclic cubical complexes and median product graphs are characterized by roots of cube polynomials. Derivatives of polynomials are also treated in this section. A multivariable polynomial generalization of a cube polynomial – a Hamming polynomial – is treated in Section 13.5. Results on a different type of one-variable Hamming polynomial and cage-amalgamation graphs are presented. In Section 13.6 some formulas are presented for counting maximal induced hypercubes in median graphs of circular split systems. In Section 13.7 we briefly describe how median graphs appear in phylogenetics and mention the applications of cube polynomials. We end the chapter with a brief summary and some suggestions for further research.

Most of the theoretical results that we survey in this chapter have been obtained by Brešar, Klavžar, Škrekovski, and their coworkers over the last decade.

13.2
Preliminaries

For $n \in \mathbb{N}$, let $[n]$ denote the set $\{1, 2, \ldots, n\}$. All graphs considered in this chapter are finite and simple. As usual with $V(G)$ and $E(G)$ we denote the vertex and edge set, respectively, of a graph G. If G does not include H as an induced subgraph, then we say that G is an *H-free graph*.

The n-dimensional hypercube, or simply *n-cube*, is a graph that has all n-tuples of 0s and 1s as its vertices, where two such tuples are adjacent if their Hamming distance is equal to 1. The *Hamming distance* between two n-tuples is defined as the number of positions in which these n-tuples differ. We denote the n-dimensional hypercube by Q_n. The two-dimensional hypercube Q_2 is commonly referred as a square, while the three-dimensional hypercube Q_3 is commonly referred as a cube.

For a graph G, let $\alpha_i(G)$, $i \geq 1$, denote the number of induced i-cubes of G. In particular, $\alpha_1(G) = |E(G)|$; in other words $\alpha_1(G)$ equals the number of edges of G. Let also $\alpha_0(G) = |V(G)|$; hence $\alpha_0(G)$ equals the number of vertices of G. The following equality holds for any hypercube (see [27]):

$$1 = \sum_{k=0}^{n} (-1)^k \alpha_k(Q_n)\,. \tag{13.3}$$

See [46] for other identities related to counting hypercubes in hypercubes and different approaches to proving (13.3). A natural generalization of hypercubes are *Hamming graphs*, whose vertices are n-tuples $u = u^{(1)} \ldots u^{(n)}$ such that $0 \leq u^{(i)} \leq n_i - 1$, where for each i we have $n_i \geq 2$, and two vertices are adjacent precisely when their Hamming distance is equal to 1.

13.2.1 Median Graphs

The length of a path is simply the number of its edges. Let u and v be vertices of a connected graph G. *The (shortest path) distance* is defined as a length of a shortest u, v-path in G and is denoted by $d_G(u, v)$, or $d(u, v)$ for short if it is clear from the context which graph is being considered. The corresponding metric space is also called the graphic metric space, associated with graph G, see [29]. See [23] for different aspects of distances in graphs.

The *interval* $I(u, v)$ between u and v is the set of all vertices on all shortest u, v-paths. A subgraph H of G is *isometric* if $d_H(u, v) = d_G(u, v)$ for all $u, v \in V(H)$ and *convex* if $I(u, v) \subseteq V(H)$ for any $u, v \in V(H)$. A graph H *isometrically embeds* into a graph G if H is isomorphic to an isometric subgraph of G. Isometric subgraphs of Hamming graphs are called *partial Hamming graphs*, and in particular isometric subgraphs of hypercubes are called *partial cubes*.

A connected graph is a *median graph* if for every triple u, v, w of its vertices $|I(u, v) \cap I(u, w) \cap I(v, w)| = 1$. The unique vertex that lies in $I(u, v) \cap I(u, w) \cap I(v, w)$ is called the *median vertex* (or simply the *median*) of vertices u, v, and w. Median graphs were first introduced in [2] and later independently in [56,59]. Prototype examples are trees and hypercubes. Further examples of families of median graphs include trees, grids, graphs of acyclic cubical complexes, superextensions, simplex graphs, Fibonacci and Lucas cubes, graphs of linear extensions, and covering graphs of distributive lattices. Besides these examples, which might also indicate the applications of median graphs, we mention also applications in location theory [55], social choice theory [28], and phylogenetics [60]. In the sequel we provide some basic characterizations for median graphs. For more information on median graphs see surveys [7,49] or the recent paper [20] and references therein.

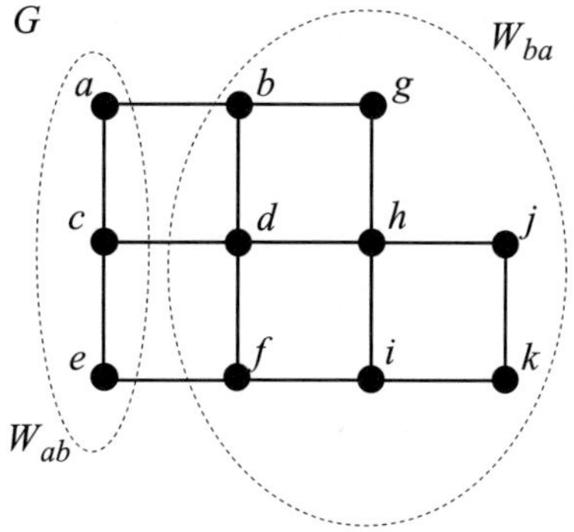

Figure 13.1 Median graph G together with its basic subsets. Here $W_{ab} = \{a, c, e\}$, $W_{ba} = \{b, d, f, g, h, i, j, k\}$, $U_{ab} = \{a, c, e\}$, $U_{ba} = \{b, d, f\}$, and $F_{ab} = \{ab, cd, ef\}$.

Let G be a connected graph. For any edge ab of G we set

$$W_{ab} = \{w \in V(G) \mid d_G(a, w) < d_G(b, w)\}\,, \tag{13.4}$$

$$U_{ab} = \{w \in W_{ab} \mid w \text{ has a neighbor in } W_{ba}\}\,, \tag{13.5}$$

$$F_{ab} = \{e \in E(G) \mid e \text{ is an edge between } W_{ab} \text{ and } W_{ba}\}\,. \tag{13.6}$$

For an example of a median graph and its basic subsets consider Figure 13.1.

Note that if G is bipartite, then for any edge ab, $V(G) = W_{ab} \cup W_{ba}$. Throughout the literature different names are used for sets W_{ab}: *semicubes* [36] or sometimes simply W-sets. The set $\{W_{ab}, W_{ba}\}$ is often called a *split*. More formally, a bipartition of $[n]$ is called a *split* [60]. Let $\overline{A} = [n] \setminus A$, then sets A and $\overline{A}$ form a split of $[n]$ and we put $S = \{A, \overline{A}\}$ and refer to this bipartition as split S. A set of splits is called a *split system*. Let G be a bipartite graph and let $|V(G)| = n$. Then we can label vertices of G with elements from $[n]$. And for any edge ij the sets W_{ij} and W_{ji} form a bipartition of the set $[n]$.

Mulder proved in [57] that median graphs are partial cubes. The first characterization of partial cubes is due to Djoković [31], and we present it in the next theorem.

Theorem 13.1 *A connected bipartite graph G is a partial cube if and only if all subgraphs W_{ab} are convex.*

For median graphs a stronger property holds as Bandelt proved in [3].

Theorem 13.2 *A connected graph G is a median graph if and only if G is bipartite and for every edge ab of G, the sets U_{ab} and U_{ba} are convex.*

The relation Θ [31, 65] is defined on the edge set of a graph G in the following way. Edges $e = xy$ and $f = uv$ of G are in relation Θ if

$$d(x, u) + d(y, v) \neq d(x, v) + d(y, u)\,.$$

The relation Θ is clearly reflexive and symmetric. The following theorem by Winkler from [65] characterizes graphs for which the relation Θ is transitive.

Theorem 13.3 *The relation Θ is transitive precisely on graphs isometrically embedable into Cartesian products of K_3. If in addition G is bipartite, then the relation Θ is transitive if and only if G is a partial cube.*

In the case of partial cubes Θ is therefore an equivalence relation and the Θ-class containing edge ab coincides with the set F_{ab}. Because of Theorems 13.1 and 13.3 the relation Θ is often called a Djoković–Winkler relation Θ. The set F_{ab} forms a matching between U_{ab} and U_{ba} that induces an isomorphism between the subgraphs induced by U_{ab} and U_{ba}. Any Θ-class also forms a minimal edge cutset, and a factor set of relation Θ is often also referred to as a cutset coloring of a graph and Θ-classes as color classes. Since partial cubes are by definition isometric subgraphs of hypercubes, the binary labeling of the vertices of hypercubes can also be used to express the distance between vertices in a partial cube with Hamming distance between labels. Moreover, Θ-classes of a partial cube G represent also the coordinates of the binary labeling induced by an isometric embedding into a hypercube. Therefore, Θ-classes are sometimes also called parallel classes. Since a Θ-class F_{ab} of a partial cube G is uniquely determined by the split W_{ab}, W_{ba}, and vice versa, labeling vertices of G that belong to W_{ab} with 0 and those that belong to W_{ba} with 1, and similarly for any other Θ-class of G, gives the mentioned labeling (at the same time assuming some ordering of Θ-classes of G). In particular, for a median graph G embedded into a hypercube it is straightforward to see that if x is the median of vertices u, v, and w, then the kth coordinate of x can be obtained by applying the so-called *majority rule* for the kth coordinates of u, v, and w: take 0 if at least two vertices among u, v, and w have 0 as the kth coordinate and 1 otherwise.

A convex amalgamation consists of gluing together two graphs along isomorphic convex subgraphs. Bandelt and van de Vel [13] characterized median graphs as connected graphs that can be obtained from hypercubes by a sequence of convex amalgamations.

Median graphs are by definition connected graphs. However, when studying the cube polynomials it turns out that it is natural to consider also a wider class of graphs, that is, a class of graphs that have connected components as median graphs. We denote by $\mathcal{M}$ the class of all median graphs and by $\mathcal{M}^*$ the class of all graphs with connected components as median graphs.

Note that ideas similar to those above can also be used for partial Hamming graphs and other related classes of graphs, see [7, 15, 42].

13.2.1.1 Expansion Procedure

The expansion procedure plays a very important role in metric graph theory. Many interesting graph classes can be characterized in this way. Also, when dealing with families of graphs, it serves as an induction tool for building proofs of statements that involve the number of vertices, edges, or some other graph characteristics that can be controlled well by considering the expansion procedures (of special type).

A *cover* $\mathcal{C}$ of a graph G is a collection $\mathcal{C} = \{G_1, \dots, G_n\}$ of induced subgraphs of G such that they cover the whole graph G, that is, $G = G_1 \cup G_2 \cup \dots G_n$.

Let $\{G_1, G_2\}$ be a cover of a connected graph G, where $G_1 \cap G_2 \neq \emptyset$ and there is no edge between $G_1 \setminus G_2$ and $G_2 \setminus G_1$. Let $\widetilde{G}_1$ and $\widetilde{G}_2$ be isomorphic copies of G_1 and G_2, respectively. For any vertex $u \in G_i$, $1 \leq i \leq 2$, let $\widetilde{u}_i$ be the corresponding vertex in $\widetilde{G}_i$. The *expansion of G with respect to* G_1 *and* G_2 *over* $G_1 \cap G_2$ is the graph $\widetilde{G}$ obtained from the disjoint union of $\widetilde{G}_1$ and $\widetilde{G}_2$, where for any $u \in G_1 \cap G_2$ the vertices $\widetilde{u}_1$ and $\widetilde{u}_2$ are joined by an edge [57]. If G_1 and G_2 are convex (isometric) sets in G, then G' is called a *convex expansion (isometric expansion)* of G. Next we state Mulder's expansion theorem for median graphs [56].

Theorem 13.4 *A connected graph G is a median graph if and only if G can be obtained from* K_1 *by a sequence of convex expansions.*

Chepoi [25] generalized Mulder's convex expansion Theorem 13.4 to all partial cubes, where the convex expansions are replaced by more general isometric expansions. When a cover of a graph consists of more than two subgraphs, one can obtain similar characterizations also for classes of nonbipartite graphs; for example, see [25] for a characterization of partial Hamming graphs. Assuming some special properties of the cover can give characterizations for some special classes of graphs and may consequently also simplify their recognition. For a discussion of different expansion procedures, the reader may consult the books [42, 57] and articles [15, 40, 58].

13.2.1.2 The Canonical Metric Representation and Isometric Dimension

The *Cartesian product* $G \square H$ of two graphs G and H is a graph with vertex set $V(G) \times V(H)$ and $(a, x)(b, y) \in E(G \square H)$ whenever either $ab \in E(G)$ and $x = y$, or $a = b$ and $xy \in E(H)$, see [43] and [42] for many references on Cartesian product and other standard products of graphs. It is well known and also follows easily from the above definition that hypercubes are the simplest example of Cartesian products; they are Cartesian products of K_2. Similarly Hamming graphs can be viewed as Cartesian products of (arbitrary) complete graphs.

One of the main reasons why Cartesian product plays a very significant role in metric graph theory is the fact that the distance function is additive on Cartesian products of graphs; see, for instance, [42]. More precisely, for (g, h) and (g', h') vertices of $G\Box H$, where G and H are arbitrary graphs, the following equality holds:

$$d_{G\Box H}((g, h), (g', h')) = d_G(g, g') + d_H(h, h') .$$

Probably one of the nicest results from metric graph theory is the so-called canonical metric representation due to Graham and Winkler [38]. Let Θ^* denote the transitive closure of the relation Θ. Furthermore, let $E_1, \ldots, E_k$ be the Θ^*-(equivalence) classes. For a connected graph G let G_i denote the graph with the same vertex set as G, that is, $V(G_i) = V(G)$, and the edge set equal to the edge set of graph G minus the edges from the Θ^*-class E_i, that is, $E(G_i) = E(G) \setminus E_i$, $1 \le i \le k$. Further let $C_1^{(i)}, \ldots, C_{r_i}^{(i)}$ denote the connected components of graph G_i. For every i, $1 \le i \le k$, graphs G_i^* are then defined as graphs having connected components $C_1^{(i)}, \ldots, C_{r_i}^{(i)}$ as vertices, that is, $V(G_i^*) = \{C_1^{(i)}, \ldots, C_{r_i}^{(i)}\}$, and $C_j^{(i)}$ and $C_k^{(i)}$ are adjacent in G_i^* if there is an edge between $C_j^{(i)}$ and $C_k^{(i)}$ in the original graph G. For every i, $1 \le i \le k$, mapping $\beta_i : V(G) \to V(G_i^*)$ is defined as simply identifying a vertex $v \in V(G)$ with the connected component $C_j^{(i)}$ to which v belongs, that is, $\beta_i(v) = C_j^{(i)}$, where $v \in C_j^{(i)}$. *The canonical metric representation of graph G* is the mapping

$$\beta : G \to G_1^*\Box \ldots \Box G_k^* ,$$

where $\beta(v) = (\beta_1(v), \ldots, \beta_k(v))$.

In [38] it is shown that β is an irredundant isometric embedding. It is irredundant in the following sense: every factor graph G_i^* has at least two vertices and each vertex of G_i^* appears as a coordinate of some vertex. Furthermore β has the largest possible number of factors among all irredundant isometric embeddings of G, and among such embeddings (with the largest possible number of factors) it is unique. Hence every graph G_i^* is prime with respect to the Cartesian product. From all the properties of the canonical metric representation it follows that the next graph invariant is well defined for any connected graph. The *isometric dimension of a graph G*, denoted by $\dim_I(G)$, is the number of factors appearing in the canonical metric representation of G. Since complete graphs are prime with respect to the Cartesian product, it follows that the isometric dimension of a partial Hamming graph G is simply the smallest number of complete graphs into which G embeds isometrically. In particular, for a partial cube G, its isometric dimension equals the smallest dimension of a hypercube into which G embeds isometrically. In the sequel we will describe the isometric dimension in other ways as well, and it will turn out to play an important role in many of the counting formulas we are going to present.

For more information and further examples see the books [30,42,43] and the paper [47] for a very nice characterization of partial Hamming graphs using canonical metric representation.

13.3 Treelike Equalities and Euler-Type Inequalities

In this section we present treelike equalities and Euler type inequalities for median graphs and their generalizations: quasi-median graphs, partial cubes, and cage-amalgamation graphs.

13.3.1 Treelike Eequalities and Euler-Type Inequalities for Median Graphs

Let G be a median graph. Two splits $\{A, \overline{A}\}$ and $\{B, \overline{B}\}$ of G are said to be *incompatible* if all four of the possible intersections $A \cap B$, $A \cap \overline{B}$, $\overline{A} \cap B$, and $\overline{A} \cap \overline{B}$ are nonempty (otherwise they are called *compatible*). The corresponding Θ-classes F_A and F_B are said to *cross* if the splits $\{A, \overline{A}\}$ and $\{B, \overline{B}\}$ are incompatible. The next theorem, probably observed for the first time by Isbell [44] (in slightly different language) and later rediscovered several times, see [55,63] characterizes hypercubes.

Theorem 13.5 *Let G be a median graph. Splits of G are pairwise incompatible if and only if G is a hypercube.*

It is also worth noting that trees are exactly median graphs with all splits pairwise compatible, as was observed already, in a slightly different language, by Buneman [24]. Dress *et al.* [32] proved the following theorem.

Theorem 13.6 *Let G be a median graph. The number of all sets of pairwise incompatible splits of G, including the empty set and sets of single splits, equals the number of vertices of median graph G.*

In the proof we follow closely the arguments given by Bandelt *et al.* in [12].

Proof. Choose a vertex u and make a BFS ordering from a root of all other vertices of G, corresponding to a root u. The crucial observation is that for a vertex v of G all edges that are incident with v and lie on a shortest path between root u and v induce a hypercube. Therefore, by Theorem 13.5 the corresponding splits are pairwise incompatible, and we associate this set of splits with vertex v. Conversely, for any set of k pairwise incompatible splits there is a unique k-dimensional hypercube, corresponding to this set, that is closest to root u. Uniqueness can be proved by using an expansion procedure to construct G by a sequence of expansion steps starting from a k-dimensional hypercube, corresponding to the set of k pairwise incompatible splits, and then applying the induction principle on the isometric dimension of G (the number of Θ-classes of G).

To count all i-dimensional hypercubes of median graph G, it is enough to count the pairs of vertices v and w, where w lies on a shortest path between v and root u and $d(v, w) = i$, and v is incident with exactly i edges that lie on a shortest path between a root u and v. It follows from the correspondence from the proof that it is equivalent to count all $(k - i)$ subsets of the k sets of pairwise incompatible splits, where $0 \le i \le k$. Let $\beta_k(G)$ denote the number of k sets of pairwise incompatible splits and, as before, $\alpha_i(G)$ the number of induced i cubes in a median graph G. Then it follows that

$$\alpha_i(G) = \sum_{k \ge i} \binom{k}{i} \beta_k(G) \qquad \text{for } i \ge 0\,, \tag{13.7}$$

see [12, 32]. Using binomial inversion, see [1], it follows that

$$\beta_i(G) = \sum_{k \ge i} \binom{k}{i} (-1)^{k-i} \alpha_k(G) \qquad \text{for } i \ge 0\,. \tag{13.8}$$

For $\beta_0(G)$, (13.8) becomes

$$\sum_{i \ge 0} (-1)^i \alpha_i(G) = 1 \tag{13.9}$$

and therefore generalizes the treelike equality (13.3) to all median graphs, as first observed by Soltan and Chepoi [62] and later independently by Škrekovski [61]. For $\beta_1(G)$, (13.8) becomes

$$\sum_{i \ge 0} (-1)^{i+1} i \alpha_i(G) = \dim_I(G)\,, \tag{13.10}$$

as observed first by Škrekovki [61]. Moreover, using the binomial theorem, we get

$$\sum_{i \ge 0} (q-1)^i \alpha_i(G) = \sum_{i \ge 0} q^i \beta_i(G) \qquad q \in \mathbb{R}\,. \tag{13.11}$$

It is also suggested in [12] that for positive choices of q the value $\iota(G) = \sum_{i \ge 0} (q-1)^i \alpha_i(G)$ can be regarded as a kind of complexity measure of the median graph. From (13.11) and Theorem 13.5 it follows that the larger the value of q, more the higher dimensional hypercubes add to the value of $\iota(G)$. If $q = 3$, then $\iota(G)$ equals the number of all hypercubes of dimension at most k in which some vertex of G is contained. In Section 13.4 another complexity measure for median graphs is discussed – cube polynomial – that is the generating function of the sequence $\alpha_i(G)$. Some results about yet another complexity measure for median graphs, the number of all maximal induced hypercubes of given dimension, are presented in Section 13.6.

The incompatibility graph of a median graph G has splits of G as vertices, two vertices $\{A, \overline{A}\}$ and $\{B, \overline{B}\}$ being adjacent if they are incompatible. The incompatibility graph is also called the *crossing graph* in [50], where in the definition, instead of splits, Θ-classes of G are taken. See [6, 17, 48] for more results on crossing graphs of median graphs (and partial cubes). By Theorem 13.5 hypercubes with a dimension of at least 2 in a median graph G correspond to complete subgraphs in their incompatibility graph, while maximal induced hypercubes in median graphs correspond to cliques in their incompatibility graph. Since any graph can be represented as an incompatibility graph of a median graph [50], counting (maximal) hypercubes with a dimension of at least 2 is equivalent to counting complete subgraphs (cliques) of an arbitrary graph, which is widely known to be a nontrivial task. However, many interesting results of counting hypercubes in median graphs have been obtained so far, and the problem has even found interesting biological applications [12, 39].

Klavžar *et al.* in [52] obtained the following Euler-type inequality for median graphs.

Theorem 13.7 *Let G be a median graph with n vertices, m edges, and $dim_I(G) = k$. Then*

$$2n - m - k \leq 2 .$$

The equality holds if and only if G is cube-free.

Combining Theorem 13.7 with Euler's Formula (13.2) the following theorem, obtained by Janaqi [45], follows.

Theorem 13.8 *Let G be a planar, cube-free median graph with n vertices and $dim_I(G) = k$. Then the number of faces in its planar embedding is equal to $n - k$.*

13.3.1.1 Cube-Free Median Graphs

Recall that cube-free median graphs are, by definition, median graphs without an induced three-dimensional hypercube. Interesting applications of cube-free median graphs can be found in the location theory [55]. A graph is a cube-free median graph if and only if it can be obtained from the one vertex graph by an expansion procedure, in which every expansion step is done with respect to a convex cover with a convex tree as intersection. This follows from the fact that a square in the intersection of the convex cover produces a Q_3 in the expansion. Klavžar and Škrekovski obtained in [53] the following result.

Theorem 13.9 *Let G be a cube-free median graph with n vertices, m edges, s squares, and* $\dim_I(G) = k$. *Then* $s = m - n + 1$ *and* $k = -m + 2n - 2$.

The following result by Brešar *et al.* is from [21].

Theorem 13.10 *Let G be a cube-free median graph different from a tree. Let* $r \geq 2$ *be the number of edges in its smallest* Θ*-class and let s denote the number of its squares. Then*

$$\dim_I(G) \geq 2r - 2$$

and

$$k^2 \geq 4s\,.$$

Moreover, both equalities hold if and only if G is the Cartesian product of two trees of the same order.

If G is a tree on more than one vertex, only the second inequality from Theorem 13.10 fails to be strict.

13.3.1.2 Q_4-Free Median Graphs

Klavžar and Škrekovski [53] also studied Q_4-free median graphs and obtained the following two results.

Theorem 13.11 *Let G be a* Q_4*-free median graph on n vertices and m edges, and let h be the number of subgraphs of G isomorphic to* Q_3 *and let* $\dim_I(G) = k$. *Then*

$$2n - m + h - k = 2\,.$$

A *plane graph* is a planar graph together with a given planar embedding. Since Q_4 is not a planar graph, all planar median graphs are Q_4-free median graphs. On the other hand, not all Q_4-free median graphs are planar. Consider, for example, $P_3 \square P_3 \square P_3$. For those that are, the following theorem holds.

Theorem 13.12 *Let G be a median plane graph with n vertices, f faces,* $\dim_I(G) = k$ *and h subgraphs isomorphic to* Q_3. *Then*

$$f = n - k + h\,.$$

13.3.1.3 Median Grid Graphs

A *grid graph* is a subgraph of a complete grid $P_n \square P_m$. Klavžar and Škrekovski [53] characterized median grid graphs in several ways and obtained the following result.

Theorem 13.13 *The length of the outer face of a plane median grid graph G is* $4n - 2m - 4 = 2\dim_I(G)$.

13.3.2
Euler-Type Inequalities for Quasi-Median Graphs

Mulder [57] introduced quasi-median graphs as a natural nonbipartite generalization of median graphs as follows. Let G be a graph, and let (u, v, w) be an ordered triple of vertices of G. A *pseudo-median* of the triple (u, v, w) is an ordered triple (x, y, z) satisfying the following three conditions:

(P1) $d(u, x) + d(x, y) + d(y, v) = d(u, v)$,
$d(v, y) + d(y, z) + d(z, w) = d(v, w)$,
$d(w, z) + d(z, x) + d(x, u) = d(w, u)$;
(P2) $d(x, y) = d(y, z) = d(z, x)$;
(P3) $d(x, y)$ is minimal under conditions (P1) and (P2).

The first equality from (P1) (and analogously the second and third equalities) says that there is a shortest u, v path on which lie both x and y. An ordered triple (x, y, z) is the *quasimedian* of the triple (u, v, w) if it is pseudo-median of (u, v, w) and if (u, v, w) has no other pseudo-medians. A connected graph is a *quasi-median graph* if it satisfies the following three conditions:

(Q1) Each ordered triple has a quasi-median.
(Q2) $K_4 - e$ is not an induced subgraph of G.
(Q3) Each induced C_6 in G has Q_3 or $K_3 \square K_3$ as a convex closure.

If in the definition of pseudo-median $d(x, y) = 0$, then $x = y = z$, and (u, v, w) has a median. Hence quasi-median graphs extend class of median graphs in a natural way. The definition of quasi-median graphs is taken from [57] and is rather long. The characterization, from the following theorem, by Bandelt *et al.* from [11] of quasi-median graphs, analogous to that of median graphs 13.2, is sometimes taken as an alternative and much shorter definition. A subgraph H of a graph G is called *gated* (in G) if for every vertex $v \in V(G)$ there exists a vertex $x \in V(H)$ that lies on a shortest path between v and u for every $u \in V(H)$. Note that such a vertex is always unique if it exists.

Theorem 13.14 *A graph G is quasi-median if every maximal complete subgraph of G is gated and* U_{ab} *is convex for any edge ab.*

In [57] it is proved that quasi-median graphs are partial Hamming graphs and an expansion type theorem is presented. For other characterizations of quasi-median graphs and further references about this class of graphs the reader may consult [19]. Next we present a result by Brešar *et al.* from [19] on counting i-regular Hamming subgraphs in quasi-median graphs. Let γ_i, where $i \geq 0$, be the number of induced i-regular Hamming subgraphs of G.

Theorem 13.15 *Let G be a quasi-median graph. Then*

$$\sum_{i\geq 0}(-1)^i\gamma_i(G) = 1$$

and

$$\sum_{i\geq 0}(-1)^{i+1}i\gamma_i(G) = \dim_I(G)\,.$$

A wider class, including quasi-medians graphs, has been introduced in [15]. *Quasi-semimedian graphs* are partial Hamming graphs with U_{ab} connected for any edge *ab*. It is shown that quasi-semimedian graphs can be obtained by a sequence of connected expansions from K_1.

The next two Euler-type inequalities are from [19].

Theorem 13.16 *Let G be a graph with n vertices, m edges, and* $\dim_I(G) = t$*, that is, it is obtained by a sequence of connected expansions from* K_1*. Then*

$$2n - m - k \leq 2\,.$$

Moreover, the equality holds if and only if G is $C_t \square K_2$*-free, where* $t \geq 3$*, and* K_4*-free.*

Combining Theorem 13.16 with Euler's Formula (13.2) gives the next result.

Theorem 13.17 *Let G be a planar graph with n vertices and* $dim_I(G) = t$*, that is, it is obtained by a sequence of connected expansions from* K_1*. Let f be the number of faces in its planar embedding. Then*

$$f \geq n - k\,.$$

Moreover, the equality holds if and only if G is $C_t \square K_2$*-free* ($t \geq 3$) *and* K_4*-free.*

13.3.3
Euler-Type Inequalities for Partial Cubes

Recall that a graph G is a *partial cube* if G is an isometric subgraph of some hypercube. In addition to the above-mentioned examples of families of median graphs we mention some more interesting families of graphs that belong to the class of partial cubes: hexagonal graphs, phenylenes, graphs of linear extensions, zonotopes, and, more generally, tope graphs of oriented matroids, (all) bipartite outerplanar graphs, etc. The examples just mentioned already point to those fields where, in addition to graph theory, partial cubes have found applications: computational geometry, topology, algebra, computer science, mathematical chemistry, mathematical biology, social sciences, and psychology. See also the book [36] for further applications of partial cubes in media theory and some more interesting examples of partial cubes.

The Euler-type inequality for median graphs 13.9 clearly does not hold for all partial cubes. Consider, for example, the six cycle C_6 to see that the inequality fails. Recently Klavžar and Shpectorov [51] obtained an Euler-type inequality for partial cubes, which we present in this subsection. A well-known characterization of Bandelt [3] says that a connected graph is a median graph if and only if the convex closure of any isometric cycle of G is a cube. Hence the only possible convex cycles in median graphs are four cycles and there are no cycles of length greater than 4. Let $\mathcal{C}(G)$ denote the set of all convex cycles of a given graph G. The *convex excess* of a graph G has been introduced in [51] as $cex(G) = \sum_{C \in \mathcal{C}(G)} \frac{|C|-4}{2}$.

Theorem 13.18 *[51] Let G be a partial cube with n vertices, m edges and $dim_I(G) = k$. Then*

$$2n - m - k - cex(G) \leq 2 .$$

13.3.4
Treelike Equality for Cage-Amalgamation Graphs

In this section we present results by Brešar and Tepeh Horvat from [22], where they introduced cage-amalgamation graphs that generalize both chordal and median graphs.

A chord in a cycle is an edge joining two vertices that are not adjacent in the cycle. A *chordal graph* (also *triangulated graph*) is a graph with the property that each of its cycles on four or more vertices has a chord. It is not hard to see that the isometric dimension of a chordal graph G equals the number of blocks of G. Let $\kappa_i(G)$ denote the number of i-cliques in a graph G. McKee [54] obtained the following two treelike equalities that hold for any connected chordal graph G:

$$\sum_{j \geq 0} (-1)^j \kappa_j(G) = 1 \tag{13.12}$$

and

$$\sum_{j \geq 0} (-1)^{j+1} j \kappa_j(G) = \dim_I(G) . \tag{13.13}$$

Note that Equalities 13.12 and 13.13 are symbolically very similar to Equalities 13.9 and 13.10, respectively.

A *C-block* in a graph G is a maximal connected, chordal-induced subgraph without a cut vertex. A *C-block graph* is a connected chordal graph without a cut vertex. The Cartesian product of arbitrary C-block graphs is called a *cage*. A graph G is a *cage-amalgamation graph* if it can be obtained by a sequence of gated amalgamations from cages.

Theorem 13.19 *For a connected graph G the following propositions are equivalent:*

(i) *G is a cage-amalgamation graph.*

(ii) *Every C-block is gated in G and every set U_{ab} is convex.*

(iii) *G can be obtained from K_1 by a sequence of gated expansions with respect to C-block graphs.*

Note that Theorem 13.19(ii) is similar in nature to Theorem 13.2, while Theorem 13.19(iii) is similar in nature to Theorem 13.4. Even more, triangle-free cage-amalgamation graphs are precisely median graphs, while square-free cage-amalgamation graphs are precisely chordal graphs and the intersection of both classes of graphs is all trees.

It follows straightforward from the definition of C-block graphs that they are prime with respect to the Cartesian product; hence the isometric dimension of a cage-amalgamation graph G equals the smallest number of factors in a Cartesian product of C-block graphs into which G can be isometrically embedded.

Recall that $\gamma_i(G)$ denotes the number of induced i-regular Hamming graphs in graph G. Note that in chordal graphs i-regular Hamming subgraphs coincide with i-cliques, while in median graphs they coincide with i-cubes. The next result generalizes both Equalities 13.9 and 13.10 for median graphs and Equalities 13.12 and 13.13 for chordal graphs.

Theorem 13.20 *Let G be a cage-amalgamation graph. Then*

$$\sum_{i\geq 0}(-1)^i\gamma_i(G) = 1$$

and

$$\sum_{i\geq 0}(-1)^{i+1}i\gamma_i(G) = \dim_I(G)\,.$$

13.4
Cube Polynomials

Many graph polynomials have been introduced so far. For different examples, see, for instance, the book [37] and recent surveys [34, 35]. Although most graph polynomials are naturally defined as generating functions of a special type, the reason they have attracted attention is that often algebraic methods allow us to decode some combinatorial information contained in graph polynomials. In this section we survey known results on the cube polynomial. In particular, we survey results on the cube polynomial of special classes of median graphs and give bounds for rational and real zeros for both classes $\mathcal{M}$ and $\mathcal{M}^*$. A special subclass of median graphs – graphs of acyclic cubical complexes – can be nicely characterized with the roots of their cube polynomials. The Cartesian product of trees of the same order also has a special cube polynomial. Results on higher derivatives are also presented.

Recall that $\alpha_i(G)$, $i \geq 1$, denotes the number of induced i-cubes of G. The *cube polynomial* $c(G, x)$ of G is defined as

$$c(G, x) = \sum_{i \geq 0} \alpha_i(G) x^i .$$

For instance, $c(Q_n, x) = (x + 2)^n$ and $c(T, x) = (n - 1)x + n$, where T denotes a tree on n vertices. Note also that

$$\alpha_k(G) = \frac{c^{(k)}(G, 0)}{k!} .$$

A cover $\mathcal{C}$ of a graph G, as defined in Subsection 13.2.1.1, is *cubical* if every induced hypercube of G is contained in at least one of the members of $\mathcal{C}$. Every cubical cover is also an isometric cover but may not be convex cover. Using the inclusion-exclusion principle the following theorem by Brešar *et al.* from [18] can be easily proved.

Theorem 13.21 *Let $\mathcal{C} = G_1, \ldots, G_n$ be a cubical cover of a graph G. Then,*

$$c(G, x) = \sum_{A \subseteq [n]} (-1)^{|A|-1} c(G_A, x) .$$

By $\mathcal{F}(G)$ we denote the set of edges of graph G consisting of representatives of the Θ-classes of a median graph G. Recall that for an edge $e = uv \in E(G)$ the set $U_e = U_{xy}$ is the subgraph of G induced by the vertices x of G incident with some edge from Θ-class F_{xy}. The *derivative* ∂G of a median graph G is defined as the disjoint union of the graphs U_e, $e \in \mathcal{F}(G)$. Let $G \in \mathcal{M}^* \setminus \mathcal{M}$, in other words $G = G_1 \cup G_2 \cup \ldots \cup G_t$ and $G_i \in \mathcal{M}$, where $1 \leq i \leq t$. The derivative of G is then defined as $G = \partial G_1 \cup \partial G_2 \cup \ldots \cup \partial G_t$.

For $k \geq 0$, higher derivatives are defined recursively in the natural way:

$$\partial_k G = \begin{cases} G & \text{for} \quad k = 0 \\ \partial(\partial_{k-1} G) & \text{for} \quad k \geq 1 . \end{cases}$$

As usual, by $c'(G, x)$ we denote the derivative of $c(G, x)$, and by $c^{(k)}(G, x)$ its kth derivative.

Brešar *et al.* obtained in [19] the following basic properties of cube polynomials.

Theorem 13.22 *Let $c(G, x)$ be the cube polynomial of a graph G.*

(i) *Let G be the expansion with respect to the cubical cover G_1, G_2 and let $G_0 = G_1 \cap G_2$. Then $c(G, x) = c(G_1, x) + c(G_2, x) + xc(G_0, x)$.*
(ii) *For every median graph G, it holds $c(G, -1) = 1$.*
(iii) *For all graphs G and H, it holds that $c(G \Box H, x) = c(G, x)c(H, x)$.*
(iv) *For every median graph G and every integer $k \geq 1$, it holds that $c^{(k)}(G, x) = c(\partial^k G, x)$.*
(v) *For every median graph G, it holds that $c'(G, -1) = \dim_I(G)$.*

Note that when the expressions on the left side of the equality from Theorem 13.22(ii) and (v) are expanded, these equalities are simply Equalities 13.9 and 13.10, respectively.

Motivated by Theorems 13.21 and 13.22(i) and (iii) we follow with next definitions. Let $\mathcal{G}$ denote the class of all finite graphs. We say that a function $f: \mathcal{G} \times \mathbb{R} \to \mathbb{R}$ has:

(i) The *amalgamation property* if $c(G, x) = \sum_{A \subseteq [n]} (-1)^{|A|-1} c(G_A, x)$.
(ii) The *product property* if for any graphs G and H, $f(G \Box H, x) = f(G, x)f(H, x)$.
(iii) The *expansion property* if $f(G, x) = f(G_1, x) + f(G_2, x) + xf(G_0, x)$ whenever G is the expansion with respect to the cubical cover $\{G_1, G_2\}$, where $G_0 = G_1 \cap G_2$.

Now we can state the characteristic properties from [19] regarding cube polynomials.

Theorem 13.23 *Let $f: \mathcal{G} \times \mathbb{R} \to \mathbb{R}$ be a function with:*

(i) *The expansion property and the product property. Then either $f \equiv 0$ or $f \equiv c$.*
(ii) *The amalgamation and the expansion property. Then for any graph G,*

$$f(G, x) = f(Q_0, x)c(G, x)\,.$$

Moreover, if $f: \mathcal{M} \times \mathbb{R} \to \mathbb{R}$ is a function with the expansion property, then for any graph G, $f(G, x) = f(Q_0, x)c(G, x)$.

13.4.1
Cube Polynomials of Cube-Free Median Graphs

In the case of cube-free median graphs, the cube polynomial is of degree two and therefore more can be said about its roots. Brešar *et al.* [21] obtained the following two results.

Theorem 13.24 *The cube polynomial of a cube-free median graph G always has real zeros. Moreover, it has a unique zero if and only if G is a tree or the Cartesian product of two trees of the same order.*

Since the cube polynomials of cube-free median graphs are of degree two, minimum points can also be described nicely.

Theorem 13.25 *Let G be a cube-free median graph that is not a tree, and let $x_{\min}$ be the minimum point of $c(G, x)$. Then*

$$x_{\min} = -1 - \frac{k}{\sum_{e \in \mathcal{F}(G)} |E(U_e)|}.$$

If G is 2-edge-connected, then $x_{\min} \geq -2$, and $x_{\min} = -2$ if and only if $G = Q_2$.

13.4.2 Roots of Cube Polynomials

It is quite natural to ask for roots of graph polynomials. The next theorem by Brešar *et al.* [21] gives an upper bound for roots of cube polynomials.

Theorem 13.26 *Let $G \in \mathcal{M}^*$ be a graph with at least one edge. Then $c(G, x)$ is a strictly increasing function on $[-1, \infty)$.*

Hence for any graph $G \in \mathcal{M}^*$, its cube polynomial $c(G, x)$ has no zeros in $[-1, \infty)$.

13.4.2.1 Rational Roots of Cube Polynomials

As was already observed, $c(T, x) = n + (n-1)x$ for any tree T on n vertices. Thus $-(n/(n-1))$ is the root of $c(T, x)$. As the next result by Brešar *et al.* [21] shows, all possible rational roots of cube polynomials of median graphs are already realized on trees.

Theorem 13.27 *Let G be a median graph. Then any rational zero of $c(G, x)$ is of the form $-((t+1)/t)$ for some $t \in \mathbb{N}$.*

Hence for median graphs all rational zeros are bounded to the interval $[-2, -1)$. The situation is different in the class $\mathcal{M}^*$, where every rational number smaller than -1 is realizable as a root of some cube polynomial. For example, let G be the disjoint union of a tree on $t+1$ vertices and $s-t-1$ additional vertices, where $s, t \in \mathbb{N}$ with $s \geq t+1$. Then, $c(G, -\frac{s}{t}) = 0$.

13.4.2.2 **Real Roots of Cube Polynomials**

For a median graph G, let $z(G)$ denote the largest real zero of its cube polynomial $c(G, x)$. That $z(G)$ is well defined follows from the following theorem from [21].

Theorem 13.28 *Let G be a nontrivial median graph. Then $c(G, x)$ has a real zero in the interval $[-2, -1)$. Moreover, for any nontrivial convex subgraph H of G, it holds that $z(H) \leq z(G)$.*

For $N \geq 1$, let G be a median graph obtained by gluing together a square and a tree on $2N$–3 vertices. Then $c(G, x) = x^2+2Nx+2N$ and the smallest zero of $c(G, x)$ equals $-N - \sqrt{N^2 - 2N}$. Hence the next theorem from [21] follows.

Theorem 13.29 *There exists a median graph with an arbitrarily small negative real zero of its cube polynomial.*

13.4.2.3 **Graphs of Acyclic Cubical Complexes**

Analogously to simplicial complexes, cubical complexes can be defined in a similar way by considering hypercubes instead of simplices. More precisely, a *cubical complex* K is a finite set of hypercubes of any dimension that is closed under taking subcubes and nonempty intersections. The (underlying) graph of cubical complex K has 0-dimensional hypercubes of K as its vertices, and two vertices being adjacent if they belong to a common 1-dimensional hypercube. The wheel W_n, $n \geq 3$, consists of the n-cycle C_n together with an extra vertex joined to all the vertices of the cycle. The *cogwheel* (also *bipartite wheel*) BW_n is obtained from the wheel W_n by subdividing all the edges of the outer cycle. Bandelt and Chepoi introduced graphs of acyclic cubical complexes [6] and characterized them in several different ways, including as follows.

Theorem 13.30 *A graph G is the graph of an acyclic cubical complex if and only if G is a median graph not containing any convex cogwheel.*

By Theorem 13.27 the only possible candidate for an integer root of the cube polynomial of a median graph is –2. It is straightforward to see that $c(BW_n, -2) = 1$. Using the characterization from Theorem 13.30 acyclic cubical complexes are characterized algebraically in [21] as follows.

Theorem 13.31 *Let G be a median graph. Then G is a graph of an acyclic cubical complex if and only if for every 2-connected convex subgraph H of G it holds that $c(H, -2) = 0$.*

For the problem of efficient recognition of graphs of acyclic cubical complexes see [41].

13.4.2.4 **Product Median Graphs**

A graph G is a *product graph* if it is a Cartesian product of nontrivial graphs. Using the observation that for H and K, two convex and nondisjoint subgraphs of a median graph G the union $H \cup K$ is an isometric subgraph of G, median graphs that are product graphs are characterized in [21] as follows.

Theorem 13.32 *Let G be a median graph. Then G is a product graph with $G = H \Box K$ if and only if G contains convex subgraphs H and K such that $|V(H) \cap V(K)| = 1$ and $c(G, x) = c(H, x)c(K, x)$.*

Special Cartesian products of trees are characterized in an algebraic way in [21] as follows.

Theorem 13.33 *Let G be a median graph with the cube polynomial $c(G, x)$ of degree p. Then, $c(G, x)$ has a p-multiple zero if and only if G is a Cartesian product of p trees all of the same order.*

Bandelt *et al.* characterized in [5] the Cartesian products of trees by a forbidden list of isometric subgraphs.

13.4.3
Higher Derivatives of Cube Polynomials

Using the induction on the number of amalgamation steps the following relation, from [19], between derivatives of cube polynomials can be proved.

Theorem 13.34 *Let G be a median graph and $s \geq 0$. Then,*

$$c^{(s)}(G, x+1) = \sum_{i \geq s} \frac{c^{(i)}(G, x)}{(i-s)!}$$

and

$$c^{(s)}(G, x) = \sum_{i \geq s} \frac{(-1)^{i-s}}{(i-s)!} c^{(i)}(G, x+1) .$$

Let $\theta_s(G)$, $s \geq 0$, denote the number of connected components in the graph $\partial^s G$. The following relations between different parameters are from [19].

Theorem 13.35 *Let G be a median graph and $s \geq 0$. Then*

$$\theta_s(G) = c^{(s)}(G,-1)$$
$$\alpha_s(G) = \frac{1}{s!}\sum_{i\geq s}\frac{\theta_s(G)}{(i-s)!}$$
$$\theta_s(G) = s!\sum_{i\geq 0}(-1)^{i-s}\binom{i}{s}\alpha_i$$
$$\sum_{i\geq 0}(-1)^i 2^i \alpha_i = \sum_{i\geq 0}(-1)^i\frac{\theta_i(G)}{i!}\,.$$

13.5 Hamming Polynomials

Brešar *et al.* introduced in [16] the Hamming polynomial $h(G)$ of a graph G as the Hamming subgraph counting polynomial. More precisely, the *Hamming polynomial of a graph G* is defined as

$$h(G) = h(G; x_2, x_3, \ldots, x_\omega) = \sum_{r_2, r_3, \ldots, r_\omega \geq 0} \alpha(G; r_2, r_3, \ldots, r_\omega) x_2^{r_2} x_3^{r_3} \ldots x_\omega^{r_\omega}\,,$$

where $\alpha(G; r_2, r_3, \ldots, r_\omega)$ denotes the number of induced subgraphs of G isomorphic to the Hamming graph $K_2^{r_2}\Box K_3^{r_3}\Box\ldots\Box K_\omega^{r_\omega}$, and $\omega = \omega(G)$, where $\omega(G)$ denotes the clique number of G.

Cube polynomials are therefore only special case of Hamming polynomials, since $c(G,x) = \sum_{r_2\geq 0}\alpha(G; r_2)x_2^{r_2}$.

Before defining the derivatives of a Hamming polynomial, we first define a relation on the set of all complete subgraphs of G on k vertices, denoted by $\mathcal{K}_k(G)$. Complete subgraphs $X, Y \in \mathcal{K}_k(G)$ on vertices $x_1, \ldots, x_k$ and $y_1, \ldots, y_k$, respectively, are in relation $\sim_k$ if the notation of vertices can be chosen in such a way that there exists an integer p such that $d(x_i, y_j) = p+1$ for $i \neq j$, and $d(x_i, y_i) = p$. Relation $\sim_k$ is an equivalence relation on $\mathcal{K}_k(G)$ for a partial Hamming graph, see [21]. Moreover, if G is a partial Hamming graph, then $2 \leq k \leq \omega$, and E is an equivalence class of relation $\sim_k$, then there exists a graph U_E such that $\langle E\rangle = K_k\Box U_E$. Let G be a partial Hamming graph and $E_1, \ldots, E_r$ the equivalence classes of relation $\sim_k$. By the above-mentioned fact, there exist graphs U_i, $1 \leq i \leq r$ such that $\langle E_i\rangle = K_k\Box U_i$. The *$K_k$-derivative* $\partial_k G$ of G (with respect to k) is defined as the disjoint union of the graphs U_i: $\partial_k G = \bigcup_{i=1}^r U_i$.

For a median graph G, $\partial_k G$ is defined only for $k = 2$; moreover, $\partial_2 G = \partial G$, where ∂G is defined as above. The relation between derivatives of Hamming polynomials and derivatives of Hamming graphs is explained in the next theorem, from [16].

Theorem 13.36 *Let G be a partial Hamming graph. Then for any k,* $2 \le k \le \omega$,

$$\frac{\partial h(G; x_2, \ldots, x_\omega)}{\partial x_k} = h(\partial_k G; x_2, \ldots, x_\omega) .$$

Cartesian product behaves nicely on Hamming polynomials as well.

Theorem 13.37 *For any graphs G and H,* $h(G \Box H) = h(G)h(H)$.

Let $\alpha_d = \alpha(G; 0, \ldots, 0, d)$ denote the number of induced subgraphs isomorphic to K_r^d. The next theorem from [16] generalizes Equality 13.3 to Hamming graphs.

Theorem 13.38 *Let G be the Hamming graph* K_r^n. *Then*

$$\sum_{k=0}^{n} (-1)^k \alpha_k = (r-1)^n$$

and

$$\sum_{k=0}^{n} (-1)^k k \alpha_k = -n(r-1)^{n-1} .$$

13.5.1
A Different Type of Hamming Polynomial for Cage-Amalgamation Graphs

Brešar and Tepeh Horvat considered in [22] a different type of Hamming polynomial, denoted by $r(G; x)$, that can be obtained by approximately merging all variables to one and summing the corresponding coefficients in the definition of the Hamming polynomial from the previous subsection. In other words, let $\gamma_i(G)$ denote the number of induced i-regular Hamming graphs in G, then

$$r(G; x) = \sum_{i \ge 0} \gamma_i(G) x_i .$$

In the rest of this section we present results from [22]. First we discuss some properties for Hamming polynomials of arbitrary graphs. For a proper cover of an arbitrary graph the next result is similar to Theorem 13.21 on cube polynomials.

Theorem 13.39 *Let* $C = G_1, \ldots, G_n$ *be a proper cover of a graph G. Then,*

$$r(G, x) = \sum_{A \subseteq [n]} (-1)^{|A|-1} r(G_A, x) .$$

Again as a consequence, when the cover is of a special type, namely both parts are gated subgraphs, the next formula follows.

Theorem 13.40 *If a graph G is a gated amalgam of graphs G_1 and G_2 along graph G_0, then*

$$r(G,x) = r(G_1,x) + r(G_2,x) - r(G_0,x)\,.$$

An expansion of a graph G is *peripheral* (also *minimal*) if one member of the cover of G includes all other members of the cover of G.

Theorem 13.41 *Let G^* be a graph obtained by the peripheral expansion of G along G_0 with respect to H. Then*

$$r(G^*,x) = r(G,x) + r(G_0,x)(r(H,x) - 1)\,.$$

Using induction on the number of vertices of a chordal graph and using Theorem 13.41 $\dim_I(G)$ times one can obtain the next result, which is a special case of 13.20.

Theorem 13.42 *For a connected chordal graph G*

$$r(G,-1) = 1$$

and

$$r'(G,-1) = \dim_I(G)\,.$$

If the expressions from the theorem are expanded, one obtains exactly Formulas 13.12 and 13.13.

13.6 Maximal Cubes in Median Graphs of Circular Split Systems

Let $\mathcal{S}$ be a nonempty collection of splits (bipartitions of $[n]$) also called a *split system*. Graph $G(\mathcal{S})$ has as vertices all sets X that contain exactly one element of each split from $\mathcal{S}$ and have the property that every pair of elements from X intersects. Two vertices U and V of $G(\mathcal{S})$ are adjacent if there exist exactly one split $S = \{A, B\}$ from $\mathcal{S}$ such that either $A \in U \cap V$ or $B \in U \cap V$. Barthélemy [14] introduced a similar construction by extending Buneman's construction of trees [24] (from a set of compatible splits). Graph $G(\mathcal{S})$ is always a median graph [14], and it can be easily seen that the isometric embedding into the corresponding hypercube together with the corresponding binary labeling of vertices is already encoded in the definition of vertices of $G(\mathcal{S})$. Moreover, every median graph can be obtained by such a construction (just consider the set of its splits). $G(\mathcal{S})$ is also called a *median network* or *Buneman graph* [32, 33, 60]. In [13, 64] the number of vertices of the median graph $G(\mathcal{S})$, where $\mathcal{S}$ is the split system of all possible splits of set $[n]$, is calculated for $n \leq 7$.

Bandelt and Dress introduced in [9] circular split systems. For $n \geq 3$, consider an n-cycle C_n and label its vertices with elements from $[n]$. For $n \geq 2$, the *full circular split system* $\mathcal{S}(n)$ on $[n]$ is defined as follows. The full circular split system $\mathcal{S}(2)$ consists of only one possible split of the set [2]; in other words, $\mathcal{S}(2) = \{\{\{1\}, \{2\}\}\}$. For $n \geq 3$, $\mathcal{S}(n)$ consists of all splits of $[n]$ that are induced by removing two edges from C_n and taking the split of $[n]$ corresponding to the two connected components.

Using Theorem 13.6 it is shown in [26] that the number of vertices of a median graph of the full circular split system $\mathcal{S}(n)$ is 2^{n-1}. Let $\varrho_i(G)$ denote the number of maximal induced i-cubes in graph G. Choe *et al.* [26] also provided the following formulas for $\varrho_i(G)$ of full circular split systems.

Theorem 13.43

$$\varrho_i(\mathcal{S}(n)) = \begin{cases} \frac{n}{n-2p}\sum_{j=0}^{p-1} 2^j \binom{p-1}{j}\binom{n-2p}{j+1} & \text{for} \quad 1 \leq p < \lfloor \frac{n}{2} \rfloor \\ 1 & \text{for} \quad p = \lfloor \frac{n}{2} \rfloor \end{cases}$$

In [26] the recursive formulas for maximal induced hypercubes of special split subsystem $\mathcal{S}(n, m)$ are presented. There, the split subsystem $\mathcal{S}(n, m)$ of $\mathcal{S}(n)$, where $1 \leq m < \lfloor \frac{n}{2} \rfloor$, is defined as a system consisting of all splits $S = \{A, B\}$ from $\mathcal{S}(n)$ with the additional property $\min\{|A|, |B|\} = m$.

13.7 Applications in Phylogenetics

Median graphs were introduced by Bandelt in 1994 as a tool for phylogenetic analysis, see [4]. Since then many new theoretical as well practical results have been obtained, that is, median graphs have also been successfully used in the analysis of population data in the form of human mitochondrial data, see [10, 12].

Median graph can be used to visualize phylogenetic relationships as follows. One way to build a median graph from a given alignment pattern of some taxa is to consider all sequences and for any position form splits in such a way as to combine in the same part all sequences that agree in this position. If this is possible, then sequences are said to be *binary* (every position has only two possible values). Once splits are determined for every position of the sequences, the construction from the previous section can be applied (the construction of the Buneman graph).

In practice, however, another approach is more appropriate. Suppose that a given taxon has a binary sequence and that its length is k. Then one can recode the alignment into binary data as follows: choose an arbitrary reference sequence L and code it by the sequence of the length k with all entries equal to 1. For any other sequence the ith position is 1 if the sequence agrees with L in the ith position and 0 otherwise. To reduce unnecessary data, remove all positions that agree on all sequences to get a set M, a set of sequences of length j. So far, sequences present vertices in a j-cube. However, they may not induce a connected graph. In the sequel, a so-called median-joining algorithm is applied that iteratively repeats the following process, for a triple of vertices find their median vertex in a j-dimensional hypercube and add it to set M. Repeat this process until it stabilizes (it is not possible to add any new median vertex to M). The resulting set is also called a *median closure* and the obtained graph is a median graph. If the sequences are not binary, then a similar median-joining algorithm can be used to produce a quasi-median graph instead of a median graph. For small data sets the construction can be done by hand, and for larger examples computer programs are freely available (see the Shareware Phylogenetic Network Software web site).

However, sometimes before the sequencing of the data or during experimental observations some phantom mutations may occur. It is observed in [12] that phantom mutations generate a pattern quite different from that of natural mutations. These mutations are then seen as high-dimensional hypercubes in the corresponding median graph. High-dimensional hypercubes are hard to visualize; therefore, one would like to avoid them if they are not necessary to be present in a median graph obtained by the above construction.

Since large sets of pairwise incompatible splits correspond to high-dimensional hypercubes in median graphs, one would like to minimize the occurrence of such situations. To this end the cube polynomial is used in [12], where it is called a *cube spectrum*. Together with some other techniques, a comparison with reliable data sets has been performed to extract the real data and to avoid errors caused by phantom mutations [12].

13.8 Summary and Conclusion

In this chapter we have surveyed almost all, to the best of our knowledge, known results on counting hypercubes in median graphs and related problems. Among special families of median graphs, other related families of graphs with interesting metric properties were also discussed, that is, quasi-median graphs, cage-amalgamation graphs, and partial cubes. The basic notions and tools from metric graph theory were presented in the first part.

All known treelike equalities and Euler-type inequalities were collected. The properties of cube and Hamming polynomials were treated and their applications to phylogenetics were mentioned. Naturally, many new questions will arise and new challenges will appear. We state some of the intriguing problems that seem promising.

Although the location of roots of cube polynomials of median graphs is quite well understood, it also seems interesting to consider the roots of Hamming polynomials.

It would be interesting to find more treelike equalities for graph classes whose members can be obtained by a sequence of amalgamations (of a special type) from families of graphs with special properties.

Perhaps one could generalize Euler-type inequalities of partial cubes to hold for partial Hamming graphs, or even more generally ℓ_1-graphs?

And finally, for an arbitrary graph G what is the role of the canonical metric representation of G in counting special subgraphs induced by products of some factors appearing in the canonical metric representation of G? Moreover, further applications would make the whole theory even more interesting and valuable.

Acknowledgment

The author would like to thank Boštjan Brešar and Sandi Klavžar for introducing him to the problems discussed herein and for their careful reading of the chapter and several useful suggestions. The author would also like to thank Martin Milanič for careful reading of the chapter and several useful remarks.

Note Added After the Editing Process

After the chapter has been submitted for the publication the author has been informed by Victor Chepoi that the class of graphs that is characterized by fulfilling the equality (13.9) has been intensively studied in [8]. These are partial cubes induced by so called lopsided sets. See [7] and [8] for more information on this very interesting class of graphs.

References

1 M. Aigner, *A Course in Enumeration Theory*, Springer, Berlin, 2007.
2 S.P. Avann, Metric ternary distributive semi-lattices, Proc. Am. Math. Soc. 12 (1961) 407–414.
3 H.-J. Bandelt, Characterizing median graphs, manuscript, 1982, University of Hamburg.

4 H.-J. Bandelt, Phylogenetic networks, Verhandl. Naturwiss. Vereins Hamburg (NF) 34 (1994) 51–71.

5 H.-J. Bandelt, G. Burosch, J.-M. Laborde, Cartesian products of trees and paths, J. Graph Theory 22 (1996) 347–356.

6 H.-J. Bandelt, V. Chepoi, Graphs of acyclic cubical complexes, Eur. J. Combin. 17 (1996) 113–120.

7 H.-J. Bandelt, V. Chepoi, Metric graph theory and geometry: a survey, in *Surveys on Discrete and Computational Geometry: Twenty Years Later* (eds. J.E. Goodman, J. Pach, R. Pollack), Contemp. Math., 453 (2008) 49–86.

8 H.-J. Bandelt, V. Chepoi, A. Dress, J. Koolen, Combinatorics of lopsided sets, European J. Combin. 27 (2006) 669–689.

9 H.-J. Bandelt, A. Dress, A canonical decomposition theory for metrics on a finite set, Adv. Math. 92 (1992) 47–105.

10 H.-J. Bandelt, P. Forster, B.C. Sykes, M.B. Richards, Mitochondrial portraits of human populations using median networks, Genetics 141 (1995) 743–753.

11 H.-J. Bandelt, H.M. Mulder, E. Wilkeit, Quasi-median graphs and algebras, J. Graph Theory 18 (1994) 681–703.

12 H.-J. Bandelt, L. Quintana-Murci, A. Salas, V. Macaulay, The fingerprint of phantom mutations in mitochondrial DNA data, Am. J. Hum. Genet. 71 (2002) 1150–1160.

13 H.-J. Bandelt, M. van de Vel, Superextensions and the depth of median graphs, J. Combin. Theory Ser. A 57 (1991) 187–202.

14 J.-P. Barthélemy, From copair hypergraphs to median graphs with latent vertices, Discrete Math. 76 (1989) 9–28.

15 B. Brešar, Partial Hamming graphs and expansion procedures, Discrete Math. 237 (2001) 13–27.

16 B. Brešar, P. Dorbec, S. Klavžar, M. Mollard, Hamming polynomials and their partial derivatives, Eur. J. Combin. 28 (2007), 1156–1162.

17 B. Brešar, S. Klavžar, Crossing graphs as joins of graphs and Cartesian products of median graphs, SIAM J. Discrete Math. 21 (2007) 26–32.

18 B. Brešar, S. Klavžar, R. Škrekovski, The cube polynomial and its derivatives: The case of median graphs, Electron. J. Combin. 10 (2003) #R3, 11.

19 B. Brešar, S. Klavžar, R. Škrekovski, Quasi-median graphs, their generalizations, and tree-like equalities, Eur. J. Combin. 24 (2003) 557–572.

20 B. Brešar, S. Klavžar, R. Škrekovski, Roots of cube polynomials of median graphs, J. Graph Theory 52 (2006) 37–50.

21 B. Brešar, S. Klavžar, R. Škrekovski, On cube-free median graphs, Discrete Math. 307 (2007) 345–351.

22 B. Brešar, A. Tepeh Horvat, Cage-amalgamation graphs, a common generalization of chordal and median graphs, Eur. J. Combin., in press.

23 F. Buckley, F. Harary, *Distance in Graphs*, Addison-Wesley, Redwood City, CA, 1990.

24 P. Buneman, The recovery of trees from measures of dissimilarity, in *Mathematics in the Archeological and Historical Sciences* (eds F. Hodson *et al.*), Edinburgh University Press, 1971, 387–395.

25 V.D. Chepoi, *d*-Convexity and isometric subgraphs of Hamming graphs, Cybernetics 1 (1988) 6–9.

26 Y.B. Choe, K. Huber, J.H. Koolen, Y.S. Kwon, V. Moulton, Counting vertices and cubes in median graphs associated to circular split systems, Eur. J. Combin. 29 (2008), 443–456.

27 D.I.A. Cohen, *Basic Techniques of Combinatorial Theory*, John Wiley & Sons, New York, 1978.

28 W.H.E. Day, F.R. McMorris, *Axiomatic Consensus Theory in Group Choice and Biomathematics*, Frontiers Appl. Math. Vol. 39 (2003).

29 M. Deza, E. Deza, *Dictionary of Distances*, Elsevier, Amsterdam, 2006.

30 M.M. Deza, M. Laurent, Geometry of cuts and metrics, *Algorithms and Combinatorics*, 15, Springer, Berlin, 1997.

31 D. Djoković, Distance preserving subgraphs of hypercubes, J. Combin. Theory Ser. B 14 (1973) 263–267.

32 A. Dress, M. Hendy, K. Huber, V. Moulton, On the number of vertices and edges of the Buneman graph, Ann. Combin. 1 (1997) 329–337.

33 A. Dress, K. Huber, V. Moulton, Some variations on a theme by Buneman, Ann. Combin. 1 (1997) 339–352.

34 J.A. Ellis-Monaghan, C. Merino, Graph Polynomials and Their Applications I: The Tutte Polynomial, http://arxiv.org/abs/0803.3079.

35 J.A. Ellis-Monaghan, C. Merino, Graph Polynomials and Their Applications II: Interrelations and Interpretations, http://arxiv.org/abs/0806.4699.
36 D. Eppstein, J.-C. Falmagne, S. Ovchinnikov, Media theory, *Interdisciplinary Applied Mathematics*, Springer, Berlin, 2008.
37 C.D. Godsil, *Algebraic Combinatorics*, Chapman & Hall, New York, 1993.
38 R.L. Graham, P.M. Winkler, On isometric embeddings of graphs, Trans. Am. Math. Soc. 288 (1985) 527–536.
39 B. Holland, F. Delsuc, V. Moulton, Visualizing conflicting evolutionary hypotheses in large collections of trees using consensus networks, Systemat. Biol. 54 (2005) 66–76.
40 W. Imrich, S. Klavžar, A convexity lemma and expansion procedures for bipartite graphs, Eur. J. Combin. 19 (1998) 677–685.
41 W. Imrich, S. Klavžar, Recognizing graphs of acyclic cubical complexes, Discrete Appl. Math. 95 (1999) 321–330.
42 W. Imrich and S. Klavžar, *Product Graphs: Structure and Recognition*, John Wiley & Sons, New York, 2000.
43 W. Imrich, S. Klavžar, D. Rall, *Topics in Graph Theory: Graphs and Their Cartesian Products*, AK Peters, Wellesley, MA, 2008.
44 J.R. Isbell, Median algebra, Trans. Ar. Math. Soc. 260 (1980) 319–362.
45 S. Janaqi, Quelques elements de la theorie des graphes, These Universite. Joseph Fourier, Mediatheque, IMAG, Grenoble, France, 1995.
46 S. Klavžar, Counting hypercubes in hypercubes, Discrete Math. 306 (2006) 2964–2967.
47 S. Klavžar, On the canonical metric representation, average distance, and partial Hamming graphs, Eur. J. Combin. 27 (2006) 68–73.
48 S. Klavžar, M. Kovše, Induced cycles in crossing graphs of median graphs, submitted.
49 S. Klavžar, H.M. Mulder, Median graphs: characterizations, location theory and related structures, J. Combin. Math. Combin. Comput. 30 (1999) 103–127.
50 S. Klavžar, H.M. Mulder, Partial cubes and crossing graphs, SIAM J. Discrete Math. 15 (2002) 235–251.
51 S. Klavžar, S. Shpectorov, Convex excess and Euler-type inequality for partial cubes, manuscript, 2007.
52 S. Klavžar, H.M. Mulder, R. Škrekovski, An Euler-type formula for median graphs, Discrete Math. 187 (1998) 255–258.
53 S. Klavžar, R. Škrekovski, On median graphs and median grid graphs, Discrete Math. 219 (2000) 287–293.
54 T.A. McKee, How chordal graphs work, Bull. Inst Comb. Appl. 9 (1993) 27–39.
55 F.R. McMorris, H.M. Mulder, F.R. Roberts, The median procedure on median graphs, Discrete Appl. Math. 84 (1998) 165–181.
56 H.M. Mulder, The structure of median graphs, Discrete Math. 24 (1978) 197–204.
57 H.M. Mulder, *The Interval Function of a Graph*, Math. Centre Tracts 132, Mathematisch Centrum, Amsterdam, 1980.
58 H.M. Mulder, The expansion procedure for graphs, in *Contemporary Methods in Graph Theory*, R. Bodendiek (ed.), Bibliographisches Institut, Mannheim, Germany (1990), 459–477.
59 L. Nebeský, Median graphs, Comment. Math. Univ. Carolinae 12 (1971) 317–325.
60 C. Semple, M. Steel, *Phylogenetics*, Oxford University Press, Oxford, 2003.
61 R. Škrekovski, Two relations for median graphs, Discrete Math 226 (2001) 351–353.
62 P.S. Soltan, V.D. Chepoi, Solution of the Weber problem for discrete median metric spaces (Russian), Trudy Tbiliss Mat Inst Razmadze Akad Nauk Gruzin SSR 85 (1987) 52–76.
63 M. van de Vel, Matching binary convexities, Topol. Appl. 16 (1983) 207–235.
64 A. Verbeek, *Superextensions of Topological Spaces* in: Math. Centre Tracts 41, Mathematisch Centrum, Amsterdam, 1972.
65 P. Winkler, Isometric embeddings in products of complete graphs, Discrete Appl. Math. 7 (1984) 221–225.

14
Elementary Elliptic (*R*, *q*)-Polycycles

Michel Deza, Mathieu Dutour Sikirić, and Mikhail Shtogrin

A (R, q)*-polycycle* is a map whose faces, besides some disjoint *holes*, are i-gons, $i \in R$, and whose vertices have a degree between 2 and q with vertices outside of holes being q-valent. This notion arises in organic chemistry and crystallography as well as in purely mathematical contexts.

A (R, q)-polycycle is called *elementary* if it cannot be cut along an edge. Every (R, q)-polycycle can be uniquely decomposed into elementary ones. This decomposition is useful for computer enumeration and determination of classes of plane graphs and (R, q)-polycycles [6, 11–14, 17, 20]. A critical step for using the decomposition theorem is to be able to list all elementary polycycles occurring in a given problem.

A (R, q)-polycycle is called *elliptic*, *parabolic*, or *hyperbolic* if $\frac{1}{q} + \frac{1}{r} - \frac{1}{2}$ (where $r = max_{i \in R} i$) is positive, zero, or negative, respectively). Here we determine all elementary elliptic (R, q)-polycycles. For parabolic and hyperbolic cases, there is a continuum of possibilities, so this method is less useful.

14.1
Introduction

Given $q \in \mathbb{N}$ and $R \subset \mathbb{N}$, a (R, q)*-polycycle* P is a nonempty 2-connected map on a closed surface S with faces partitioned in two nonempty sets F_1 and F_2, so that:

(1) all elements of F_1 (called *proper faces*) are combinatorial i-gons with $i \in R$;
(2) all elements of F_2 (called *holes*) are pairwisely disjoint, that is, have no common vertices;
(3) all vertices have degree within $\{2, \ldots, q\}$ and all *interior* (i.e., not on the boundary of a hole) vertices are q-valent.

The map P can be finite or infinite and some of the faces of the set F_2 can be i-gons with $i \in R$ or have a countable number of edges. In practice almost all the maps occurring here will be plane graphs, which most often will be finite plane graphs. Note that while any finite plane graph has a unique ex-

Analysis of Complex Networks: From Biology to Linguistics. Edited by Matthias Dehmer and Frank Emmert-Streib

ISBN: 978-3-527-32345-6

terior face, an infinite plane graph can have any number of exterior faces, including 0 and infinity. The exterior faces will always be holes.

An *isomorphism* between two maps, G_1 and G_2, is a function ϕ mapping vertices, edges, and faces of G_1 to those of G_2 and preserving inclusion relations. Two (R,q)-polycycles, P_1 and P_2, are *isomorphic* if there is an isomorphism ϕ of corresponding plane graphs that maps the set of holes of P_1 to the set of holes of P_2. The *automorphism group* Aut(G) of a map G is the group of all its *automorphisms*, that is, isomorphisms of G to G. The *automorphism group* Aut(P) of a polycycle P, considered below, consists of all automorphisms of map G preserving the pair (F_1, F_2); thus, Aut(P) is the stabilizer of the pair (F_1, F_2) in Aut(G).

If a (R,q)-polycycle is a finite plane graph, then its automorphism group is isomorphic to a group of isometries of $\mathbb{R}^3$. Such groups are classified and we use the Schoenflies notation explained, for example, in [18]. If the (R,q)-polycycle is infinite, then there is no general classification. However, if the (R,q)-polycycle is "infinite in only one direction," then the groups are identified as "frieze groups"; the seven possible patterns are described in [2]. It turns out that the (R,q)-polycycles described here belong to those cases. But, in general, many other types of groups could occur.

We now explain some notions of topology that allow us to reduce the problem to the case of (R,q)-polycycles, which are plane graphs. If one is not interested in those notions, then one needs only know that a simply connected (R,q)-polycycle, called an $(R,q)_{\text{simp}}$-polycycle, is a plane graph with the set of holes being the set of exterior faces. If P is an (R,q)-polycycle, then the group $\pi_1(P)$ is a *fundamental group* (for details see, for example, [22]) of the map obtained by removing the faces F_2 from P, that is, by considering them as boundaries. An (R,q)-polycycle P is *simply connected* if $\pi_1(P) = \{\text{Id}\}$, and in this case we call P an $(R,q)_{\text{simp}}$-polycycle. For example, *Prism*$_m$ (m-gonal prism) is a $(\{4\},3)$-polycycle with F_2 consisting of two m-gonal faces. The map *Prism*$_m$, drawn on a plane with one m-gon being an exterior face, is simply connected. But after removal of F_2 from the face set, it becomes nonsimply connected, since a cycle around the faces will no longer be contractible to a point. An (R,q)-polycycle is simply connected if and only if it is a plane graph with F_2 being exactly the set of exterior faces. An automorphism ϕ of an (R,q)-polycycle is called *fixed-point-free* if $\phi = \text{Id}$ or ϕ does not fix any vertex, edge, or proper face. Any (R,q)-polycycle P admits a *universal cover*, that is, an (R,q)-polycycle $\widetilde{P}$, which is simply connected and such that P is obtained as quotient of $\widetilde{P}$ by a group G of fixed point free automorphisms, which is isomorphic to $\pi_1(P)$.

The notion of an (R,q)-polycycle is a large generalization of *(r,q)-polycycles*, that is, $(\{r\},q)_{\text{simp}}$-polycycles introduced by Deza and Shtogrin in [7] and studied in their papers [7–16, 23, 24]. The case $|R| = 1$, that is, *(r,q)-polycycles with holes*, was considered in [3].

A *boundary* of an (R, q)-polycycle P is the boundary of any of its holes.

A *bridge* of an (R, q)-polycycle is an edge that is not on a boundary and goes from a hole to a hole (possibly the same one). An (R, q)-polycycle is called *elementary* if it has no bridges. See below for an illustration of these concepts:

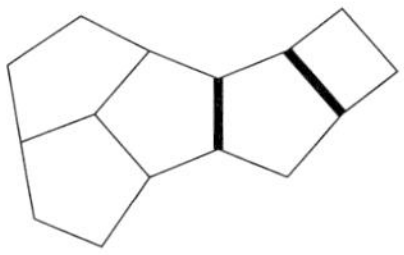

A nonelementary $(\{4, 5\}, 3)_{\text{simp}}$-polycycle with its bridges

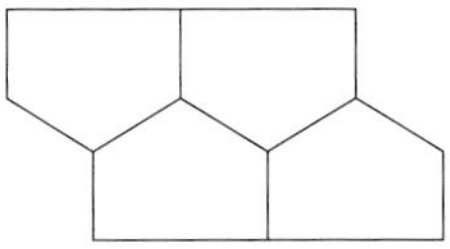

An elementary $(\{5\}, 3)_{\text{simp}}$-polycycle

An *open edge* of an (R, q)-polycycle is an edge on a boundary such that each of its end vertices have a degree less than q. See below the open edges of some (R, q)-polycycles:

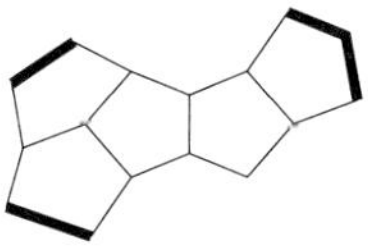

The open edges of a $(\{5\}, 3)_{\text{simp}}$-polycycle

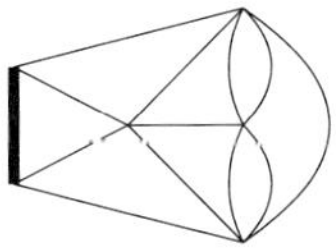

The open edge of a $(\{2, 3\}, 5)_{\text{simp}}$-polycycle

Theorem 14.1 *Every (R, q)-polycycle is uniquely formed by the agglomeration of elementary (R, q)-polycycles along open edges or, in other words, it can be uniquely cut, along the bridges, into elementary (R, q)-polycycles.*

See below for an example of a decomposition of a $(\{5\}, 3)$-polycycle:

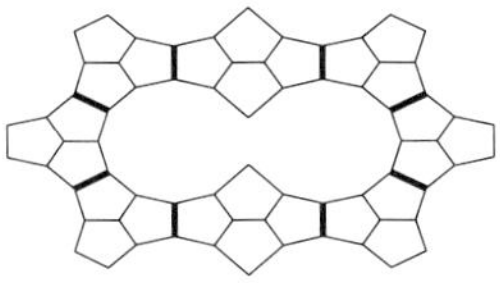

A $(\{5\}, 3)$-polycycle with its bridges being overlined

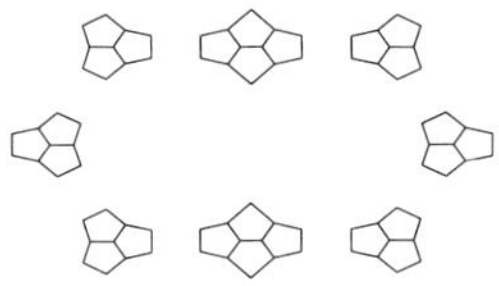

The elementary components of this polycycle

Theorem 14.1 gives a simple way to describe an (r, q)-polycycle: give the names of its elementary components and use the symbol +. In some cases, this is ambiguous, that is, the same elementary component can be used to form an (r, q)-polycycle in different ways; in the same way as the formula of a molecule, giving its number of atoms does not define it in general. For example, with D denoting the $(5, 3)$-polycycle formed by a 5-gon, $D + D + D$ refers unambiguously to the following $(5, 3)$-polycycle:

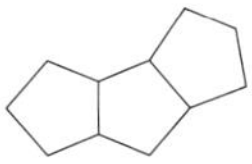

There is another $(5, 3)$-polycycle with three 5-gons sharing a vertex; but this one is elementary. On the other hand $D+D+D+D$ is ambiguous, since there are two $(5, 3)$-polycycles having four elementary components D.

Hence, the interesting question is to enumerate, if possible, those elementary (R, q)-polycycles. Call an (R, q)-polycycle *elliptic, parabolic,* or *hyperbolic* if the number $\frac{1}{q}+\frac{1}{r}-\frac{1}{2}$ (where $r = max_{i\in R}i$) is positive, zero, or negative, respectively. The number of elementary $(\{r\}, q)_{\text{simp}}$-polycycles is uncountable for any parabolic or hyperbolic pairs (r, q) (for example, [5,14]). But in [12,14], all elliptic elementary $(\{r\}, q)_{\text{simp}}$-polycycles were determined. That is to say, the countable sets of all elementary $(\{5\}, 3)_{\text{simp}}$- and $(\{3\}, 5)_{\text{simp}}$-polycycles were described; the cases of $(\{3\}, 3)_{\text{simp}}$-, $(\{4\}, 3)_{\text{simp}}$- and $(\{3\}, 4)_{\text{simp}}$-polycycles are easy. We generalize this classification to all elliptic (R, q)-polycycles. In fact, we will consider the case $R = \{i : 2 \le i \le r\}$ covering all elliptic possibilities: $(\{2, 3, 4, 5\}, 3)$-, $(\{2, 3\}, 4)$- and $(\{2, 3\}, 5)$-polycycles in Sections 2, 3, and 4, respectively.

Given an (R, q)-polycycle P, one can define another (R, q)-polycycle P' by removing a face f from F_1. If f has no common vertices with other faces from F_1, then removing it leaves unchanged the plane graph G and only changes the pair (F_1, F_2). If f has some edges in common with a hole, then we remove them and merge it with the hole. If f has a vertex v in common with a hole and if v does not belong to a common edge, then we split v into two vertices. See below for two examples of this operation:

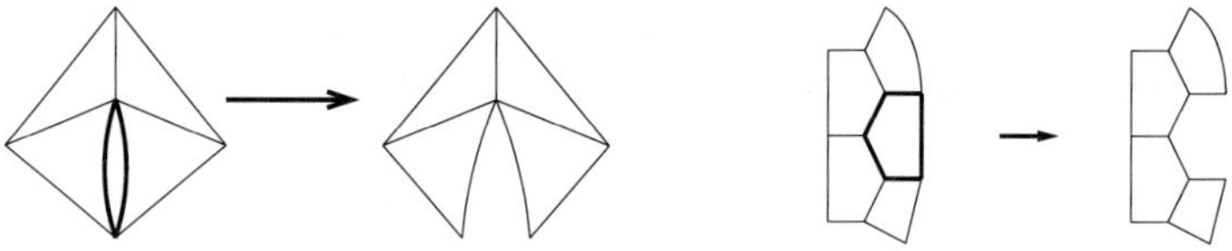

Removal of a 2-, 5-gon having a boundary vertex, edge, respectively.

The reverse operation is the *addition* of a face. A (R, q)-polycycle P is called *extensible* if there exists another (R, q)-polycycle P' such that the removal of a face of P' yields P, that is, if one can add a face to it.

Theorem 14.1, together with the determination of the elementary $(\{r\}, q)_{\text{simp}}$-polycycles, has been the main tool for the following applications: determination of $(\{r\}, q)_{\text{simp}}$-polycycles having the maximal number of interior vertices for a fixed number of faces [12, 14], determination of nonextensible finite $(\{r\}, q)_{\text{simp}}$-polycycles [6, 14], classification of 2-embeddable $(\{r\}, q)_{\text{simp}}$-polycycles [11, 13], determination of $(\{5\}, 3)_{\text{simp}}$-polycycles not uniquely characterized by their boundary [17]. But (R, q)-polycycles are also useful for the enumeration of plane graphs; actually we came to this notion in working on classification questions of *face-regular two-faced* maps [20].

In all pictures below, we put under an (R, q)-polycycle P, its symmetry group Aut(P), and mark *nonext.* for nonextensible P. Also, we put in parentheses the group Aut(G) of the corresponding graph G if Aut(P) $\neq$ Aut(G) and no other polycycle with the same graph G exists. In fact, the same plane graph G can admit several realizations as a (R, q)-polycycle; see examples below mentioned in Appendices 1 and 2.

The computations and graphics presented here were done with the GAP computer algebra program [21], the PlanGraph package [19], and the CaGe drawing program [1].

We thank Gil Kalai for a question that led us to this study.

14.2 Kernel Elementary Polycycles

Call *kernel Ker*(P) of an $(R, q)_{\text{simp}}$-polycycle P the cell complex formed by its vertices, edge, and faces that do not contain a boundary vertex. Call an (R, q)-polycycle *kernel-elementary* if it is an r-gon or if it has a nonempty connected kernel such that the deletion of any face from the kernel will diminish it (i.e., any face of the polycycle is incident to its kernel).

Theorem 14.2

(i) If an $(R, q)_{simp}$-polycycle is kernel-elementary, then it is elementary.

(ii) If (R, q) is elliptic, then any elementary $(R, q)_{simp}$-polycycle is also kernel-elementary.

Proof. (i) Take a kernel-elementary (R, q)-polycycle P; one can assume it to be different from an r-gon. Let $P_1, \ldots, P_m$ be the elementary components of this polycycle. The connectedness condition on the kernel gives that all P_i but one are r-gons with $r \in R$. But removing the components P_i that are r-gons does not change the kernel; thus, $m = 1$ and P is elementary.

(ii) Consider any two vertices of an r-gon of an elliptic (R, q)-polycycle that belongs to the kernel of this polycycle. The shortest edge path between these vertices lies inside the union of two stars of r-gons with the centers at these two vertices; this result can easily be verified in each particular case for any elliptic parameters $(R, q) = (\{2, 3, 4, 5\}, 3)$, $(\{2, 3\}, 4)$, and $(\{2, 3\}, 5)$. Hence, any r-gon of an elliptic (R, q)-polycycle is incident with only one simply connected component of its kernel. All r-gons that are incident with the same nonempty connected component of the kernel constitute a nontrivial elementary component. Since the polycycle is elementary, this is its totality and the kernel is connected.

In [14] the notion of kernel-elementary was called *elementary*. See below for an example of a $(\{6\}, 3)$-polycycle that is elementary but not kernel-elementary, since its kernel is not connected:

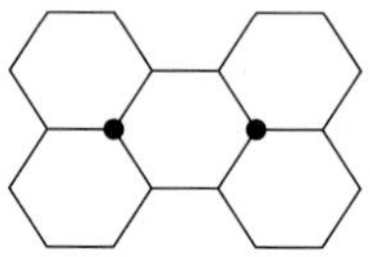

The decomposition Theorem 14.1 (of (r, q)-polycycles into elementary polycycles) is the main reason why we prefer to call the property elementary rather than kernel-elementary. Another reason is that if an (R, q)-polycycle is elementary, then its universal cover is also elementary.

The notion of kernel is used as a technical tool in the classification results obtained below. If P is an $(R, q)_{\text{simp}}$-polycycle, then we denote by $Ker(P)$ its kernel and by $G(P)$ the cell complex obtained by removing from $Ker(P)$ all edges and vertices that are not contained in any face. See below for an example:

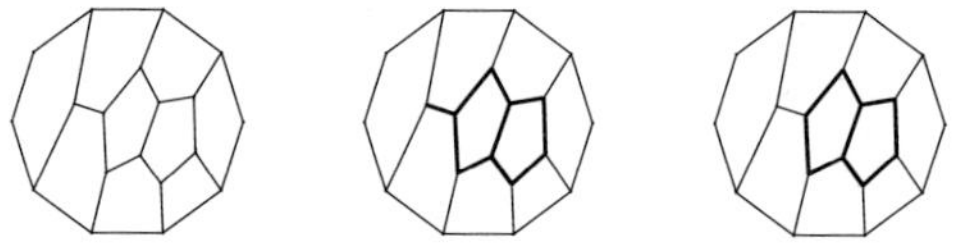

P $\quad$ $Ker(P)$ $\quad$ $G(P)$

We will prove that $G(P)$ is itself an $(R, q)_{\text{simp}}$-polycycle if (R, q) is elliptic and P is an elementary $(R, q)_{\text{simp}}$-polycycle. Call an $(R, q)_{\text{simp}}$-polycycle P' *kernelable* if there is an $(R, q)_{\text{simp}}$-polycycle P such that $P' = G(P)$.

14.3
Classification of Elementary ({2, 3, 4, 5}, 3)-Polycycles

For $2 \le m \le \infty$, denote by $Barrel_m$ the $(\{5\}, 3)$-polycycle with two m-gonal holes separated by two m-rings of 5-gonal proper faces. This polycycle is nonextensible if and only if $m \ge 6$. Its symmetry group, D_{md}, coincides with the symmetry group of the underlying 3-valent plane graph if and only if $m \neq 5$. See below for pictures of $Barrel_m$ for $m = 2, 3, 4, 5$, and ∞:

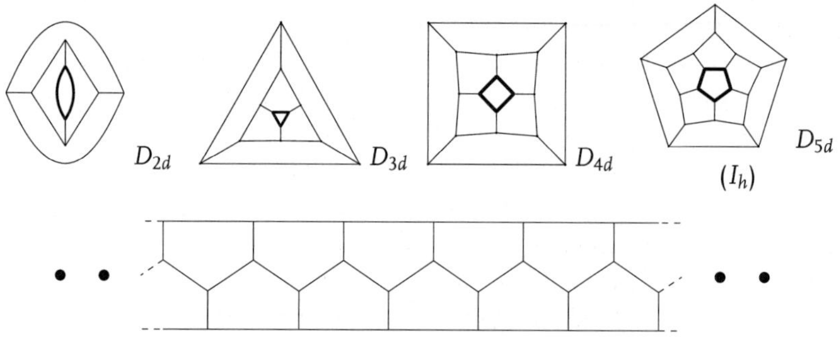

$D_{\infty d}$=pma2, nonext.

Theorem 14.3 *The list of elementary $(\{2,3,4,5\},3)$-polycycles consists of:*

(i) 204 *sporadic $(\{2,3,4,5\},3)_{simp}$-polycycles, given in Appendix 1.*
(ii) Six $(\{3,4,5\},3)_{simp}$-polycycles, infinite in one direction:

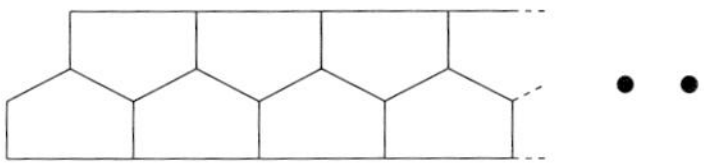

α: C_1

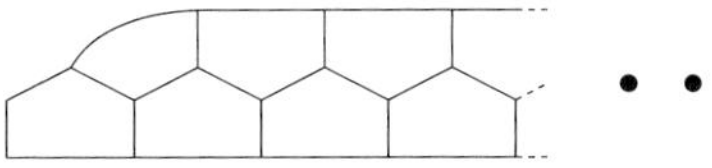

δ: C_1

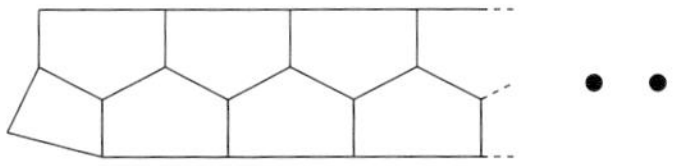

β: C_1

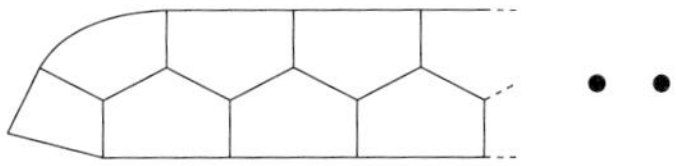

ε: C_1, *nonext.*

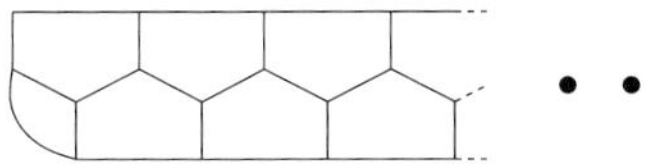

γ: C_1, *nonext.*

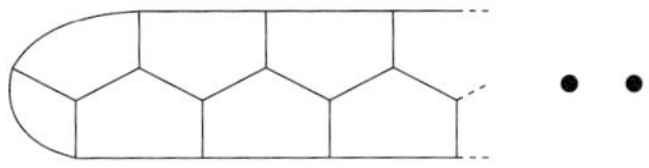

μ: C_1, *nonext.*

(iii) $21 = \binom{6+1}{2}$ *infinite series obtained by taking two ends of the infinite polycycles from (ii) above and concatenating them.*
For example, merging α with itself produces the infinite series of elementary $(\{5\},3)_{simp}$-polycycles, denoted by E_n in [12]. See Figure 14.1 for the first 3 members (starting with 6 faces) of two such series: $\alpha\alpha$ and $\beta\varepsilon$.
(iv) The infinite series of Barrel$_m$, $2 \le m \le \infty$, and its nonorientable quotient for m odd.

Proof. Take an elementary $(\{2,3,4,5\},3)_{\rm simp}$-polycycle P that, by Theorem 14.2, is kernel-elementary. If its kernel is empty, then P is simply an r-gon with $r \in R$. If the kernel is reduced to a vertex, then P is simply a triple of r-gons with $r \in R - \{2\}$. If each r-gon of P has at most three vertices from the kernel that are arranged in succession along the perimeter, then the kernel does not contain any face and has the form of a *geodesic* or a *propeller*, that is, a vertex and three adjacent vertices. If $Ker(P)$ is a geodesic, then P is one of $Barrel_\infty$, (ii) or (iii). If it is a propeller, then there is a finite number of possibilities, which occur in (i).

Suppose now that $Ker(P)$ contains some faces. Then any face not contained in $Ker(P)$ contains at most 3 vertices of $Ker(P)$. Now, if two faces F and F' of $Ker(P)$ are related by a sequence of kernel edges, then one can see that F and F' are also related by a sequence of faces. Similarly, if $e_1 = \{v_0, v_1\}$, $e_2 = \{v_1, v_2\}$ are kernel edges with v_0 belonging to a kernel face F'' and $e_1 \notin F''$, then e_1 belongs to another face of the kernel. So, $G(P)$ is a $(\{2,3,4,5\},3)$-polycycle and $Ker(P)$ is obtained from $G(P)$ by adding edges sharing a vertex with $G(P)$. The method for enumerating the kernelable $(\{2,3,4,5\},3)_{\rm simp}$-polycycles is then as follows:

1. Denote by $\mathcal{L}(N)$ a list of kernelable $(\{2,3,4,5\},3)_{\text{simp}}$-polycycles, which is the complete list of such polycycles for those with at most N interior faces.
2. Start with $\mathcal{L}(1)$ being the list of r-gons for $r \in \{2,3,4,5\}$.
3. For every N take any element in $\mathcal{L}(N)$ and add to it the kernelable $(\{2,3,4,5\},3)_{\text{simp}}$-polycycles obtained by adding one or two faces.
4. Reduce by isomorphism and obtain $\mathcal{L}(N+1)$.

It turns out that this algorithm stops at $N = 6$. Let P be a kernelable $(\{2,3,4,5\},3)_{\text{simp}}$-polycycle. If P is infinite, then one can find finite subpolycycles of arbitrary size that are also kernelable. If P is finite, then one can remove one or two faces and still have a kernelable $(\{2,3,4,5\},3)_{\text{simp}}$-polycycle. This proves that there are no kernelable $(\{2,3,4,5\},3)$-polycycles with more than 6 proper faces, and so we have the complete list.

Given a kernelable $(\{2,3,4,5\},3)_{\text{simp}}$-polycycle, we consider all possible ways of adding edges to it to form a kernel and obtain therefore all elementary $(\{2,3,4,5\},3)_{\text{simp}}$-polycycles. Let us now determine all elementary $(\{2,3,4,5\},3)$-polycycles that are not simply connected. The universal cover $\widetilde{P}$ of such a polycycle P is an elementary $(\{2,3,4,5\},3)_{\text{simp}}$-polycycle, which has a nontrivial fixed-point-free automorphism group in $\text{Aut}(\widetilde{P})$. Consideration of the above list of polycycles yields $Barrel_\infty$ as the only possibility. The polycycle $Barrel_m$ and its nonorientable quotients arise in this process.

Infinite series $\alpha\alpha$ of elementary $(\{2,3,4,5\},3)_{\text{simp}}$-polycycles:

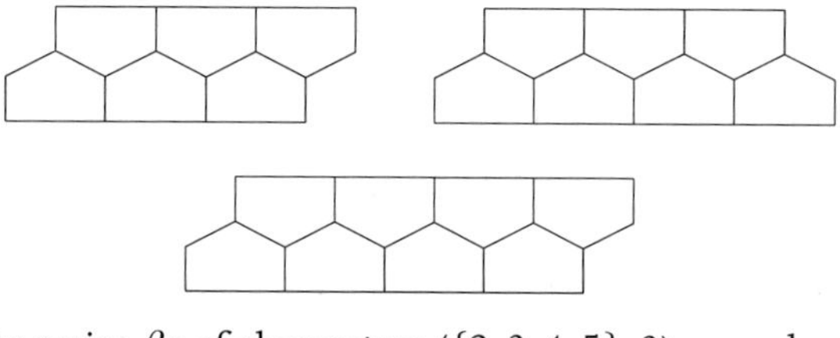

Infinite series $\beta\varepsilon$ of elementary $(\{2,3,4,5\},3)_{\text{simp}}$-polycycles:

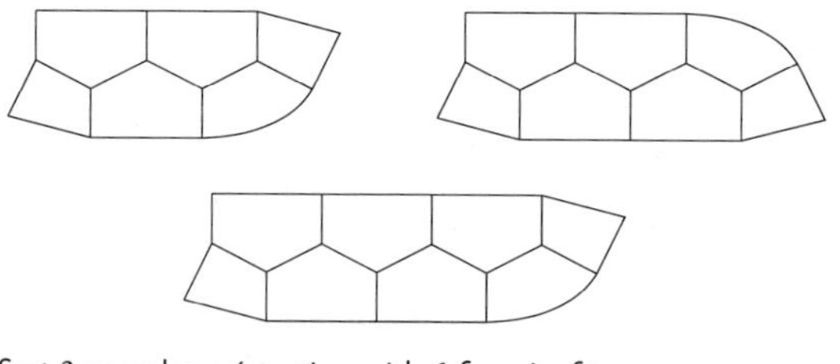

Figure 14.1 The first 3 members (starting with 6 faces) of two infinite series, among 21 series of $(\{2,3,4,5\},3)_{\text{simp}}$-polycycles in Theorem 14.3(iii).

14.4 Classification of Elementary $(\{2,3\},4)$-Polycycles

Theorem 14.4 *Any elementary $(\{2,3\},4)$-polycycle is one of the following eight:*

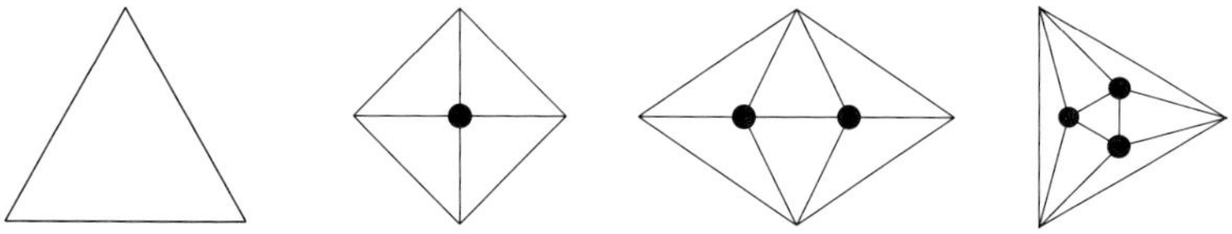

C_{3v} (D_{3h}) $\quad$ C_{4v} $\quad$ C_{2v} $\quad$ C_{3v}, *nonext.* (O_h)

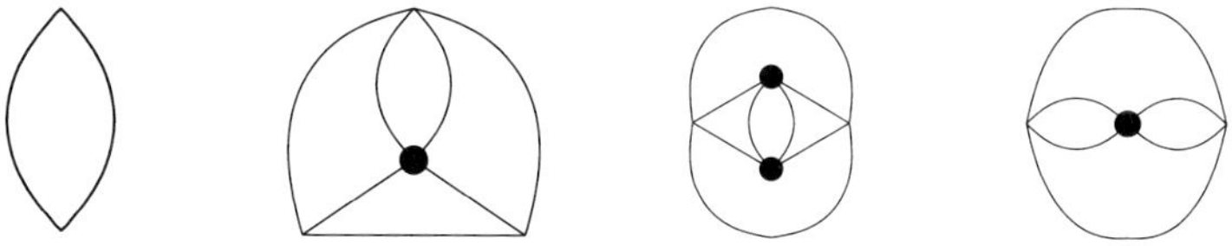

C_{2v} (D_{2h}) $\quad$ C_s $\quad$ C_{2v}, *nonext.* (D_{2d}) $\quad$ C_{2v}, *nonext.* (D_{3h})

Proof. The list of elementary $(\{3\},4)_{\text{simp}}$-polycycles is determined in [3] and consists of the first four graphs of this theorem. Let P be a $(\{2,3\},4)$-polycycle containing a 2-gon. If $|F_1| = 1$, then it is the 2-gon. Clearly, the case where two 2-gons share one edge is impossible. Assume that P contains two 2-gons that share a vertex. Then we should add a triangle on both sides and thus obtain the second polycycle given above. If there is a 2-gon that does not share a vertex with a 2-gon, then P contains the following pattern:

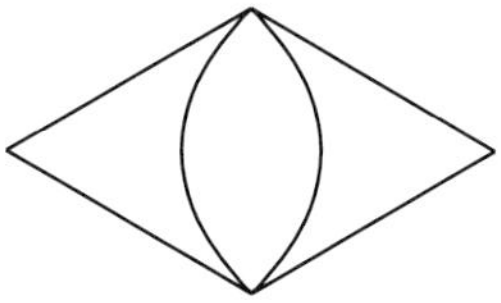

Thus, clearly, P is one of the last two possibilities above.

Note that the seventh and fourth polycycles in Theorem 14.4 are, respectively, 2- and 3-antiprisms; here the exterior face is the unique hole. The m-antiprism for any $m \geq 2$ can also be seen as a $(\{2,3\},4)$-polycycle with F_2 consisting of the exterior and interior m-gons; this polycycle is not elementary.

14.5 Classification of Elementary $(\{2,3\},5)$-Polycycles

For $2 \leq m \leq \infty$, call *snub m-antiprism* the $(\{3\},5)_{\text{simp}}$-polycycle with two m-gonal holes separated by $4m$ 3-gonal proper faces. This polycycle is nonextensible if and only if $m \geq 4$. Its symmetry group, D_{md}, coincides with the symmetry group of an underlying 5-valent plane graph if and only if $m \neq 3$. See below for pictures of a snub m-antiprism for $m = 2, 3, 4, 5$, and ∞ (see [4], p. 119 for a formal definition):

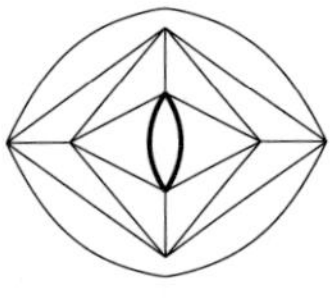

D_{2d}

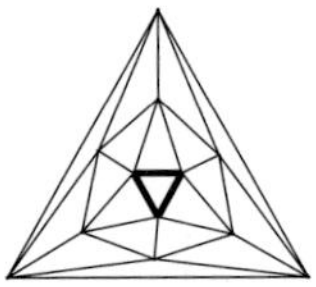

D_{3d} (I_h)

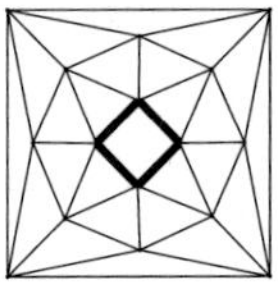

D_{4d}

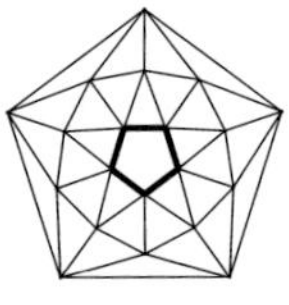

D_{5d}

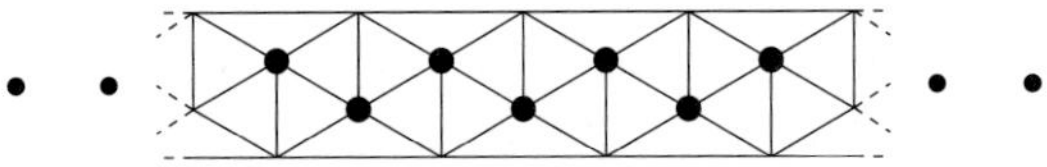

$D_{\infty d}$=pma2, nonext.

Theorem 14.5 *The list of elementary* $(\{2, 3\}, 5)$*-polycycles consists of:*

(i) 57 *sporadic* $(\{2, 3\}, 5)_{simp}$*-polycycles, given in Appendix 2.*
(ii) The following 3 *infinite* $(\{2, 3\}, 5)_{simp}$*-polycycles:*

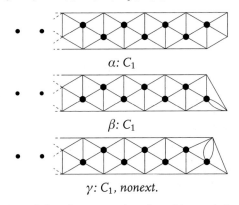

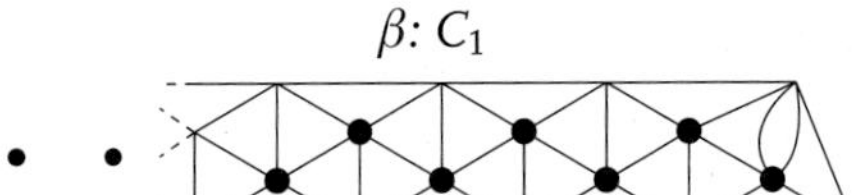

α: C_1

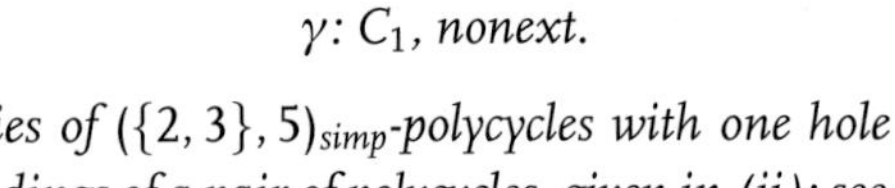

β: C_1

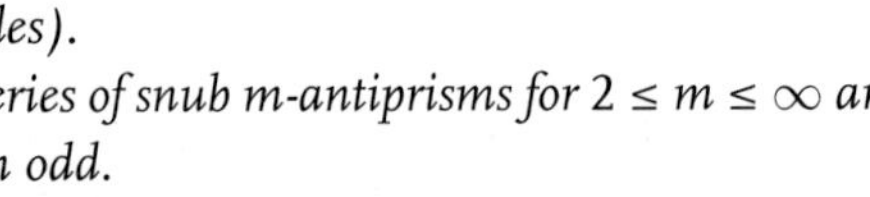

γ: C_1, *nonext.*

(iii) 6 *infinite series of* $(\{2, 3\}, 5)_{simp}$*-polycycles with one hole (obtained by concatenating endings of a pair of polycycles, given in (ii); see Figure 14.2 for the first* 4 *polycycles).*
(iv) The infinite series of snub m-antiprisms for $2 \leq m \leq \infty$ *and its nonorientable quotient for m odd.*

Proof. Take an elementary $(\{2, 3\}, 5)_{\text{simp}}$-polycycle P that, by Proposition 14.2, is kernel-elementary. If its kernel is empty, then P is simply an r-gon with

$r \in \{2, 3\}$. If the kernel is reduced to a vertex, then P is simply a 5-tuple of 2-, 3-gons. If three edges $e_i = \{v_{i-1}, v_i\}$, $i = 1, \ldots, 3$ are part of the kernel with e_1, e_2 and e_2, e_3 not part of an r-gon of the kernel, then we have the following local configurations:

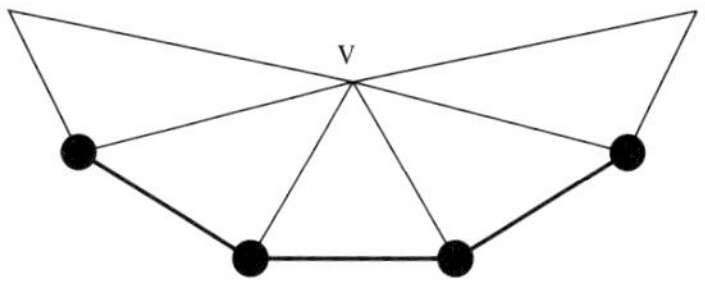

Configuration 1

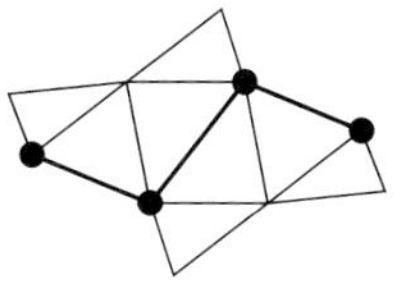

Configuration 2

In both of the above pictures, we used 3-gons; if 2-gons occur, then the degree of involved vertices only increases. Therefore, configuration 1 is not possible since it involves v of degree at least 6. A *geodesic* is a sequence (finite, or infinite in one or two directions) of edges in configuration 2. From this we conclude that if the kernel $Ker(P)$ does not contain any face, then it is a geodesic and it is a snub ∞-antiprism, α, β, γ, and the polycycles obtained in (iii).

If $Ker(P)$ contains some faces, then one can prove, using the nonexistence of configuration 1, that any two faces of $Ker(P)$ are related by a sequence of kernel faces. If $e_1 = \{v_0, v_1\}$, $e_2 = \{v_1, v_2\}$ are kernel edges with v_0 belonging to a kernel face F and $e_1 \notin F$, then e_1 belongs to another kernel face. Thus, $Ker(P)$ is formed by $G(P)$ and, possibly, some edges sharing a vertex with them. $G(P)$ is actually a kernelable $(\{2, 3\}, 5)_{\text{simp}}$-polycycle. The enumeration of kernelable $(\{2, 3\}, 5)_{\text{simp}}$-polycycles is done in a way similar to the $(\{2, 3, 4, 5\}, 3)$ case. One gets that $G(P)$ has at most 10 faces and then, after adding the edges, all 57 sporadic $(\{2, 3\}, 5)$-polycycles.

If P is an elementary $(\{2, 3\}, 5)$-polycycle that is not simply connected, then its universal cover $\widetilde{P}$ is an elementary $(\{2, 3\}, 5)_{\text{simp}}$-polycycle that has a nontrivial fixed-point-free automorphism group included in $\text{Aut}(\widetilde{P})$. The only $(\{2, 3\}, 5)$-polycycle with a non-trivial fixed-point-free automorphism is snub ∞-antiprism. It yields the infinite series of snub m-antiprisms and its non-orientable quotients.

14.6 Conclusion

The classification, developed here, is an extension of previous work of Deza and Shtogrin. It allows for various possible numbers of sides of interior faces and several possible holes. A natural question is if one can further enlarge the class of polycycles.

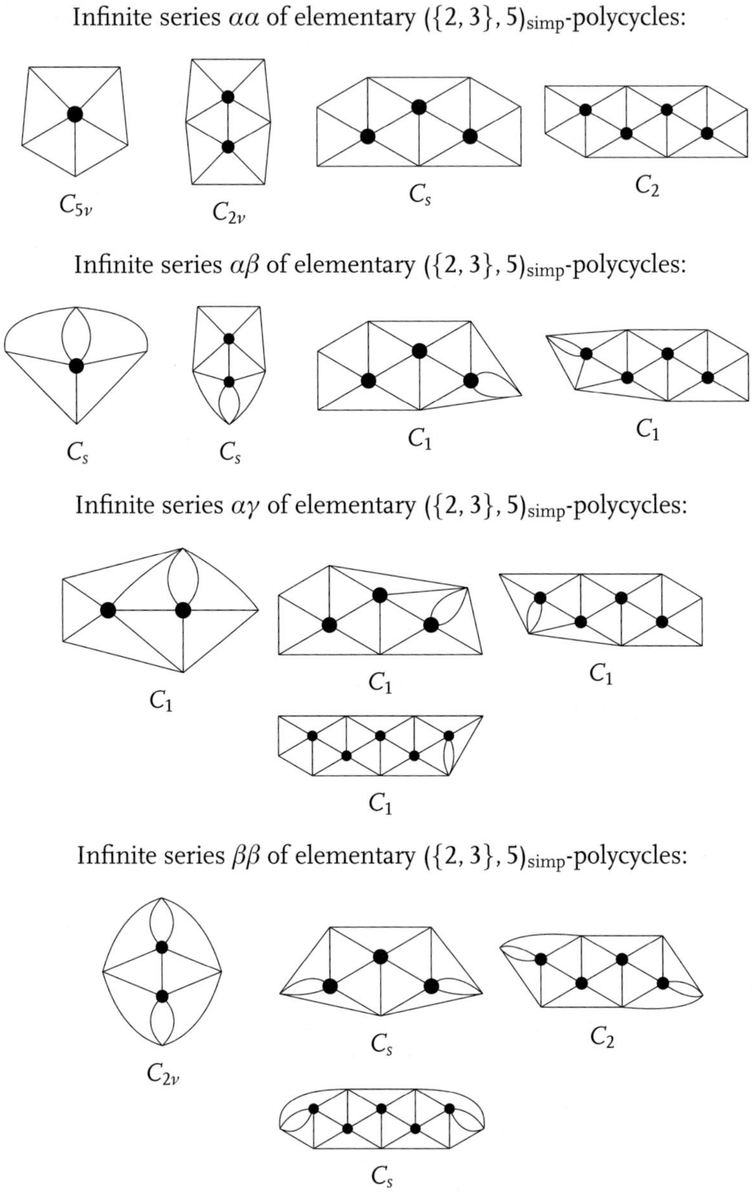

Figure 14.2 The first 4 members of the six infinite series of $(\{2, 3\}, 5)_{\text{simp}}$-polycycles from Theorem 14.5(iii).

Infinite series $\beta\gamma$ of elementary $(\{2,3\},5)_{\text{simp}}$-polycycles:

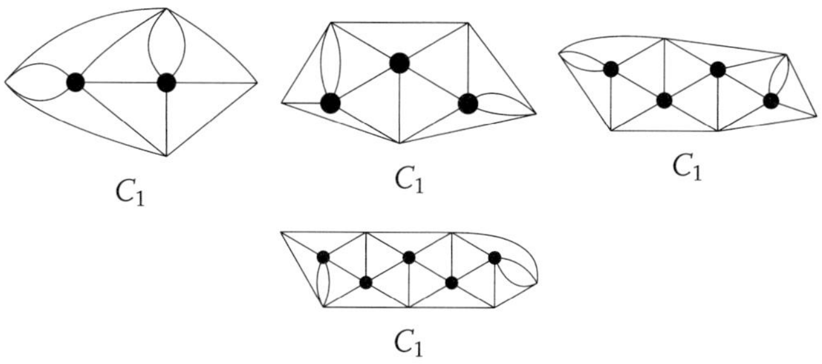

Infinite series $\gamma\gamma$ of elementary $(\{2,3\},5)_{\text{simp}}$-polycycles:

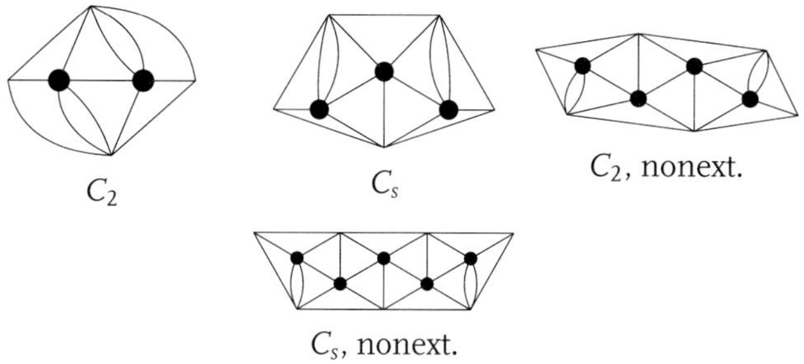

Figure 14.2 (continued)

There will be only some technical difficulties if one tries to obtain the catalog of elementary (R, Q)-*polycycles,* that is, the generalization of the (R, q)-polycycle allowing the set Q for values of a degree of interior vertices. Such a polycycle is called an *elliptic,* a *parabolic,* or a *hyperbolic* if $\frac{1}{q}+\frac{1}{r}-\frac{1}{2}$ (where $r = max_{i\in R}i$, $q = max_{i\in Q}i$) is positive, zero, or negative, respectively. The decomposition and other main notions could be applied directly.

We required 2-connectivity and that any two holes not share a vertex. If one removes those two conditions, then too many other graphs appear.

The omitted cases $(R, q) = (\{2\}, q)$ are not interesting. In fact, consider the infinite series of $(\{2\}, 6)$-polycycles, *m-bracelets,* $m \geq 2$ (i.e., m-circle with each edge being tripled). The central edge is a bridge for those polycycles, for both 2-gons of the edge triple. But if one removes those two digons, then the resulting plane graph has two holes sharing a face, that is, it violates the crucial point (ii) of the definition of the (R, q)-polycycle. For even m, each even edge (for some order $1, \ldots, m$ of them) can be duplicated t times (for fixed t, $1 \leq t \leq 5$) and each odd edge duplicated $6 - t$ times; thus the degrees of all vertices will still be 6. On the other hand, two holes (m-gons inside and outside of the m-bracelet) have common vertices; thus, it is again not our polycycle.

Acknowledgment

This work was supported by the Croatian Ministry of Science, Education, and Sports under contract 098-0982705-2707.

Appendix 1: 204 Sporadic Elementary $(\{2, 3, 4, 5\}, 3)$-Polycycles

Always below (11 cases), when several elementary sporadic $(\{2, 3, 4, 5\}, 3)_{\text{simp}}$-polycycles correspond to the same plane graph, we add the sign **x** with $1 \leq x \leq 11$.

List of 4 sporadic elementary $(\{2, 3, 4, 5\}, 3)_{\text{simp}}$-polycycles with 1 proper face:

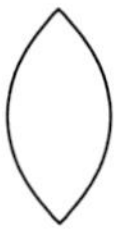

C_{2v} (D_{2h})

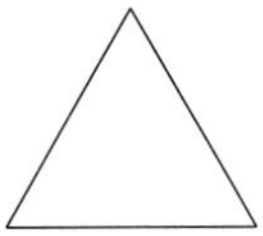

C_{3v} (D_{3h})

C_{4v} (D_{4h})

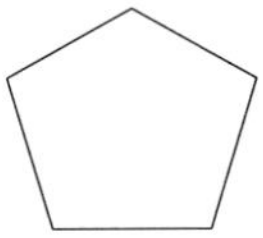

C_{5v} (D_{5h})

List of 13 sporadic elementary $(\{2, 3, 4, 5\}, 3)_{\text{simp}}$-polycycles with 3 proper faces:

C_{2v}, nonext. (D_{2h})

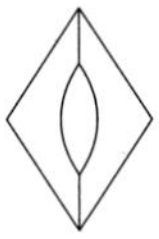

C_{2v}

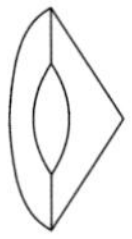

C_s, nonext.

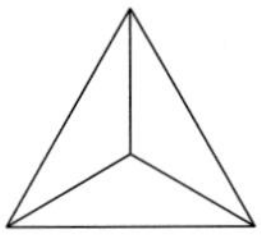

C_{3v}, nonext. (T_d)

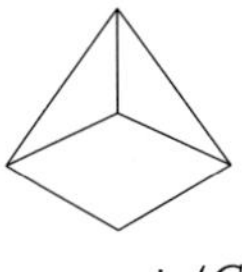

C_s, nonext. (C_{2v})

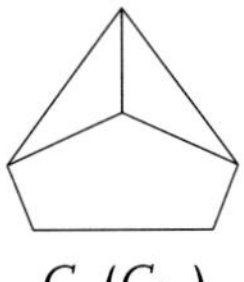

C_s (C_{2v})

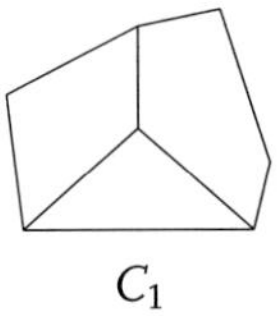

C_1

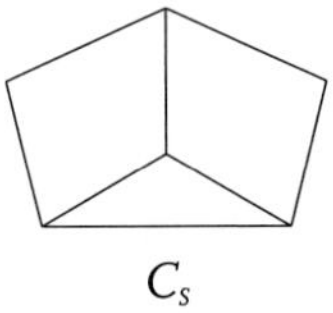

C_s

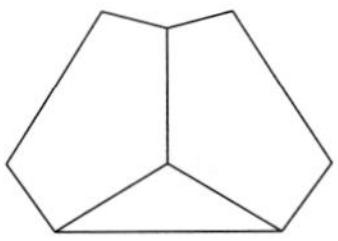

C_s

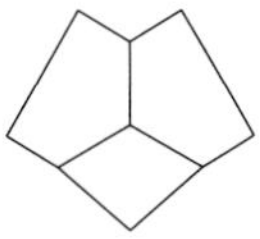

C_s

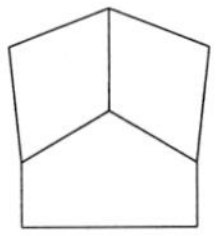

C_s

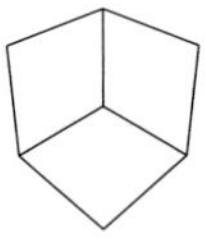

C_{3v}

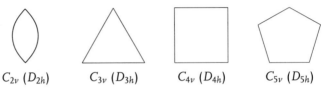

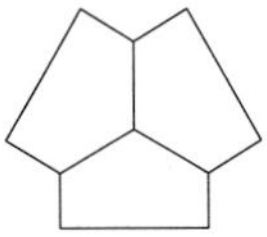

C_{3v}

List of 26 sporadic elementary $(\{2,3,4,5\},3)_{\text{simp}}$-polycycles with 4 proper faces:

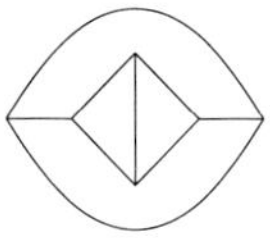
C_{2v}, nonext. **1**

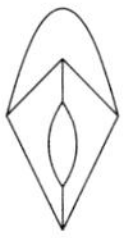
C_s, nonext. (C_{2v}) **1**

C_s, nonext. (C_{2v})

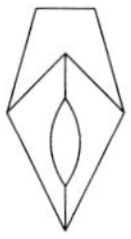
C_s (C_{2v})

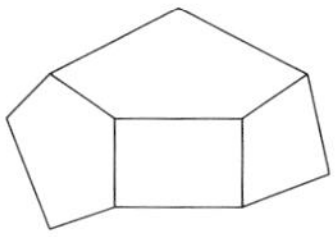
C_1

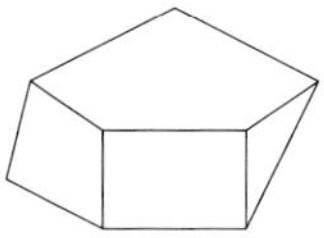
C_1

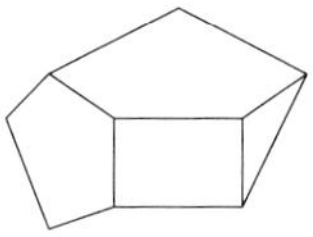
C_1

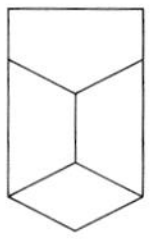
C_s

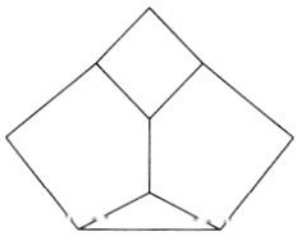
C_s

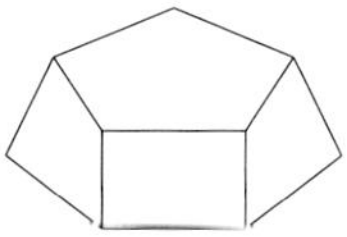
C_s

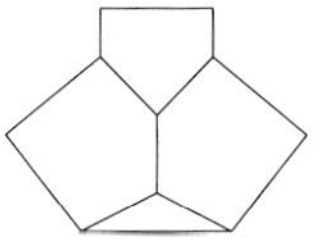
C_s

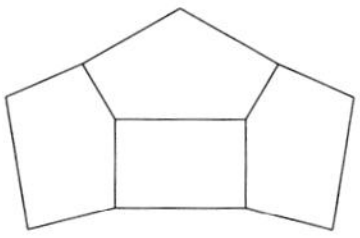
C_s

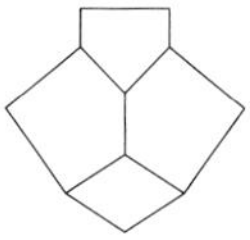
C_s

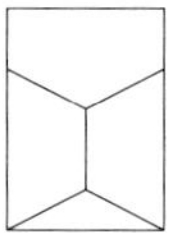
C_s

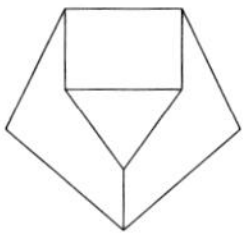
C_s

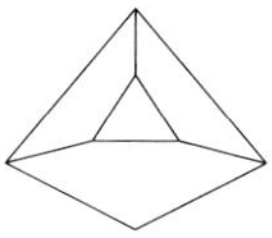
C_s, nonext. **2**

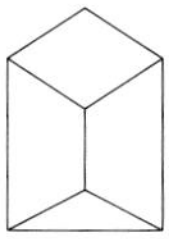
C_s, nonext. **2**

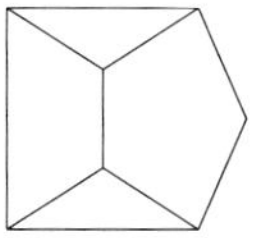
C_s, nonext. (C_{2v})

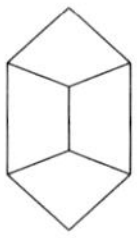
C_{2v}

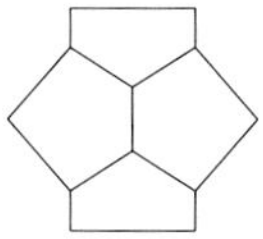
C_{2v}

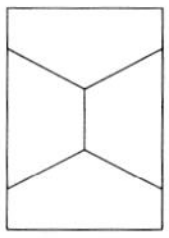
C_{2v}

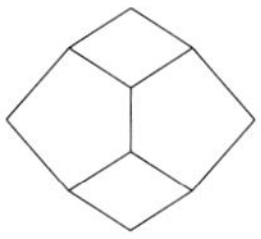
C_{2v}

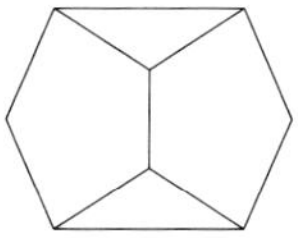
C_{2v}

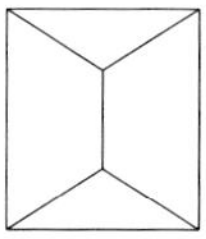
C_{2v}, nonext. **3**

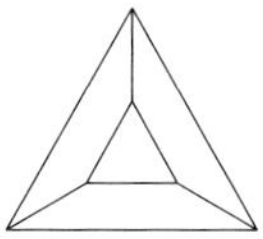
C_{3v}, nonext. **3**

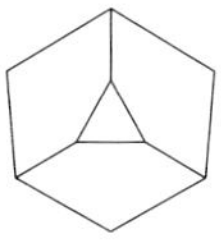
C_{3v}

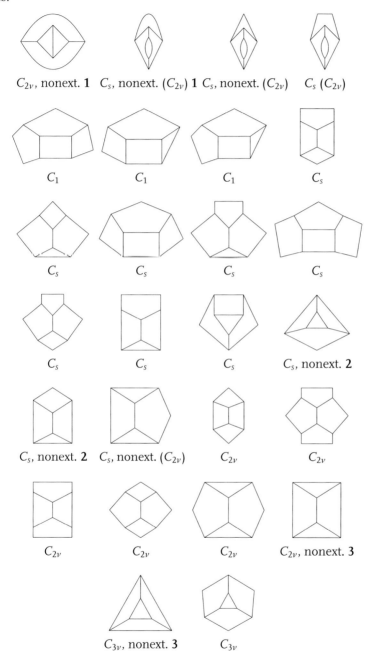

List of 36 sporadic elementary $(\{2,3,4,5\},3)_{\text{simp}}$-polycycles with 5 proper faces:

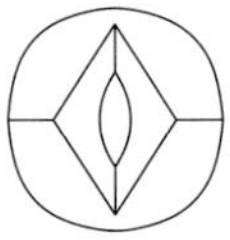

C_{2v}, nonext. (D_{2d})

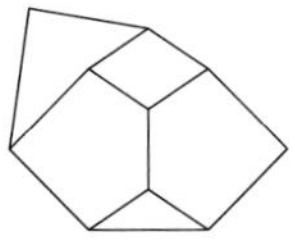

C_1

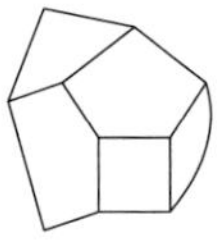

C_1

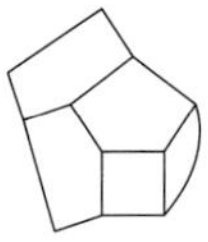

C_1

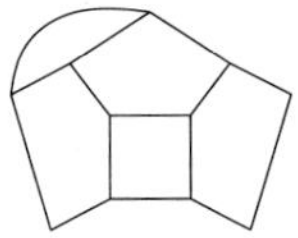

C_1

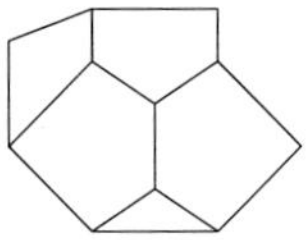

C_1

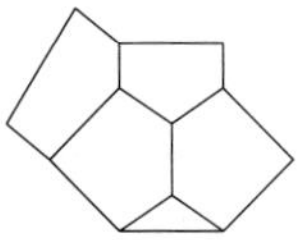

C_1

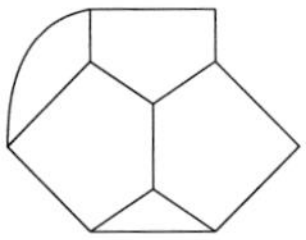

C_1

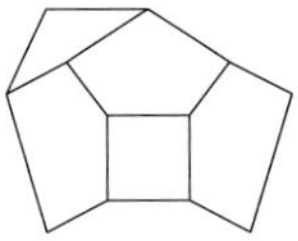

C_1

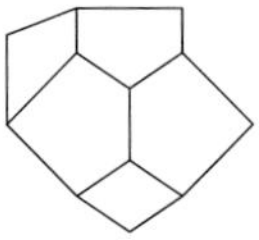

C_1

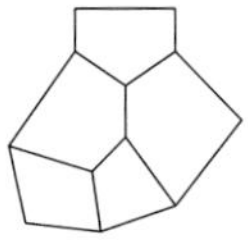

C_1

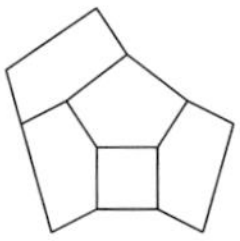

C_1

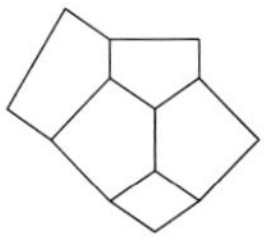

C_1

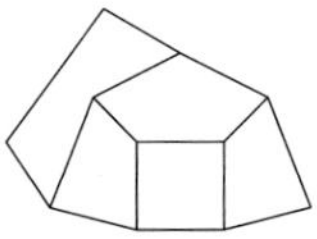

C_1

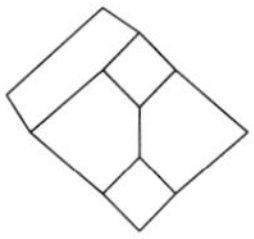

C_1

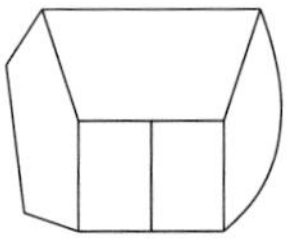

C_1

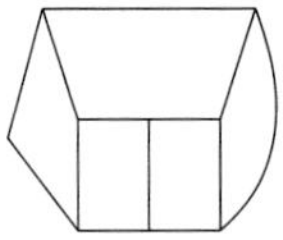

C_1, nonext.

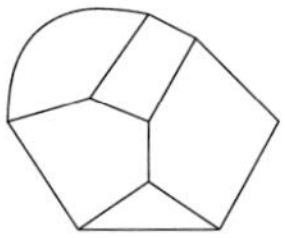

C_1, nonext.

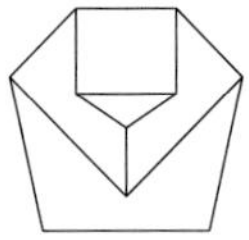

C_s

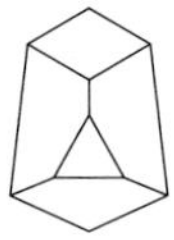

C_s

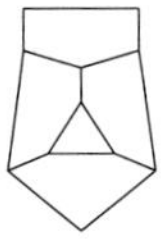

C_s

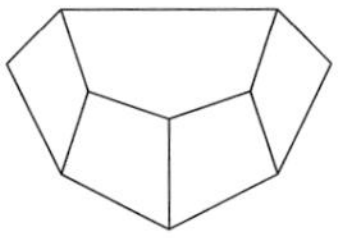

C_s

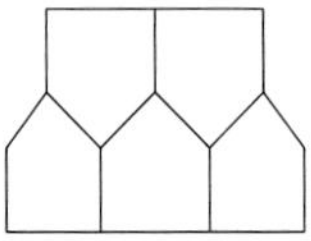

C_s

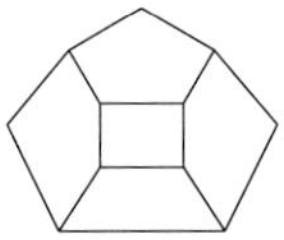

C_s

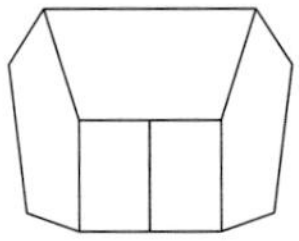

C_s

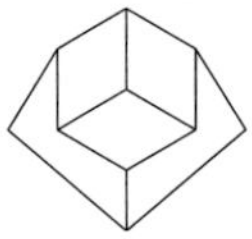

C_s

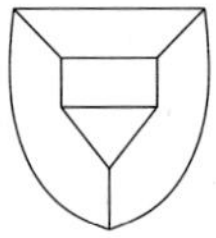

C_s, nonext. **4**

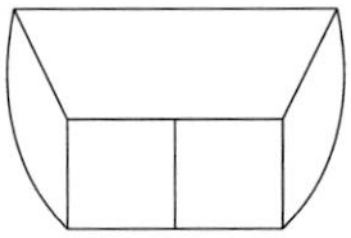

C_s, nonext. **4**

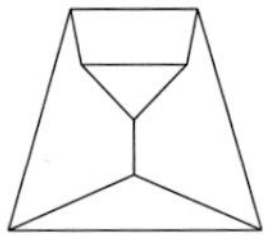

C_s, nonext. **4**

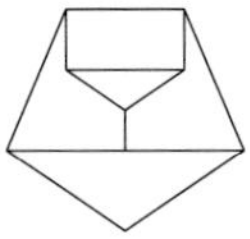

C_s, nonext. **5**

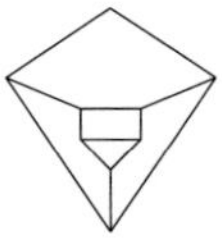

C_s, nonext. **5**

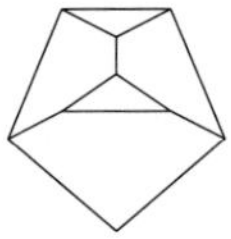

C_s, nonext. (C_{2v})

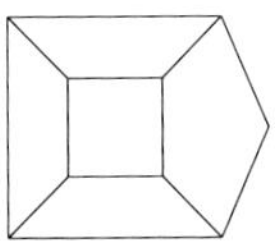
C_s, nonext. (C_{2v})

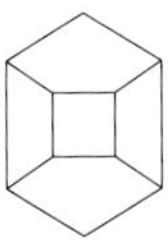
C_{2v}

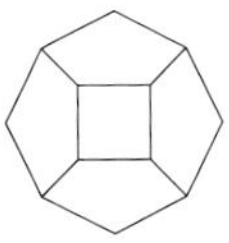
C_{4v}

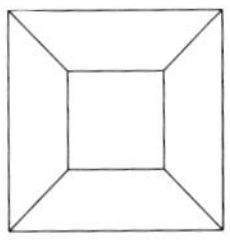
C_{4v}, nonext. (O_h)

List of 34 sporadic elementary $(\{2,3,4,5\},3)_{simp}$-polycycles with 6 proper faces:

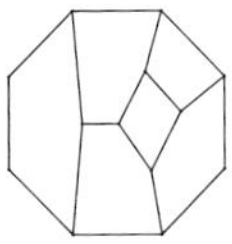
C_1

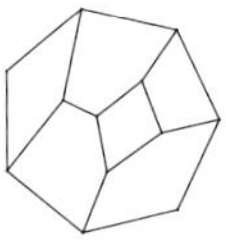
C_1

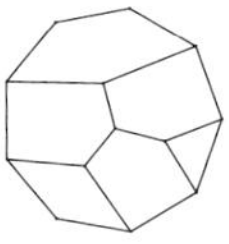
C_1

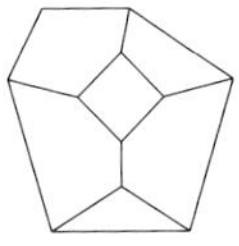
C_1, nonext.

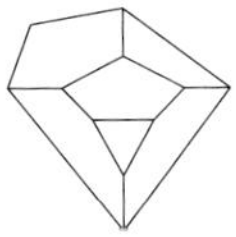
C_1, nonext.

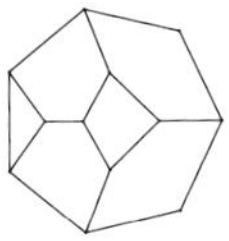
C_s

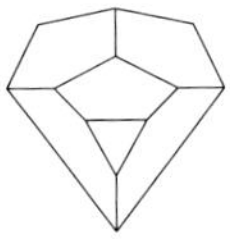
C_s

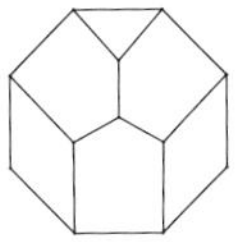
C_s

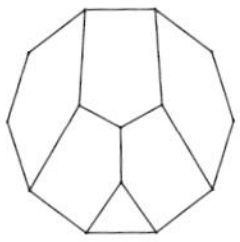
C_s

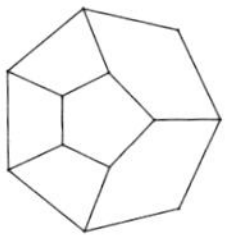
C_s

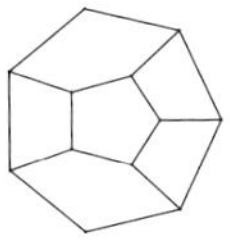
C_s

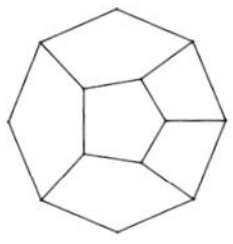
C_s

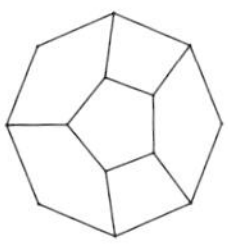
C_s

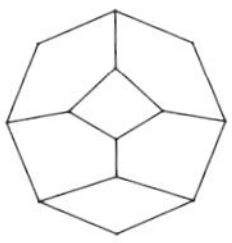
C_s

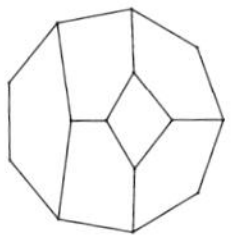
C_s

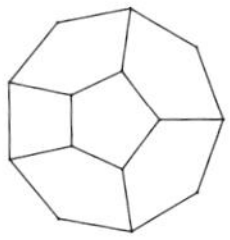
C_s

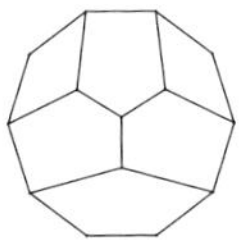
C_s

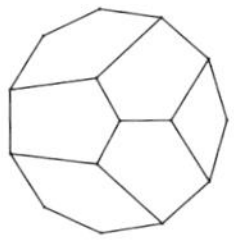
C_s

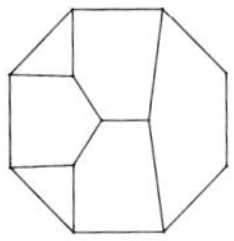
C_s

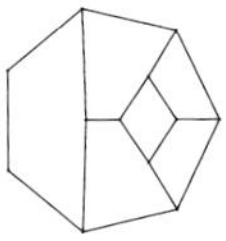
C_s

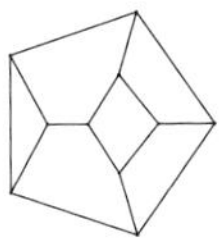
C_s, nonext. **6**

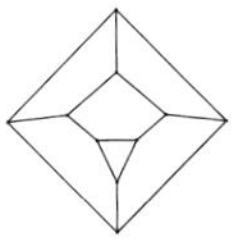
C_s, nonext. **6**

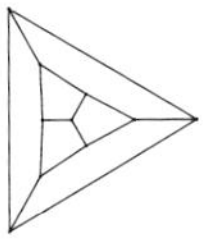
C_{3v}, nonext. **6**

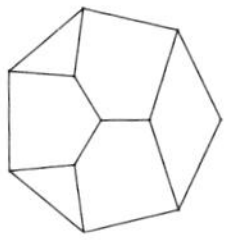
C_s, nonext.

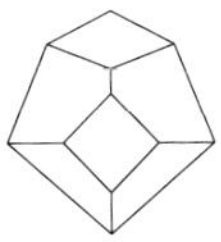
C_s, nonext.

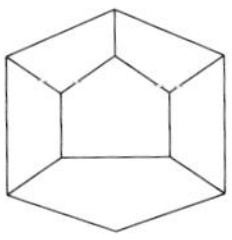
C_s, nonext.

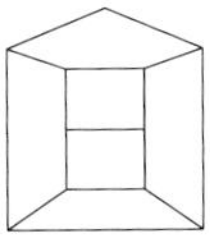
C_s, nonext. (C_{2v})

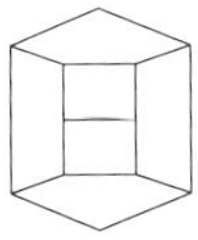
C_{2v}

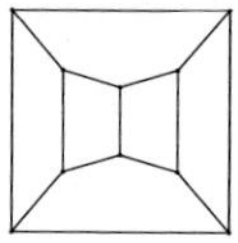
$C_{2\nu}$, nonext. **7**

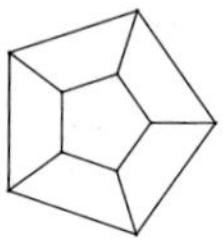
$C_{5\nu}$, nonext. **7**

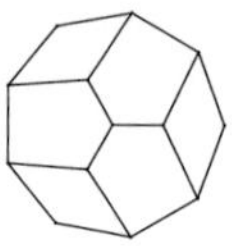
$C_{3\nu}$

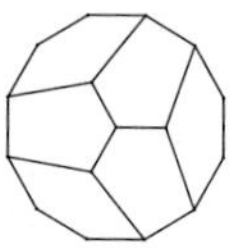
$C_{3\nu}$

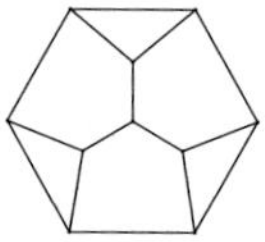
$C_{3\nu}$, nonext.

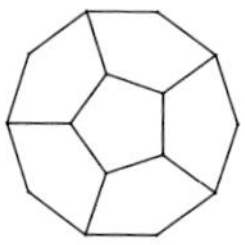
$C_{5\nu}$

List of 36 sporadic elementary $(\{2, 3, 4, 5\}, 3)_{\text{simp}}$-polycycles with 7 proper faces:

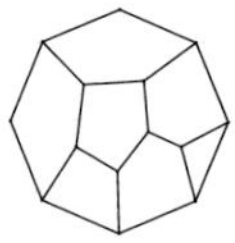
C_1

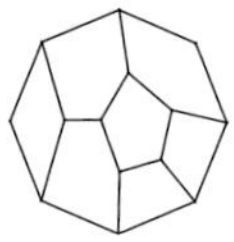
C_1

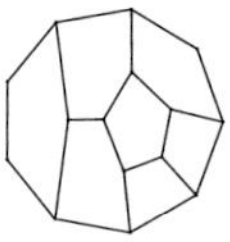
C_1

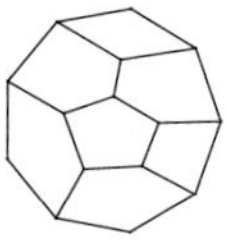
C_1

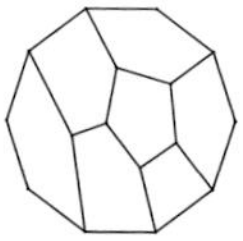
C_1

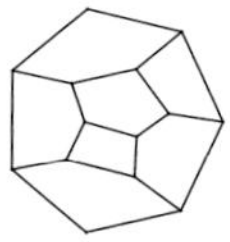
C_1

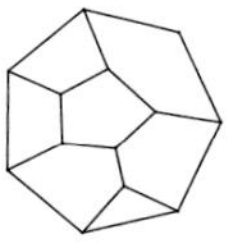
C_1, nonext.

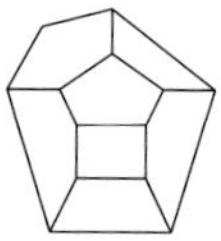
C_1, nonext.

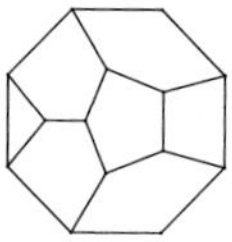
C_s

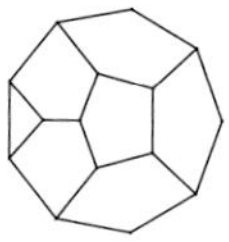
C_s

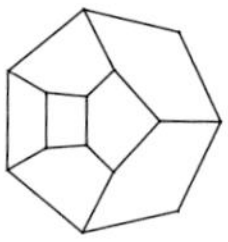
C_s

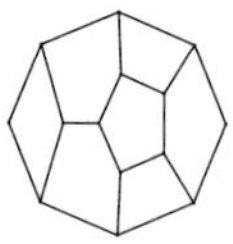
C_s

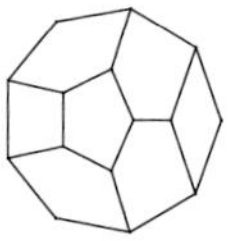
C_s

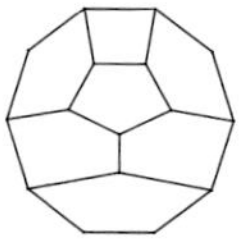
C_s

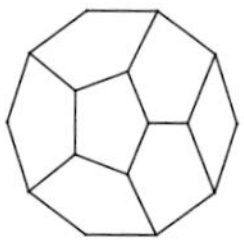
C_s

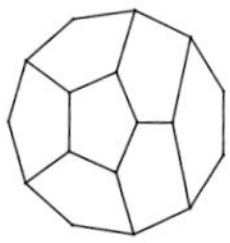
C_s

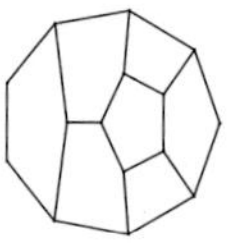
C_s

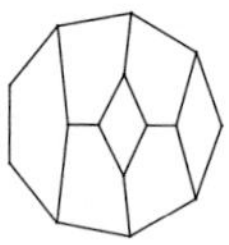
C_s

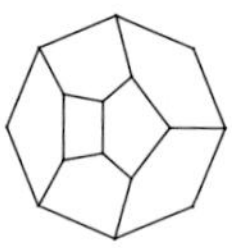
C_s

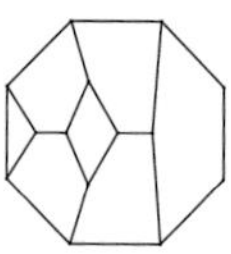
C_s

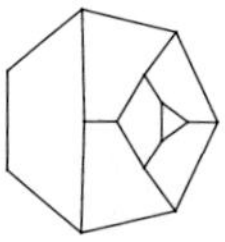
C_s

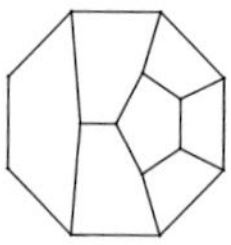
C_s

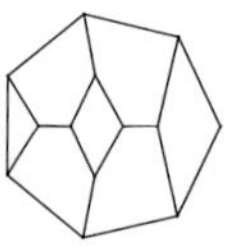
C_s, nonext.

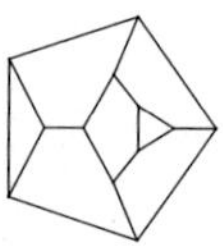
C_s, nonext. **8**

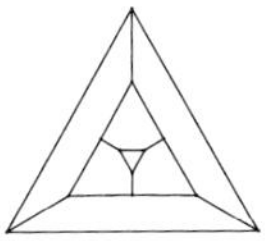

C_{3v}, nonext. **8**

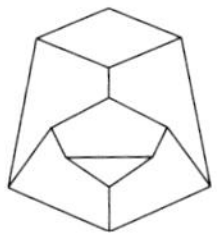

C_s, nonext.

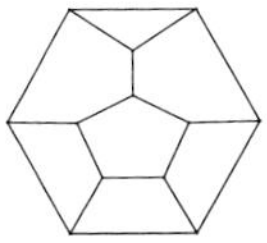

C_s, nonext.

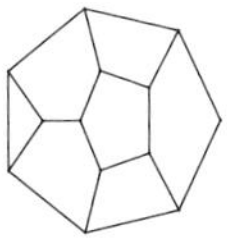

C_s, nonext.

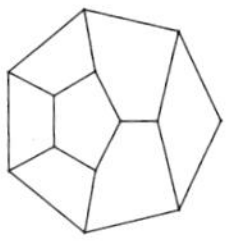

C_s, nonext.

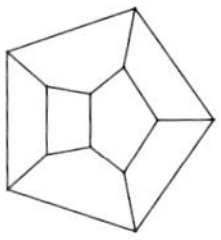

C_s, nonext. **9**

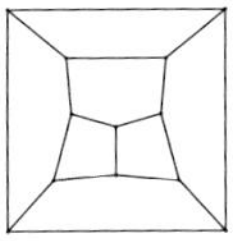

C_s, nonext. **9**

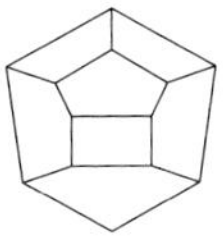

C_s, nonext.

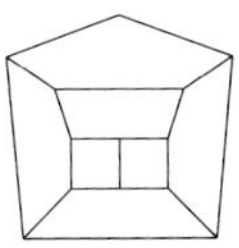

C_s, nonext. (C_{2v})

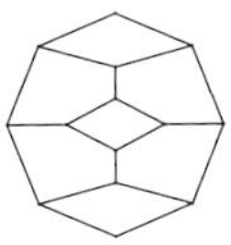

C_{2v}

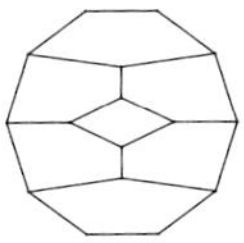

C_{2v}

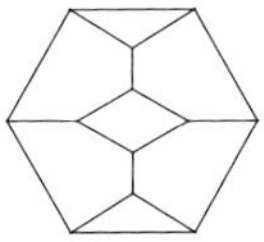

C_{2v}, nonext.

List of 29 sporadic elementary $(\{2, 3, 4, 5\}, 3)_{\text{simp}}$-polycycles with 8 proper faces:

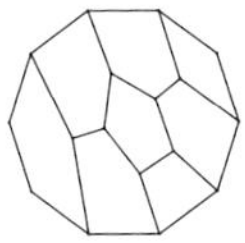

C_1

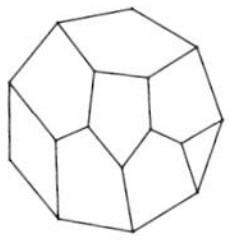

C_1

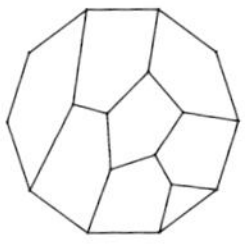

C_1

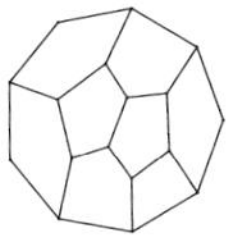

C_1

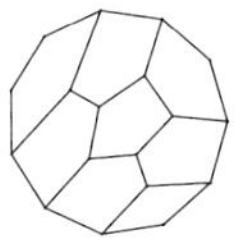

C_1

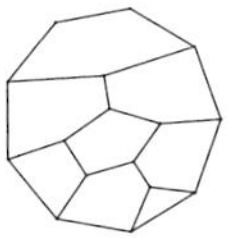

C_1

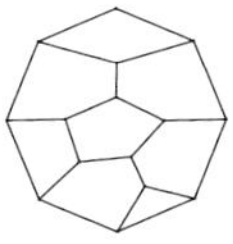

C_1, nonext.

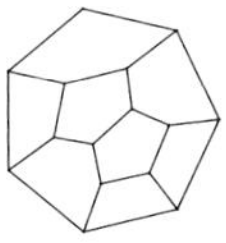

C_1, nonext.

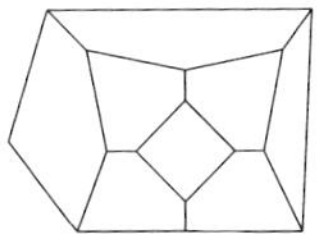

C_1, nonext.

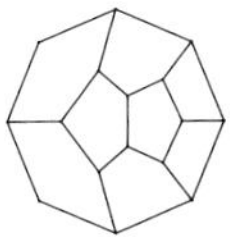

C_s

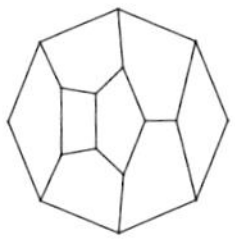

C_s

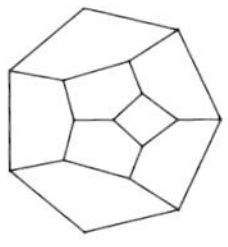

C_s

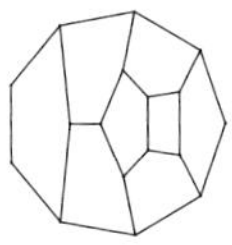

C_s

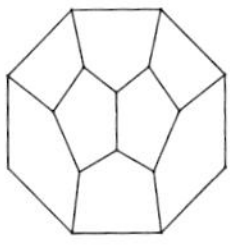

C_s

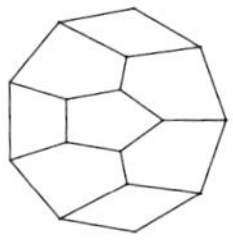

C_s

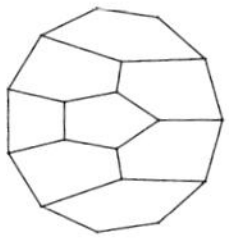

C_s

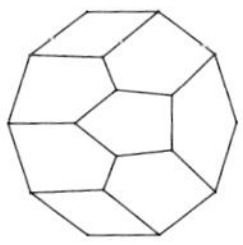

C_s

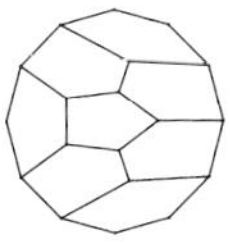

C_s

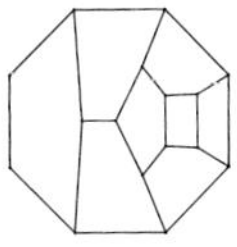

C_s

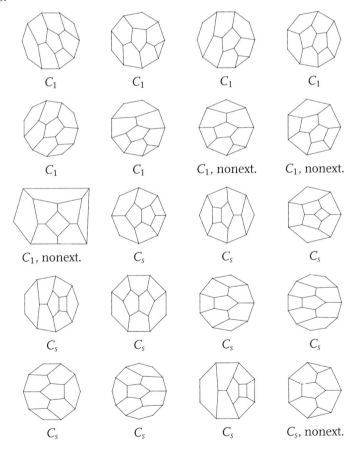

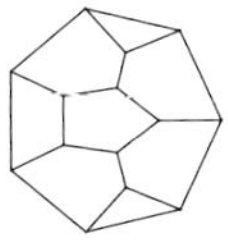

C_s, nonext.

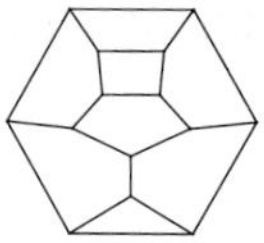
C_s, nonext.

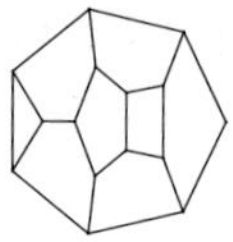
C_s, nonext.

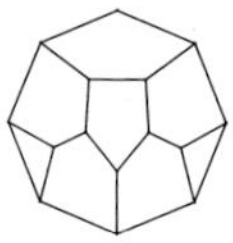
C_s, nonext.

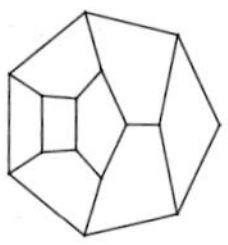
C_s, nonext.

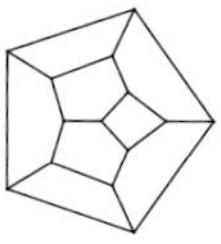
C_s, nonext. (D_{3h}) **10**

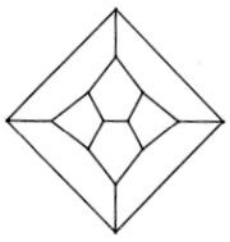
C_{2v}, nonext. (D_{3h}) **10**

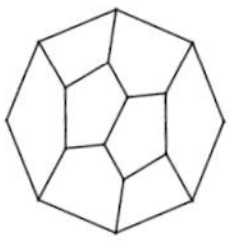
C_2

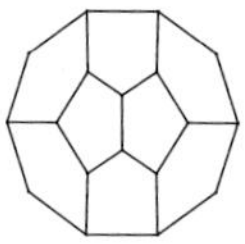
C_{2v}

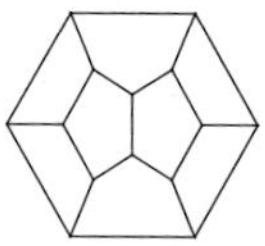
C_{2v}, nonext.

List of 16 sporadic elementary $(\{2, 3, 4, 5\}, 3)_{\text{simp}}$-polycycles with 9 proper faces:

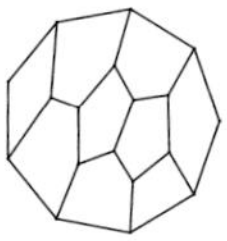
C_1

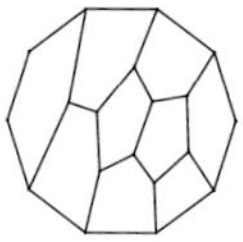
C_1

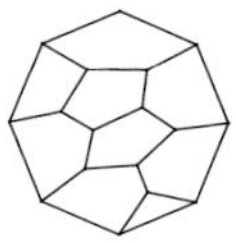
C_1, nonext.

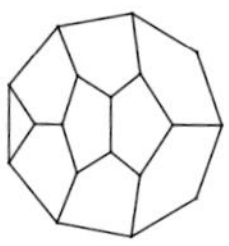
C_s

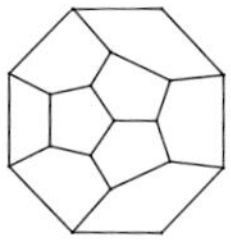
C_s

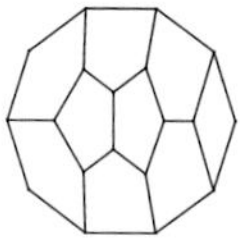
C_s

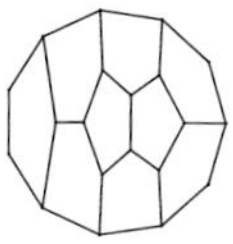
C_s

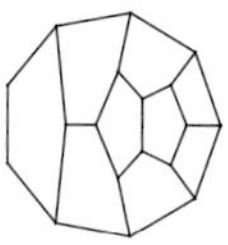
C_s

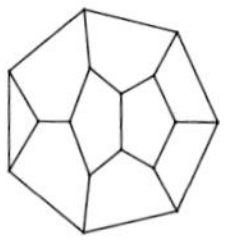
C_s, nonext.

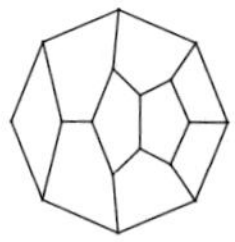
C_s, nonext.

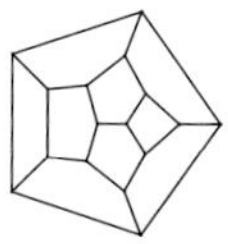
C_s, nonext. **11**

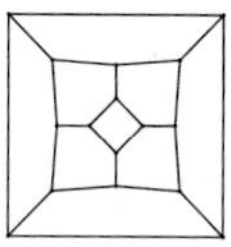
C_{4v}, nonext. **11**

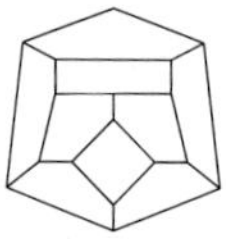
C_s, nonext.

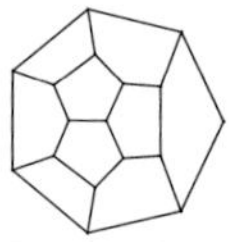
C_s, nonext.

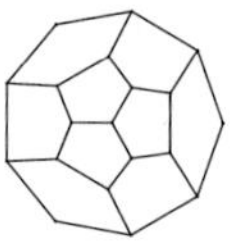
C_{3v}

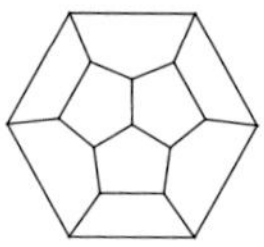
C_{3v}, nonext.

List of 9 sporadic elementary $(\{2, 3, 4, 5\}, 3)_{\text{simp}}$-polycycles with 10 proper faces:

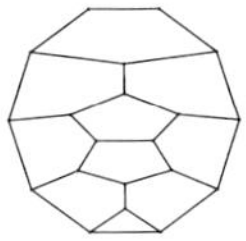

C_s

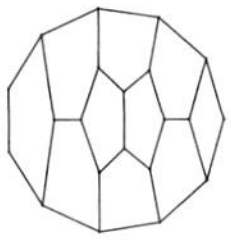

C_s

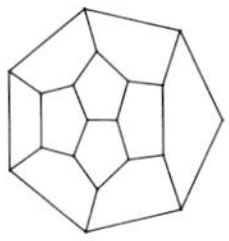

C_s, nonext.

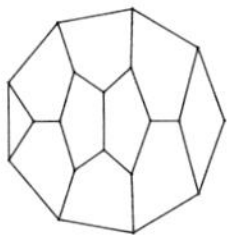

C_s, nonext.

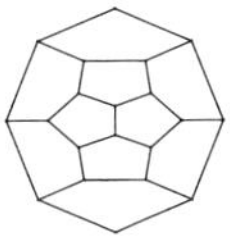

C_{2v}

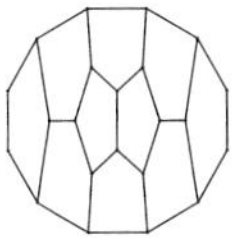

C_{2v}

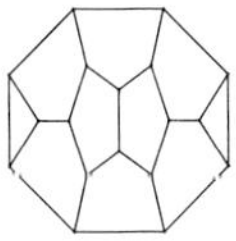

C_{2v}, nonext.

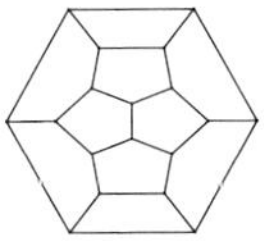

C_{2v}, nonext.

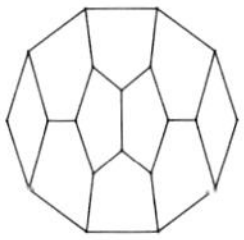

C_{2v}, nonext.

Unique sporadic elementary $(\{2, 3, 4, 5\}, 3)_{\text{simp}}$-polycycle with at least 11 proper faces:

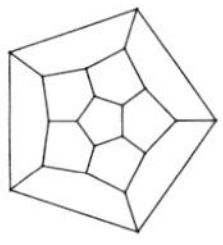

C_{5v}, nonext. (I_h)

Appendix 2: 57 Sporadic eLementary ({2, 3}, 5)-polycycles

Always below (three cases) when several elementary sporadic $(\{2, 3\}, 5)_{\text{simp}}$-polycycles correspond to the same plane graph, we add the sign **A**, **B**, or **C**.

List of 2 sporadic elementary $(\{2, 3\}, 5)_{\text{simp}}$-polycycles without interior vertices:

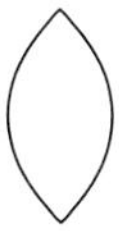

C_{2v} (D_{2h})

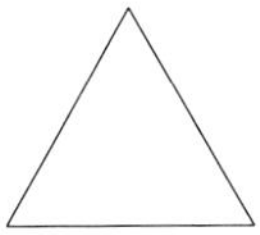

C_{3v} (D_{3h})

List of 3 sporadic elementary $(\{2,3\},5)_{simp}$-polycycles with 1 interior vertex:

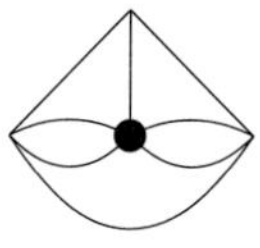

C_s

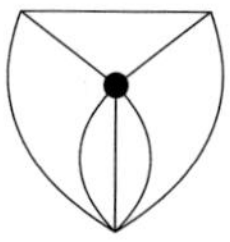

C_s (C_{2v})

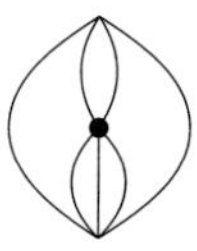

C_s, nonext. (C_{2v})

List of 6 sporadic elementary $(\{2,3\},5)_{simp}$-polycycles with 2 interior vertices:

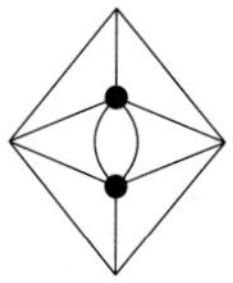

C_{2v}

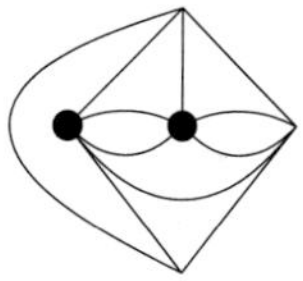

C_1 (C_s)

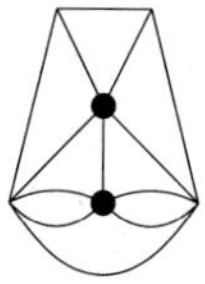

C_s

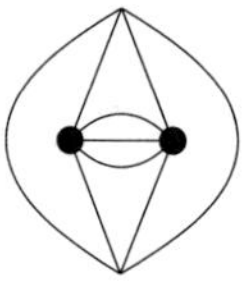

C_{2v}, nonext.

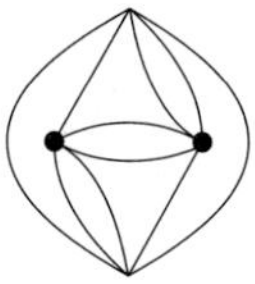

C_2, nonext. (D_{2d})

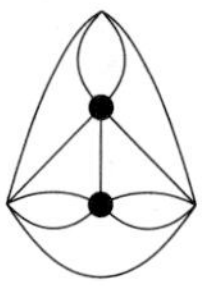

C_s, nonext. **A**

List of 10 sporadic elementary $(\{2,3\},5)_{simp}$-polycycles with 3 interior vertices:

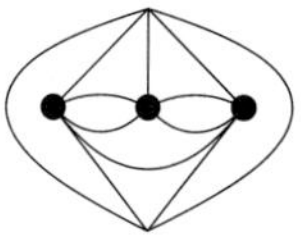

C_s, nonext. **A**

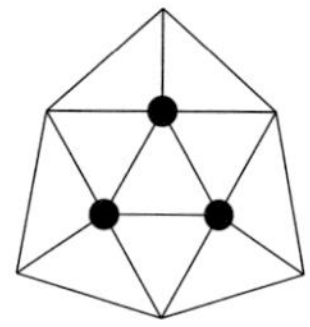

C_{3v}

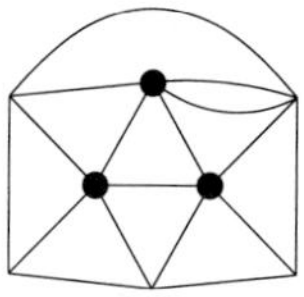

C_1

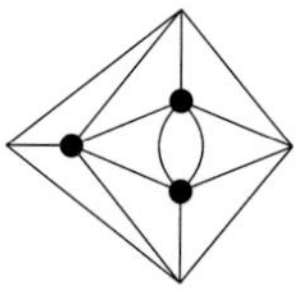

C_s

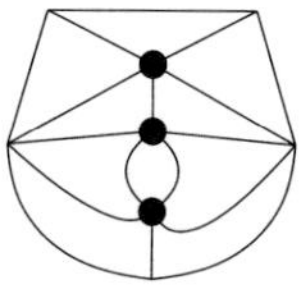

C_s

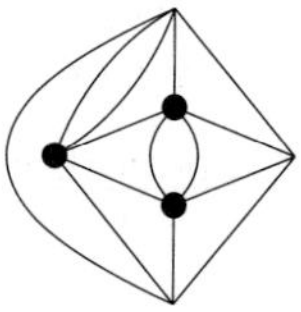

C_1 (C_2)

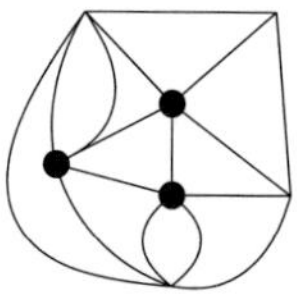

C_1

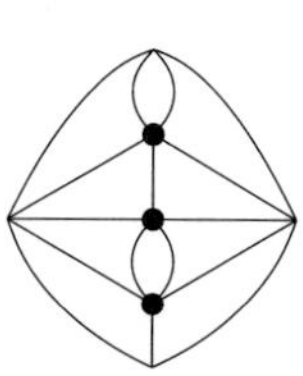

C_s

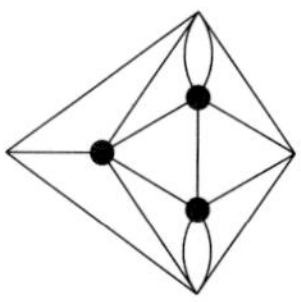

C_s

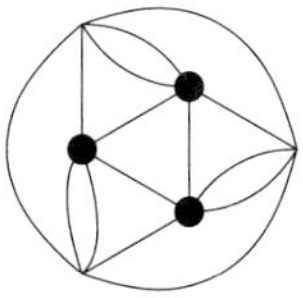

C_3, nonext. (D_3) **B**

List of 14 sporadic elementary $(\{2,3\},5)_{\text{simp}}$-polycycles with 4 interior vertices:

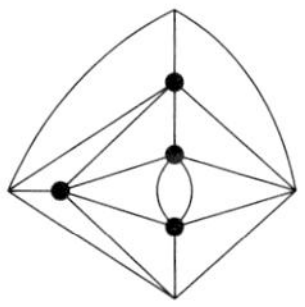

C_1

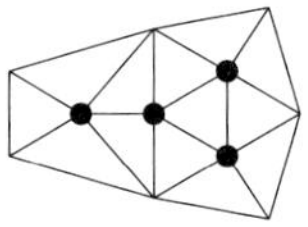

C_s

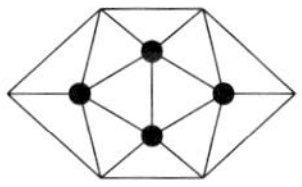

$C_{2\nu}$

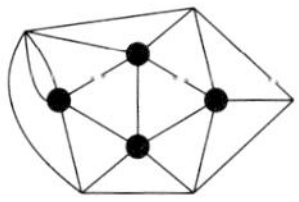

C_1

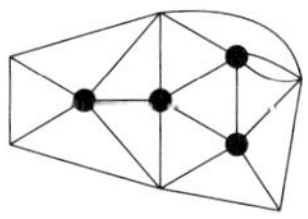

C_1

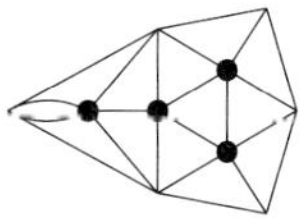

C_s

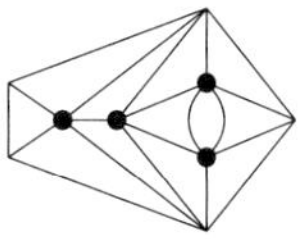

C_s

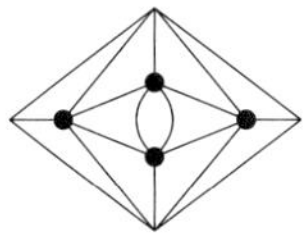

$C_{2\nu}$

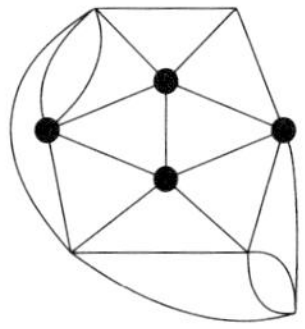

C_1

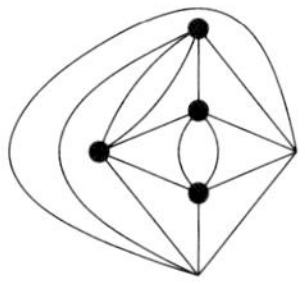

C_1, nonext. **B**

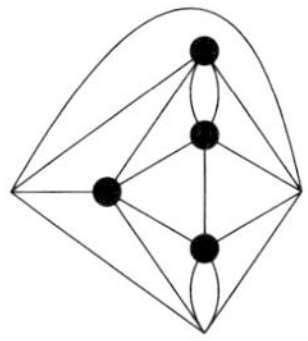

C_1, nonext. ($C_{2\nu}$)

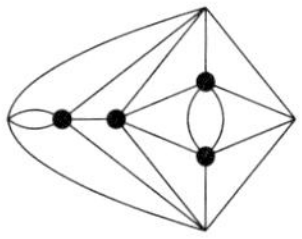

C_s

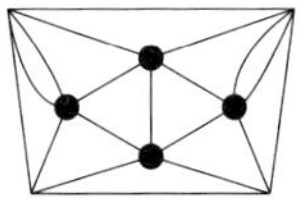

C_s

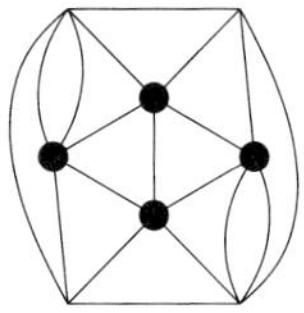

C_2

List of 10 sporadic elementary $(\{2,3\},5)_{\text{simp}}$-polycycles with 5 interior vertices:

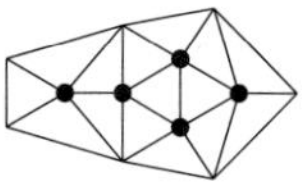

C_s

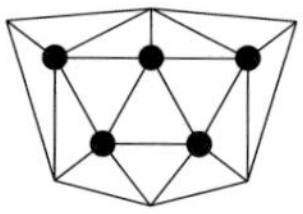

C_s

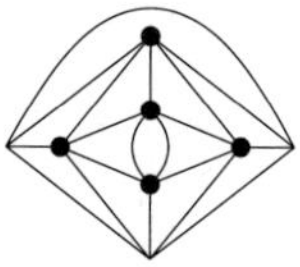

C_1 (C_{2v})

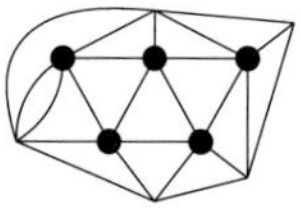

C_1

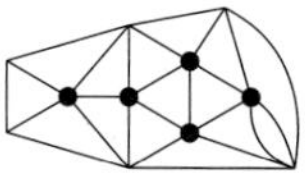

C_1

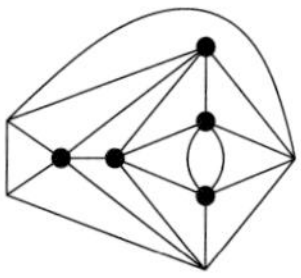

C_1

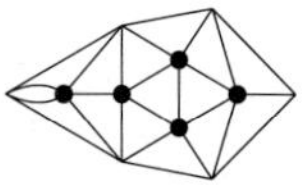

C_s

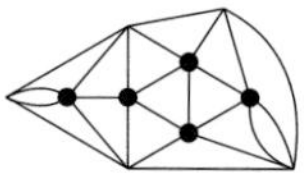

C_1

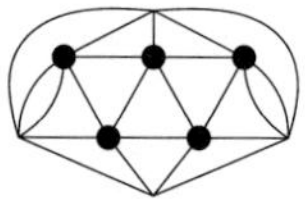

C_s, nonext.

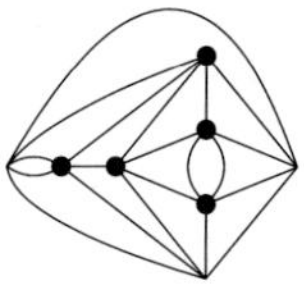

C_1, nonext. **C**

List of 9 sporadic elementary $(\{2,3\},5)_{\text{simp}}$-polycycles with 6 interior vertices:

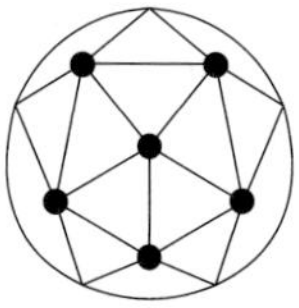

C_{5v}

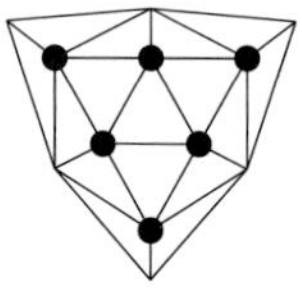

C_{3v}

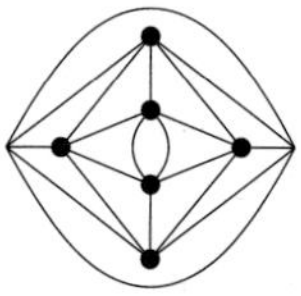

C_{2v}, nonext. **C**

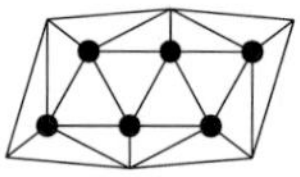

C_2

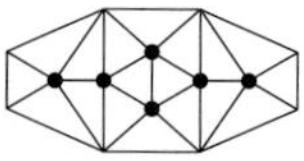

C_{2v}

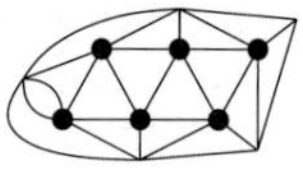

C_1

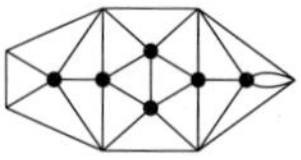

C_s

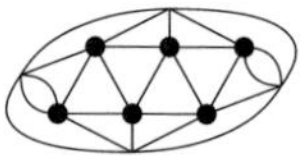

C_2, nonext.

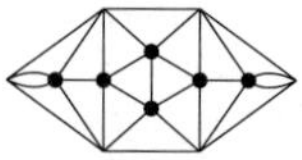

C_{2v}, nonext.

Sporadic elementary $(\{2,3\},5)_{simp}$-polycycles with 7, 8, or 9 interior vertices:

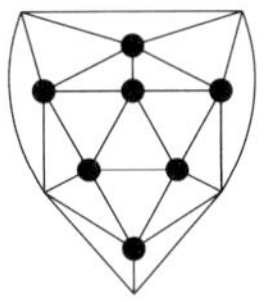

C_s

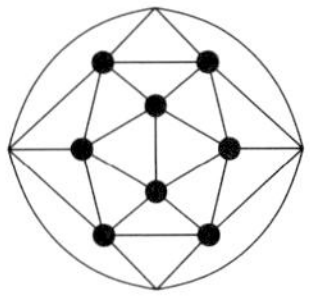

$C_{2\nu}$

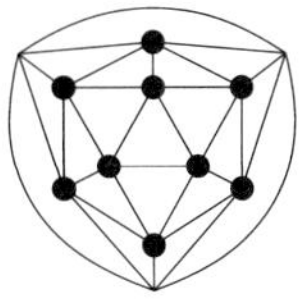

$C_{3\nu}$, nonext. (I_h)

References

1 G. Brinkmann, O. Delgado Friedrichs, A. Dress, T. Harmuth, *CaGe – a virtual environment for studying some special classes of large molecules*, MATCH: Communications in Mathematical and in Computer Chemistry, **36** (1997), 233–237.

2 B. Clair, *Frieze groups*, http://euler.slu.edu/~clair/escher/friezehandout.pdf.

3 M. Deza, M. Dutour, M.I. Shtogrin, *Elliptic polycycles with holes*, Uspekhi Mat. Nauk. **60–2** (2005), 157–158 (in Russian). English translation in Russian Math. Surveys **60–2**.

4 M. Deza, V.P. Grishukhin, M.I. Shtogrin, *Scale-Isometric Polytopal Graphs in Hypercubes and Cubic Lattices*, World Scientific and Imperial College Press, 2004.

5 M. Deza, M. Dutour Sikirić, *Geometry of Chemical Graphs: Polycycles and Two-faced Maps*, Cambridge University Press, 2008.

6 M. Deza, S. Shpectorov, M.I. Shtogrin, *Non-extendible finite polycycles*, Izv. Ross. Akad. Nauk Ser Mat. **70–1** (2006), 3–22 (in Russian). English translation in Izvestia of Russian Academy of Sciences, Sec. Math. **70–1** (2006), 1–18.

7 M. Deza, M.I. Shtogrin, *Polycycles*, Voronoi Conference on Analytic Number Theory and Space Tilings (Kyiv, September 7–14, 1998), Abstracts, 19–23.

8 M. Deza, M.I. Shtogrin, *Primitive polycycles and helicenes*, Uspekhi Mat. Nauk. **54–6** (1999), 159–160 (in Russian). English translation in Russian Math. Surveys **54–6**, 1238–1239.

9 M. Deza, M.I. Shtogrin, *Infinite primitive polycycles*, Uspekhi Mat. Nauk. **55–1** (2000), 179–180 (in Russian). English translation in Russian Math. Surveys **55–1**, 169–170.

10 M. Deza, M.I. Shtogrin, *Embedding of chemical graphs into hypercubes*, Math. Zametki **68–3** (2000), 339–352 (in Russian). English translation in Mathematical Notes **68–3**, 295–305.

11 M. Deza, M.I. Shtogrin, *Polycycles: symmetry and embeddability*, Uspekhi Mat. Nauk. **55–6** (2000), 129–130 (in Russian). English translation in Russian Math. Surveys **55–6**, 1146–1147.

12 M. Deza, M.I. Shtogrin, *Clusters of Cycles*, Journal of Geometry and Physics **40–3, 4** (2001), 302–319.

13 M. Deza, M.I. Shtogrin, *Criterion of embedding of (r, q)-polycycles*, Uspekhi Mat. Nauk. **57–3** (2002), 149–150 (in Russian). English translation in Russian Math. Surveys **57–3**, 589–591.

14 M. Deza, M.I. Shtogrin, *Extremal and non-extendible polycycles*, Proceedings of Steklov Mathematical Institute, **239** (2002), 117–135. (Translated from Trudy of Steklov Math. Institut **239** (2002), 127–145).

15 M. Deza, M.I. Shtogrin, *Archimedean polycycles*, Uspekhi Mat. Nauk. **59–3** (2004), 165–166 (in Russian). English translation in Russian Math. Surveys **59-3**, 564–566.

16 M. Deza, M.I. Shtogrin, *Metrics of constant curvature on polycycles*, Mat. Zametki **78–2** (2005), 223–233 (in Russian). English translation in Math. Notes **78–1&2** (2005), 204–212.

17 M. Deza, M.I. Shtogrin, *Types and boundary unicity of polypentagons*, Uspekhi Math. Nauk **61–6** (2006), 183–184. English translation in Russian Math. Surveys **61–6**, 1170–1172.

18 M. Dutour, *Point Groups*, http://www.liga.ens.fr/~dutour/PointGroup/

19 M. Dutour, *PlanGraph, a GAP package for Planar Graph*, http://www.liga.ens.fr/~dutour/PlanGraph/

20 M. Dutour, M. Deza, *Face-regular 3-valent two-faced spheres and tori*, Research Memorandum Nr. 976 of Institute of Statistical Mathematics, 2006 (see also http://www.liga.ens.fr/~dutour/).

21 The GAP Group, GAP — Groups, Algorithms, and Programming, Version 4.3, 2002. http://www.gap-system.org.

22 A. Hatcher, *Algebraic topology*, http://www.math.cornell.edu/~hatcher/AT/ATpage.html.

23 M.I. Shtogrin, *Primitive polycycles: criterion*, Uspekhi Mat. Nauk. **54–6** (1999), 177–178 (in Russian). English translation in Russian Math. Surveys **54–6**, 1261–1262.

24 M.I. Shtogrin, *Non-primitive polycycles and helicenes*, Uspekhi Mat. Nauk. **55–2** (2000), 155–156 (in Russian). English translation in Russian Math. Surveys **55-2**, 358–360.

15
Optimal Dynamic Flows in Networks and Algorithms for Finding Them

Dmitrii Lozovanu and Maria Fonoberova

15.1
Introduction

Dynamic flow problems are among the most important and challenging problems in network optimization due to the large size of these models in real-world applications. Dynamic flows are widely used in modelling of control processes from different technical, economic, biological and informational systems. Road traffic assignment, evacuation planning, production and distribution, schedule planning, telecommunications, modeling of biological and ecological systems, and management problems can be formulated and solved as single-commodity or multicommodity flow problems [1–3, 36].

The field of network flows blossomed in the 1940s and 1950s with interest in transportation planning and has developed rapidly since then. There is a significant body of literature devoted to this subject (see, for example, [1, 3, 11, 26, 34]). However, it has largely ignored a crucial aspect of transportation: transportation occurs over time. In the 1960s, Ford and Fulkerson [19, 20] introduced flows over time to include time in the network model and proposed scheme for the reduction of the dynamic maximum flow problem with fixed transit times and capacities of arcs to the static maximum flow problem.

The following two aspects of dynamic flows distinguish them from the traditional model. Firstly, the flow value on an arc may change over time. This feature is important in applications where supply and demand are not given as fixed measures; instead, they change over time. Naturally, the flow value on each arc should adjust to these changes. Secondly, there is a transit time on every arc that specifies the amount of time flow units need to traverse the arc.

Two basic network flow problems are concerned with determining maximum flows and with finding minimum cost flows. These problems have large implementation for many practical problems and have theoretical im-

Analysis of Complex Networks: From Biology to Linguistics. Edited by Matthias Dehmer and Frank Emmert-Streib

ISBN: 978-3-527-32345-6

portance for investigating and solving various optimization problems on graphs. The techniques for solving such problems are similar, so in this chapter we will mainly study the minimum cost flow problem.

Linear models of optimal dynamic flows have been studied by Cai, Sha, and Wong [6], Carey and Subrahmanian [7], Fleischer [15, 16], Glockner and Nemhauser [21] Hoppe and Tardos [25], Klinz and Woeginger [27, 28], Lozovanu [29, 30], and Ma, Cui, and Cheng [33]. In this chapter the classical optimal flow problems on networks are extended and generalized for the cases of nonlinear cost functions on arcs, multicommodity flows, and time- and flow-dependent transactions on arcs of the network. Dynamic networks are considered with time-varying capacities of arcs and the demand–supply function that depends on time. It is assumed that cost functions, defined on arcs, are nonlinear and depend on time and flow. Moreover, the dynamic model with transit time functions that depend on flow value and entering time-moment of flow in the arc is considered. Algorithms for solving such problems are proposed and justified on the basis of the time-expanded network method [18–20, 31, 32].

15.2 Optimal Dynamic Single-Commodity Flow Problems and Algorithms for Solving Them

In this section we consider two basic models concerning dynamic flows: the minimum cost flow problem and the maximum flow problem on dynamic networks. We formulate and investigate the minimum cost dynamic flow problem on a network with nonlinear cost functions, defined on arcs, that depend on time and on flow, and demand–supply and capacity functions that depend on time. The maximum dynamic flow problem is also considered in the case where all network parameters depend on time. To solve the considered problems, we propose algorithms based on the reduction of dynamic problems to classical static problems on auxiliary time-expanded networks. Generalized problems with transit time functions that depend on the amount of flow and the entering time-moment of flow in the arc are analyzed and algorithms for solving such problems are developed and grounded.

15.2.1 The Minimum Cost Dynamic Flow Problem

A dynamic network $N = (V, E, \tau, d, u, \varphi)$ is determined by directed graph $G = (V, E)$ with set of vertices V, $|V| = n$, set of arcs E, $|E| = m$, transit time function $\tau\colon E \to R_+$, demand-supply function $d\colon V \times \mathbb{T} \to R$, capacity function $u\colon E \times \mathbb{T} \to R_+$, and cost function $\varphi\colon E \times R_+ \times \mathbb{T} \to R_+$. We consider the discrete time model, in which all times are integral and bounded by horizon T. Time is measured in discrete steps, so that if one unit of flow leaves vertex z at time t on arc $e = (z, v)$, then one unit of flow arrives at vertex v at time $t + \tau_e$, where τ_e is the transit time on arc e. The time horizon is the time until which the flow can travel in the network, and it defines the set $\mathbb{T} = \{0, 1, \ldots, T\}$ of time moments we consider.

In order for the flow to exist, it is required that $\sum_{t \in \mathbb{T}} \sum_{v \in V} d_v(t) = 0$. It is evident that this condition is necessary, but it is not a sufficient one. If for an arbitrary node $v \in V$ at a moment of time $t \in \mathbb{T}$ the condition $d_v(t) > 0$ holds, then we treat this node v at time moment t as a source. If at a moment of time $t \in \mathbb{T}$ the condition $d_v(t) < 0$ holds, then we regard the node v at time moment t as a sink. In the case $d_v(t) = 0$ at a moment of time $t \in \mathbb{T}$, we consider the node v at time moment t as an intermediate node. In this way, the same node $v \in V$ at different moments of time can serve as a source, a sink, or an intermediate node.

Without loss of generality we consider that the set of vertices V is divided into three disjoint subsets V_+, V_-, V_*, such that:

1. V_+ consists of nodes $v \in V$, for which $d_v(t) \geq 0$ for $t \in \mathbb{T}$, and there exists at least one moment of time $t_0 \in \mathbb{T}$ such that $d_v(t_0) > 0$;
2. V_- consists of nodes $v \in V$, for which $d_v(t) \leq 0$ for $t \in \mathbb{T}$ and there exists at least one moment of time $t_0 \in \mathbb{T}$ such that $d_v(t_0) < 0$;
3. V_* consists of nodes $v \in V$, for which $d_v(t) = 0$ for every $t \in \mathbb{T}$.

Thus V_+ is a set of sources, V_- is a set of sinks, and V_* is a set of intermediate nodes of network N.

A feasible dynamic flow in network N is a function x: $E \times \mathbb{T} \rightarrow R_+$ that satisfies the following conditions:

$$\sum_{e \in E^-(v)} x_e(t) - \sum_{\substack{e \in E^+(v) \\ t - \tau_e \geq 0}} x_e(t - \tau_e) = d_v(t), \ \forall t \in \mathbb{T}, \ \forall v \in V; \tag{15.1}$$

$$0 \leq x_e(t) \leq u_e(t), \ \ \forall t \in \mathbb{T}, \ \forall e \in E; \tag{15.2}$$

$$x_e(t) = 0, \ \forall e \in E, \ t = \overline{T - \tau_e + 1, T}. \tag{15.3}$$

Here the function x defines the value $x_e(t)$ of flow entering arc e at time t. It is easy to observe that the flow does not enter arc e at time t if it has to leave the arc after time T; this is ensured by (15.3). Capacity constraints (15.2) mean that in a feasible dynamic flow, at most $u_e(t)$ units of flow can enter arc e at time moment t. Conditions 15.1 represent flow conservation constraints.

To model transit costs, which may change over time, we define the cost function $\varphi_e(x_e(t), t)$ with the meaning that the flow of value $\varrho = x_e(t)$ entering arc e at time t will incur a transit cost of $\varphi_e(\varrho, t)$. We assume that $\varphi_e(0, t) = 0$ for all $e \in E$ and $t \in \mathbb{T}$.

The total cost of the dynamic flow x in network N is defined as follows:

$$F(x) = \sum_{t \in \mathbb{T}} \sum_{e \in E} \varphi_e(x_e(t), t) . \tag{15.4}$$

The minimum cost dynamic flow problem consists in finding a feasible dynamic flow that minimizes the objective function (15.4).

It is easy to observe that if $\tau_e = 0, \forall e \in E$, and $T = 0$, then the formulated problem becomes the classical minimum cost flow problem on a static network.

15.2.2
The Maximum Dynamic Flow Problem

The concept of dynamic flows can be extended for the maximum flow problem in the following way. We introduce the dynamic network $N_m = (V, E, \tau, u, V_s, V_f)$, which consists of directed graph $G = (V, E)$ with set of vertices V and set of arcs E, transit time function $\tau: E \to R_+$, capacity function $u: E \times \mathbb{T} \to R_+$, set of sources V_s, and set of sinks V_f. In an analogous way as in the previous subsection we consider the discrete time model with $\mathbb{T} = \{0, 1, 2, \ldots, T\}$.

A feasible dynamic flow in the network N_m is a function $x: E \times \mathbb{T} \to R_+$ that satisfies Conditions (15.2)–(15.3) and the following conditions:

$$\sum_{e \in E^-(v)} x_e(t) - \sum_{\substack{e \in E^+(v) \\ t-\tau_e \geq 0}} x_e(t - \tau_e) = \begin{cases} y_v(t), & \forall v \in V_s; \\ 0, & \forall v \in V \setminus (V_s \cup V_f); \\ -y_v(t), & \forall v \in V_f, \end{cases} \qquad \forall t \in \mathbb{T};$$

$$y_v(t) \geq 0, \ \forall t \in \mathbb{T}, \ \forall v \in V .$$

The maximum dynamic flow problem consists in finding a feasible dynamic flow that maximizes the total value of the flow:

$$Z = \sum_{t \in \mathbb{T}} \sum_{v \in V_s} y_v(t) .$$

15.2.3
Algorithms for Solving the Optimal Dynamic Flow Problems

To solve the minimum cost dynamic flow problem, we propose an approach based on the reduction of the dynamic problem to a corresponding static problem. We show that the minimum cost dynamic flow problem on network $N = (V, E, \tau, d, u, \varphi)$ can be reduced to a minimum cost static flow problem on an auxiliary network $N^T = (V^T, E^T, d^T, u^T, \varphi^T)$, which is called the time-expanded network. The advantage of such an approach is that it turns the problem of determining an optimal flow over time into a classical static network flow problem.

The essence of the time-expanded network is that it contains a copy of the vertex set of the dynamic network for each moment of time $t \in \mathbb{T}$, and the transit times and flows are implicit in arcs linking those copies. We define the network N^T as follows:

1. $V^T: = \{v(t) \mid v \in V,\ t \in \mathbb{T}\}$;
2. $E^T: = \{e(t) = (v(t), z(t+\tau_e)) \mid e \in E,\ 0 \le t \le T-\tau_e\}$;
3. ${d_{v(t)}}^T: = d_v(t)$ for $v(t) \in V^T$;
4. ${u_{e(t)}}^T: = u_e(t)$ for $e(t) \in E^T$;
5. ${\varphi_{e(t)}}^T({x_{e(t)}}^T): = \varphi_e(x_e(t), t)$ for $e(t) \in E^T$.

In what follows we state a correspondence between feasible flows in the dynamic network N and feasible flows in the time-expanded network N^T. Let $x_e(t)$ be a flow in the dynamic network N; then the function x^T defined as follows:

$$ {x_{e(t)}}^T = x_e(t),\ \forall\, e(t) \in E^T \tag{15.5} $$

represents a flow in the time-expanded network N^T.

Lemma 15.1 *Correspondence (15.5) is a bijection from the set of feasible flows in the dynamic network N onto the set of feasible flows in the time-expanded network N^T.*

Proof. It is obvious that Correspondence (15.5) is a bijection from the set of T-horizon functions in the dynamic network N onto the set of functions in the time-expanded network N^T. In what follows we have to show that each dynamic flow in the dynamic network N is put into the correspondence with a static flow in the time-expanded network N^T and vice versa.

Let $x_e(t)$ be a dynamic flow in N, and let ${x_{e(t)}}^T$ be a corresponding function in N^T. Let us prove that ${x_{e(t)}}^T$ satisfies the conservation constraints in the static network N^T. Let $v \in V$ be an arbitrary vertex in N and t, and let $0 \le t \le T-\tau_e$ be an arbitrary moment of time:

$$ \begin{aligned} d_v(t) &\overset{(i)}{=} \sum_{e \in E^-(v)} x_e(t) - \sum_{\substack{e \in E^+(v) \\ t-\tau_e \ge 0}} x_e(t-\tau_e) = \\ &= \sum_{e(t) \in E^-(v(t))} {x_{e(t)}}^T - \sum_{e(t-\tau_e) \in E^+(v(t))} {x_{e(t-\tau_e)}}^T \overset{(ii)}{=} {d_{v(t)}}^T. \end{aligned} \tag{15.6} $$

Note that, according to the definition of the time-expanded network, the set of arcs $\{e(t-\tau_e) | e(t-\tau_e) \in E^+(v(t))\}$ consists of all arcs that enter $v(t)$, while the set of arcs $\{e(t)|e(t) \in E^{-}(v(t))\}$ consists of all arcs that originate from $v(t)$. Therefore, all necessary conditions are satisfied for each vertex $v(t) \in V^T$. Hence, ${x_{e(t)}}^T$ is a flow in the time-expanded network N^T.

Let ${x_{e(t)}}^T$ be a static flow in the time-expanded network N^T, and let $x_e(t)$ be a corresponding function in the dynamic network N. Let $v(t) \in V^T$ be an arbitrary vertex in N^T. The conservation constraints for this vertex in the static network are expressed by Equality (ii) from (15.6), which holds for all $v(t) \in V^T$ at all times t, $0 \le t \le T - \tau_e$. Therefore, Equality (i) holds for all $v \in V$ at all moments of time t, $0 \le t \le T - \tau_e$. In this way $x_e(t)$ is a flow in the dynamic network N.

It is easy to verify that a feasible flow in the dynamic network N is a feasible flow in the time-expanded network N^T and vice versa. Indeed,

$$0 \le {x_{e(t)}}^T = x_e(t) \le u_e(t) = {u_{e(t)}}^T .$$

The lemma is proved.

The total cost of the static flow x^T in the time-expanded network N^T is determined as follows:

$$F^T(x^T) = \sum_{t \in \mathbb{T}} \sum_{e(t) \in E^T} {\varphi_{e(t)}}^T({x_{e(t)}}^T) .$$

Theorem 15.1 *If x is a flow in the dynamic network N and x^T is a corresponding flow in the time-expanded network N^T, then*

$$F(x) = F^T(x^T) .$$

Moreover, for each minimum cost flow x^* in the dynamic network N there is a corresponding minimum cost flow x^{*T} in the static network N^T such that

$$F(x^*) = F^T(x^{*T})$$

and vice versa.

Proof. Let $x : E \times T \to R_+$ be an arbitrary dynamic flow in the dynamic network N. Then according to Lemma 15.1 the unique flow x^T in N^T corresponds to the flow x in N, and therefore we have:

$$F(x) = \sum_{t \in \mathbb{T}} \sum_{e \in E} \varphi_e(x_e(t), t) = \sum_{t \in \mathbb{T}} \sum_{e(t) \in E^T} {\varphi_{e(t)}}^T({x_{e(t)}}^T) = F^T(x^T) .$$

Thus the first part of the theorem is proved.

To prove the second part of the theorem we again use Lemma 15.1. Let $x^* : E \times T \to R_+$ be the optimal dynamic flow in N and x^{*T} the corresponding optimal flow in N^T. Then

$$F(x^*) = \sum_{t \in \mathbb{T}} \sum_{e \in E} \varphi_e(x_e^*(t), t) = \sum_{t \in \mathbb{T}} \sum_{e(t) \in E^T} {\varphi_{e(t)}}^T({x^*_{e(t)}}^T) = F^T(x^{*T}) .$$

The converse proposition is proved in an analogous way.
The theorem is proved.

The results described above allow us to propose the following algorithm for solving the minimum cost dynamic flow problem.

1. To build the time-expanded network N^T for the dynamic network N.
2. To solve the classical minimum cost flow problem on the static network N^T, using one of the known algorithms [1,4,12,13,22,23,26,34].
3. To reconstruct the solution of the static problem on network N^T to the dynamic problem on network N.

In what follows let us examine the complexity of this algorithm including the time necessary to solve the resulting problem on the static time-expanded network. Building the time-expanded network and reconstructing the solution of the minimum cost static flow problem to the dynamic one has complexity $O(nT + mT)$, where n is the number of vertices in the dynamic network and m is the number of arcs in this network. The complexity of step 2 depends on the complexity of the algorithm used for the minimum cost flow problem on static networks. If such an algorithm has complexity $O(f(n', m'))$, where n' is a number of vertices and m' is a number of arcs in the network, then the complexity of solving the minimum cost flow problem on the time-expanded network employing the same algorithm is $O(f(nT, mT))$.

To solve the maximum dynamic flow problem we can also use the time-expanded network method. We define the auxiliary network $N_m^T = (V^T, E^T, u^T, V_s^T, V_f^T)$ in the following way:

1. $V^T: = \{v(t) \mid v \in V,\ t \in \mathbb{T}\}$;
2. $E^T: = \{(v(t), z(t + \tau_e)) \mid e = (v, z) \in E,\ 0 \le t \le T - \tau_e\}$;
3. $u_{e(t)}{}^T: = u_e(t)$ for $e(t) \in E^T$;
4. $V_s^T: = \{v(t) \mid v \in V_s,\ t \in \mathbb{T}\}$;
5. $V_f^T: = \{v(t) \mid v \in V_f,\ t \in \mathbb{T}\}$.

In a similar way as above we can prove that if x is a flow in the dynamic network N_m and x^T is a corresponding flow in the time-expanded network N_m^T, then $Z = Z^T$. Moreover, for each maximum flow x^* in the dynamic network N_m there is a corresponding maximum flow x^{*T} in the static network N_m^T such that $Z^* = Z^{*T}$ and vice versa.

In this way, to solve the maximum flow problem on dynamic network N_m we have to construct the time-expanded network N_m^T, then solve the classical maximum flow problem on the static network N_m^T, and, finally, reconstruct the solution of the static problem to the dynamic problem.

15.2.4
The Dynamic Model with Flow Storage at Nodes

In the above mathematical model it is assumed that flow cannot be stored at nodes. Such a model can be extended for the case of flow storage at nodes if we associate to each node $v \in V$ a transit time τ_v, which means that the flow passage through this node takes τ_v units of time. If in addition we associate to each node v the capacity function $u_v(t)$ and the cost function $\varphi_v(x_v(t), t)$, we obtain a more general mathematical model. It is easy to observe that in this case the problem can be reduced to the previous one by simple transformation of the network where each node v is changed by a couple of vertices v' and v'' connected with directed arc $e_v = (v', v'')$. Here v' preserves all entering arcs and v'' preserves all leaving arcs of the previous network. To arc e_v we associate the transit time $\tau_{e_v} = \tau_v$, the cost function $\varphi_{e_v}(x_{e_v}(t), t) = \varphi_v(x_v(t), t)$, and the capacity function $u_{e_v}(t) = u_v(t)$.

Another mathematical model with unlimited flow storage at nodes can be obtained by introducing loops in those nodes in which there is flow storage. The flow that was stored at the nodes passes through these loops. Moreover, by introducing transit times for the loops and the costs we can formulate the problem with flow storage at nodes and storage costs at nodes.

An important particular case of the considered problem is when all flow is dumped into the network from sources $v \in V_+$ at time moment $t = 0$ and arrives at sinks $v \in V_-$ at time moment $t = T$. This means that the supply–demand function $d : V \times T \to R$ satisfies the following conditions:

(a) $d_v(0) > 0$, $d_v(t) = 0$, $t = 1, 2, \ldots, T$, for $v \in V_+$;
(b) $d_v(T) < 0$, $d_v(t) = 0$, $t = 0, 1, 2, \ldots, T-1$, for $v \in V_-$.

In what follows we show that in this case another dynamic mathematical model can be used.

15.2.5
The Dynamic Model with Flow Storage at Nodes and Integral Constant Demand–Supply Functions

We consider the minimum cost flow problem on the dynamic network with flow storage at nodes and integral constant demand–supply functions. Suppose a dynamic network $N = (V, E, \tau, d, u, \varphi)$, where the demand–supply function $d : V \to R$ does not depend on time. Without loss of generality, we assume that no arcs enter sources or exit sinks. According to this dynamic model, all flow is dumped into the network at zero time moment and arrive in its entirety at the final moment of time T. We note that in order for a flow to exist, supply must equal demand: $\sum_{v \in V} d_v = 0$.

The mathematical model of the minimum cost flow problem on this dynamic network is the following:

$$\sum_{e \in E^-(v)} \sum_{t=0}^{T} x_e(t) - \sum_{e \in E^+(v)} \sum_{t=\tau_e}^{T} x_e(t-\tau_e) = d_v, \ \forall v \in V; \tag{15.7}$$

$$\sum_{e \in E^-(v)} \sum_{t=0}^{\theta} x_e(t) - \sum_{e \in E^+(v)} \sum_{t=\tau_e}^{\theta} x_e(t-\tau_e) \le 0, \ \forall v \in V_*, \ \forall \theta \in \mathbb{T}; \tag{15.8}$$

$$0 \le x_e(t) \le u_e(t), \ \ \forall t \in \mathbb{T}, \ \forall e \in E; \tag{15.9}$$

$$x_e(t) = 0, \ \forall e \in E, \ t = \overline{T - \tau_e + 1, T}. \tag{15.10}$$

Condition 15.10 ensures that there is no flow in the network after time horizon T. Condition 15.9 is a capacity constraint. As flow travels through the network, we allow unlimited flow storage at the nodes, but prohibit any deficit by Constraint 15.8. Finally, all demands must be met, flow must not remain in the network after time T, and each source must not exceed its supply. This is ensured by Constraint 15.7.

We are seeking a feasible dynamic flow x that minimizes the total cost:

$$F(x) = \sum_{t \in \mathbb{T}} \sum_{e \in E} \varphi_e(x_e(t), t) \, .$$

To solve the formulated problem we use the time-expanded network method. We construct the auxiliary static network N^T as follows:

1. $V^T := \{v(t) | v \in V, t \in \mathbb{T}\}$;
2. $V_+^T := \{v(0) | v \in V_+\}$ and $V_-^T := \{v(T) | v \in V_-\}$;
3. $E^T := \{(v(t), z(t+\tau_e)) \mid e = (v, z) \in E, \ 0 \le t \le T - \tau_e\} \cup$
   $\{v(t), v(t+1) \mid v \in V, \ 0 \le t < T\}$;
4. ${d_{v(t)}}^T := d_v$ for $v(t) \in V_+^T \cup V_-^T$; otherwise ${d_{v(t)}}^T := 0$;
5. ${u_{(v(t),z(t+\tau_{(v,z)}))}}^T : = u_{(v,z)}(t)$ for $(v(t), z(t+\tau_{(v,z)})) \in E^T$;
   ${u_{(v(t),v(t+1))}}^T : = \infty$ for $(v(t), v(t+1)) \in E^T$;
6. ${\varphi_{(v(t),z(t+\tau_{(v,z)}))}}^T({x_{(v(t),z(t+\tau_{(v,z)}))}}^T) : = \varphi_{(v,z)}(x_{(v,z)}(t), t)$
   for $(v(t), z(t+\tau_{(v,z)})) \in E^T$;
   ${\varphi_{(v(t),v(t+1))}}^T({x_{(v(t),v(t+1))}}^T) : = 0$ for $(v(t), v(t+1)) \in E^T$.

If we define a flow correspondence to be ${x_{e(t)}}^T := x_e(t)$, where ${x_{(v(t),v(t+1))}}^T$ in N^T corresponds to the flow in N stored at node v in the period of time from t to $t+1$, then the minimum cost flow problem on dynamic networks can be solved by solving the minimum cost static flow problem on the time-expanded network.

15.2.6
Approaches to Solving Dynamic Flow Problems with Different Types of Cost Functions

Next we analyze the following cases of the minimum cost flow problem on a dynamic network.

Linear Cost Functions on Arcs

If cost functions $\varphi_e(x_e(t), t)$ are linear with regard to $x_e(t)$, then cost functions of the time-expanded network are linear. In this case we can apply well-established methods for minimum cost flow problems, including linear programming algorithms [19, 20, 24, 25] and combinatorial algorithms [20], as well as other developments such as [14, 17].

Convex Cost Functions on Arcs

If cost functions $\varphi_e(x_e(t), t)$ are convex with regard to $x_e(t)$, then cost functions of the time-expanded network are convex. Algorithms for solving the dynamic version of the minimum cost flow problem with convex cost functions can be obtained by using the time-expanded network method. We construct the auxiliary time-expanded network, solve the minimum cost flow problem in the static network with convex cost functions on arcs, and then reconstruct the obtained solution. To solve the static problem we can apply methods from convex programming and the specification of such methods for the minimum cost flow problem.

Concave Cost Functions on Arcs

If there is exactly one source, and cost functions $\varphi_e(x_e(t), t)$ are concave with regard to $x_e(t)$, then cost functions in the time-expanded network are concave. If the dynamic network is acyclic, then the time-expanded network is acyclic. Therefore, we can solve the static problem using classical algorithms for minimum cost flow problems in acyclic networks with concave cost functions [30, 35].

In what follows we present an approach for dynamic networks with cost functions that are concave with regard to flow value and do not change over time. Relying on concavity, we reduce the problem to the minimum cost flow problem on a static network of equal size, not the time-expanded network.

Suppose a dynamic network $\mathcal{N} = (V, E, \tau, u, \varphi, d)$ constant in time capacities of arcs, constant in time demand–supply of nodes and the possibility of flow storage at nodes. As above, we consider that no arcs enter sources or exit sinks. The corresponding static network $\mathcal{N}^0$ of $\mathcal{N}$ is obtained by discarding all time-related information: $\mathcal{N}^0 = (V, E, u, \varphi^0, d)$, where $\varphi_e^0(\varrho) = \varphi_e(\varrho, 0)$.

Lemma 15.2 *Let $\mathcal{N}$ be an uncapacitated dynamic network with cost functions concave with regard to flow and constant in time. If x is a flow on $\mathcal{N}$, then $y_e = \sum_{t\in\mathbb{T}} x_e(t)$ is a flow in the corresponding static network $\mathcal{N}^0$ and $F^0(y) \le F(x)$.*

Proof. Note that if $\phi : R_+ \to R_+$ is a concave function, then $\phi(\alpha + \beta) \le \phi(\alpha) + \phi(\beta)$ for all $\alpha, \beta \in R_+$. Since F and F^0 are concave with regard to flow value, we obtain:

$$F^0(y) = \sum_{e\in E} \varphi_e^0(y_e) = \sum_{e\in E} \varphi_e^0\left(\sum_{t\in\mathbb{T}} x_e(t)\right)$$
$$\le \sum_{e\in E}\sum_{t\in\mathbb{T}} \varphi_e^0(x_e(t)) = \sum_{e\in E}\sum_{t\in\mathbb{T}} \varphi_e(x_e(t), t) = F(x)\,.$$

Moreover, $y_e = \sum_{t=0}^{T} x_e(t) = \sum_{t=\tau_e}^{T} x_e(t-\tau_e)$, since flow x obeys Constraint 15.10. Hence, by substituting y_e in dynamic conservation Constraint 15.7, we obtain the corresponding static conservation constraint. Therefore, y is a flow in $\mathcal{N}^0$. The lemma is proved.

Further we will need the following definitions.

Definition 15.1 The graph $\mathcal{G}_x = (V_x, E_x)$ that consists of arc set $E_x = \{e \in E | x_e > 0\}$ and node set $V_x = \{v | \exists z$ such that $(v, z) \in E_x$ or $(z, v) \in E_x\}$ is called the base graph of flow x in network $\mathcal{N}$.

Definition 15.2 The graph $\mathcal{G}_x = (V_x, E_x)$ consisting of arc set $E_x = \left\{e \in E | \sum_{t\in\mathbb{T}} x_e(t) > 0\right\}$ and node set $V_x = \left\{v | \exists z \text{ such that } (v, z) \in E_x \text{ or } (z, v) \in E_x\right\}$ is called the base graph of dynamic flow x in $\mathcal{N}$.

Lemma 15.3 *Let $\mathcal{N}$ be an infinite-horizon dynamic network with cost functions constant in time. If y is a static flow in $\mathcal{N}^0$ such that its base graph $\mathcal{G}_y$ is a forest, then there exists a dynamic flow x in $\mathcal{N}$ such that $F(x) = F^0(y)$.*

Proof. Let $x_{(v,z)}(t) = y_{(v,z)}$ if $t = t_v$, and $x_{(v,z)}(t) = 0$ otherwise, where:

$$t_v = \begin{cases} 0, & \text{if } v \in V_+, \\ \max\{t_z + \tau_{(z,v)} | (z, v) \in E_y\}, & \text{otherwise.} \end{cases} \tag{15.11}$$

Since $\mathcal{G}_y = (V_y, E_y)$ is a forest, the constants t_v are well defined and finite. To prove that x is a flow in $\mathcal{N}$, we have to show that it satisfies Constraints 15.10, 15.7, and 15.8.

Because $T = +\infty$, it follows that $T \geq t_v, \forall v \in V_y$. Therefore, for any $e = (v, z) \in E_y$, we obtain $T \geq t_z \geq t_v + \tau_e$, hence $T - \tau_e \geq t_v$. Since $t \neq t_v$ implies $x_e(t) = 0$, it follows that $x_e(t) = 0$ for all $t_v > T - \tau_e \geq t_v$, hence Constraint 15.10 is obeyed.

Condition 15.11 means that flow starts leaving $\forall v \in V_* \cap V_y$ only after all inbound flow has arrived. Thus for $\theta < t_v$ we have $\sum_{e \in E^-(v)} \sum_{t=0}^{\theta} x_e(t) = 0$, hence Constraint 15.8 holds for $\theta < t_v$. For $\theta \geq t_v$ flow summed over time on any arc is the same as the flow on that arc in the static network:

$$\sum_{e \in E^+(v)} \sum_{t=\tau_e}^{\theta} x_e(t - \tau_e) = \sum_{e \in E^+(v)} y_e = \sum_{e \in E^-(v)} y_e = \sum_{e \in E^-(v)} \sum_{t=0}^{\theta} x_e(t).$$

Therefore, Constraint 15.8 holds for $\theta \geq t_v$. We have established that Constraint 15.8 is obeyed.

By taking $\theta = T \geq t_v$ in the previous argument, we obtain that Constraint 15.7 holds for all $v \in V_* \cap V_y$. For all sources $v \in V_+$ incoming flow is zero: $\sum_{e \in E^+(v)} \sum_{t \in \mathbb{T}} x_e(t) = 0$, since no arcs enter a source. On the other hand, outgoing flow equals supply: $\sum_{e \in E^-(v)} \sum_{t \in \mathbb{T}} x_e(t) = \sum_{e \in E^-(v)} y_e = d_v$. Therefore, Constraint 15.7 holds for all sources. The proof for sinks is similar, taking into account no arcs exit sinks. Therefore, Constraint 15.7 is obeyed.

Having proved that x is a flow, it is easy to see that it is feasible, since $0 \leq x_e(t) \leq y_e \leq u_e$. Finally, $F(x) = \sum_{e \in E} \sum_{t \in \mathbb{T}} \varphi_e(x_e(t), t) = \sum_{e \in E_y} \varphi_e(x_e(t_v), t_v) = \sum_{e \in E} \varphi_e^0(y_e) = F^0(y)$. The lemma is proved.

In the above proof we employ the fact that $T = +\infty$ only to maintain that $t_v \leq T, \forall v \in V_y$. However, if we denote by $|L| = \sum_{e \in L} \tau_e$ the time length of a path in $\mathcal{N}$, then we immediately obtain that $t_v \leq \max_{L \in \mathcal{N}} \{|L|\}$. Hence $\max_{L \in \mathcal{N}} \{|L|\}$ is an upper bound for the makespan of flow x as constructed in the above lemma, and we can broaden the class of networks we examine.

Lemma 15.4 *Let $\mathcal{N}$ be a dynamic network with cost functions constant in time such that $T \geq \max_{L \in \mathcal{N}} \{|L|\}$. If y is a static flow in $\mathcal{N}^0$ such that its base graph $\mathcal{G}_y$ is a forest, then there exists a dynamic flow x in $\mathcal{N}$ such that $F(x) = F^0(y)$.*

To make the connection between Lemma 15.2, Lemma 15.4, and minimum cost flows in dynamic networks, we will employ the following property of minimum cost flows in static networks with concave cost functions [29, 30].

Lemma 15.5 *Let $\mathcal{N}^0$ be an uncapacitated static network with concave nondecreasing cost functions. If there exists a flow in $\mathcal{N}^0$, then there exists a minimum cost flow y in $\mathcal{N}^0$ such that its base graph $\mathcal{G}_y$ is a forest.*

We are now able to prove the main result of this subsection. Denote by y^T the dynamic flow in $\mathcal{N}$ obtained from a forestlike flow y in $\mathcal{N}^0$ such that $y^T_{(v,z)}(t) = y_{(v,z)}$ if $t = t_v$, and $y^T_{(v,z)}(t) = 0$ otherwise, where t_v are defined as in (15.11).

Theorem 15.2 *Let $\mathcal{N}$ be an uncapacitated dynamic network with cost functions concave with regard to flow and constant in time such that $T \geq \max_{L \in \mathcal{N}}\{|L|\}$. If there exists a flow in $\mathcal{N}$, then there exists a minimum cost forestlike flow ξ in $\mathcal{N}^0$, and the flow ξ^T is a minimum cost flow in $\mathcal{N}$.*

Proof. Since there exists a flow in $\mathcal{N}$, a flow can be constructed in $\mathcal{N}^0$ according to Lemma 15.2. Hence, according to Lemma 15.5 there exists a minimum cost forestlike flow in $\mathcal{N}^0$; denote this flow by ξ. Flow ξ^T is a minimum cost flow in $\mathcal{N}$. Indeed, for any flow x in $\mathcal{N}$ we have:

$$F(\xi^T) = F^0(\xi) \leq F^0(y) \leq F(x),$$

where y is a static flow in $\mathcal{N}^0$ such that $y_e = \sum_{t \in \mathbb{T}} x_e(t)$. Equality $F(\xi^T) = F^0(\xi)$ follows from Lemma 15.4, inequality $F^0(\xi) \leq F^0(y)$ from the fact that ξ is a minimum cost flow in $\mathcal{N}^0$, and inequality $F^0(y) \leq F(x)$ from Lemma 15.2. The theorem is proved.

Therefore, a dynamic minimum cost flow can be computed using the following procedure:

(1) Find a forestlike minimum cost flow ξ in the corresponding static network.
(2) Compute the constants t_v and construct the dynamic minimum cost flow ξ^T.

To solve the minimum cost flow problem on dynamic networks that meet the conditions of Theorem 15.2, and have exactly 1 source, we use the approach that represents an extension of the method for solving the static minimum cost flow problem proposed in [29, 30]. In detailed form the approach can be found in [32]. In general, the algorithm can be developed for the case where the cost functions on arcs are concave with respect to flow for every fixed moment of time $t \in \mathbb{T}$.

15.2.7
Determining the Optimal Dynamic Flows in Networks with Transit Functions That Depend on Flow and Time

In the above dynamic models the transit time functions are assumed to be constant on each arc of the network. In this setting, the time it takes to traverse an arc does not depend on the current flow situation on the arc and the entering time moment of flow in the corresponding arc. Intuitively, it is clear that in many applications the amount of time needed to traverse an arc of the network increases as the arc becomes more congested and it also depends on the entering time moment of flow in the arc. If we take into account these assumptions, we obtain a more general minimum cost dynamic flow problem. In this model we consider two-sided restrictions on arc capacities. We assume that the transit time function $\tau_e(x_e(t), t)$ is a nonnegative, nondecreasing, left-continuous step function with respect to the amount of flow $x_e(t)$ for every fixed time moment $t \in \mathbb{T}$ and an arbitrary given arc $e \in E$. We denote by $P_{e,t}$ the set of numbers of steps of the transit time function for the fixed moment of time t and the given arc e.

Thus we consider a dynamic network $N = (V, E, \tau, d, u', u'', \varphi)$ determined by directed graph $G = (V, E)$ with set of vertices V and set of arcs E, transit time function $\tau\colon E \times \mathbb{T} \times R_+ \to R_+$, demand–supply function $d\colon V \times \mathbb{T} \to R$, lower and upper capacity functions $u', u''\colon E \times \mathbb{T} \to R_+$, and cost function $\varphi\colon E \times R_+ \times \mathbb{T} \to R_+$. As above, we consider the discrete time model, in which all times are integral and bounded by a time horizon T, which defines the set $\mathbb{T} = \{0, 1, \dots, T\}$ of time moments we consider. The supply is equal to the demand, that is, $\sum_{t\in\mathbb{T}} \sum_{v\in V} d_v(t) = 0$.

A dynamic flow in network N is represented by a function $x\colon E \times \mathbb{T} \to R_+$, which determines the value $x_e(t)$ of flow entering arc e at time t. Since we require that all arcs must be empty after time horizon T, the following implication must hold for all $e \in E$ and $t \in \mathbb{T}$: if $x_e(t) > 0$, then $t + \tau_e(x_e(t), t) \le T$. The dynamic flow x must satisfy the flow conservation constraints, which means that at any time moment $t \in \mathbb{T}$ for every vertex $v \in V$ the difference between the total amount of flow that leaves node v and the total amount of flow that enters node v is equal to $d_v(t)$.

The dynamic flow x is called feasible if it satisfies the following capacity constraints: $u'_e(t) \le x_e(t) \le u''_e(t), \ \forall t \in \mathbb{T}, \ \forall e \in E$.

The total cost of the dynamic flow x in network N is determined as follows:

$$F(x) = \sum_{t\in\mathbb{T}} \sum_{e\in E} \varphi_e(x_e(t), t) .$$

The minimum cost dynamic flow problem consists in finding a feasible dynamic flow that minimizes this objective function.

In [18] it is shown that the minimum cost flow problem on dynamic network with transit time functions that depend on the amount of flow and the entering time moment of flow in the arc can be reduced to a static problem on a special modified time-expanded network $N^T = (V^T, E^T, d^T, u'^T, u''^T, \varphi^T)$, which is defined as follows:

1. $\overline{V}^T: = \{v(t) \mid v \in V,\ t \in \mathbb{T}\}$;
2. $\widetilde{V}^T: = \{e(v(t)) \mid v(t) \in \overline{V}^T,\ e \in E^-(v),\ t \in \mathbb{T} \setminus \{T\}\}$;
3. $V^T: = \overline{V}^T \cup \widetilde{V}^T$;
4. $\widetilde{E}^T: = \{\widetilde{e}(t) = (v(t), e(v(t))) \mid v(t) \in \overline{V}^T$ and corresponding $e(v(t)) \in \widetilde{V}^T,\ t \in \mathbb{T} \setminus \{T\}\}$;
5. $\overline{E}^T: = \{e^p(t) = (e(v(t)), z(t + \tau_e^p(x_e(t), t))) \mid e(v(t)) \in \widetilde{V}^T,\ z(t + \tau_e^p(x_e(t), t)) \in \overline{V}^T,\ e = (v, z) \in E,\ 0 \le t \le T - \tau_e^p(x_e(t), t),\ p \in P_{e,t}\}$;
6. $E^T: = \overline{E}^T \cup \widetilde{E}^T$;
7. ${d_{v(t)}}^T: = d_v(t)$ for $v(t) \subset \overline{V}^T$;
   ${d_{e(v(t))}}^T: = 0$ for $e(v(t)) \in \widetilde{V}^T$;
8. ${u'_{\widetilde{e}(t)}}^T: = u'_e(t)$ for $\widetilde{e}(t) \in \widetilde{E}^T$;
   ${u''_{\widetilde{e}(t)}}^T: = u''_e(t)$ for $\widetilde{e}(t) \in \widetilde{E}^T$;
   ${u'_{e^p(t)}}^T: = \overline{x_e^{p-1}}(t)$ for $e^p(t) \in \overline{E}^T$, where $\overline{x_e^0}(t) = u'_e(t)$;
   ${u''_{e^p(t)}}^T: = \overline{x_e^p}(t)$ for $e^p(t) \in \overline{E}^T$;
9. ${\varphi_{\widetilde{e}(t)}}^T({x_{\widetilde{e}(t)}}^T): = \varphi_e(x_e(t), t)$ for $\widetilde{e}(t) \in \widetilde{E}^T$;
   ${\varphi_{e^p(t)}}^T({x_{e^p(t)}}^T): = \varepsilon_p$ for $e^p(t) \in \overline{E}^T$, where $\varepsilon_1 < \varepsilon_2 < \cdots < \varepsilon_{|P_{e,t}|}$ are small numbers.

The most complicated moment in solving the considered problem is the construction of the auxiliary time-expanded network N^T. The solution of the dynamic problem can be found on the basis of the following results.

Lemma 15.6 *Let $x^T: E^T \to R_+$ be a flow in the static network N^T. Then the function $x: E \times \mathbb{T} \to R_+$ defined as follows:*

$$x_e(t) = {x_{\widetilde{e}(t)}}^T = {x_{e^p(t)}}^T$$

for $e = (v, z) \in E,\ \widetilde{e}(t) = (v(t), e(v(t))) \in \widetilde{E}^T$,

$$e^p(t) = (e(v(t)), z(t + \tau_e^p(x_e(t), t))) \in \overline{E}^T,$$

$p \in P_{e,t}$ *is such that* ${x_{\widetilde{e}(t)}}^T \in (\overline{x_e^{p-1}}(t), \overline{x_e^p}(t)],\ t \in \mathbb{T}$,

represents a flow in the dynamic network N.

Let x: $E \times \mathbb{T} \to R_+$ be a flow in the dynamic network N. Then the function x^T: $E^T \to R_+$ defined as follows:

$$x_{\tilde{e}(t)}{}^T = x_e(t) \textit{ for } \tilde{e}(t) = (v(t), e(v(t))) \in \widetilde{E}^T,\ e = (v, z) \in E,\ t \in \mathbb{T};$$

$$x_{e^p(t)}{}^T = x_e(t) \textit{ for such } p \in P_{e,t} \textit{ that } x_e(t) \in (\overline{x_e^{p-1}}(t), \overline{x_e^p}(t)]$$

$$\textit{and } x_{e^p(t)}{}^T = 0 \textit{ for all other } p \in P_{e,t}$$

$$\textit{for } e^p(t) = (e(v(t)), z(t + \tau_e^p(x_e(t), t))) \in \overline{E}^T,\ e = (v, z) \in E,\ t \in \mathbb{T},$$

represents a flow in the static network N^T.

Theorem 15.3 *If x^{*T} is a static minimum cost flow in the static network N^T, then the corresponding according to Lemma 15.6 dynamic flow x^* in the dynamic network N is also a minimum cost flow and vice versa.*

In such a way, to solve the minimum cost flow problem on dynamic networks with transit time functions that depend on the amount of flow and the entering time moment of flow in the arc, we construct a time-expanded network, then solve the minimum cost static flow problem, and reconstruct the solution of the static problem to the dynamic problem.

15.3
Optimal Dynamic Multicommodity Flow Problems and Algorithms for Solving Them

In this section we formulate and investigate the optimal multicommodity flow problems on dynamic networks ([18, 31]). The multicommodity flow problem consists in shipping several different commodities from their respective sources to their sinks through a given network satisfying certain objectives in such a way that the total flow going through arcs does not exceed their capacities. No commodity ever transforms into another commodity, so that each one has its own flow conservation constraints, but they compete for the resources of the common network. We consider the minimum cost multicommodity flow problem on dynamic networks with time-varying capacities of arcs and transit times on arcs that depend on a kind of commodity entering them. We assume that cost functions, defined on arcs, are nonlinear and depend on time and flow, and demand–supply functions depend on time. For solving the considered problem, we propose algorithms based on the time-expanded network method.

15.3.1
The Minimum Cost Dynamic Multicommodity Flow Problem

The minimum cost dynamic multicommodity flow problem is a problem of finding the flow of a set of commodities through a network with a given time horizon, satisfying all supplies and demands with minimum total cost such that arc capacities are not exceeded. We consider the discrete time model, where all times are integral and bounded by horizon T, which defines the set $\mathbb{T} = \{0, 1, \ldots, T\}$ of time moments.

We consider a dynamic network $N = (V, E, K, \tau, d, u, w, \varphi)$, determined by directed graph $G = (V, E)$, where V is a set of vertices and E is a set of arcs, set of commodities $K = \{1, 2, \ldots, q\}$ that must be routed through the same network, transit time function $\tau: E \times K \to R_+$, demand–supply function $d: V \times K \times \mathbb{T} \to R$, mutual capacity function $u: E \times \mathbb{T} \to R_+$, individual capacity function $w: E \times K \times \mathbb{T} \to R_+$, and cost function $\varphi: E \times R_+ \times \mathbb{T} \to R_+$. Thus $\tau_e = (\tau_e^1, \tau_e^2, \ldots, \tau_e^q)$ is a vector, each component of which reflects the transit time on arc $e \in E$ for commodity $k \in K$.

In order for the flow to exist, it is required that $\sum_{t\in\mathbb{T}} \sum_{v\in V} d_v^k(t) = 0, \forall k \in K.$ If for an arbitrary node $v \in V$ at a moment of time $t \in \mathbb{T}$ the condition $d_v^k(t) > 0$ holds, then we treat this node v at the time moment t as a source for commodity $k \in K$. If at a moment of time $t \in \mathbb{T}$ the condition $d_v^k(t) < 0$ holds, then we regard node v at time moment t as a sink for commodity $k \in K$. In the case $d_v^k(t) = 0$ at a moment of time $t \in \mathbb{T}$, we consider node v at the time moment t as an intermediate node for commodity $k \in K$. In such a way, the same node $v \in V$ at different moments of time can serve as a source, a sink on an intermediate node for commodity $k \in K$.

Without loss of generality we consider that for every commodity $k \in K$ the set of vertices V is divided into three disjoint subsets V_+^k, V_-^k, V_*^k, such that:

- V_+^k consists of nodes $v \in V$, for which $d_v^k(t) \geq 0$ for $t \in \mathbb{T}$, and there exists at least one moment of time $t_0 \in \mathbb{T}$ such that $d_v^k(t_0) > 0$;
- V_-^k consists of nodes $v \in V$, for which $d_v^k(t) \leq 0$ for $t \in \mathbb{T}$, and there exists at least one moment of time $t_0 \in \mathbb{T}$ such that $d_v^k(t_0) < 0$;
- V_*^k consists of nodes $v \in V$, for which $d_v^k(t) = 0$ for every $t \in \mathbb{T}$;
- Thus V_+^k is a set of sources, V_-^k is a set of sinks, and V_*^k is a set of intermediate nodes for the commodity $k \in K$ in the network N.

A feasible dynamic multicommodity flow in network N is determined by a function $x: E \times K \times \mathbb{T} \to R_+$ that satisfies the following conditions:

$$\sum_{e\in E^-(v)} x_e^k(t) - \sum_{\substack{e\in E^+(v)\\ t-\tau_e^k\geq 0}} x_e^k(t-\tau_e^k) = d_v^k(t), \ \forall t \in \mathbb{T}, \ \forall v \in V, \ \forall k \in K; \tag{15.12}$$

$$\sum_{k\in K} x_e^k(t) \leq u_e(t), \ \forall t \in \mathbb{T}, \ \forall e \in E; \tag{15.13}$$

$$0 \leq x_e^k(t) \leq w_e^k(t), \ \ \forall t \in \mathbb{T}, \ \forall e \in E, \ \forall k \in K; \tag{15.14}$$

$$x_e^k(t) = 0, \ \forall e \in E, \ t = \overline{T - \tau_e^k + 1, T}, \ \forall k \in K. \tag{15.15}$$

Here the function x defines the value $x_e^k(t)$ of flow of commodity k entering arc e at moment of time t. Condition 15.15 ensures that the flow of commodity k does not enter arc e at time t if it has to leave the arc after time horizon T. Individual capacity Constraints 15.14 mean that at most $w_e^k(t)$ units of flow of commodity k can enter arc e at time t. Mutual capacity Constraints 15.13 mean that at most $u_e(t)$ units of flow can enter arc e at time t. Constraints 15.14 and 15.13 are called weak and strong forcing constraints, respectively. Conditions 15.12 represent flow conservation constraints.

The total cost of the dynamic multicommodity flow x in network N is defined as follows:

$$F(x) = \sum_{t \in \mathbb{T}} \sum_{e \in E} \varphi_e(x_e^1(t), x_e^2(t), \ldots, x_e^q(t), t). \tag{15.16}$$

The minimum cost dynamic multicommodity flow problem consists in finding a feasible dynamic multicommodity flow that minimizes the objective Function 15.16.

It is important to note that in many practical cases cost functions are presented in the following form:

$$\varphi_e(x_e^1(t), x_e^2(t), \ldots, x_e^q(t), t) = \sum_{k \in K} \varphi_e^k(x_e^k(t), t)\,. \tag{15.17}$$

The case where $\tau_e^k = 0$, $\forall\, e \in E$, $\forall\, k \in K$ and $T = 0$ can be considered as the minimum cost static multicommodity flow problem on network.

15.3.2
Algorithm for Solving the Minimum Cost Dynamic Multicommodity Flow Problem

To solve the formulated problem, we propose an approach based on the reduction of the dynamic problem to a static problem. We show that the minimum cost multicommodity flow problem on network N can be reduced to a static problem on a special auxiliary network N^T.

In the case of the minimum cost multicommodity flow problem on a dynamic network with different transit times on an arc for different commodities we define the auxiliary time-expanded network $N^T = (V^T, E^T, K, d^T, u^T, w^T, \varphi^T)$ in the following way:

1 $\overline{V}^T := \{v(t) \mid v \in V,\ t \in \mathbb{T}\}$;
2 $\widetilde{V}^T := \{e(v(t)) \mid v(t) \in \overline{V}^T,\ e \in E^-(v),\ t \in \mathbb{T} \setminus \{T\}\}$;
3 $V^T := \overline{V}^T \cup \widetilde{V}^T$;
4 $\widetilde{E}^T := \{\widetilde{e}(t) = (v(t), e(v(t))) \mid v(t) \in \overline{V}^T$ and corresponding $e(v(t)) \in \widetilde{V}^T$, $t \in \mathbb{T} \setminus \{T\}\}$;

5 $\overline{E}^T := \{e^k(t) = (e(v(t)), z(t+\tau_e^k)) \mid e(v(t)) \in \widetilde{V}^T,\ z(t+\tau_e^k) \in \overline{V}^T,\ e = (v,z) \in E,\ 0 \le t \le T - \tau_e^k,\ k \in K\}$;

6 $E^T := \overline{E}^T \cup \widetilde{E}^T$;

7 $d_{v(t)}^{k\ T} := d_v^k(t)$ for $v(t) \in \overline{V}^T,\ k \in K$;

$d_{e(v(t))}^{k\ T} := 0$ for $e(v(t)) \in \widetilde{V}^T,\ k \in K$;

8 $u_{\widetilde{e}(t)}^{\ T} := u_e(t)$ for $\widetilde{e}(t) \in \widetilde{E}^T$;

$u_{e^k(t)}^{\ T} := \infty$ for $e^k(t) \in \overline{E}^T$;

9 $w_{e^k(t)}^{l\ T} := \begin{cases} w_e^{k(t)}, & \text{if } l = k \text{ for } e^k(t) \in \overline{E}^T,\ l \in K; \\ 0, & \text{if } l \ne k \text{ for } e^k(t) \in \overline{E}^T,\ l \in K; \end{cases}$

$w_{\widetilde{e}(t)}^{l\ T} = \infty$ for $\widetilde{e}(t) \in \widetilde{E}^T,\ l \in K$;

10 $\varphi_{\widetilde{e}(t)}^{\ T}(x_{\widetilde{e}(t)}^{1\ T}, x_{\widetilde{e}(t)}^{2\ T}, \ldots, x_{\widetilde{e}(t)}^{q\ T}) := \varphi_e(x_e^1(t), x_e^2(t), \ldots, x_e^q(t), t)$ for $\widetilde{e}(t) \in \widetilde{E}^T$;

$\varphi_{e^k(t)}^{\ T}(x_{e^k(t)}^{1\ T}, x_{e^k(t)}^{2\ T}, \ldots, x_{e^k(t)}^{q\ T}) := 0$ for $e^k(t) \in \overline{E}^T$.

The following lemma represents the relationship between flows in dynamic network N and flows in the time-expanded network N^T.

Lemma 15.7 *Let $x^T: E^T \times K \to R_+$ be a multicommodity flow in the static network N^T. Then the function $x: E \times K \times \mathbb{T} \to R_+$ defined in the following way:*

$$x_e^k(t) = x_{e^k(t)}^{k\ T} = x_{\widetilde{e}(t)}^{k\ T}$$

for $e = (v,z) \in E,\ e^k(t) = (e(v(t)), z(t+\tau_e^k)) \in \overline{E}^T,$

$\widetilde{e}(t) = (v(t), e(v(t))) \in \widetilde{E}^T,\ k \in K,\ t \in \mathbb{T},$

represents a multicommodity flow in the dynamic network N.

Let $x: E \times K \times \mathbb{T} \to R_+$ be a multicommodity flow in the dynamic network N. Then the function $x^T: E^T \times K \to R_+$ defined in the following way:

$$x_{\widetilde{e}(t)}^{k\ T} = x_e^k(t) \text{ for } \widetilde{e}(t) = (v(t), e(v(t))) \in \widetilde{E}^T,\ e = (v,z) \in E,\ k \in K,\ t \in \mathbb{T};$$

$$x_{e^k(t)}^{k\ T} = x_e^k(t);\ x_{e^k(t)}^{l\ T} = 0,\ l \ne k$$

for $e^k(t) = (e(v(t)), z(t+\tau_e^k)) \in \overline{E}^T,\ e = (v,z) \in E,\ l, k \in K,\ t \in \mathbb{T},$

represents a multicommodity flow in the static network N^T.

Proof. To prove the first part of the lemma, we have to show that Conditions 15.12–15.15 for the x defined above in the dynamic network N are true. These conditions evidently result from the following definition of multicommodity flows in the static network N^T:

$$\sum_{e(t)\in E^-(v(t))} x^{k\;T}_{e(t)} - \sum_{e(t-\tau^k_e)\in E^+(v(t))} x^{k\;T}_{e(t-\tau^k_e)} = d^{k\;T}_{v(t)}, \ \forall v(t) \in V^T, \ \forall k \in K; \tag{15.18}$$

$$\sum_{k\in K} x^{k\;T}_{e(t)} \le u_{e(t)}{}^T, \ \forall e(t) \in E^T; \tag{15.19}$$

$$0 \le x^{k\;T}_{e(t)} \le w^{k\;T}_{e(t)}, \ \forall\, e(t) \in E^T, \ \forall k \in K; \tag{15.20}$$

$$x^{k\;T}_{e(t)} = 0, \ \forall\, e(t) \in E^T, \ t = \overline{T - \tau^k_e + 1, T}, \ \forall k \in K, \tag{15.21}$$

where by $v(t)$ and $e(t)$ we denote $\overline{v}(t)$ or $\widetilde{v}(t)$ and $\overline{e}(t)$ or $\widetilde{e}(t)$, respectively, against context.

To prove the second part of the lemma it is sufficient to show that Conditions 15.18–15.21 hold for x^T defined above. The correctness of these conditions results from the procedure of constructing the time-expanded network, the correspondence between flows in static and dynamic networks, and the satisfied Conditions 15.12–15.15.

The lemma is proved.

The following theorem holds.

Theorem 15.4 *If x^{*T} is a minimum cost multicommodity flow in the static network N^T, then, according to Lemma 15.7, the corresponding multicommodity flow x^* in the dynamic network N is also a minimum cost one and vice versa.*

Proof. Taking into account the correspondence between static and dynamic multicommodity flows on the basis of Lemma 15.7, we obtain that costs of the static multicommodity flow in the time-expanded network N^T and the corresponding dynamic multicommodity flow in the dynamic network N are equal. To solve the minimum cost multicommodity flow problem on the static time-expanded network N^T, we have to solve the following problem:

$$F^T(x^T) = \sum_{t\in\mathbb{T}} \sum_{e(t)\in E^T} \varphi_{e(t)}{}^T(x^{1\;T}_{e(t)}, x^{2\;T}_{e(t)}, \ldots, x^{q\;T}_{e(t)}) \to \min$$

subject to (15.18)–(15.21).

In the case of the minimum cost multicommodity flow problem on dynamic network with separable cost functions (15.17) and without mutual capacities of arcs we can simplify the procedure of constructing the time-expanded network. In this case we don't have to add a new set of vertices $\widetilde{V}^T$ and a new set of arcs $\widetilde{E}^T$. In this way the time-expanded network N^T is defined as follows:

1. $V^T: = \{v(t) \mid v \in V,\ t \in \mathbb{T}\}$;
2. $E^T: = \{e^k(t) = (v(t), z(t+\tau_e^k)) \mid v(t) \in V^T,\ z(t+\tau_e^k) \in V^T,\ e = (v,z) \in E,$ $0 \le t \le T - \tau_e^k,\ k \in K\}$;
3. ${d_{v(t)}^k}^T: = d_v^k(t)$ for $v(t) \in V^T,\ k \in K$;
4. ${w_{e^k(t)}^l}^T: = \begin{cases} w_e^k(t), & \text{if } l = k \text{ for } e^k(t) \in E^T,\ l \in K; \\ 0, & \text{if } l \ne k \text{ for } e^k(t) \in E^T,\ l \in K; \end{cases}$
5. ${\varphi_{e^k(t)}^l}^T({x_{e^k(t)}^l}^T): = \begin{cases} \varphi_e^k(x_e^k(t), t), & \text{if } l = k \text{ for } e^k(t) \in E^T,\ l \in K; \\ 0, & \text{if } l \ne k \text{ for } e^k(t) \in E^T,\ l \in K. \end{cases}$

As corollaries of Lemma 15.7 and Theorem 15.4 we can obtain the following results.

Lemma 15.8 *Let x^T: $E^T \times K \to R_+$ be a multicommodity flow in the static network N^T. Then the function x: $E \times K \times \mathbb{T} \to R_+$ defined as follows:*

$$x_e^k(t) = {x_{e^k(t)}^k}^T \textit{ for } e \in E,\ e^k(t) \in E^T,\ k \in K,\ t \in \mathbb{T},$$

represents the multicommodity flow in the dynamic network N.

Let x: $E \times K \times \mathbb{T} \to R_+$ be a multicommodity flow in the dynamic network N. Then the function x^T: $E^T \times K \to R_+$ defined as follows:

$${x_{e^k(t)}^k}^T = x_e^k(t);\ {x_{e^k(t)}^l}^T = 0,\ l \ne k \textit{ for } e^k(t) \in E^T,\ e \in E,\ l, k \in K,\ t \in \mathbb{T},$$

represents the multicommodity flow in the static network N^T.

Theorem 15.5 *If x^{*T} is a minimum cost multicommodity flow in the static network N^T, then the corresponding according to Lemma 15.8 multicommodity flow x^* in the dynamic network N is also a minimum cost one and vice versa.*

In the case of the minimum cost multicommodity flow problem on dynamic network with common transit times on an arc for different commodities, the time-expanded network N^T can be constructed even more simply and is defined in the following way:

1. $V^T: = \{v(t) \mid v \in V,\ t \in \mathbb{T}\}$;
2. $E^T: = \{e(t) = (v(t), z(t+\tau_e)) \mid v(t) \in V^T,\ z(t+\tau_e) \in V^T,\ e = (v,z) \in E,$ $0 \le t \le T - \tau_e\}$;
3. ${d_{v(t)}^k}^T: = d_v^k(t)$ for $v(t) \in V^T,\ k \in K$;
4. ${u_{e(t)}}^T: = u_e(t)$ for $e(t) \in E^T$;
5. ${w_{e(t)}^k}^T: = w_e^k(t)$ for $e(t) \in E^T,\ k \in K$;
6. ${\varphi_{e(t)}}^T({x_{e(t)}^1}^T, {x_{e(t)}^2}^T, \dots, {x_{e(t)}^q}^T)\ mbox: = \varphi_e(x_e^1(t), x_e^2(t), \dots, x_e^q(t), t)$ for $e(t) \in E^T$.

The following lemma and theorem can be considered as particular cases of Lemma 15.7 and Theorem 15.4.

Lemma 15.9 *Let x^T: $E^T \times K \to R_+$ be a multicommodity flow in the static network N^T. Then the function x: $E \times K \times \mathbb{T} \to R_+$ defined as follows:*

$$x_e^k(t) = x_{e(t)}^{k\ \ T} \text{ for } e \in E,\ e(t) \in E^T,\ k \in K,\ t \in \mathbb{T},$$

represents the multicommodity flow in the dynamic network N.

Let x: $E \times K \times \mathbb{T} \to R_+$ be a multicommodity flow in the dynamic network N. Then the function x^T: $E^T \times K \to R_+$ defined as follows:

$$x_{e(t)}^{k\ \ T} = x_e^k(t) \text{ for } e(t) \in E^T,\ e \in E,\ k \in K,\ t \in \mathbb{T},$$

represents the multicommodity flow in the static network N^T.

Theorem 15.6 *If x^{*T} is a minimum cost multicommodity flow in the static network N^T, then the corresponding according to Lemma 15.9 multicommodity flow x^* in the dynamic network N is also a minimum cost one and vice versa.*

In this way, to solve the minimum cost multicommodity flow problem on dynamic networks, we must build the time-expanded network N^T for the given dynamic network N, after which we must solve the classical minimum cost multicommodity flow problem on the static network N^T, using one of the known algorithms [5, 8–10, 14], and then reconstruct the solution of the static problem on N^T to the dynamic problem on N.

The minimum cost dynamic multicommodity flow problems can be extended for the case with transit time functions that depend on the amount of flow and the entering time moment of flow in the arc. The time-expanded network method for such s class of problems can be developed and specified. Similar algorithms based on the time-expanded network method can be derived for maximum dynamic multicommodity flow problems.

15.4
Conclusion

The mathematical models for finding optimal flows in dynamic networks represent the extension and generalization of the classical optimal flow problems on static networks. Such dynamic models can be used for studying and solving a large class of practical problems as well as problems from the theory of graphs and combinatorics. The mathematical apparatus for determining optimal solutions of the network flow problems based on the time-expanded network method has been elaborated and grounded. New efficient algorithms for finding minimum cost and maximum dynamic flows have been derived. The time-expanded network method has been specified for the multicommodity case of optimal dynamic flow problems, and algorithms for solving such problems have been developed.

References

1 Ahuja, R., Magnati, T., Orlin, J. *Network flows*. Prentice-Hall, Englewood Cliffs, 1993.

2 Aronson, J. A survey of dynamic network flows. *Annals of Operations Research*, 1989, 20, 1–66.

3 Assad, A. Multicommodity network flows: a survey. *Networks*, 1978, 8, 37–92.

4 Bland, R.G., Jensen, D.L. *On the computational behavior of a polynomial-time network flow algorithm*. Technical Report 661, School of Operations Research and Industrial Engineering, Cornell University, 1985.

5 McBride, R. Progress made in solving the multicommodity flow problems. *SIAM Journal on Optimization*, 1998, 8(4), 947–955.

6 Cai, X., Sha, D., Wong, C.K. Time-varying minimum cost flow problems. *European Journal of Operational Research*, 2001, 131, 352–374.

7 Carey, M., Subrahmanian, E. An approach to modelling time-varying flows on congested networks. *Transportation Research Part B*, 2000, 34, 157–183.

8 Castro, J., Nabona, N. An implementation of linear and nonlinear multicommodity network flows. *European Journal of Operational Research Theory and Methodology*, 1996, 92, 37–53.

9 Castro J. A specialized interior-point algorithm for multicommodity network flows. *SIAM Journal on Optimization*, 2000, 10(3), 852–877.

10 Castro, J. Solving difficult multicommodity problems with a specialized interior-point algorithm. *Annals of Operations Research*, 2003, 124, 35–48.

11 Christofides, N. *Graph Theory: An Algorithmic Approach*. Academic Press, New York, London, San Francisco, 1975.

12 Edmonds, J., Karp, R.M. Theoretical improvements in algorithmic efficiency for network flow problems. *Journal of the Association for Computing Machinery*, 1972, 19, 248–264.

13 Ermoliev, Iu., Melnic, I. *Extremal Problems on Graphs*. Naukova Dumka, Kiev, 1968.

14 Fleisher, L. Approximating multicommodity flow independent of the number of commodities. *SIAM Journal of Discrete Mathematics*, 2000, 13(4), 505–520.

15 Fleischer, L. Universally maximum flow with piecewise-constant capacities. *Networks*, 2001, 38(3), 115–125.

16 Fleischer, L. Faster algorithms for the quickest transshipment problem. *SIAM Journal on Optimization*, 2001, 12(1), 18–35.

17 Fleisher, L., Skutella, M. The quickest multicommodity flow problem. *Integer Programming and Combinatorial Optimization*. Springer, Berlin, 2002, pp. 36–53.

18 Fonoberova, M., Lozovanu, D. Minimum cost multicommodity flows in dynamic networks and algorithms for their finding. *Bulletin of the Academy of Sciences of Moldova, Mathematics*, 2007, 1(53), 107–119.

19 Ford, L., Fulkerson, D. Constructing maximal dynamic flows from static flows. *Operations Research*, 1958, 6, 419–433.

20 Ford, L., Fulkerson, D. *Flows in Networks*. Princeton University Press, Princeton, NJ, 1962.

21 Glockner, G., Nemhauser, G. A dynamic network flow problem with uncertain arc capacities: formulation and problem structure. *Operations Research*, 2000, 48(2), 233–242.

22 Goldberg, A.V., Tarjan, R.E. Solving minimum-cost flow problems by successive approximation. *Proceedings of the 19th ACM STOC*, 1987, 7–18.

23 Goldberg, A.V., Tarjan, R.E. *Finding minimum-cost circulations by canceling negative cycles*. Technical Report CS-TR 107-87, Department of Computer Science, Princeton University, 1987.

24 Goljshtein, E., Iudin, D. *Linear programming problems of transport type*. Nauka, Moscow, 1969.

25 Hoppe, B., Tardos, E. The quickest transshipment problem. *Mathematics of Operations Research*, 2000, 25, 36–62.

26 Hu, T. *Integer Programming and Network Flows*. Addison-Wesley Publishing Company, Reading, MA, 1970.

27 Klinz, B., Woeginger, C. Minimum cost dynamic flows: the series parallel case, in *Integer Programming and Combinatorial Optimization* (eds W.J. Cook and

A.S. Schulz), Springer, Berlin, 1995, pp. 329–343.

28 Klinz, B., Woeginger, C. One, two, three, many, or: complexity aspects of dynamic network flows with dedicated arcs. *Operations Research Letters*, 1998, 22, 119–127.

29 Lozovanu, D. Properties of optimal solutions of a grid transportation problem with concave cost functions of the flows on the arcs. *Engineering Cybernetics*, 1983, 20, 34–38.

30 Lozovanu, D. *Extremal-combinatorial problems and algorithms for their solving*, Stiinta, Chisinau, 1991.

31 Lozovanu, D., Fonoberova, M. Optimal dynamic multicommodity flows in networks. *Electronic Notes in Discrete Mathematics*, 2006, 25, 93–100.

32 Lozovanu, D., Stratila, D. Optimal flow in dynamic networks with nonlinear cost functions on edges. *Analysis and optimization of differential systems*. Kluwer Academic Publishers, 2003, pp. 247–258.

33 Ma, Z., Cui, D., Cheng, P. Dynamic network flow model for short-term air traffic flow management. *IEEE Transactions on Systems, Man, and Cybernetics – Part A: Systems and Humans*, 2004, 34(3), 351–358.

34 Papadimitrou, C., Steiglitz, K. *Combinatorial Optimization: Algorithms and Complexity*. Prentice-Hall, Englewood Cliffs, NJ, 1982.

35 Pardalos, P.M., Guisewite, G. Global search algorithms for minimum concave cost network flow problem. *Journal of Global Optimization*, 1991, 1(4), 309–330.

36 Powell, W., Jaillet, P., Odoni, A. Stochastic and dynamic networks and routing, in *Network Routing*, Vol. 8 of *Handbooks in Operations Research and Management Science*, chapter 3 (eds M.O. Ball, T.L. Magnanti, C.L. Monma, and G.L. Nemhauser), North Holland, Amsterdam, 1995, pp. 141–295.

16
Analyzing and Modeling European R&D Collaborations: Challenges and Opportunities from a Large Social Network

Michael J. Barber, Manfred Paier, and Thomas Scherngell

16.1
Introduction

Networks have attracted a burst of attention in the last decade (useful reviews include references [1, 9, 15, 28]), with applications to natural, social, and technological systems. While networks provide a powerful abstraction for investigating relationships and interactions, the preparation and analysis of complex real-world networks nonetheless presents significant challenges. In particular, social networks are characterized by a large number of different properties and generation mechanisms that require a rich set of indicators. The objective of the current study is to analyze large social networks with respect to their community structure and mechanisms of network formation. As a case study, we consider networks derived from the European Union's Framework Programs (FPs) for Research and Technological Development.

The EU FPs were implemented to follow two main strategic objectives: first, strengthening the scientific and technological bases of European industry to foster international competitiveness and, second, the promotion of research activities in support of other EU policies. In spite of their different scopes, the fundamental rationale of the FPs has remained unchanged. All FPs share a few common structural key elements. First, only projects of limited duration that mobilize private and public funds at the national level are funded. Second, the focus of funding is on multinational and multiactor collaborations that add value by operating at the European level. Third, project proposals are to be submitted by self-organized consortia, and the selection for funding is based on specific scientific excellence and socioeconomic relevance criteria [33]. By considering the constituents of these consortia, we can represent and analyze the FPs as networks of projects and organizations. The resulting networks are of substantial size, including over 50,000 projects and over 30,000 organizations.

We have a general interest in studying a real-world network of large size and high complexity from a methodological point of view. Furthermore, so-

Analysis of Complex Networks: From Biology to Linguistics. Edited by Matthias Dehmer and Frank Emmert-Streib

ISBN: 978-3-527-32345-6

cioeconomic research emphasizes the central importance of collaborative activities in R&D for economic competitiveness (see, for instance, reference [16], among many others). Mainly for reasons of data availability, attempts to evaluate quantitatively the structure and function of the large social networks generated in the EU FPs have begun only in the last few years, using social network analysis and complex network methodologies [2, 6–8, 34]. Studies to date point to the presence of a dense and hierarchical network. A highly connected core of frequent participants, taking leading roles within consortia, is linked to a large number of peripheral actors, forming a giant component that exhibits the characteristics of a small world.

We augment the earlier studies by applying a battery of methods to the most recent data. We begin by constructing the network, discussing the need for processing the raw data in section 16.2, and continuing with the network definition in section 16.3. We next examine the overall network structure in section 16.4, showing that the networks for each FP feature a giant component with a highly skewed degree distribution and small-world properties. We follow this with an exploration of community structure in sections 16.5 and 16.6, showing that the networks are made of heterogeneous subcommunities with strong topical differentiation. Finally, we investigate determinants of network formation with a binary choice model in section 16.7; this is similar to a recent analysis of Spanish firms [4], but with a focus on the European level and on geographic and network effects. Results are summarized in section 16.8.

16.2 Data Preparation

We draw on the latest version of the sysres EUPRO database. This database includes all information publicly available through the CORDIS projects database[1)] and is maintained by ARC systems research (ARC sys). The sysres EUPRO database presently comprises data on funded research projects of the EU FPs (complete for FP1–FP5, and about 70% complete for FP6) and all participating organizations. It contains systematic information on project objectives and achievements, project costs, project funding, and contract type, as well as information on the participating organizations including the full name, full address, and type of organization.

For purposes of network analysis, the main challenge is the inconsistency of the raw data. Apart from incoherent spelling in up to four languages per country, organizations are labeled inhomogeneously. Entries may range from

1) http://cordis.europa.eu

large corporate groupings, such as EADS, Siemens, and Philips, or large public research organizations like CNR, CNRS, and CSIC, to individual departments and labs.

Due to these shortcomings, the raw data are of limited use for meaningful network analysis. Further, any fully automated standardization procedure is infeasible. Instead, a labor-intensive, manual data-cleaning process is used in building the database. The data-cleaning process is described in reference [34]; here, we restrict our discussion to the steps of the process relevant to the present work. These are as follows.

1. Identification of unique organization name. Organizational boundaries are defined by legal control. Entries are assigned to appropriate organizations using the most recently available organization name. Most records are easily identified, but, especially for firms, organization names may have changed frequently due to mergers, acquisitions, and divestitures.

2. Creation of subentities. This is the key step for mitigating the bias that arises from the different scales at which participants appear in the data set. Ideally, we use the actual group or organizational unit that participates in each project, but this information is only available for a subset of records, particularly in the case of firms. Instead, subentities that operate in fairly coherent activity areas are pragmatically defined. Wherever possible, subentities are identified at the second lowest hierarchical tier, with each subentity comprising one further hierarchical sublayer. Thus, universities are broken down into faculties/schools, consisting of departments; research organizations are broken down into institutes, activity areas, etc., consisting of departments, groups, or laboratories; and conglomerate firms are broken down into divisions, subsidiaries, etc. Subentities can frequently be identified from the contact information even in the absence of information on the actual participating organizational unit. Note that subentities may still vary considerably in scale.

3. Regionalization. The data set has been regionalized according to the European Nomenclature of Territorial Units for Statistics (NUTS) classification system,[2)] where possible to the NUTS3 level. This has been done mostly via information on postal codes.

Due to resource limitations, only organizations appearing more than thirty times in the standardization table for FP1–FP5 have thus far been processed.

2) NUTS is a hierarchical system of regions used by the statistical office of the European Community for the production of regional statistics. At the top of the hierarchy are NUTS-0 regions (countries), below which are NUTS-1 regions and then NUTS-2 regions, etc.

This could bias the results; however, the networks have a structure such that the size of the bias is quite low (see reference [34]).

Additionally, we make use of a representative survey[3] of FP5 participants.[4] The survey focuses on the issues of partner selection, intraproject collaboration, and output performance of EU projects on the level of bilateral partnerships, including individuals as well as organizations. As the survey was restricted to small collaborative projects (specifically, projects with a minimum of 2 and a maximum of 20 partners), the survey addresses a subset of 9,107 relevant (59% of all FP5) projects. It yielded 1,686 valid responses, representing 3% of all (relevant) participants, and covering 1,089 (12% of all relevant) projects.

16.3 Network Definition

Using the sysres EUPRO database, for each FP we construct a network containing the collaborative projects and all organizational subentities that are participants in those projects. An organization is linked to a project if and only if the organization is a member of the project. Since an edge never exists between two organizations or two projects, the network is bipartite. The network edges are unweighted; in principle, the edges could be assigned weights to reflect the strength of the participation, but the data needed to assign the network weights are not available.

We will also consider, for each FP, the projections of the bipartite networks onto unipartite networks of organizations and projects. The organization projections are constructed by taking the organizations as the vertices, with edges between any organizations that are at distance two in the corresponding bipartite network. Thus, organizations are neighbors in the projection network if they take part in one or more projects together. The project projections are similar, with project vertices linked when they have one or more participants in common. While the construction of the projection networks intrinsically loses information available in the bipartite networks, they can nonetheless be useful.

For the binary choice model, we construct another network using cross-section data on 191 organizations that are selected from the survey data. We employ the collaboration network of the respondents on the organiza-

3) This survey was conducted in 2007 by the Austrian Research Centers GmbH, Vienna, Austria, and operated by b-wise GmbH, Karlsruhe, Germany.

4) We chose FP5 (1998–2002) for the survey in order to cover some of the developments over time, including prior as well as subsequent bilateral collaborations, and effects of the collaboration with respect to both scientific and commercial outcomes. Thus, the survey complements the sysres EUPRO database.

tion level (this network comprises 1,173 organizations collaborating in 1,089 projects) and extract the 2-core [14] of its largest component (203 organizations representing 17% of all vertices).[5] Finally, another 12 organizations are excluded due to nonavailability of geographical distance data, so that we end up with a sample of 191 organizations.

16.4 Network Structure

We first consider the bipartite networks for each of the FP networks. Call the size of an organization the number of projects in which it takes part, and similarly call the size of a project the number of constituent organizations taking part in the project. These sizes correspond directly to the degrees of the relevant vertices in the bipartite networks. Both parts – organizations (Figure 16.1) and projects (Figure 16.2) – of each of the networks feature strongly skewed, heavy-tailed size distributions. The sizes of vertices can differ by orders of magnitude, pointing toward the existence of high degree hubs in the networks; hubs of this sort can play an important role in determining the network structure.

The organization size distributions are similar for each of the FPs. The underlying research activities thus have not altered the mix of organizations participating in a particular number of projects in each FP, despite changes in the nature of those research activities over time. In contrast, the rule changes in FP6 that favor larger project consortia are clearly seen in the project size distributions.

Turning to the projection networks, we see that both the organization projection (Table 16.1) and the project projection (Table 16.2) show small-world properties [37]. First, note that the great majority of the (N) vertices and (M) edges are in the largest connected component of the networks. In light of this, we focus on paths in only the largest component. The average path length (l) in each projection network is short, as is the diameter. However, the clustering coefficient [37], which ranges between zero and one, is high. The combination of short path length and high clustering is characteristic of small-world networks. The small-world character is expected to be beneficial in the FP networks, as small-world networks have been shown to encourage the spread of knowledge in model systems [12].

Additionally, the heavy-tailed size distributions of the bipartite networks have a visible effect on the degrees of the projection networks. In each case, the data are quite asymmetric about the mean degree, as seen by examin-

5) This technical trick ensures optimal utilization of observed collaborations in the estimation model, while keeping the size of the model small. It is important to note that it does not make use of the network properties on this somewhat arbitrary subnetwork.

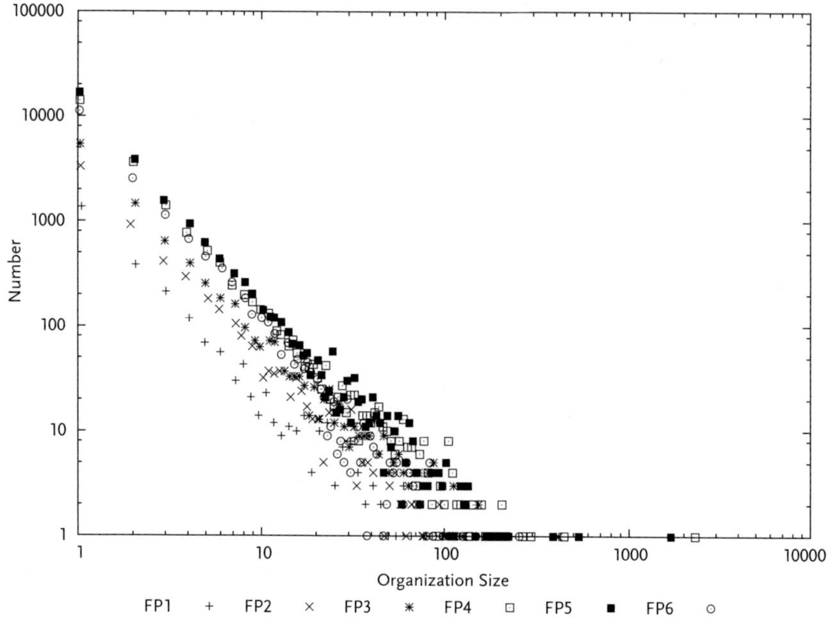

Figure 16.1 Organization sizes.

Table 16.1 Organization projection properties.

Measure	FP1	FP2	FP3	FP4	FP5	FP6
No. of vertices (N)	2116	5758	9035	21599	25840	17632
No. of edges (M)	9489	62194	108868	238585	385740	392879
No. of components	53	45	123	364	630	26
N for largest component	1969	5631	8669	20753	24364	17542
Share of total (%)	93.05	97.79	95.95	96.08	94.29	99.49
(M) for largest component	9327	62044	108388	237632	384316	392705
Share of total (%)	98.29	99.76	99.56	99.60	99.63	99.96
(N) for 2nd-largest component	8	6	9	10	12	9
(M) for 2nd-largest component	44	30	72	90	132	72
Diameter of largest component	9	7	8	11	10	7
(l) largest component	3.62	3.21	3.27	3.45	3.30	3.03
Clustering coefficient	0.65	0.74	0.74	0.78	0.76	0.80
Mean degree	9.0	21.6	24.1	22.1	29.9	44.6
Fraction of (N) above the mean (%)	29.4	28.0	23.6	22.4	23.5	26.1

ing what fraction of vertices have degrees above the mean. The fractions are between 20% and 30%, consistent with the skewed degree distributions (the distributions are shown in references [7,34]; the relation between the degrees in the bipartite networks and the projections is explored in reference [7]).

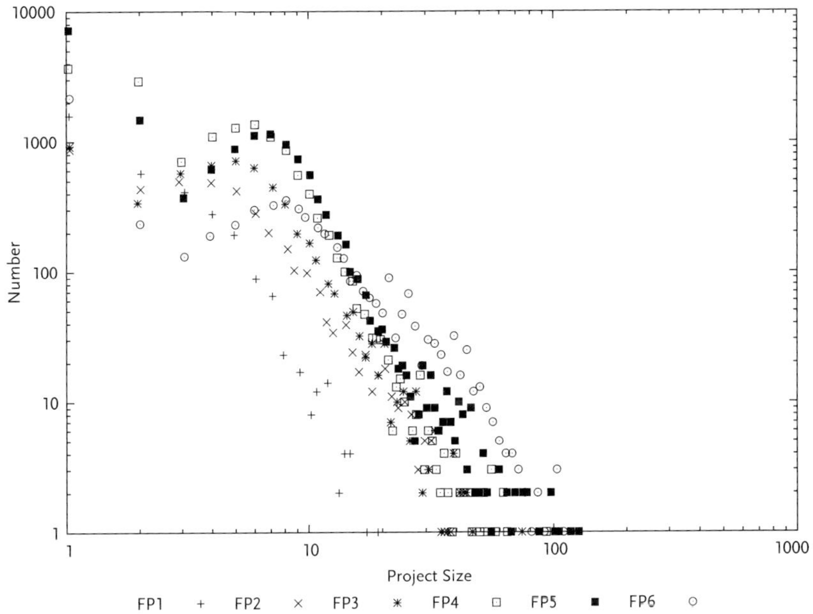

Figure 16.2 Project sizes.

Table 16.2 Project projection properties.

Measure	FP1	FP2	FP3	FP4	FP5	FP6
No. of vertices (*N*)	2116	5758	9035	21599	25840	17632
No. of edges (*M*)	9489	62194	108868	238585	385740	392879
No. of components	53	45	123	364	630	26
(*N*) for largest component	1969	5631	8669	20753	24364	17542
Share of total (%)	93.05	97.79	95.95	96.08	94.29	99.49
(*M*) for largest component	9327	62044	108388	237632	384316	392705
Share of total (%)	98.29	99.76	99.56	99.60	99.63	99.96
(*N*) for 2nd-largest component	8	6	9	10	12	9
(*M*) for 2nd-largest component	44	30	72	90	132	72
Diameter of largest component	9	7	8	11	10	7
(*l*) largest component	3.62	3.21	3.27	3.45	3.30	3.03
Clustering coefficient	0.65	0.74	0.74	0.78	0.76	0.80
Mean degree	9.0	21.6	24.1	22.1	29.9	44.6
Fraction of (*N*) above the mean (%)	29.4	28.0	23.6	22.4	23.5	26.1

16.5
Community Structure

Of great current interest is the identification of community groups, or modules, within networks. Stated informally, a community group is a portion of a network whose members are more tightly linked to one another than to other members of the network. A variety of approaches [3,11,19,20,22,26,27, 30,32] have been taken to explore this concept; see references [13,24] for useful reviews. Detecting community groups allows quantitative investigation of relevant subnetworks. Properties of the subnetworks may differ from the aggregate properties of the network as a whole, for example, modules on the World Wide Web are sets of topically related web pages. Thus, identification of community groups within a network is a first step toward understanding the heterogeneous substructures of the network.

Methods for identifying community groups can be categorized into distinct classes of networks, such as bipartite networks [5,21]. This is immediately relevant for our study of FP networks, allowing us to examine the community structure in the bipartite networks. Communities are expected to be formed of groups of organizations engaged in similar R&D activities and the projects in which those organizations take part.

16.5.1 Modularity

To identify communities, we take as our starting point the modularity, introduced by [26]. Modularity makes intuitive notions of community groups precise by comparing network edges to those of a null model. The modularity (Q) is proportional to the difference between the number of edges within communities (c) and those for a null model:

$$Q \equiv \frac{1}{2M} \sum_{c} \sum_{i,j \in c} \left(A_{ij} - P_{ij}\right) . \tag{16.1}$$

Along with Eq. (16.1), it is necessary to provide a null model, defining $\left(P_{ij}\right)$.

The standard choice for the null model constrains the degree distribution for the vertices to match the degree distribution in the actual network. Random graph models of this sort are obtained [10] by putting an edge between vertices (i) and (j) at random, with the constraint that on average the degree of any vertex (i) is (d_i). This constrains the expected adjacency matrix such that

$$d_i = E\left(\sum_{j} A_{ij}\right) . \tag{16.2}$$

Denote $\left(E\left(A_{ij}\right)\right)$ by $\left(P_{ij}\right)$ and assume further that $\left(P_{ij}\right)$ factorizes into

$$P_{ij} = p_i p_j , \tag{16.3}$$

leading to

$$P_{ij} \equiv \frac{d_i d_j}{2M} . \tag{16.4}$$

A consequence of the null model choice is that $(Q = 0)$ when all vertices are in the same community.

The goal now is to find a division of the vertices into communities such that the modularity (Q) is maximal. An exhaustive search for a decomposition is out of the question: even for moderately large graphs there are far too many ways to decompose them into communities. Fast approximate algorithms do exist (see, for example, references [25, 31]).

16.5.2
Finding Communities in Bipartite Networks

Specific classes of networks have additional constraints that can be reflected in the null model. For bipartite graphs, the null model should be modified to reproduce the characteristic form of bipartite adjacency matrices:

$$A = \begin{bmatrix} O & M \\ M^{\mathrm{T}} & O \end{bmatrix} . \tag{16.5}$$

Recently, specialized modularity measures and search algorithms have been proposed for finding communities in bipartite networks [5, 21]. These measures and methods have not been studied as extensively as the versions with the standard null model shown above, but many of the algorithms can be adapted to the bipartite versions without difficulty. Limitations of modularity-based methods (e.g., the resolution limit described in reference [17]) are expected to hold as well.

We make use of the algorithm called BRIM: bipartite, recursively induced modules [5]. BRIM is a conceptually simple, greedy search algorithm that capitalizes on the separation between the two parts of a bipartite network. Starting from some partition of the vertices of type 1, it is straightforward to identify the optimal partition of the vertices of type 2. From there, optimize the partition of vertices of type 1, and so on. In this fashion, modularity increases until a (local) maximum is reached. However, the question remains: is the maximum a "good" one? At this level, then, a random search is called for, varying the composition and number of communities, with the goal of reaching a better maximum after a new sequence of searching using the BRIM algorithm.

16.6
Communities in the Framework Program Networks

A popular approach in social network analysis – where networks are often small, consisting of a few dozen nodes – is to visualize the networks and identify community groups by eye. However, the Framework Program net-

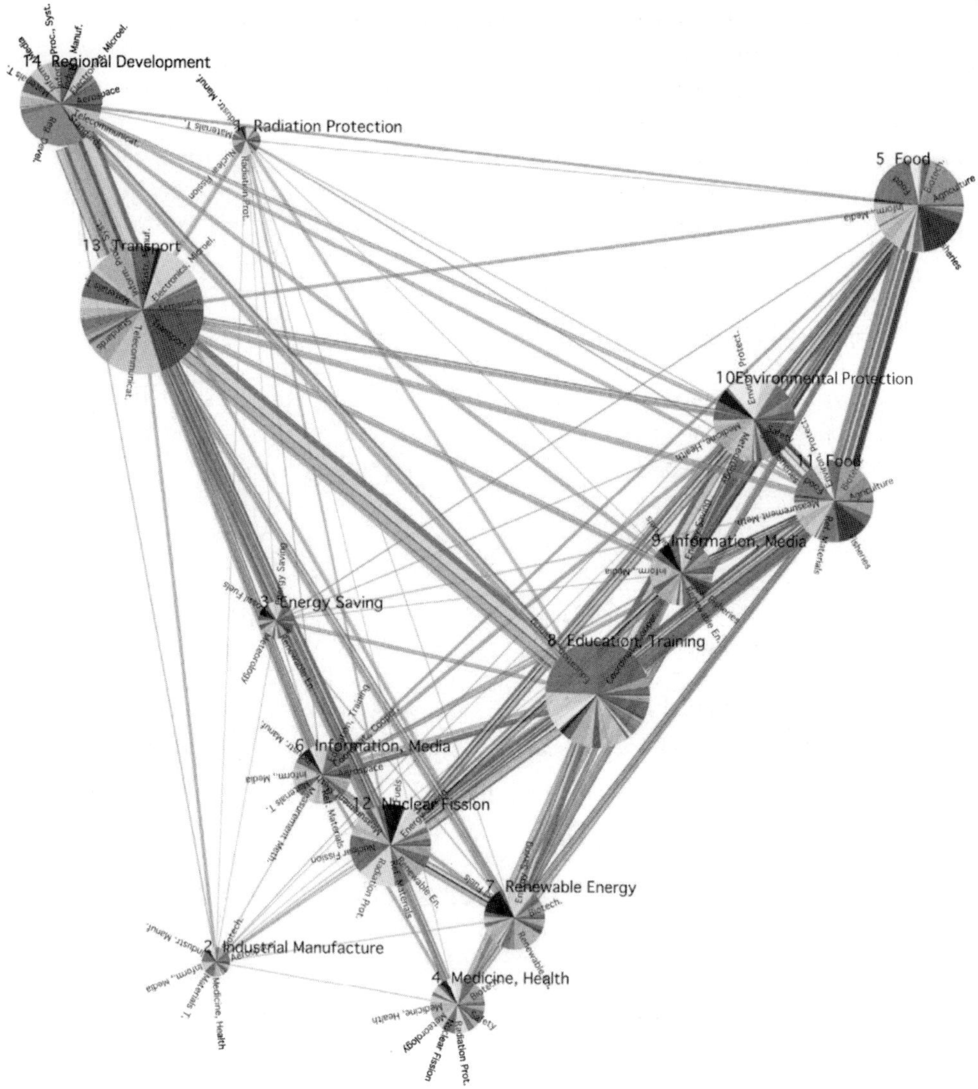

Figure 16.3 Community groups in the network of projects and organizations for FP3.

works are much larger: can we "see" the community groups in these networks?

Structural differences or similarities of such networks are not obvious at a glance. For a graphical representation of the organizations and/or projects by dots on an A4 sheet of paper, we would need to put these dots at a distance of about (1 mm) from each other, and even then we still would not have drawn the links (collaborations) that connect them.

Previous studies used a list of coarse graining recipes to compact the networks into a form that would lend itself to a graphical representation [8]. As an alternative we have attempted to detect communities just using BRIM, that is, purely on the basis of relational network structure, ignoring any additional information about the nature of agents.

In Figure 16.3, we show a community structure for FP3 found using the BRIM algorithm, with a modularity of $(Q = 0.602)$ for 14 community groups. The communities are shown as vertices in a network, with the vertex positions determined using spectral methods [35]. The area of each vertex is proportional to the number of edges from the original network within the corresponding community. The width of each edge in the community network is proportional to the number of edges in the original network connecting community members from the two linked groups. The vertices and edges are shaded to provide additional information about their topical structure, as described in the next section. Each community is labeled with the most frequently occurring subject index.

16.6.1
Topical Profiles of Communities

Projects are assigned one or more standardized subject indices. There are 49 subject indices in total, ranging from *Aerospace* to *Waste Management*. We denote by

$$f(t) > 0 \tag{16.6}$$

the frequency of occurrence of the subject index (t) in the network, with

$$\sum_t f(t) = 1\,. \tag{16.7}$$

Similarly we consider the projects within one community (c) and the frequency

$$f_c(t) \geq 0 \tag{16.8}$$

of any subject index (t) appearing in the projects only of that community. We call (f_c) the topical profile of community (c) to be compared with that of the network as a whole.

Topical differentiation of communities can be measured by comparing their profiles, among each other or with respect to the overall network. This can be done in a variety of ways [18], such as by the Kullback "distance"

$$D_c = \sum_t f_c(t) \ln \frac{f_c(t)}{f(t)}\,. \tag{16.9}$$

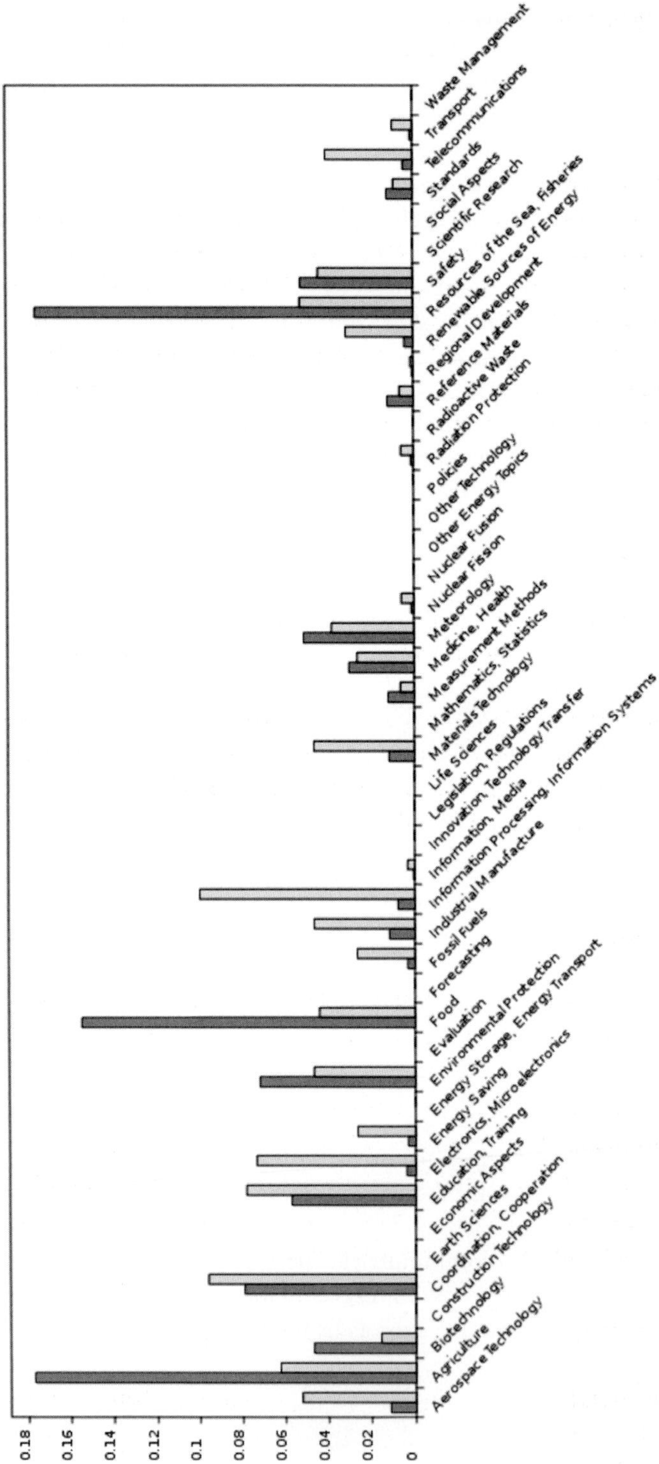

Figure 16.4 Topical differentiation in a network community. The histogram shows the difference between the topical profile ($f_c(t)$) for a specific community (dark bars) and the overall profile ($f(t)$) for the network as a whole (light bars). The community-specific profile shown is for the community labeled "11. Food" in Figure 16.3. The community has ($d_c = 0.90$).

A true metric is given by

$$d_c = \sum_t |f_c(t) - f(t)| , \tag{16.10}$$

ranging from zero to two.

Topical differentiation is illustrated in Figure 16.4. In the figure, example profiles are shown, taken from the network in Figure 16.3. The community-specific profile corresponds to the community labeled '11. Food" in Figure 16.3. Based on the most frequently occurring subject indices – *Agriculture, Food,* and *Resources of the Seas, Fisheries* – the community consists of projects and organizations focused on R&D related to food products. The topical differentiation is ($d_c = 0.90$) for the community shown.

16.7 Binary Choice Model

We now turn to modeling organizational collaboration choices in order to examine how specific individual characteristics, spatial effects, and network effects determine the choice of collaboration (the theoretical underpinnings are described in Ref. [29]). We will build upon the survey data and the sub-network constructed therefrom (Section 16.3). While this restricts us to only 191 organizations, we have considerably more information about these organizations than for the complete networks.

16.7.1 The Empirical Model

In our analytical framework, the constitution of a collaboration (Y_{ij}) between two organizations (i) and (j) will depend on an unobserved continuous variable (Y^*_{ij}) that corresponds to the profit that two organizations (i) and (j) receive when they collaborate. Since we cannot observe (Y^*_{ij}) but only its dichotomous realizations (Y_{ij}), we assume $(Y_{ij} = 1)$ if $(Y^*_{ij} > 0)$ and $(Y_{ij} = 0)$ if $(Y^*_{ij} \leq 0)$. (Y_{ij}) is assumed to follow a Bernoulli distribution so that (Y_{ij}) can take the values one and zero with probabilities (π_{ij}) and $(1 - \pi_{ij})$, respectively. The probability function can be written as

$$\Pr(Y_{ij}) = \pi_{ij}^{Y_{ij}} (1 - \pi_{ij})^{1-Y_{ij}} , \tag{16.11}$$

with $(E[Y_{ij}] = \mu_{ij} = \pi_{ij})$ and $(Var[Y_{ij}] = \sigma^2_{ij} = \pi_{ij}(1 - \pi_{ij}))$, where (μ_{ij}) denotes some mean value.

The next step in defining the model concerns the systematic structure – we would like the probabilities (π_{ij}) to depend on a matrix of observed covariates. Thus, we let the probabilities (π_{ij}) be a linear function of the covariates:

$$\pi_{ij} = \sum_{k=1}^{K} \beta_k X_{ij}^{(k)}, \tag{16.12}$$

where the $\left(X_{ij}^{(k)}\right)$ are elements of the $(\boldsymbol{X}^{(k)})$ matrix containing a constant and $(K-1)$ explanatory variables, including geographical effects, relational effects, and FP experience characteristics of organizations (i) and (j). $(\boldsymbol{\beta}_K = (\beta_0, \boldsymbol{\beta}_{K-1}))$ is the $(K \times 1)$ parameter vector, where (β_0) is a scalar constant term and $(\boldsymbol{\beta}_{K-1})$ is the vector of parameters associated with the $(K-1)$ explanatory variables.

However, estimating this model using ordinary least-squares procedures is not convenient since the probability (π_{ij}) must be between zero and one, while the linear predictor can take any real value. Thus, there is no guarantee that the predicted values will be in the correct range without imposing any complex restrictions [23]. A very promising solution to this problem is to use the logit transform of (π_{ij}) in the model, that is, replacing Eq. (16.12) by the following *ansatz*:

$$Logit\,(\pi_{ij}) = \log \frac{\pi_{ij}}{1-\pi_{ij}} = h_{ij}, \tag{16.13}$$

where we have introduced the abbreviation (h_{ij}), defined as

$$h_{ij} = \beta_0 + \beta_1 X_{ij}^{(1)} + \beta_2 X_{ij}^{(2)} + \cdots + \beta_K. \tag{16.14}$$

This leads to the binary logistic regression model to be estimated given by

$$\Pr\left(Y_{ij} = 1 \mid X_{ij}^{(k)}\right) = \pi_{ij} = \frac{\exp(h_{ij})}{1+\exp(h_{ij})}. \tag{16.15}$$

The focus of interest is on estimating the parameters $(\boldsymbol{\beta})$. The standard estimator for the logistic model is the maximum-likelihood estimator. The reduced log-likelihood function is given by [23]

$$\log L(\boldsymbol{\beta} \mid Y_{ij}) = -\sum_{i,j} \log\left(1+\exp\left((1-2Y_{ij})\, h_{ij}\right)\right), \tag{16.16}$$

assuming independence over the observations (Y_{ij}). The resulting variance matrix $(\boldsymbol{V}(\hat{\boldsymbol{\beta}}))$ of the parameters is used to calculate standard errors. $(\hat{\boldsymbol{\beta}})$ is consistent and asymptotically efficient when the observations of (Y_{ij}) are stochastic and in the absence of multicollinearity among the covariates.

16.7.2
Variable Construction

16.7.2.1 The Dependent Variable

To construct the dependent variable (Y_{ij}) that corresponds to observed collaborations between two organizations (i) and (j), we construct the $(n \times n)$ collaboration matrix $(\boldsymbol{Y})$ that contains the collaborative links between the $((i),(j))$ organizations. One element (Y_{ij}) denotes the existence of collaboration between two organizations (i) and (j) as measured in terms of the existence of a common project. $(\boldsymbol{Y})$ is symmetric by construction, so that $(Y_{ij} = Y_{ji})$. Note that the matrix is very sparse. The number of observed collaborations is 702, so that the proportion of zeros is approximately 98%. The mean collaboration intensity between all $((i),(j))$ organizations is 0.02.

16.7.2.2 Variables Accounting for Geographical Effects

We use two variables, $\left(x_{ij}^{(1)}\right)$ and $\left(x_{ij}^{(2)}\right)$, to account for geographical effects on the collaboration choice. The first step is to assign specific NUTS-2 regions to each of the 191 organizations that are given in the sysres EUPRO database. Then we take the great circle distance between the economic centers of the regions where organizations (i) and (j) are located to measure the geographical distance variable $\left(x_{ij}^{(1)}\right)$. The second variable, $\left(x_{ij}^{(2)}\right)$, controls for country border effects and is measured in terms of a dummy variable that takes a value of zero if two organizations (i) and (j) are located in the same country, and one otherwise, in order to get empirical insight on the role of country borders for collaboration choice of organizations.

16.7.2.3 Variables Accounting for FP Experience of Organizations

This set of variables controls for the experience of the organizations with respect to participation in the European FPs. First, thematic specialization within FP5 is expected to influence the potential to collaborate. We define a measure of thematic distance $\left(x_{ij}^{(3)}\right)$ between any two organizations that is constructed in the following way. Each organization is associated with a unit vector of specialization $(\boldsymbol{s}_i)$ that relates to the number of project participations $(N_{i,1}, \ldots, N_{i,7})$ of organization (i) in the seven subprograms of FP5:[6)]

$$\boldsymbol{s}_i = (N_{i,1}, \ldots, N_{i,7}) \, / \sqrt{N_{i,1}^2 + \cdots + N_{i,7}^2} \, . \tag{16.17}$$

The thematic distance of organizations (i) and (j) is then defined as the Euclidean distance of their respective specialization vectors $(\boldsymbol{s}_i)$ and $(\boldsymbol{s}_j)$, giving

6) EESD, GROWTH, HUMAN POTENTIAL, INCO 2, INNOVATION-SME, IST, and LIFE QUALITY

$\left(x_{ij}^{(3)} = x_{ji}^{(3)}\right)$ and $\left(0 \leq x_{ij}^{(3)} \leq \sqrt{2}\right)$. The second variable accounting for FP experience focuses on the individual (or research group) level and takes into account the respondents inclination or openness to FP research. As a proxy for openness of an organization (i) to FP research, we choose the total number (P_i) of FP5 projects in the respondent's own organization that they are aware of.[7] Then we define

$$x_{ij}^{(4)} = P_i + P_j \tag{16.18}$$

as a measure for the aggregated openness of the respective pair of organizations to FP research. The third variable related to FP experience is the overall number of FP5 project participations an organization is engaged in. Denoting, as above, $(N_i = N_{i,1} + \cdots + N_{i,7})$ as the total number of project participations of organization (i) in FP5, we define

$$x_{ij}^{(5)} = \left|N_i - N_j\right| \tag{16.19}$$

as the difference in the number of participations of organizations (i) and (j) in FP5. This is taken from the sysres EUPRO database and is an integer ranging from $\left(0 \leq x_{ij}^{(5)} \leq 1{,}228\right)$, resulting from the minimal value of one participation and the maximum of 1,229 participations among the sample of 191 organizations.

16.7.2.4 **Variables Accounting for Relational Effects**

We consider a set of three variables accounting for potential relational effects on the decision to collaborate. In this way we distinguish between joint history and network effects. The first factor to be taken into account is prior acquaintance of two organizations and is measured by a binary variable denoting acquaintance on the individual (research group) level before the FP5 collaboration started. It is taken from the survey.[8] By convention, $\left(x_{ij}^{(6)} = 1\right)$ if at least one respondent from organization (i) nominated organization (j) as being a prior acquaintance, $\left(x_{ij}^{(5)} = 0\right)$ otherwise. All other relational factors we take into account in the model are network effects. For conceptual reasons we must look at the global FP5 network, where we make use of the structural embeddedness of our 191 sample organizations.

One of the most important centrality measures is betweenness centrality. Betweenness is a centrality concept based on the question of the extent to

7) The exact wording of the question was, "How many FP5 projects of your organization are you aware of?" For multiple responses from an organization, the numbers of known projects are summarized. In cases of missing data, this number is set to zero.

8) The exact wording of the question was, "Which of your [project acronym] partners (i.e., persons from which organization) did you know before the project began?"

which a vertex in a network is able to control the information flow through the whole network [36]. Organizations that are high in betweenness may thus be especially attractive as collaboration partners. More formally, the betweenness centrality of a vertex can be defined as the probability that a shortest path between a pair of vertices of the network passes through this vertex. Thus, if $(B(k, l; i))$ is the number of shortest paths between vertices (k) and (l) passing through vertex (i), and $(B(k, l))$ is the total number of shortest paths between vertices (k) and (l), then

$$b(i) = \sum_{k \neq l} \frac{B(k, l; i)}{B(k, l)} \tag{16.20}$$

is called the betweenness centrality of vertex (i) [15]. We calculate the betweenness centralities in the global FP5 network and include

$$x_{ij}^{(7)} = b(i)\, b(j) \tag{16.21}$$

as a combined betweenness measure.

The third variable accounting for relational effects is local clustering. Due to social closure, we may assume that within densely connected clusters organizations are mutually quite similar, so that it might be strategically advantageous to search for complementary partners from outside. In this way, communities with lower clustering may be easier to access. We use the clustering coefficient $(CC_1(i))$, which is the share of existing links in the number of all possible links in the direct neighborhood (at a distance of $(d = 1)$) of a vertex (i). Thus, let (k_i) be the number of direct neighbors and (T_i) the number of existing links among these direct neighbors; then

$$CC_1(i) = \frac{2T_i}{k_i(k_i - 1)} \tag{16.22}$$

is the relevant clustering coefficient [37]. We employ the difference in the local clustering coefficients within the global FP5 network for inclusion in the statistical model by setting

$$x_{ij}^{(8)} = |CC_1(i) - CC_1(j)| \tag{16.23}$$

in order to obtain a symmetric variable in (i) and (j).

16.7.3 Estimation Results

This section discusses the estimation results of the binary choice model of R&D collaborations as given by Eq. (16.15). The binary dependent variable corresponds to observed collaborations between two organizations (i) and

(j), taking a value of one if they collaborate and zero otherwise. The independent variables are geographical separation variables, variables capturing FP experience of the organizations and relational effects (joint history and network effects). We estimate three model versions. The standard model includes one variable for geographical effects and FP experience, respectively, and two variables accounting for relational effects. In the extended model version we add country border effects as an additional geographical variable, in order to isolate country border effects from geographical distance effects, and openness to FP research as an additional FP experience variable. The full model additionally includes balance variables accounting for FP experience and network effects, respectively.

Table 16.3 presents the sample estimates derived from maximum likelihood estimation for the model versions. The number of observations is equal to 36,481; asymptotic standard errors are given in parentheses. The statistics given in Table 16.4 indicate that the selected covariates show a quite high predictive ability. The Goodman–Kruskal–Gamma statistic ranges from 0.769 for the basic and 0.782 for the extended model to 0.786 for the full model, indicating that more than 75% fewer errors are made in predicting interorganizational collaboration choices by using the estimated probabilities than by the probability distribution of the dependent variable alone. The Somers (D) statistic and the (C) index confirm these findings. The Nagelkerke (R)-Squared is 0.391 for the basic model, 0.395 for the extended model, and 0.397 for the full model version.[9] A likelihood ratio test for the null hypothesis of $(\beta_k = 0)$ yields a (χ^2_4) test statistic of 2,565.165 for the basic model, a (χ^2_6) test statistic of 2,582.421 for the extended model, and a (χ^2_8) test statistic of 2,597.911 for the full model. These are statistically significant and we reject the null hypothesis that the model parameters are zero for all model versions.

The model reveals some promising empirical insight in the context of the relevant literature on innovation as well as on social networks. The results provide a fairly remarkable confirmation of the role of geographical effects, FP experience effects and network effects for interorganizational collaboration choice in EU FP R&D networks. In general, the parameter estimates are statistically significant and quite robust over different model versions.

The results of the basic model show that geographical distance between two organizations significantly determines the probability of collaboration. The parameter estimate of $(\beta_1 = -0.145)$ indicates that for any additional

9) Nagelkerke's R-squared is an attempt to imitate the interpretation of multiple R-squared measures from linear regressions based on the log-likelihood of the final model versus log likelihood of the null model. It is defined as $\left(R^2_{\text{Nag}} = \left[1 = (L_0/L_1)^{2/n}\right] / \left[1 - L_0^{2/n}\right]\right)$ where (L_0) is the log-likelihood of the null model, (L_1) is the log-likelihood of the model to be evaluated, and (n) is the number of observations.

Table 16.3 Maximum-likelihood estimation results for the collaboration model based on $\binom{n}{2} = 36{,}481$ observations. Asymptotic standard errors are given parenthetically.

Coefficient	Basic model	Extended model	Full model
β_0	−1.882 (***) (0.313)	−1.951 (***) (0.342)	−1.816 (***) (0.385)
β_1	−0.145 (***) (0.038)	−0.116 (***) (0.039)	−0.128 (***) (0.040)
β_2	—	−0.103 (***) (0.034)	−0.094 (**) (0.034)
β_3	−1.477 (***) (0.110)	−1.465 (***) (0.114)	−1.589 (***) (0.117)
β_4	—	0.004 (***) (0.001)	0.003 (***) (0.001)
β_5	—	—	0.001 (0.000)
β_6	4.224 (***) (0.089)	4.189 (***) (0.089)	4.194 (***) (0.089)
β_7	0.161 (***) (0.023)	0.135 (***) (0.025)	0.119 (***) (0.027)
β_8	—	—	0.070 (**) (0.025)

(***) Significant at the 0.001 significance level.
(**) Significant at the 0.01 significance level.
(*) Significant at the 0.05 significance level.

Table 16.4 Performance of the three collaboration model versions based on $\binom{n}{2} = 36{,}481$ observations.

Performance	Basic model	Extended model	Full model
Somers (D)	0.733	0.746	0.753
Goodman–Kruskal–Gamma	0.769	0.782	0.786
(C) index	0.876	0.873	0.875
Nagelkerke (R)-squared index	0.391	0.395	0.397
Log-likelihood	−2,190.151	−2,176.768	−2,169.578
Likelihood ratio test	2,565.156 (***)	2,582.421 (***)	2,597.911 (***)

(***) Significant at the 0.001 significance level.
(**) Significant at the 0.01 significance level.
(*) Significant at the 0.05 significance level.

100 km between two organizations the mean collaboration frequency decreases by about 15.6%. Geographical effects matter, but effects of the FP experience of organizations are more important. As evidenced by the estimate ($\beta_3 = -1.477$) it is most likely that organizations choose partners that are located closely in thematic space. A 1% increase in thematic distance reduces the probability of collaboration by more than 3.25%. The most important determinants of collaboration choice are network effects. The estimate of ($\beta_6 = 4.224$) tells us that the probability of collaboration between two organizations increases by 68.89% when they are prior acquaintances. Also network embeddedness matters as given by the estimate for ($\beta_7 = 0.161$), indicating that choice of collaboration is more likely between organizations that are central players in the network with respect to betweenness centrality.

Turning to the results of the extended model version it can be seen that taking into account country border effects decreases geographical distance effects by about 24% (($\beta_1 = -0.116$)). The existence of a country border between two organizations has a significant negative effect on their collaboration probability; the effect is slightly smaller than geographical distance effects (($\beta_2 = -0.103$)). Adding openness to FPs as an additional variable to capture FP experience does not influence the other model parameters much. Openness to FPs, though statistically significant, shows only a small impact on collaboration choice.

In the full model version we add one balance variable accounting for FP experience and network effects. The difference in the number of submitted FP projects has virtually no effect on the choice of collaboration, as given by the estimate of (β_5). An interesting result from a social network analysis perspective provides the integration of the difference between two organizations with respect to the clustering coefficient. The estimate of ($\beta_8 = 0.070$) tells us that it is more likely that two organizations collaborate when the difference of their clustering coefficients is higher. This result points to the existence of strategic collaboration choices for organizations that are highly cross-linked searching for organizations to collaborate with lower clustering coefficients, and the other way round. The effect is statistically significant but smaller than other network effects and geographical effects.

16.8 Summary

We have presented an investigation of networks derived from the European Union's Framework Programs for Research and Technological Development. The networks are of substantial size, complexity, and economic importance. We have attempted to provide a coherent picture of the complete process, beginning with data preparation and network definition, then continuing with analysis of the network structure and modeling of network formation.

We first considered the challenges involved in dealing with a large amount of imperfect data, detailing the tradeoffs made to clean the raw data into a usable form under finite resource constraints. The processed data were then used to define bipartite networks with vertices consisting of all the projects and organizations involved in each FP. To provide alternative views of the data, we defined projection networks for each part (organizations or projects) of the bipartite networks. Additionally, we used results of a survey of FP5 participants to define a smaller network about which we have more detailed information than we have for the networks as a whole.

Next we examined the structural properties of the bipartite and projection networks. We found that the vertex degrees in the FP networks have a highly skewed, heavy tailed distribution. The networks further show characteristic features of small-world networks, having both high clustering coefficients and short average path lengths. We followed this with an analysis of the community structure of the Framework Programs. Using a modularity measure and search algorithm adapted to bipartite networks, we identified communities from the networks and found that the communities were topically differentiated based on the standardized subject indices for Framework Program projects.

In the final stage of analysis, we constructed a binary choice model to explore determinants of interorganizational collaboration choice. The model parameters were estimated using logistic regression. The model results show that geographical effects matter but are not the most important determinants. The strongest effect comes from relational characteristics, in particular prior acquaintance, and, to a minor extent, network centrality. Also, thematic similarity between organizations highly favors a partnership.

By using a variety of networks and analyses, we have been able to address several different questions about the Framework Programs. The results complement one another, giving a more complete picture of the Framework Programs than the results from any one method alone. We are confident that our understanding of collaborative R&D in the European Union can be improved by extending the analyses presented in this chapter and by expanding the types of analyses we undertake.

Acknowledgment

The authors gratefully acknowledge financial support from the European FP6-NEST-Adventure Program, under Contract No. 028875 (Project NEMO: Network Models, Governance, and R&D Collaboration Networks).

References

1 R. Albert and A.-L. Barabási. Statistical mechanics of complex networks. *Reviews of Modern Physics*, 74(1):47, 2002.

2 J.A. Almendral, J.G. Oliveira, L. López, M.A.F. Sanjuán, and J.F.F. Mendes. The interplay of universities and industry through the fp5 network. *New Journal of Physics*, 9(183), 2007.

3 L. Angelini, S. Boccaletti, D. Marinazzo, M. Pellicoro, and S. Stramaglia. Identification of network modules by optimization of ratio association. *Chaos: An Interdisciplinary Journal of Nonlinear Science*, 17(2):023114, 2007.

4 N. Arranz and J.C. Fernández de Arroyabe. The choice of partners in R&D cooperation: An empirical analysis of Spanish firms. *Technovation*, 28:88–100, 2008.

5 M.J. Barber. Modularity and community detection in bipartite networks. *Physical Review E (Statistical, Nonlinear, and Soft Matter Physics)*, 76(6):066102, 2007.

6 M.J. Barber, M. Faria, L. Streit, and O. Strogan. Searching for communities in bipartite networks. In Christo-

pher C. Bernido and M. Victoria Carpio-Bernido (eds.), *Proceedings of the 5th Jagna International Workshop: Stochastic and Quantum Dynamics of Biomolecular Systems*. AIP, 2008.

7 M.J. Barber, A. Krueger, T. Krueger, and T. Roediger-Schluga. Network of European Union–funded collaborative research and development projects. *Physical Review E (Statistical, Nonlinear, and Soft Matter Physics)*, 73(3):036132, 2006.

8 S. Breschi and L. Cusmano. Unveiling the texture of a European Research Area: Emergence of oligarchic networks under EU Framework Programmes. *International Journal of Technology Management*, 27(8):747–772, 2004.

9 C. Christensen and R. Albert. Using graph concepts to understand the organization of complex systems. *International Journal of Bifurcation and Chaos*, 17(7):2201–2214, 2007. Special Issue "Complex Networks' Structure and Dynamics".

10 Fan Chung and Linyuan Lu. Connected components in random graphs with given expected degree sequences. *Annals of Combinatorics*, 6(2):125–145, 2002.

11 A. Clauset, M.E.J. Newman, and C. Moore. Finding community structure in very large networks. *Physical Review E (Statistical, Nonlinear, and Soft Matter Physics)*, 70(6):066111, 2004.

12 R. Cowan and N. Jonard. Network structure and the diffusion of knowledge. *Journal of Economic Dynamics and Control*, 28(8):1557–1575, June 2004.

13 L. Danon, A. Díaz-Guilera, J. Duch, and A. Arenas. Comparing community structure identification. *J. Stat. Mech.*, P09008, 2005.

14 W. deNooy, A. Mrvar, and V. Batagelj. *Exploratory Social Network Analysis with Pajek*. Cambridge University Press, 2004.

15 S.N. Dorogovtsev and J.F.F. Mendes. The shortest path to complex networks. In N. Johnson, J. Efstathiou, and F. Reed-Tsochas (eds.), *Complex Systems and Interdisciplinary Science*. World Scientific, 2004.

16 J. Fagerberg, D.C. Mowery, and R.R. Nelson. *The Oxford Handbook of Innovation*. Oxford University Press, 2005.

17 S. Fortunato and M. Barthelemy. Resolution limit in community detection. *PNAS*, 104(1):36–41, 2007.

18 A.L. Gibbs and F.E. Su. On choosing and bounding probability metrics. *International Statistical Review*, 70(3):419–435, 2002.

19 M. Girvan and M.E.J. Newman. Community structure in social and biological networks. *PNAS*, 99(12):7821–7826, 2002.

20 V. Gol'dshtein and G.A. Koganov. An indicator for community structure. Preprint, July 2006.

21 R. Guimerà, M. Sales-Pardo, and L.A. Nunes Amaral. Module identification in bipartite and directed networks. *Physical Review E (Statistical, Nonlinear, and Soft Matter Physics)*, 76(3):036102, 2007.

22 M.B. Hastings. Community detection as an inference problem. *Physical Review E (Statistical, Nonlinear, and Soft Matter Physics)*, 74(3):035102, 2006.

23 J. Johnston and J. Dinardo. *Econometric Methods*. McGraw-Hill, Inc., New York, 2007.

24 M.E.J. Newman. Detecting community structure in networks. *Eur. Phys. J. B*, 38:321–330, 2004.

25 M.E.J. Newman. Fast algorithm for detecting community structure in networks. *Physical Review E (Statistical, Nonlinear, and Soft Matter Physics)*, 69(6):066133, 2004.

26 M.E.J. Newman and M. Girvan. Finding and evaluating community structure in networks. *Physical Review E (Statistical, Nonlinear, and Soft Matter Physics)*, 69(2):026113, 2004.

27 M.E.J. Newman and E.A. Leicht. Mixture models and exploratory data analysis in networks. *PNAS*, 104(23):9564–9569, 2007.

28 M.E.J. Newman. The structure and function of complex networks. *SIAM Review*, 45:167–256, 2003.

29 M. Paier and T. Scherngell. Determinants of collaboration in European R&D networks: Empirical evidence from a binary choice model perspective. SSRN Working Paper Series No. 1120081, Rochester, NY, 2008

30 G. Palla, I. Derenyi, I. Farkas, and T. Vicsek. Uncovering the overlapping community structure of complex networks in nature and society. *Nature*, 435:814–818, June 2005.

31 J.M. Pujol, J. Bejar, and J. Delgado. Clustering algorithm for determining community structure in large networks. *Physical Review E (Statistical, Nonlinear, and Soft Matter Physics)*, 74(1):016107, 2006.

32 J. Reichardt and S. Bornholdt. Statistical mechanics of community detection. *Physical Review E (Statistical, Nonlinear, and Soft Matter Physics)*, 74(1):016110, 2006.

33 T. Roediger-Schluga and M.J. Barber. The structure of R&D collaboration networks in the European Framework Programmes. Working Paper 2006-036, UNU-MERIT, 2006.

34 T. Roediger-Schluga and M.J. Barber. R&D collaboration networks in the European Framework Programmes: Data processing, network construction and selected results. *IJFIP*, 4(3/4):321–347, 2008. Special Issue on "Innovation Networks".

35 A.J. Seary and W.D. Richards. Spectral methods for analyzing and visualizing networks: an introduction. In R. Breiger, K. Carley, and P. Pattison (eds.), *Dynamic Social Network Modeling and Analysis: Workshop Summary and Papers*, pp. 209–228, The National Academic Press, Washington, D.C., 2003.

36 S. Wasserman and K. Faust. *Social Network Analysis – Methods and Applications*. Cambridge University Press, 1994.

37 D.J. Watts and S.H. Strogatz. Collective dynamics of 'small-world' networks. *Nature*, 393:440–2, 1998.

17
Analytic Combinatorics on Random Graphs

Michael Drmota and Bernhard Gittenberger

17.1
Introduction

In this chapter we present some results on various types of random graphs that can be obtained by methods of analytic combinatorics. Our journey through random graphs starts with the simplest type, random trees, and goes on to more and more complex random graphs. The term "analytic combinatorics" summarizes the combinatorial and asymptotic analysis of certain properties of discrete structures that can be treated by means of a generating function; see the monograph by Flajolet and Sedgewick [24]. In fact, many combinatorial constructions have their counterpart in generating functions (GFs). In particular, if some strucure has a recursive description, then there is usually a GF approach to the counting problem for this structure. Trees, in particular rooted trees, are one of the most prominent structures that can be analyzed in that way. There are at least two major advantages of a GF approach. First, it is usually not only possible to count the number of objects of a given size but several parameters at the same time by introducing several variables. Second, GFs can be considered as analytic objects, more precisely as analytic functions, so that asymptotic methods like saddle point methods or a singularity analysis provide asymptotic expansions for the coefficients and also probabilistic limiting distributions.

This chapter is organized as follows. In Section 17.2 we discuss (rooted) trees. From a graph-theoretic viewpoint, trees have a very simple structure. Therefore, a wealth of information is known about them and we had to make a choice. We decided to focus on the class of simply generated trees that can be viewed as realizations of Galton–Watson branching processes. Furthermore, we chose four shape characteristics of trees, namely, degree distribution, height, profile, and width. The recursive structure of such trees allows for a translation into functional equations for the corresponding GFs that can be treated with analytic tools. In Section 17.3 we turn our attention to random mappings from a finite set into itself. Such mappings give rise to

Analysis of Complex Networks: From Biology to Linguistics. Edited by Matthias Dehmer and Frank Emmert-Streib

ISBN: 978-3-527-32345-6

directed graphs that themselves can be decomposed into trees arranged in cycles. Due to this decomposition, random mappings are a natural generalization of trees. In view of analytic combinatorics the study of random mappings is similar to that of trees. As an example we briefly discuss the degree distribution in random mappings. The next section deals with the random graph model of Erdős and Rényi, which is the classical model in random graph theory. One of the most exciting events is the emergence of a giant component when random graphs become more and more dense and that appears – at first glance – to occur suddenly. It turned out that a different scaling than the one used originally by Erdős and Rényi allows a precise observation of this phase transition. As a preparation for the study of the phase transition, we present Wright's method to count connected graphs. Then we give a brief overview of the analysis of the emergence of the giant component where we focus on the analytic arguments used there. The last section deals with various classes of planar graphs. Here the basic counting problem is already very difficult. We present the functional equations relating the GFs for 2-connected, connected, and all planar graphs. These lead to the numbers of such graphs. Finally, we discuss the degree distribution for random planar graphs.

17.2
Trees

Trees are a fundamental object in graph theory and combinatorics as well as a basic object for data structures and algorithms in computer science. In recent years research related to (random) trees has been constantly increasing and several asymptotic and probabilistic techniques have been developed to describe characteristics of interest of large trees in different settings.

A basic class of rooted trees are *planted plane trees*. Starting from the root, every node has an arbitrary number of successors with a natural left to right order (Figure 17.1). In particular, the subtrees of the root vertex are again planted plane trees.

This example is also very instructive in order to give a flavor of analytic combinatorics. Let $\mathcal{P}$ denote the set of planted plane trees. Then from the

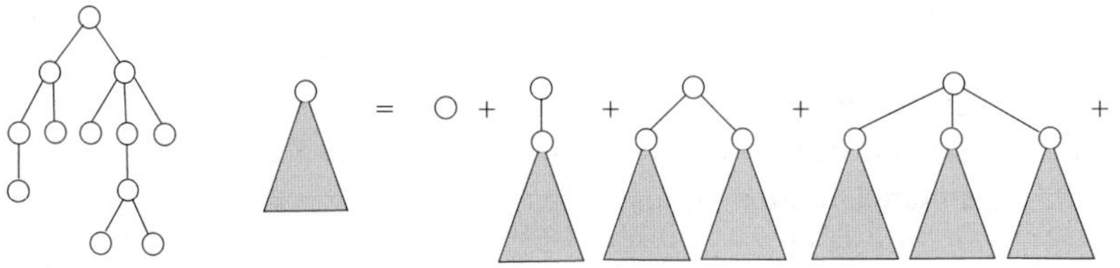

Figure 17.1 Planted plane tree.

above description we obtain the recursive relation

$$\mathcal{P} = \circ + \circ \times \mathcal{P} + \circ \times \mathcal{P}^2 + \circ \times \mathcal{P}^3 + \cdots$$

(see again the schematic Figure 17.1). With the GF

$$p(z) = \sum_{n\geq 1} p_n z^n$$

this translates to

$$p(z) = z + z\,p(z) + z\,p(z)^2 + z\,p(z)^3 = \frac{z}{1-p(z)}\,. \tag{17.1}$$

Hence

$$p(z) = \frac{1-\sqrt{1-4z}}{2}\,, \tag{17.2}$$

and consequently

$$p_n = \frac{1}{n}\binom{2n-2}{n-1} \sim \frac{4^{n-1}}{\sqrt{\pi}n^{3/2}}\,.$$

Note that $z_0 = 1/4$ is the dominating singularity of $p(z)$ and the singular behavior of $p(z)$ is given by (17.2), which can also be used to obtain the asymptotic expansion of p_n.

Interestingly enough, there is an intimate relation to Galton–Watson branching processes. A Galton–Watson process is a discrete stochastic process $(Z_0, Z_1, \ldots)$ that can be defined as follows. Consider "particles" that can give birth to a random number of "children." The number of children is governed by the so-called offspring distribution ξ, and all particles are assumed to behave independently and with identical offspring distributions. More precisely, $(Z_0, Z_1, \ldots)$ is given by $Z_0 = 1$, and for $k \geq 1$ by

$$Z_k = \sum_{j=1}^{Z_{k-1}} \xi_j^{(k)}\,,$$

where the $(\xi_j^{(k)})_{k,j}$ are iid random variables distributed as ξ. The random variable Z_i is precisely the number of particles in the ith generation. We also assume that $\mathbb{P}\{\xi = 0\} > 0$, so that eventually the process terminates and the resulting object (interpreted as a family tree) can be viewed as a rooted plane tree. The size of this tree is precisely $Z_0 + Z_1 + \cdots$, the so-called total progeny.

Simply generated random trees (according to Meir and Moon [32]) are precisely random rooted plane trees of size n, where the distribution is determined by a Galton–Watson branching process conditioned to have total progeny n. For example, if the offspring distribution ξ is a geometric distribution, then every tree of size n appears with equal probability. Thus, the probability model in this case is precisely the combinatorial counting model.

In the sequel we assume that an exponential moment of the offspring distribution exists. Under this assumption the offspring distribution ξ can be written in the form

$$\mathbb{P}\{\xi = k\} = \frac{\tau^k \varphi_k}{\varphi(\tau)}, \tag{17.3}$$

where $(\varphi_k; k \geq 0)$ is a sequence of nonnegative numbers such that $\varphi_0 > 0$ and $\varphi(t) = \sum_{k\geq 0} \varphi_k t^k$ has a positive or infinite radius of convergence R and τ is a positive number within the circle of convergence of $\varphi(t)$. In particular, these conditions imply that all moments of ξ exist. Without loss of generality we assume that the Galton–Watson process is critical, that is, we have $\mathbb{E}\xi = 1$ which equivalently means that τ satisfies $\tau\varphi'(\tau) = \varphi(\tau)$. The variance of ξ can also be expressed in terms of $\varphi(t)$ and is given by

$$\sigma^2 = \frac{\tau^2 \varphi''(\tau)}{\varphi(\tau)}. \tag{17.4}$$

Note that the offspring distribution (17.3) can be interpreted as assigning weights to all trees defined by

$$\omega(T) = \prod_{k\geq 0} \varphi_k^{n_k(T)}$$

for a tree T having n nodes, n_k of which have out-degree k, $k \geq 0$. Denote by $|T|$ the number of nodes of such a tree and let y_n be the (weighted) number of all trees with n nodes, that is,

$$y_n = \sum_{T:|T|=n} \omega(T).$$

Then the corresponding GF

$$y(z) = \sum_{n\geq 0} y_n z^n$$

satisfies the functional equation

$$y(z) = z\varphi(y(z)).$$

Obviously, the tree class discussed at the beginning of this chapter, namely, planted plane trees, is a special case of simply generated trees: the set $\phi(t) = 1/(1-t)$ returns the functional Equation (17.1). In the aperiodic case (i.e., if $\gcd\{i|\varphi_i > 0\} = 1$), which we will assume for the rest of this chapter, $y(z)$ has only one singularity ϱ on the circle of convergence. Then it can be shown that $y(z)$ admits the asymptotic expansion

$$y(z) = \tau - \frac{\tau\sqrt{2}}{\sigma}\sqrt{1-\frac{z}{\varrho}} + O\left(\left|1-\frac{z}{\varrho}\right|\right).$$

The singularity analysis of Flajolet and Odlyzko [21] is a powerful method for obtaining asymptotic expansions for the coefficients of power series, if their behavior near the singularites on the circle of convergence is known (as asymptotic expansion in terms of "simple" functions). Using this method we get the asymptotic number of simply generated trees $y_n \sim (\tau/\sqrt{2\pi\sigma^2})\varrho^{-n}n^{-3/2}$.

17.2.1
The Degree Distribution

As a first application of the GF approach we describe the degree distribution of simply generated resp. Galton–Watson trees. Let $X_n^{(k)}$ denote the number of nodes of out-degree k in a random simply generated tree of size n; the out-degree of a node is the number of successors. Then the corresponding GF

$$y_k(x,u) = \sum_{n\ge1} y_n \mathbb{E}\, u^{X_n^{(k)}} z^n$$

satisfies the functional equation

$$y_k(x,u) = x(u-1)\varphi_k y_k(x,u)^k + x\varphi(y_k(x,u)).$$

If follows, then, that $y_k(x,u)$ has a local expansion of the form

$$y_k(z,u) = g(u) - h(u)\sqrt{1-\frac{z}{\varrho(u)}} + O\left(\left|1-\frac{z}{\varrho(u)}\right|\right)$$

that directly leads to a central limit theorem for $X_n^{(k)}$, that is,

$$\frac{X_n^{(d)} - \mathbb{E}X_n^{(d)}}{\mathbb{V}\text{ar}\, X_n^{(k)}} \xrightarrow{w} N(0,1),$$

where the expected value and variance are asymptotically proportional to n (cf. [15]. For example, we have

$$\mathbb{E}\, X_n^{(k)} = \frac{\varphi_k\tau^k}{\varphi(\tau)}n + O(1).$$

This shows that the offspring distribution of the critial Galton–Watson branching process (which is related to the simply generated tree family) can be recovered just by looking at the degree distribution statistics.

17.2.2 The Height

A very important paramater of rooted trees is the height, that is, the maximal distance between a node and the root. The first contribution to the height of simply generated trees was made by de Bruijn, Knuth, and Rice [12], who considered the special case of planted plane trees ($\varphi(t) = 1/(1-t)$). The height of simply generated trees has been studied by Flajolet *et al.* [22, 23]. They showed the following results, where we again assume that an exponential moment of the offspring distribution exists.

Theorem 17.1 (Flajolet and Odlyzko [23]) *Let H_n denote the height of a simply generated random tree with n vertices. Then we have*

$$\mathbb{E}(H_n^r) = 2^{r/2}\sigma^{-r} r(r-1)\Gamma(r/2)\zeta(r) \cdot n^{r/2}\left(1 + O(n^{-\frac{1}{4}+\eta})\right),$$

where $\zeta(s)$ denotes the Riemann zeta function and $(r-1)\zeta(r) = 1$ for $r = 1$ and η is any positive number.

If $y_n^{(h)}$ denotes the weighted number of simply generated trees with n vertices and height equal to h, then $\mathbb{P}\{H_n = h\} = y_n^{(h)}/y_n$.

Theorem 17.2 (Flajolet *et al.* [22]) *Let $\delta > 0$ and $\beta = 2\sqrt{n}/h$. Then, as $n \to \infty$, we have*

$$\mathbb{P}\{H_n = h\} = \frac{y_n^{(h)}}{y_n} \sim 4b\sqrt{\frac{\varrho\pi^5}{n}}\beta^4 \sum_{m\ge 1} m^2(2(m^2\pi^2\beta^2 - 3)e^{-m^2\pi^2\beta^2} \tag{17.5}$$

uniformly for $\frac{1}{\delta\sqrt{\log n}} \le \frac{h}{\sqrt{n}} \le \delta\sqrt{\log n}$.

The assumption that an exponential moment of the offspring distribution exists can be weakened to the assumption that just the second moment of the offspring distribution exists. For example, this can be deduced from the concept of continuum random trees that was introduced by Aldous [1]. A continuum random tree is, in a proper sense, the weak limit of scaled Galton–Watson trees. Since the height is a continuous functional (in this context), one directly gets a weak limit theorem for the height.

The approach of Aldous is quite general, but it does not give an error term. The only known method that provides an error term is that of Flajolet and Odlyzko [23] and is based on GFs. We quickly sketch their approach.

Let $y_k(z)$ denote the GFs

$$y_h(z) = \sum_{n\ge 1} y_n^{(h)} z^n.$$

Then

$$y_0(z) = \varphi_0 z, \quad y_{k+1}(z) = z\varphi(y_k(z)), \qquad (k \geq 0).$$

A subtle analysis of the above recurrence yields (with $\alpha(z) = z\varphi'(z)$)

$$y(z) - y_k(z) = \frac{2\tau}{\sigma^2} \cdot \frac{\alpha(z)^k}{\alpha(z)^k/(1-\alpha(z)^k) + \left(\log|1/(1-\alpha(z))|\right)},$$

and by singularity analysis it is then possible to extract the coefficients of $y(z) - y_k(z)$, which eventually leads to (17.5).

17.2.3 The Profile

The height gives a first impression of the "expected shape" of a rooted tree. The so-called profile of a rooted tree provides much more precise shape characteristics.

Consider a rooted tree T of size n. We denote by $L_T(k)$ the number of nodes of T at distance k from the root. The sequence $(L_T(k))_{k\geq 0}$ is called the profile of T. If T is a random tree of size n, for example a simply generated tree, then we denote the profile by $(L_n(k))_{k\geq 0}$, which is now a sequence of random variables.

It is convenient to consider the continuous version of this stochastic process obtained by defining the values for noninteger arguments by linear interpolation, that is, we set

$$L_n(t) = (\lfloor t \rfloor + 1 - t)L_n(\lfloor t \rfloor) + (t - \lfloor t \rfloor)L_n(\lfloor t \rfloor + 1), \quad t \geq 0.$$

Of course, $(L_n(t))_{t\geq 0}$ is a stochastic process and is called the profile process of a random tree. By definition, the sample paths of the profile are continuous functions on $[0, \infty)$.

Since the height of a simply generated tree is of order $\sqrt{n}$, we consider the rescaled process

$$l_n(t) = \frac{1}{\sqrt{n}} L_n(t\sqrt{n}), \quad t \geq 0.$$

This process describes in some sense the local behavior of the contour process obtained by recording the depth during a depth-first search traversal of the tree. Aldous [1] showed that the contour process converges weakly to a (properly scaled) Brownian excursion. The standard Brownian excursion $e(t)$ is a one-dimensional Brownian motion W_t on the unit interval $[0, 1]$ and conditioned to be nonnegative with zeroes only at $t = 0$ and $t = 1$. The convergence of the contour to $e(t)$ led to the conjecture ([1], Conjecture 4) that $l_n(t)$

weakly converges to the (properly rescaled) total local time of the Brownian excursion, which is the process defined by

$$l(t) = \lim_{\varepsilon \to 0} \frac{1}{\varepsilon} \int_0^1 I_{[t,t+\varepsilon]}(e(s))\, ds\,,$$

where $I_{[t,t+\varepsilon]}$ denotes the indicator function of the interval $[t, t+\varepsilon]$. This conjecture was first proved in [16]:

Theorem 17.3 *Let $\varphi(t)$ be the defining series of a family of simply generated random trees. Furthermore, suppose that the equation $t\varphi'(t) = \varphi(t)$ has a minimal positive solution $\tau < R$ and that σ^2 defined by (17.4) is finite. Then the process $l_n(t)$ converges weakly to Brownian excursion local time, that is,*

$$l_n(t) \xrightarrow{w} \frac{\sigma}{2} l\left(\frac{\sigma}{2} t\right)$$

in $C([0,\infty))$, as $n \to \infty$.

Remark 17.1 The weak convergence above is the usual weak convergence of probability measures on the space $C([0,\infty))$ equipped with the supremum norm $\|.\|_\infty$ (for details see [6]).

We will now lay out the idea of the proof of the above theorem. The density of the distribution of $l(\varrho)$ for some fixed level ϱ is well studied. Several approaches are used in the literature that lead to different representations of this density. We use the one obtained by Cohen and Hooghiemstra [11]. If $f_\varrho(x) = \frac{d}{dx}\mathbb{P}\{l(\varrho) \le x\}$, then we have

$$f_\varrho(x) = \frac{1}{i\sqrt{2\pi}} \int_\gamma \frac{-se^{-s}}{\sinh^2(\varrho\sqrt{-2s})} \exp\left(-\frac{x}{\sqrt{2}} \frac{\sqrt{-s}e^{\varrho\sqrt{-2s}}}{\sinh(\varrho\sqrt{-2s})}\right) ds\,, \tag{17.6}$$

where γ is the straight line $\{z : \Re z = -1\}$.

Remark 17.2 Note that this equation describes only the continuous part of the local time density. Since $\mathbb{P}\{\sup_{0\le t\le 1} W_t < \varrho\} > 0$, there is a jump of this magnitude at 0. This quantity is well known (see (17.10)).

In order to prove Theorem 17.3, two facts have to be shown (cf. [6], Theorem 12.3). First, weak convergence of the finite dimensional distributions (fdd's) of $l_n(t)$ to those of Brownian excursion local time, that is,

$$(l_n(t_1), \ldots, l_n(t_d)) \xrightarrow{w} (l(t_1), \ldots, l(t_d)),$$

for any choice of d and $t_1, \ldots, t_d$. Second, the process must be tight, which means, roughly speaking, that its sample paths have only moderate variation. Tightness proofs are necessary to complete functional limit theorems but are usually very technical. We will therefore confine ourselves to stating a sufficient condition for tightness, namely, that one has to show that there exists a constant $C > 0$ such that

$$\mathbb{E}\left(L_n(r) - L_n(r+h)\right)^4 \le C h^2 n \tag{17.7}$$

holds for all nonnegative integers n, r, h. So let us now turn to the weak limit theorem.

Consider a random tree T and set

$$y_{dmn} = \sum_{T: L_n(d)=m} \omega(T).$$

Thus the distribution of $L_n(d)$ is given by $\mathbb{P}\{L_n(d) = m\} = y_{dmn}/y_n$. The GF of this sequence satisfies

$$\sum_{n,m\ge 0} y_{dmn} u^m z^n = y_d(z, u y(z)),$$

where

$$y_0(z,u) = u, \quad y_{i+1}(z,u) = z\varphi(y_i(z,u)), \quad i \ge 0. \tag{17.8}$$

In order to prove the weak limit theorem (Theorem 17.3) it suffices to obtain enough knowledge of the characteristic function of the distribution of $L_n(d)$ that is encoded in the GF. In fact, the characteristic function of $L_n(k)/\sqrt{n}$ is

$$\phi_{kn}(t) = \frac{1}{y_n}[z^n] y_k\left(z, e^{it/\sqrt{n}} y(z)\right)$$

and that of the fdd, that is, the distribution of $\left(L_n(k_1)/\sqrt{n}, \ldots, L_n(k_p)/\sqrt{n}\right)$, is

$$\phi_{k_1\cdots k_p n}(t_1, \ldots, t_p) =$$
$$\frac{1}{y_n}[z^n] y_{k_1}\left(z, e^{it_1/\sqrt{n}} y_{k_2-k_1}\left(z, \ldots y_{k_p-k_{p-1}}\left(z, e^{it_p/\sqrt{n}} y(z)\right)\ldots\right)\right).$$

In order to extract the desired coefficient, we will use Cauchy's integral formula with a suitably chosen integration contour and approximate the integrand there. Therefore, we need a detailed knowledge of the behavior of the recursion (17.8). Note that $y_k(z, y(z)) = y(z)$, and knowing the behavior of the error $w_k(z,u)$ in $y_k(z,u) = y(z) + w_k(z,u)$ is enough. It is possible to show the following lemma.

Lemma 17.1 *Set* $z = \varrho\left(1 + \frac{x}{n}\right)$ *and* $\alpha(z) = z\varphi'(y(z))$. *Furthermore, assume that* $|u - y(z)| = O\left(\frac{1}{\sqrt{n}}\right)$ *and* $\frac{x}{n} \to 0$ *in such a way that* $|\arg(-x)| < \pi$ *and* $|1 - \sqrt{-x/n}| = 1 + O\left(\sqrt{n}\right)$ *are satisfied. Then we have*

$$w_k(z,u) = \frac{2\sqrt{-x/n}(u - y(z))\alpha(z)^k}{(1+\alpha(z)^k)\sqrt{-x/n} + (\tau - u)(1 - \alpha(z)^k)\sigma/\tau\sqrt{2} + O(|x|/n)},$$

uniformly for $k = O\left(\sqrt{n}\right)$.

By means of this lemma it is possible to derive the characteristic functions of the limiting distributions $L_n(k)/\sqrt{n}$ as well as of the multivariate sequence of random variables $\left(L_n(k_1)/\sqrt{n}, \ldots, L_n(k_p)/\sqrt{n}\right)$:

Theorem 17.4 *Let $k_i = \kappa_i\sqrt{n}, i = 1, \ldots, p$ where $0 < \kappa_1 < \cdots < \kappa_p$. Then the characteristic function $\phi_{\kappa_1\ldots\kappa_p}(t_1, \ldots, t_p) = \lim_{n\to\infty} \phi_{k_1\cdots k_p n}(t_1, \ldots, t_p)$ of the limiting distribution of $\left(\frac{1}{\sqrt{n}}L_n(k_1), \ldots, \frac{1}{\sqrt{n}}L_n(k_p)\right)$ satisfies*

$$\phi_{\kappa_1\ldots\kappa_p}(t_1, \ldots, t_p) = 1 + \frac{\sigma}{i\sqrt{2\pi}} \int_\gamma f_{\kappa_1,\ldots,\kappa_p,\sigma}(x, t_1, \ldots, t_p)e^{-x}\,dx\,,$$

where

$$f_{\kappa_1,\ldots,\kappa_p,\sigma}(x, t_1, \ldots, t_p) = \Psi_{\kappa_1\sigma}(x, it_1 + \Psi_{\kappa_2-\kappa_1,\sigma}(\ldots \Psi_{\kappa_{p-1}-\kappa_{p-2},\sigma}(x, it_{p-1} + \Psi_{\kappa_p-\kappa_{p-1},\sigma}(x, it_p))\cdots)$$

with

$$\Psi_{\kappa\sigma}(x, t) = \frac{t\sqrt{-x}e^{-\kappa\sigma\sqrt{-x/2}}}{\sqrt{-x}e^{\kappa\sigma\sqrt{-x/2}} - t\frac{\sigma}{\sqrt{2}}\sinh\left(\kappa\sigma\sqrt{-\frac{x}{2}}\right)}\,,$$

and γ is the Hankel contour $\gamma_1 \cup \gamma_2 \cup \gamma_3$ defined by

$$\gamma_1 = \{s||s| = 1 \text{ and } \Re s \le 0\}\,, \quad \gamma_2 = \{s|\Im s = 1 \text{ and } \Re s \ge 0\}\,, \quad \gamma_3 = \overline{\gamma}_2\,.$$

This limiting distribution has to be identified as the characteristic function of Brownian excursion local time, which was computed in [11] for $p = 1$ (cf. (17.6)) and $p = 2$. One way to do this is to compute the characteristic function of Brownian excursion occupation time for the sets $[\kappa_1, \kappa_1+\eta]\cup\cdots\cup[\kappa_p, \kappa_p+\eta]$ and do a kind of differentiation process afterwards. Another approach would be the use of Itô's excursion theory by computing the expected value of a suitably chosen random variable with respect to Itô's measure and then taking the inverse Laplace transform. Both approaches are presented in [16].

17.2.4
The Width

We are interested in the width of such a tree, which is defined by

$$w_n = \max_{t\ge 0} L_n(t).$$

This quantity has attracted the interest of many authors. First, Odlyzko and Wilf [33] became interested in this tree parameter when studying the bandwidth of a tree. Regarding the width, they showed that there are positive constants c_1 and c_2 such that

$$c_1\sqrt{n} < \mathbb{E}w_n < c_2\sqrt{n\log n} \tag{17.9}$$

holds. The exact order of magnitude was left as an open problem.

Biane and Yor [5] showed that $\sup_{t\geq 0} l(t)$ and $2\sup_{0\leq t\leq 1} W_t$ have the same distribution, and due to Kennedy [29] this is

$$\mathbb{P}\{\sup_{0\leq t\leq 1} W_t \leq x\} = 1 - 2\sum_{k\geq 1}(4x^2k^2 - 1)e^{-2x^2k^2}\,. \tag{17.10}$$

Hence Theorem 17.3 implies $\sup_{t\geq 0} l_n(t) \xrightarrow{w} \sigma \sup_{0\leq t\leq 1} W_t$ as $n \to \infty$. This suggests, but does not imply, $\sqrt{n}$ as the correct order of magnitude in (17.9). A moment convergence theorem of $\sup_{t\geq 0} l_n(t)$ to $\sup_{t\geq 0} l(t)$ was shown by Chassaing and Marckert [10] for Cayley trees and in [17] for all simply generated trees. Formulated in terms of the tree width $w_n = \max_{t\geq 0} L_n(t) = (\sigma/2)\sqrt{n}\,\sup_{t\geq 0} l_n(t)$ it reads as follows.

Theorem 17.5 *Suppose that there exists a minimal positive solution $\tau < R$ of $t\varphi'(t) = \varphi(t)$. Then the width w_n satisfies*

$$\mathbb{E}\left(w_n^p\right) = \sigma^p 2^{-p/2} p(p-1)\Gamma\left(\frac{p}{2}\right)\zeta(p)\cdot n^{p/2}\cdot(1 + o(1))$$

as $n \to \infty$.

A weak limit theorem for the joint law of height and width of simply generated trees was given by Chassaing, Marckert, and Yor [9].

The proof of Theorem 17.5 relies on the notion of polynomial convergence.

Definition 17.1 Let x_n be a sequence of stochastic processes in $C[0,\infty)$. Then we say that x_n is *polynomially convergent* to $x \in C([0,\infty))$ if for every continuous functional $F : C[0,\infty) \to \mathbb{R}$ of polynomial growth (i.e., $|F(y)| = O\left((1 + \|y\|_\infty)^r\right)$ for some $r \geq 0$) we have $\lim_{n\to\infty} \mathbb{E}F(x_n) = \mathbb{E}F(x)$.

Theorem 17.6 *Let x_n be a sequence of stochastic processes in $C[0,\infty)$ that converges weakly to x. Assume that for any choice of fixed positive integers p and d there exist positive constants c_0, c_1, c_2, c_3 such that*

$$\sup_{n\geq 0} \mathbb{E}|x_n(t)|^p \leq c_0 e^{-c_1 t} \text{ for all } t \geq 0\,,$$

and

$$\sup_{n\geq 0} \mathbb{E}|x_n(t+s) - x_n(t)|^{2d} \leq c_2 e^{-c_3 t} s^d \text{ for all } s, t \geq 0\,. \tag{17.11}$$

Then x_n is polynomially convergent to x.

The quite technical proof of this theorem is given in [17]. Theorem 17.5 is now a consequence of the weak limit theorem for the profile as well as the previous theorem. In order to show the exponential estimates needed to apply Theorem 17.6, one must conduct a careful singularity analysis similar to that used to prove tightness. Equation (17.11) resembles the sufficient condition for tightness (17.7) and is in fact a much stronger statement.

Finally, we mention that the method allows one to derive moment convergence theorems for other combinatorial problems as well. Examples are the degree distribution in layers of random trees, strata of random mappings, or the height of random trees (*cf.* [17]).

17.3
Random Mappings

While in the previous section we considered a particularly simple kind of random graphs, we now turn to slightly more complex graphs. Random mappings are mappings from an n element set M into itself where uniform distribution is assumed. Obviously, any such mapping gives rise to a directed graph: Draw n points in the plane and label them with the elements of M. Then draw a directed edge from i to j ($i, j \in M$) whenever the mapping maps i onto j. This graph is called the functional digraph of the mapping and allows an easy decomposition. Each component contains exactly one cycle (which may be a loop) and to each vertex of the cycle is attached a Cayley tree (i.e., a labeled rooted tree). So a functional digraph decomposes into a multiset of cycles of labeled tree (see Figure 17.2 for an example).

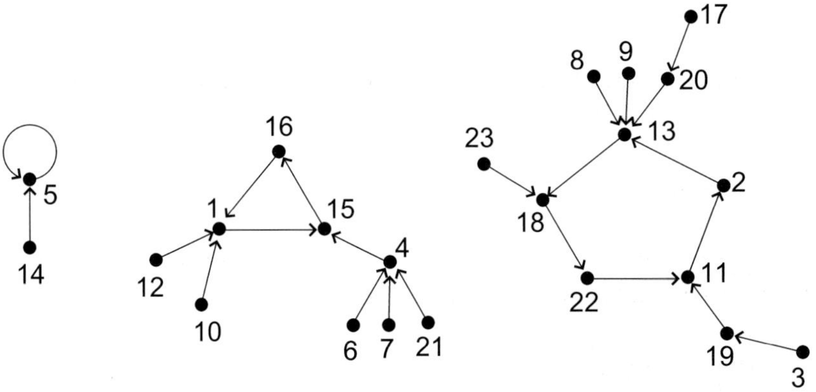

Figure 17.2 A random mapping on $M = \{1, 2, \ldots, 23\}$.

From this decomposition it is easy to derive the GF for the numbers of such mappings. It is well known (cf. [24]) that the multiset and cycle construction for labeled objects correspond to e^z and $\log(1/(1-z))$ for the exponential GFs. Hence the exponential GF is

$$A(z) = \exp\left(\log \frac{1}{1-a(z)}\right) = \frac{1}{1-a(z)},$$

where $a(z)$ is the GF for Cayley trees, given by the functional equation

$$a(z) = ze^{a(z)}. \tag{17.12}$$

Of course, elementary enumeration shows that $[z^n]A(z) = n^n/n!$ such that the above decomposition seems unnecessary. But it is useful to study more complex features of random mappings. As an example we will sketch the analysis of the degree distribution in random mappings that was first done by Arney and Bender [3] by enumerative arguments and later in a more general context by Drmota and Soria [18] via GFs. The analysis of the degree distribution amounts to a study of the in-degree distribution since obviously every vertex in a random mapping has an out-degree one. The bivariate GF Cayley trees encoding tree size and number of nodes of out-degree r satisfies the functional equation

$$a_r(z, u) = ze^{a_r(z,u)} + z(u-1)\frac{a_r(z, u)}{r!}.$$

When building functional digraphs in each tree one of the incoming edges of the root stems from the cycle, that is, from a node outside the tree. Taking this into account yields the GF (for mappings)

$$A_r(z, u) = \frac{1}{1 - a_r(z, u) + z(u-1)\left(\frac{a_r(z,u)}{r!} - \frac{a_{r-1}(z,u)}{(r-1)!}\right)}.$$

It can be shown that $a_r(z, u)$ has a single algebraic singularity. More precisely, it admits (locally at $(1/e, 1)$ a representation of the form

$$a_r(z, u) = g(z, u) - h(z, u)\sqrt{1 - \frac{z}{\varrho(u)}},$$

where $g(z, u)$ and $h(z, u)$ are analytic functions and $\varrho(u)$ is analytic at $u = 1$. The expected number of nodes with in-degree r is

$$\frac{n!}{n^n}[z^n]\left[\frac{\partial}{\partial u}A_r(z, u)\right]_{u=1}.$$

One easily obtains

$$\left[\frac{\partial}{\partial u}A_r(z, u)\right]_{u=1} \sim \frac{\varrho'(u)}{\varrho(u)}\frac{1}{2^{3/2}(1-ez)^{3/2}}.$$

Applying the transfer lemma ([21]) we obtain that the asymptotic number of nodes of in-degree r is $n/er!$. Even more can be shown:

Theorem 17.7 *Let* $\mu = 1/(er!)$ *and* $\sigma^2 = 1/(er!) + (r^2 - 2r + 2)/(er!)^2$. *Then the number of nodes with r preimages in a random mapping on an n-element set* *is asymptotically normally distributed with asymptotic mean μn and asymptotic variance $\sigma^2 n$. Moreover, a local limit theorem holds: Let a_{nk} denote the number of mappings on an n-element set with exactly k nodes of in-degree r. Then, as $n \to \infty$ and uniformly for $k \geq 0$, we have*

$$\frac{a_{nk}}{n^n} = \frac{1}{\sqrt{2\pi\sigma^2 n}} \left(\exp\left(\frac{(k-\mu n)^2}{2\sigma^2 n} \right) + O\left(n^{-1/4}\right) \right).$$

17.4 The Random Graph Model of Erdős and Rényi

In this section we want to present some results on the random graph model of Erdős and Rényi [19, 20]. This model is defined as follows. Given are n vertices with labels $1, \ldots, n$ and a probability p. Then each possible edge is included in the edge set of the graph with probability p, where all edges are treated independently. The resulting graph is denoted by $G(n, p)$. A second model is considering the set of all graphs with n vertices and m edges and choosing one graph uniformly at random. This graph is called $G(n, m)$ and it is known that the two models are equivalent in many respects, provided that $p\binom{n}{2}$ is close to m (see [28, Section 1.4] for details).

One interesting phenomenon of random graphs is the emergence of a giant component. If we let $n \to \infty$ and set $p = c/n$, then for $c < 1$ a typical graph consists of small and simple components, that is, each component has a typical size of $O(\log n)$ and does not contain many cycles. If $c > 1$, then a typical graph consists of one giant component that comprises roughly two thirds of all vertices and many small and simple components. The phase transition occurring around $c = 1$ was thoroughly studied by Janson *et al.* [27].

17.4.1 Counting Connected Graphs with Wright's Method

When analyzing the phase transition at the emergence of a giant component, it is of prime importance to understand the behavior of the connected components of the graph. A systematic treatment of the enumeration of connected graphs according to their complexity was done by Wright [38–41]. We briefly sketch his method.

Let $f(n, n+k)$ denote the number of connected graphs with n vertices and $n+k$ edges. Of course, $f(n, n+k) = 0$ for $k < -1$ and $k > N = \binom{n}{2}$. The number of trees is the well-known Cayley's formula, $f(n, n-1) = n^{n-2}$. The goal is to derive a recurrence relation for the GFs

$$W_k(z) = \sum_{n \geq 1} f(n, n+k) z^n / n!.$$

We call a graph with n vertices and q edges an (n, q)-graph. A connected $(n, q+1)$-graph can be generated from an (n, q)-graph in two ways. First, one could add an edge to a connected (n, q)-graph. In this case there are $N-q$ possibilities to add the edge. The second way to generate a connected $(n, q+1)$-graph is to add an edge to an (n, q)-graph that consists of exactly two connected components, an (s, t)-graph and an $(n-s, q-t)$-graph. Thus there are $s(n-s)/2$ possibilities to connect these two components by an edge and so we generate

$$Q(n, q) = \frac{1}{2}\left(\sum_{s=1}^{n-1}\binom{n}{s}s(n-s)\sum_{t=s-1}^{q-n+s+1} f(s, t)f(n-s, q-t)\right)$$

connected $(n, q+1)$-graphs in that way. The two procedures above generate each connected $(n, q+1)$-graph $q+1$ times since there are $q+1$ ways to choose the edge to be added to an (n, q)-graph. Thus

$$(q+1)f(n, q+1) = (N-q)f(n, q) + Q(n, q).$$

This recurrence relation for $f(n, n+k)$ can be translated into a recurrence relation for $W_k(z)$,

$$a(z)^{k+1}W_{k+1}(z) = \int_0^z J_k(x)a(x)^k a'(x)\,\mathrm{d}x\,, \tag{17.13}$$

where

$$J_k(x) = \frac{1}{2}\left(\left(\frac{\partial^2}{\partial x^2} - \frac{\partial}{\partial x} - 2k\right)W_k(x) + \sum_{h=0}^{k}\left(\frac{\partial}{\partial x}W_h(x)\right)\left(\frac{\partial}{\partial x}W_{k-h}(x)\right)\right)$$

and $a(z)$ is the GF for Cayley trees defined in (17.12). Using (17.13) and the well-known relation

$$W_{-1}(z) = a(z) - \frac{a(z)^2}{2}$$

we obtain

$$W_0(z) = \frac{1}{2}\log\frac{1}{1-a(z)} - \frac{1}{2}a(z) - \frac{1}{4}a(z)^2$$

and

$$W_1(z) = \frac{1}{24}\frac{a(z)^4(6-a(z))}{(1-a(z))^3}.$$

Similarly, $W_2(z)$ can be expressed in terms of finitely many powers of $a(z)$ and no logarithmic term. In fact, we have the following result.

Theorem 17.8 *For $k \geq 1$ there are rational numbers w_{ks} such that*

$$W_k(z) = \frac{1}{(1-a(z))^k} \sum_{s=0}^{2k-1} w_{ks} \left(\frac{a(z)}{1-a(z)} \right)^{2k-s}.$$

Since $1 - a(z) \sim \sqrt{2}\sqrt{1-ez}$ as $z \to 1/e$, the transfer theorems of Flajolet and Odlyzko [21] can be used to obtain asymptotic expressions for $f(n, n+k)$. Note that using the first term suffices, since this is clearly the dominant one. Equation (17.12) allows an approach by Langrange's inversion formula as well and one obtains exact expressions. Asymptotic results were given by Wright as well. The next two theorems summarize these results.

Theorem 17.9 *For $k \geq 1$ there are rational numbers p_{ks} and q_{ks} such that*

$$f(n, n+k) = (-1)^k \left(h(n) \sum_{s=0}^{\lfloor 3k/2 \rfloor - 1} p_{ks} n^s - (n-1)n^{n-2} \sum_{s=0}^{\lfloor (3k+1)/2 \rfloor} q_{ks} n^s \right),$$

where

$$h(n) = \sum_{s=0}^{n} \binom{n}{s} s^s (n-s)^{n-s}.$$

Theorem 17.10 *For $k \geq 0$ there are rational numbers w_k such that*

$$f(n, n+k) \sim w_k n^{n-\frac{1}{2}+\frac{3}{2}k}, \quad \text{as } n \to \infty.$$

Furthermore, the constants w_k satisfy, as $k \to \infty$,

$$w_k \sim \left(\frac{e}{12(k+1)} (1 + o(1)) \right)^{(k+1)/2}.$$

17.4.2 Emergence of the Giant Component

As mentioned at the beginning of this section, the component structure undergoes a phase transition when a graph acquires more and more edges. In order to describe this, we have to define a suitable graph evolution model. Two models are considered in the literature. First, the classical model is the graph process. Start with an empty graph with n vertices and a permutation of the $N = \binom{n}{2}$ possible edges, which is chosen uniformly at random among all $N!$ permutations. Then add succesively the edges according to the order given by the permutation. In this way, the graph at "time" m has exactly m edges and is distributed as $G(n, m)$. The second, which was introduced and analyzed by Janson *et al.* [27] in their thorough investigation of the evolution of random graphs, is the multigraph model. Here, at each instance an edge $\langle x, y \rangle$ is generated uniformly at random among all n^2 possible edges and then added to the graph. Obviously, in general this process produces multiple edges and self-loops. With respect to the component structure this process behaves very similarly to the graph process but is simpler to analyze.

The evolution process has three stages. First, the stage where the graph has few edges, precisely $m = o(n)$ or $c < 1$, is called the subcritical range. There the maximal component consists of $O(\log n)$ vertices. Second is the critical range at $m = n/2$, where one component dominates and has $\Theta(n^{2/3})$ vertices while all other component sizes are of logarithmic order. Third is the supercritical range with many edges, $n = o(m)$. Such a graph typically has one component with $const \cdot n$ vertices and several small components. In this range the giant component constantly grows and swallows the small components until the graph becomes connected.

In the first stage the component structure can be analyzed by choosing a vertex v and determining its component by a greedy algorithm. First, mark v and look for all neighbors of v and mark them, too. Then look for the neighbors of neighbors that are not already marked and mark them. Continuing like this eventually leads the component containing v. It turns out that this procedure almost behaves like a branching process. Therefore, the theory of branching processes can be used to obtain the results such as that there are only small components of size $O(\log n)$ and most of them are trees (for details see Spencer [35]).

Now let us turn to the critical range where the phase transition occurs. The critical range was first studied by Erdős and Rényi [19, 20] in the region $m = \frac{1}{2}n + \omega(n)\sqrt{n}$ ($\omega(n) \to \infty$ arbitrarily slowly), which turned out to be the wrong scaling. In fact, the correct parameterization is

$$m = \frac{1}{2}n + \lambda n^{2/3}. \tag{17.14}$$

One of the first detailed studies of the behavior of the component sizes inside the phase transition was done by Łuczak [30]. The method uses the enumeration formulae of trees (Cayley's formula) and those of Theorem 17.10 in conjunction with probabilistic arguments resembling ideas of the so-called probabilistic method (see [2]). Indeed, if we fix λ and some $\kappa > 0$ and let Y denote the number of tree components of size $k = \kappa n^{2/3}$ in $G(n, p)$ with $p = n^{-1} + \lambda n^{-4/3}$ (which corresponds to (17.14)), then

$$\mathbb{E}Y = \binom{n}{k} k^{k-2} p^{k-1} (1-p)^{k(n-k)+\binom{k}{2}-(k-1)}.$$

This holds, since we have to choose the k vertices of the components first. Then there are k^{k-2} possibilities to construct a tree with these vertices. If we fix one of these trees, then each of the $k-1$ edges must be in $G(n,p)$ and each of the $k(n-k)$ pairs of a node of the tree and some node outside the tree, as well as all the $\binom{k}{2}-(k-1)$ pairs of nodes of the tree that are not neighbors, must not be connected by an edge in $G(n,p)$. Evaluating this expression asymptotically we obtain

$$\mathbb{E}X \longrightarrow \frac{1}{\sqrt{2\pi}}\int_a^b \exp\left(-\frac{t^3}{6}-\frac{\lambda^2 t}{2}+\frac{\lambda t^2}{2}+o(1)\right)\frac{dt}{t^{5/2}}, \qquad \text{as } n \to \infty, \tag{17.15}$$

where X denotes the number of tree components of size between $an^{2/3}$ and $bn^{2/3}$. If we look at components with excess ℓ, that is, components where the number of edges exceeds the number of vertices by exactly ℓ, a similar formula holds. Let X_ℓ denote the number of components of excess ℓ and with size between $an^{2/3}$ and $bn^{2/3}$ (hence $X = X_{-1}$). Then

$$\mathbb{E}X_\ell \longrightarrow \frac{1}{\sqrt{2\pi}}\int_a^b \exp\left(-\frac{t^3}{6}-\frac{\lambda^2 t}{2}+\frac{\lambda t^2}{2}+o(1)\right) w_\ell t^{3(\ell+1)/2}\frac{dt}{t^{5/2}}, \tag{17.16}$$

as $n \to \infty$, where w_ℓ is the constant of Theorem 17.10. Note that with setting $w_{-1} = 1$ (17.16) and (17.15) coincide. Note that

$$g(t) = \sum_{\ell \ge -1} w_\ell t^{\frac{3}{2}(\ell+1)}$$

converges for all t and is related to the number of all components. In fact, the probability that a random component of size $tn^{2/3}$ has excess ℓ is $w_\ell t^{3(\ell+1)/2}/g(t)$. In particular, if t is very close to zero, then $g(t) \approx 1$ and the probability of having excess -1 (and thus being a tree component) is $1/g(t) \approx 1$. Thus small components are likely to be trees.

A more detailed study can be done with GFs, which were the key tool in [27], to which we refer the reader for details. The paper contains a wealth of profound results on the phase transition that cannot even be mentioned here. We can only excerpt the paper very briefly.

It turns out that it is useful to decompose a graph into its simple part and its complex part. The simple part consists of all components that are trees or unicyclic (excess -1 or 0) while components with an excess of at least 2 are called complex components. If $G(w,z)$ is the GF for graphs (w and z keeping track of the number of edges and the number of vertices, respectively) and $F(w,z)$ the GF of cyclic graphs, that is, graphs without tree components, and $E(w,z)$ the GF of complex graphs, then the identities

$$G(w,z) = e^{W_{-1}(wz)/w}F(w,z) \quad \text{and} \quad F(w,z) = e^{W_0(wz)}E(w,z)$$

hold. Further relations can be obtained by looking at the graph process: adding an edge means choosing two vertices that are not connected by an edge and then connecting them with the new edge. Equivalently, this can be regarded as a graph with a marked edge that is not counted. In the world of GFs this corresponds to the derivative w.r.t. w. On the other hand, we can distinguish two vertices and add a new edge if they are not connected and mark the edge otherwise. We then have to subtract these cases afterwards. Again this corresponds to differential operators for GFs. We get

$$\frac{1}{w}\vartheta_w G(w,z) = \left(\frac{\vartheta_z^2 - \vartheta_z}{2} - \vartheta_w\right) G(w,z)$$

with $\vartheta_w = d/dw$ and $\vartheta_z = d/dz$. Using these techniques it is possible to derive a differential equation for $E(z)$:

$$\frac{1}{w}\left(\vartheta_w - a(z)\vartheta_z\right) E(w,z) = e^{-W_0(wz)}\left(\frac{\vartheta_z^2 - \vartheta_z}{2} - \vartheta_w\right) e^{W_0(wz)} E(w,z).$$

Solving this equation and the one for the analogous GF for the multigraph process and splitting the functions according to the excess of the graphs counted, that is, letting $E(w,z) = 1 + \sum_{\ell\geq 1} w^\ell E_\ell(wz)$, we get formulae similar to those of Wright (Theorem 17.8):

$$E_\ell(z) = 1 + \sum_{s\geq 0} e_{\ell s}\frac{a(z)^{5\ell-s}}{(1-a(z))^{3\ell-s}} \text{ and } E_\ell^{(M)}(z) = \sum_{s=0}^{2\ell} e_{\ell s}^{(M)}\frac{a(z)^{2\ell-s}}{(1-a(z))^{3\ell-s}}, \quad (17.17)$$

where $E_\ell^{(M)}(z)$ is the GF for the multigraph process. The coefficients $e_{\ell s}^{(M)}$ turn out to have nice algebraic properties. For instance, they satisfy a fairly simple recurrence relation that allows for the derivation of exact expressions involving only polynomials and factorials. Moreover, they are asymptotically equal to Wright's coefficients of Theorem 17.10, that is, $w_{\ell s} \sim e_{\ell s}^{(M)}$, as $\ell \to \infty$. The expansions (17.17) are amenable to contour integration à la Flajolet and Odlyzko [21]. In this respect, it is obvious that the first terms are the dominant ones. Fortunately, it can be shown that $e_{\ell 0}^{(M)} = e_{\ell 0}$. Therefore, the analysis of components is the same for graphs and multigraphs. The following theorem gives the joint distribution of all kinds of components (according to excess).

Theorem 17.11 *The probability that a random graph (or multigraph) with n vertices and $\frac{n}{2} + O\left(n^{1/3}\right)$ edges has exactly r_i components with excess i ($i = 1, 2, \ldots, q$) and no components of higher excess is*

$$\left(\frac{4}{3}\right)^r \sqrt{\frac{2}{3}}\frac{w_{10}^{r_1}}{r_1!}\frac{w_{20}^{r_2}}{r_2!}\cdots\frac{w_{q0}^{r_q}}{r_q!}\frac{(r_1+2r_2+\cdots+qr_q)!}{(2(r_1+2r_2+\cdots+qr_q))!} + O\left(n^{-1/3}\right).$$

The excess is one of the crucial concepts in the analysis of the phase transition. The theorem above tells us the probability that a random graph has the configuration $[r_1, r_2, \ldots, r_q]$. The evolution process induces a stochastic process with state space equal to the set of all possible configurations. With the help of the following theorem it is possible to compute the transition probabilities of this process (see Figure 17.3 as well).

Theorem 17.12 *Set* $r = r_1 + 2r_2 + \cdots + qr_q$ *and let* $\delta_1 + 2\delta_2 + \cdots + q\delta_q = 1$. *Then the probability that a random graph (or multigraph) of configuration* $[r_1, r_2, \ldots, r_q]$ *with n vertices and having no tree components will change to configuration* $[r_1 + \delta_1, r_2 + \delta_2, \ldots, r_q + \delta_q, \delta_{q+1}, \ldots]$ *when a random edge is added can be computed as follows:*

Probability	*If the nonzero δs are*
$\frac{5}{(6r+1)(6r+5)} + O\left(n^{-1/2}\right)$	$\delta_1 = 1$
$\frac{36j(j+1)r_j}{(6r+1)(6r+5)} + O\left(n^{-1/2}\right)$	$\delta_j = -1, \delta_{j+1} = 1$
$\frac{36j^2 r_j(r_j-1)}{(6r+1)(6r+5)} + O\left(n^{-1/2}\right)$	$\delta_j = -2, \delta_{2j+1} = 1$
$\frac{72jkr_jr_k}{(6r+1)(6r+5)} + O\left(n^{-1/2}\right)$	$\delta_j = -1, \delta_k = -1, \delta_{j+k+1} = 1, j < k$
0	*otherwise*

It can be further shown that the evolution process is nearly a Markov process. The Markov process on the set of possible configurations with transition probabilities as stated in the previous theorem describes in a certain sense almost all evolutions of random graphs. This gives a precise picture of the phase transition. Looking at these numbers shows that the evolution of the form $[0] \to [1] \to [0,1] \to [0,0,1] \to \ldots$ is the most probable one. If the Markov process were to reflect the exact behavior, the probability of this evolution would be

$$\prod_{r\geq 1} \frac{36r(r+1)}{(6r+1)(6r+5)} = \frac{5\pi}{18} \approx 0.8726646 .$$

Janson *et al.* [27] showed that the probability that a graph on n vertices evolves like this tends to $5\pi/18$ as $n \to \infty$. This means that a typical graph never has more than one complex component during its evolution.

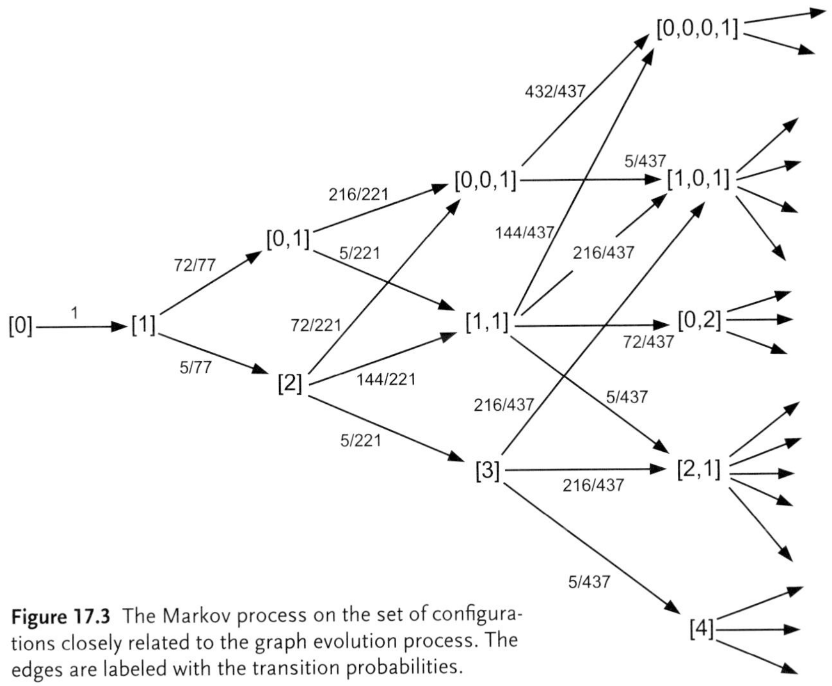

Figure 17.3 The Markov process on the set of configurations closely related to the graph evolution process. The edges are labeled with the transition probabilities.

17.5 Planar Graphs

The counting problem of several classes of planar graphs resp. planar maps goes back to Tutte [8, 36, 37]. Interestingly enough, the study of random vertex labeled planar graphs is a recent one. Random planar graphs were introduced by Denise *et al.* [13], and since then they have been widely studied. Several natural parameters defined on $\mathcal{R}_n$ have been studied, starting with the number of edges, which is probably the most basic one. Partial results were obtained in [7, 13, 25, 34], until it was shown by Giménez and Noy [26] that the number of edges in random planar graphs obeys asymptotically a normal limit law with linear expectation and variance. The expectation is asymptotically κn, where $\kappa \approx 2.21326$ is a well-defined analytic constant. This implies that the average degree of the vertices is $2\kappa \approx 4.42652$. McDiarmid *et al.* showed that with high probability a planar graph has a linear number of vertices of degree k, for each $k \geq 1$ [32].

In what follows we present here an approach to random (vertex) labeled planar graphs that is based on GFs and indicate how corresponding counting problems and distributional results can be obtained. We recall that a graph is 2-connected if it is connected and one has to remove at least two vertices (and all incident edges) to disconnect it. Similarly, a graph is 3-connected if it is 2-connected and one has to remove at least three vertices to disconnect.

We first provide a system of equations for the corresponding GFs (see [4, 26]).

Theorem 17.13 *Let $b_{n,m}$ denote the number of 2-connected labeled planar graphs, $c_{n,m}$ the number of connected labeled planar graphs, and $g_{n,m}$ the number of all labeled planar graphs with n vertices and m edges. Furthermore, let*

$$B(x,y) = \sum_{m,n\geq 0} b_{n,m} \frac{x^n}{n!} y^m, \qquad C(x,y) = \sum_{m,n\geq 0} c_{n,m} \frac{x^n}{n!} y^m,$$

$$G(x,y) = \sum_{m,n\geq 0} g_{n,m} \frac{x^n}{n!} y^m$$

be the corresponding GFs. Then these functions are determined by the following system of equations:

$$G(x,y) = \exp\left(C(x,y)\right),$$

$$\frac{\partial C(x,y)}{\partial x} = \exp\left(\frac{\partial B}{\partial x}\left(x\frac{\partial C(x,y)}{\partial x}, y\right)\right),$$

$$\frac{\partial B(x,y)}{\partial y} = \frac{x^2}{2}\frac{1+D(x,y)}{1+y}, \tag{17.18}$$

$$\frac{M(x,D(x,y))}{2x^2 D(x,y)} = \log\left(\frac{1+D(x,y)}{1+y}\right) - \frac{xD(x,y)^2}{1+xD(x,y)}, \tag{17.19}$$

$$M(x,y) = x^2y^2\left(\frac{1}{1+xy} + \frac{1}{1+y} - 1 - \frac{(1+U)^2(1+V)^2}{(1+U+V)^3}\right),$$

$$U(x,y) = xy(1+V(x,y))^2,$$

$$V(x,y) = y(1+U(x,y))^2.$$

Note that the number of edges have to be taken into account, too, as Equations 17.18 and 17.19 could not be stated without the variable y. Nevertheless, we can set $y = 1$ after all and obtain GFs for the numbers $b_n = \sum_{k\geq 0} b_{n,m}$ etc. [4, 26].

Theorem 17.14 *The numbers b_n, c_n, and g_n of labeled 2-connected resp. connected resp. all planar graphs are asymptotically given by*

$$b_n = b \cdot n^{-\frac{7}{2}} \varrho_1^n n! \left(1 + O\left(\frac{1}{n}\right)\right),$$

$$c_n = c \cdot n^{-\frac{7}{2}} \varrho_2^n n! \left(1 + O\left(\frac{1}{n}\right)\right),$$

$$g_n = g \cdot n^{-\frac{7}{2}} \varrho_2^n n! \left(1 + O\left(\frac{1}{n}\right)\right),$$

where $\varrho_1 = 0.03819\ldots$, $\varrho_2 = 0.03672841\ldots$ *and*

$$b = 0.3704247487\ldots \cdot 10^{-5},$$
$$c = 0.4104361100\ldots \cdot 10^{-5},$$
$$g = 0.4260938569\ldots \cdot 10^{-5}$$

are positive constants.

It is much more difficult to get a precise description of the singular behavior of the above GFs than in previous examples. Nevertheless, it turns out that $B(x, y)$, $C(x, y)$, and $G(x, y)$ have a respresentation of the form

$$g(x, y) + h(x, y)\left(1 - \frac{x}{\varrho(y)}\right)^{5/2}$$

with certain analytic functions $g(x, y)$, $h(x, y)$, $\varrho(y)$. Of course, this implies Theorem 17.14. Furthermore, we directly obtain a central limit theorem for the number X_n of edges where the expected value and variance are asymptotically proportional to n:

$$\mathbb{E}\, X_n = \mu n + O(1) \quad \text{and} \quad \mathbb{V}\, X_n = \sigma^2 n + O(1).$$

For example, for connected resp. all planar graphs we have $\mu = 2.2132652\ldots$ and $\sigma^2 = 0.4303471\ldots$ (compare with [26]).

By extending the above procedure one can also get a description of the degree distribution of planar graphs. This has been worked out recently by Drmota, Giménez, and Noy [14].

Theorem 17.15 *Let $d_{n,k}$ be the probability that a random node in a random planar graph $\mathcal{R}_n$ has degree k. Then the limit*

$$d_k := \lim_{n\to\infty} d_{n,k}$$

exists. The probability GF $p(w) = \sum_{k\geq 1} d_k w^k$ can be explicitly computed. The first few values are given in the following table, and asymptotically we have

$$d_k \sim c k^{-1/2} q^k,$$

where $c \approx 3.0826285$ and $q \approx 0.6734506$ are computable constants.

d_1	d_2	d_3	d_4	d_5	d_6
0.0367284	0.1625794	0.2354360	0.1867737	0.1295023	0.0861805

The proof is based on the GFs $B^\bullet(x, y, w)$, $C^\bullet(x, y, w)$, and $G^\bullet(x, y, w)$ that correspond to 2-connected, connected, resp. all planar graphs, where one vertex is marked and the exponent of w counts the degree of the rooted vertex. The corresponding system of equations is now the following one (see [14]):

$$\begin{aligned}
G^\bullet(x,y,w) &= \exp\left(C(x,y,1)\right) C^\bullet(x,y,w)\,,\\
C^\bullet(x,y,w) &= \exp\left(B^\bullet\left(xC^\bullet(x,y,1),y,w\right)\right)\,,\\
w\frac{\partial B^\bullet(x,y,w)}{\partial w} &= xyw\exp\left(S(x,y,w)+\frac{1}{x^2D(x,y,w)}\right.\\
&\qquad \left.\times\, T^\bullet\left(x, D(x,y,1), \frac{D(x,y,w)}{D(x,y,1)}\right)\right)\\
D(x,y,w) &= (1+yw)\exp\left(S(x,y,w)+\frac{1}{x^2D(x,y,w)}\right.\\
&\qquad \left.\times\, T^\bullet\left(x, D(x,y,1), \frac{D(x,y,w)}{D(x,y,1)}\right)\right)-1\\
S(x,y,w) &= xD(x,y,1)\left(D(x,y,w)-S(x,y,w)\right)\,,\\
T^\bullet(x,y,w) &= \frac{x^2y^2w^2}{2}\left(\frac{1}{1+wy}+\frac{1}{1+xy}-1\right.\\
&\qquad \left.-\frac{(U+1)^2\left(-w_1(U,V,w)+(U-w+1)\sqrt{w_2(U,V,w)}\right)}{2w(Vw+U^2+2U+1)(1+U+V)^3}\right)\,,\\
U(x,y) &= xy(1+V(x,y))^2, \quad V(x,y) = y(1+U(x,y))^2
\end{aligned}$$

with polynomials $w_1 = w_1(U,V,w)$ and $w_2 = w_2(U,V,w)$ given by

$$\begin{aligned}
w_1 =& -UVw^2 + w(1+4V+3UV^2+5V^2+U^2+2U+2V^3\\
&+3U^2V+7UV)+(U+1)^2(U+2V+1+V^2)\,,\\
w_2 =& U^2V^2w^2 - 2wUV(2U^2V+6UV+2V^3+3UV^2+5V^2+U^2\\
&+2U+4V+1)+(U+1)^2(U+2V+1+V^2)^2\,.
\end{aligned}$$

It turns out that the singular behavior of the functions $B^\bullet(x,y,w)$, $C^\bullet(x,y,w)$, and $G^\bullet(x,y,w)$ is of the form

$$g(x,y,w)+h(x,y,w)\left(1-\frac{x}{\varrho(y)}\right)^{\frac{3}{2}},$$

that is, the singularity does not depend on w. With the help of these kinds of representations the degree distributions can be characterized.

Acknowledgment

The authors gratefully acknowledge financial support from the European FP6-NEST-Adventure Program, under Contract No. 028875 (Project NEMO: Network Models, Governance, and R&D Collaboration Networks).

References

1 D. Aldous. The continuum random tree. II. An overview. In *Stochastic analysis (Durham, 1990)*, Vol. 167 of *London Math. Soc. Lecture Note Ser.*, pp. 23–70. Cambridge Univ. Press, Cambridge, 1991.

2 N. Alon and J.H. Spencer. *The probabilistic method.* Wiley-Interscience Series in Discrete Mathematics and Optimization. Wiley-Interscience, New York, 2nd ed., 2000. With an appendix on the life and work of Paul Erdős.

3 J. Arney and E.A. Bender. Random mappings with constraints on coalescence and number of origins. *Pacific J. Math.*, 103(2):269–294, 1982.

4 E.A. Bender, Zhicheng Gao, and N.C. Wormald. The number of labeled 2-connected planar graphs. *Electron. J. Combin.*, 9(1):Research Paper 43, 13 pp. (electronic), 2002.

5 Ph. Biane and M. Yor. Valeurs principales associées aux temps locaux browniens. *Bull. Sci. Math. (2)*, 111(1):23–101, 1987.

6 P. Billingsley. *Convergence of probability measures.* John Wiley & Sons, New York, 1968.

7 N. Bonichon, C. Gavoille, N. Hanusse, D. Poulalhon, and G. Schaeffer. Planar graphs, via well-orderly maps and trees. *Graphs Combin.*, 22(2):185–202, 2006.

8 W.G. Brown and W.T. Tutte. On the enumeration of rooted non-separable planar maps. *Canad. J. Math.*, 16:572–577, 1964.

9 P. Chassaing, J.F. Marckert, and M. Yor. The height and width of simple trees. *Mathematics and computer science (Versailles, 2000)*, Trends Math., pp. 17–30. Birkhäuser, Basel, 2000.

10 Ph. Chassaing and J.-F. Marckert. Parking functions, empirical processes, and the width of rooted labeled trees. *Electron. J. Combin.*, 8(1):Research Paper 14, 19 pp. (electronic), 2001.

11 J.W. Cohen and G. Hooghiemstra. Brownian excursion, the $M/M/1$ queue and their occupation times. *Math. Oper. Res.*, 6(4):608–629, 1981.

12 N.G. de Bruijn, D.E. Knuth, and S.O. Rice. The average height of planted plane trees. In *Graph theory and computing.* Academic Press, New York, 1972, pp. 15–22.

13 A. Denise, M. Vasconcellos, and D.J.A. Welsh. The random planar graph. *Congr. Numer.* Festschrift for C. St. J. A. Nash-Williams.

14 M. Drmota, O. Gimenez, and M. Noy. Degree distribution in random planar graph, 2008, submitted.

15 M. Drmota. Asymptotic distributions and a multivariate Darboux method in enumeration problems. *J. Combinatorial Th. Ser. A*, 67:169–184, 1994.

16 M. Drmota and Bernhard Gittenberger. On the profile of random trees. *Random Structures Algorithms*, 10(4):421–451, 1997.

17 M. Drmota and B. Gittenberger. The width of Galton-Watson trees conditioned by the size. *Discrete Math. Theor. Comput. Sci.*, 6(2):387–400 (electronic), 2004.

18 M. Drmota and M. Soria. Images and preimages in random mappings. *SIAM J. Discrete Math.*, 10(2):246–269, 1997.

19 P. Erdős and A. Rényi. On random graphs. I. *Publ. Math. Debrecen*, 6:290–297, 1959. Reprinted in P. Erdős, *The Art of Counting.* MIT Press, Cambridge, MA, 1973, pp. 561–568.

20 P. Erdős and A. Rényi. On the evolution of random graphs. *Magyar Tud. Akad. Mat. Kutató Int. Közl.*, 5:17–61, 1960. Reprinted in P. Erdős (ed.), *The Art of Counting.* MIT Press, Cambridge, MA, 1973, pp. 574–618.

21 P. Flajolet and A.M. Odlyzko. Singularity analysis of generating functions. *SIAM J. Discrete Math.*, 3:216–240, 1990.

22 Ph. Flajolet, Zhicheng Gao, A. Odlyzko, and B. Richmond. The distribution of heights of binary trees and other simple trees. *Combin. Probab. Comput.*, 2(2): 145–156, 1993.

23 Ph. Flajolet and A. Odlyzko. The average height of binary trees and other simple trees. *J. Comput. System Sci.*, 25(2): 171–213, 1982.

24 Ph. Flajolet and R. Sedgewick. *Analytic Combinatorics.* Cambridge University Press, 2009.

25 St. Gerke and C. McDiarmid. On the number of edges in random planar graphs. *Combin. Probab. Comput.*, 13(2):165–183, 2004.

26 O. Giménez and M. Noy. Asymptotic enumeration and limit laws of planar

graphs. *J. Amer. Math. Soc., to appear.* arXiv:math/0501269.

27 S. Janson, D.E. Knuth, T. Łuczak, and B. Pittel. The birth of the giant component. *Random Structures Algorithms,* 4(3):231–358, 1993. With an introduction by the editors.

28 S. Janson, T. Łuczak, and A. Rucinski. *Random graphs.* Wiley-Interscience Series in Discrete Mathematics and Optimization. Wiley-Interscience, New York, 2000.

29 D.P. Kennedy. The distribution of the maximum Brownian excursion. *J. Appl. Probability,* 13(2):371–376, 1976.

30 T. Łuczak. Component behavior near the critical point of the random graph process. *Random Structures Algorithms,* 1(3):287–310, 1990.

31 C. McDiarmid, A. Steger, and D.J.A. Welsh. Random planar graphs. *J. Combin. Theory Ser. B,* 93:187–205, 2005.

32 A. Meir and J.W. Moon. On the altitude of nodes in random trees. *Canadian Journal of Mathematics,* 30:997–1015, 1978.

33 A.M. Odlyzko and H.S. Wilf. Bandwidths and profiles of trees. *J. Combin. Theory Ser. B,* 42(3):348–370, 1987.

34 D. Osthus, H.J. Prömel, and A. Taraz. On random planar graphs, the number of planar graphs and their triangulations. *J. Combin. Theory Ser. B,* 88(1):119–134, 2003.

35 J. Spencer. Nine lectures on random graphs. In *École d'Été de Probabilités de Saint-Flour XXI—1991,* Vol. 1541 of *Lecture Notes in Math.* Springer, Berlin, 1993, pp. 293–347.

36 W.T. Tutte. A census of planar triangulations. *Canad. J. Math.,* 14:21–38, 1962.

37 W.T. Tutte. A census of planar maps. *Canad. J. Math.,* 15:249–271, 1963.

38 E.M. Wright. The number of connected sparsely edged graphs. *J. Graph Theory,* 1(4):317–330, 1977.

39 E.M. Wright. The number of connected sparsely edged graphs. II. Smooth graphs and blocks. *J. Graph Theory,* 2(4):299–305, 1978.

40 E.M. Wright. The number of connected sparsely edged graphs. III. Asymptotic results. *J. Graph Theory,* 4(4):393–407, 1980.

41 E.M. Wright. The number of connected sparsely edged graphs. IV. Large nonseparable graphs. *J. Graph Theory,* 7(2): 219–229, 1983.

Index

a
A*-based optimal GED computation 119
Ackermann's function 188
activity spreading 254
acyclic cubical complex 342
addition 354
adjacency matrix 4ff., 25ff., 145
AIDS database 127
Albert–Barabasi limit 35
algorithm
– breadth-first search (BES) 236
– Dijkstra's 190
– dynamic programming (DP) 278
– Hungarian 122f.
– Kuhn-Munkres' 122f.
– minimum cost dynamic multicommodity flow problem 394
– optimal 118
– optimal dynamic flow 377ff.
– spanning peripheral arcs 201
– spanning peripheral edge 186
– suboptimal 121
Alon–Milman theorem 64
analytic combinatorics 425ff.
m-antiprism 359ff.
arc 199, 247, 386
– downward 199
– kernel 199
– lateral 199
– peripheral 199
– reflexive 199
– upward 199
assignment problem 121
authority, network 13
automatic protein prediction 114
automorphism 1ff., 352
– computation 7
automorphism group 4, 352
– abelian 5
average shortest path (ASP) 247, 256ff.

b
BABEL 231f.
Bacillus subtilis 282ff.
balanced incomplete block design (BIBD) 153
base graph 387
basic interval 302
basis problem 318
basis system
– complete 318
– problem 296
batch machine 301
batch machine scheduling (BMS) problem 301f.
Beamsearch(s), Beam(s) 121, 133
benzenoid hydrocarbon 154
betweenness 206
betweenness centrality (BC) 15ff., 52, 58
– vector 17f.
binary choice model 413
biological network 68
biometric person identification 114
bipartite edit distance 133
bipartite graph 149
bipartite graph matching 121
bipartite (BP) method 125ff.
bipartite model 213
bipartite network, community 409
bipartite substructure, complex network 77
bipartite wheel 341
Boltzmann factor, generalized 40
bound
– graph energy 147
– lower 154
– upper 147
Bourgas Indices (BI) 49ff.

Analysis of Complex Networks: From Biology to Linguistics. Edited by Matthias Dehmer and Frank Emmert-Streib

ISBN: 978-3-527-32345-6

bow-tie model 176
m-bracelet 363
branching process 427, 441
breadth-first search (BES) algorithm 236
bridge 353
Brownian excursion 431
Buneman graph 346
bypass deletion 281

c
C-block graph 337
CACTVS 231f.
Caenorhabditis elegans 245ff.
cage-amalgamation graph 337ff.
Cahn-Ingold-Prelog 223
canonical discriminant analysis (CDA) 68
canonical metric representation 328
capacity function 378
cardiac defibrillation 261
cardiac system 261
cardioversion 261
CARMEN Neuroinformatics project 266
Cartesian product 3, 327f.
Cauchy's integral formula 433
Cauchy–Schwarz inequality 147
Cayley graph 87ff.
Cayley tree 435ff.
Cayley's formula 438
CCDC 232
centrality 58
centrality measure 52
– integrated 52
ChEBI 232
Chebyshev's inequality 99
chemical database 232
chemical graph format 231
Chemical Markup Language (CML) 231
chemical software package 232
Chemistry Development Kit (CDK) 232
chemoinformatics, graph theory 221ff.
Cheng, Harrison, and Zelikovsky theorem 278
chordal graph 336
chromatic decomposition 19
chromatic information content 20
chromatic number 19
circular split system 345
class
– CLC($\mathcal{X}$) 304
– complex network 66
classification
– elementary ({2,3},4)-polycycle 359
– elementary ({2,3},5)-polycycle 359
– elementary ({2,3,4,5},3)-polycycle 356
– GED-based nearest-neighbor 129
classifier
– *k*-nearest-neighbor 129
– NN 133f.
CLC, *see* connected list coloring
closed walk (CW) 57, 77
closeness centrality (CC) 19, 52, 58
cluster 247
clustering coefficient 15, 30
cogwheel 341
– convex 341
collaboration model 419
color
– initial 311
– label 316
color class 19
coloring
– complete 19
– connected 301
– graph 19
– proper 301
– proper interval edge 301
communicability 70
– complex network 69
– function 77f.
– network community 71
communicability graph 73ff.
community 71f.
– detection 73
– identification 75
– structure 407
– topical profile 411
complementary geraph 181, 199
complete basis system 318f.
complex network 23, 65
– class 66
– communicability 69
– global topological organization 62
– relational 48
– statistical mechanics 23ff.
– structure 55
complexity 304
component 209f.
computation, graph edit distance 118f.
computational geometry 61

computational tractability 295ff.
computing minimum cost homomorphism 277
concave cost function on arcs 386
conceptual domain 204
conceptual graph (CG) 209ff.
– semiotic system 212
conceptual space 204ff., 216
conceptualistic interpretation 213
connected graph 151ff., 182, 326, 345, 438ff.
connected group 66
connected list coloring (CLC) 296
Connected List Coloring (CLC) problem 295ff.
– CLC($\mathcal{X}$) problem 296
– CLL-negative input 312
– CLL-positive input 312
connected list labeling (CLL) 309
– problem 309
connected service area, problem 298
connected solution 299
connection 253
connectivity 49, 90
– average nearest-neighbor 30
– local 110
– neutral network 90
– peripheral 195
constraint satisfaction problem (CSP) 234
– finite domain (FCSP) 234
convex amalgamation 327
convex cost function on arcs 386
convex excess 336
convex expansion 328
core-periphery 66
correspondence 381
cortical network 245
– property 246
cortical system 264
cost function 378f., 393
cost matrix 123
Coulson integral formula 162f.
counting connected graph 438
cover 328
– cubical 338
cross reference 193
crossing graph 332
crystal graph 230
crystal packing 231
cube
– counting 323ff.
– spectrum 347
n-cube 85ff., 324
– binary 94
cube polynomial 324ff., 337ff.
– root 340
cycle
– graph 235
– handling in pattern 280

d
Daylight 232
defibrillation shock (DS) 261
degree centrality (DC) 52, 58
degree distribution 25f., 39ff.
– cumulative 26
– Poissonian 36
degree vector 13
deletion 116
demand–supply function 378, 393
density matrix 40
dependency tree (DT) 202
depth-first search (DFS) 279
digraph 8, 200
– functional 436
Dijkstra's algorithm 190
dimension, isometric 328
directed generalized dependency tree (DiGDT) 202
directed generalized spanning tree (DiGST) 200
directed generalized tree (DGT) 198ff.
directed minimum spanning generalized tree (DiMSGT) 202
directed spanning tree 200
discrete mathematical model 295
discrete optimization problem 295
disjoint hole 351
dissimilarity computation 114
dissimilarity embedding graph kernel 132ff.
– suboptimal GED 136
distance 34, 49
– computation 116
– *n*-cube 105
– geodesic 179
– graph 114
– graph energy 169
– matrix 169
– shortest path 325
distortion operation 116
distribution, linking probability 30
Djokoviæ–Winkler relation 327
domain formation 216
domain networking 216

dynamic flow 385ff.
dynamic model, flow storage at node 384
dynamic multicommodity flow 394
dynamic network 378ff., 393
– uncapacitated 387
dynamic programming (DP)
– algorithm 278
– table 279
dynamics, structural and functional 245

e
EC (Enzyme Commission) number 275
eccentricity 51
ecological network 68
edge
– coloring 301
– cross-reference 196
– disengaged 230
– engaged 230
– kernel 181ff.
– lateral 181ff.
– open 353
– peripheral 190
– proper interval 301
– reflexive 181
– removal 256
– short cut 196
– transverse 196
– vertical 181ff.
edit distance 117
edit path 117
eigenvalue 57
– principal 64
eigenvector 57
– centrality (EC) 58
– principal 64
elementary elliptic (R,*q*)-polycycle 351ff.
elementary ({2,3},4)-polycycle 359
elementary ({2,3},5)-polycycle 359
– sporadic 371ff.
elementary ({2,3,4,5},3)-polycycle 356
– sporadic 364ff.
empirical model 413
energy, graph 146
ensemble
– canonical 32ff.
– random network 39
ensemble average 28
entropy 2, 37
– graph 20
– group-based 4
– network 25ff.
– nonextensive 31
– principle 25
epilepsy 254
epileptogenesis 254
equidistance 207
equienergetic noncospectral connected graph 158
Erdős–Rényi classical random graph 28
– model 438
Erdős–Rényi (ER) network 30, 39
Escherichia coli 282ff.
essential set of essential rings (ESER) 235
estimation result 417
Estrada index 59
Euclidean distance 133, 259
Euler-type inequality 324, 330, 332
– partial cube 335
European Bioinformatics Institute (EBI) 222
European Nomenclature of Territorial Units for Statistics (NUTS)
classification 403
European Union's framework program (FP) 401ff.
excitable medium, spreading 260
expansion factor 63f.
expansion procedure 328
expansion property 339
extended set of smallest rings (ESSR) 235

f
face-regular two faced map 354
family
– chain 300
– CLC($\mathcal{X}$) 296
– fixed 107
– induced 106
– parameterized 318ff.
– problem 320
FCSP, *see* constraint satisfaction problem
feasible dynamic flow 379f.
feasible dynamic multicommodity flow 393
finite probability scheme 2
fingerprint database 126
Fisher's linear discriminant analysis (LDA) 131ff.

flexible graph distance measure 114
flow correspondence 385
flow storage at nodes 384
food web of Canton Creek 80
framework program (FP) network
– community 409
frieze group 352
functional brain network 248
functional dynamics 260
fundamental group 352

g
GABA (γ-aminobutyric acid) receptor 254
Galton–Henry classification system 127
Galton–Watson process 427f.
Galton–Watson tree 430
– scaled 430
general embedding procedure 130
generalized forest (GF) 208ff.
generalized shortest path tree (GSPT) 175ff., 190ff., 212
generalized shortest paths tree (GPST) 175ff., 195, 212
generalized spanning tree 184
– directed 200
generalized subtree 207
– type-restricted 183
generalized tree (GT) 175ff., 204
– directed 199
– minimum spanning (MSGT) 186ff.
– orientating 197
– undirected 180
– weighted undirected 181, 199
geodesic 361
geodesic betweenness 206
geodesic distance 179
geodesic equidistance 207
geodesic path 179
giant component 438ff.
global corticocortical connectivity 263
i-gon 351ff.
good expansion (GE) 63
graph 115, 295
– 2-connected 446
– 3-connected 446
– acyclic 50f.
– automorphism group 11
– bipartite 76, 149ff., 326
– bipartite unicyclic 166
– class 175ff.
– complexity 47ff.
– centrality 19
– characteristic polynomial 145
– chordal 336, 345
– circulant 156
– coloring 19
– complementary 181, 199
– connected 151ff., 182, 326, 345, 438ff.
– connected (p_x,p_y)-pseudo-semiregular bipartite graph 151
– cycle 237
– cyclic 50f.
– data set 125
– data set characteristics 128
– distance 114
– embedding 129
– entropy 20
– equienergetic 157
– equienergetic noncospectral 158
– extremal 162
– *H*-free 324
– Hamming, *see* Hamming graph
– hyperenergetic 156f
– hypoenergetic 157
– information content 2
– isometric dimension 329
– maximum-energy unicyclic *n*-vertex 167
– minimum energy 166f.
– minimum-energy *n*-vertex 162
– non-bipartite connected 151
– non-bipartite connected *p*-pseudoregular 151
– non-isomorphic 157
– non-trivial with identity group 3
– planar 445
– polynomial 6, 145
– quasi-median 334
– quasi-semimedian 335
– second-minimum energy 166
– semiregular bipartite 149
– *k-th* spectral moment 155
– third-minimum energy 166
– tree-like 178
– type 302
– underlying 200
– unicyclic 166f.
– *n*-vertex noncospectral equienergetic 159
– *n*-vertex regular 168
(*n*, *m*)-graph, minimum energy 167

graph edit distance (GED) 113ff.
– computation 118ff., 133
– dissimilarity embedding graph kernel based on suboptimal graph edit distance 136
– optimal and suboptimal 133ff.
– optimal and suboptimal algorithm 118ff.
graph element 47
– weighted distribution 47
graph energy 145ff.
– bound 147
graph kernel
– dissimilarity-based embedding 129ff.
– method 131
graph matching method 116, 226
– erroro-tolerant 116
graph matching paradigm 116
graph spectral theory 81
graph spectrum 4, 55ff.
– background 56
graph structure 47
– analysis 295ff.
graph theoretic approaches 221
graph theory
– bioinformatics 221ff.
– chemoinformatics 221ff.
graph vertices 2
Green's function, thermal 71
group 4

h

H-theorem 25
Hamiltonian 27ff., 60
Hamming distance 86, 110, 324
Hamming graph 325
Hamming polynomial 324, 343f.
Hankel contour 434
Heaviside step function 73
height 430
Heuristic-A* 120, 133f.
hexagonal system 154
hidden variable distribution 40ff.
high-cost edge 190
hole 351
homeomorphism 276
homomorphism 276f.
hub 48
Hückel graph 156
Hückel molecular orbital (HMO) 145
Hungarian algorithm 122f.
Hurwitz generalized zeta function 42
hypercube 323ff.
– *n*-dimensional 324
hyperenergetic graph 156
hypoenergetic graph 157
hypotactic unfolding 216

i

ictiogenesis 254
identity graph 7
idleness 301
INChi 224
incompatibility graph 332
independence 318
– number 297
individual capacity function 393
information
– mean 47
– sequence specific 106
– total 47
information content, graph 2
information measure 49
information-theoretic entropy 60
informational network 68
initial color 311
initial vertex 311
– CLL-negative 312
– CLL-positive 312
insertion 116
integral constant demand–supply function 384
integral dimension 205
integral schedule 302
integrated centrality index 52
interaction 60
– strength 60
intercluster communicability 72
internal metric 32
interval 301, 325
– basic 302
interval edge coloring problem 301
– hypergraph 301
intracluster communicability 72
isometric expansion 328
isomorphism 352

j

Janson's inequality 90ff.
Jordan canonical form 4

k

KEGG 232
kernel 181, 361
– geodesic 357
– propeller 357
kernel edge 181ff.

kernel elementary polycycle 355
kernel function 132
kernel machine 132
kernel minimum spanning tree 188f.
key parameter 296
Kim *et al.* limits 35
Kneser graph 156
Kolmogorov complexity 1
König theorem 76f.
Koolen-Moulton upper bound 168
Kuhn-Munkres' algorithm 122f.
Kullback distance 411
Kuratowski's theorem 237

l
label 309
labeling 309
– connected list 309
– tentative 310
Laguerre polynomial 42
Laplacian graph energy 169
Laplacian matrix 169
Laplacian spectrum of graph 57
largest component 93
letter database 125
limit probability 89
line, covered and uncovered 122
linear chain 55
linear cost function on arcs 386
linear discriminant analysis (LDA) 131ff.
– Fisher 136
link 35ff.
linking probability 33ff.
– distribution 40
– microscopic 42
list 358ff.
local expansion 429
lopsided set 348
low-price edge 190

m
macro-level structure 210
macromolecular assembly
– crystal packing 229
Macromolecular Structure Database (MSD) 222
macroscopics 31
majority rule 327
mapping
– formylTHF biosynthesis 284ff.
– glutamate degradation VII pathway 288f.
– interconversion of arginine, ornithine, and proline pathway 288
– metabolic pathway 282
– pentose phosphate pathway 284
– statistical significance 283
Markov process 444
Markov's inequality 102
maximum flow 383
maximum dynamic flow problem 380
maximum matching problem 307
median 52
median closure 347
median graph 321ff.
– cube polynomial 339
– cube-free 332ff.
– maximal cube 345
– Q_4-free 333
median grid graph 333
median network 346
median vertex 325
medical diagnosis 114
meso-level coherence 210
metabolic pathway 271ff.
– filling hole 286
metric basic structure 210
metric space 32, 205
micro-level coherence 210
microscopics 35
microstate 60
middle complexity problem 319
minimality 206
minimum cost dynamic flow problem 378
minimum cost dynamic multicommodity flow problem 392
minimum cost homomorphism problem 277
minimum spanning generalized tree (MSGT) 186ff.
– revisited 187
minimum spanning tree (MST) 187
– kernel 188
modelling metabolic pathway mappings 275
modularity 247, 408
molecular graph 222
– common problem 223
Moon–Moser graph 73
Mulder's convex expansion 328
multicommodity flow 395
multidimensional conceptual space 211
multidomain conceptual space 208

multiparametric complexity analysis 296
multiple cluster 259
multiple dicriminant analysis (MDA) 136
– MDA* 139
– transformation 138
multisource tree 276ff.
– pattern 278
Munkres' algorithm 122f.
mutual capacity function 393

n
Nagelkerke (*R*)-squared 418
nearest-neighbor, GED-based 129
k-nearest-neighbor, classifier 129
neighbor connectivity 30
neighborhood degree vector, point-deleted 13
neighborhood betweenness centrality vector, point-deleted 18
network, *see also* graph
– almost bipartite 76
– bipartivity 76f.
– broad-scale 65
– complexity 47ff.
– connectivity 49
– definition 404
– distance 49
– entropy 25ff.
– growing 26ff.
– homogeneous 65
– macroscopic parameter 23ff.
– measure 30
– microscopic rule 23
– modular 63
– non-growing 26
– nonhomogeneous 65
– optimal dynamic flow 377ff.
– random, *see* random network
– real-world 24, 65
– scale-free 23f., 65
– single-scale 65
– small-world 23f., 258ff.
– state 25
– structure 405
– thermodynamics 25, 35
– time-expanded 381ff., 394
– universal topological class 65
– universality class 29
network change, development 258
network cluster 259
network community 71
– communicability 71
network ensemble 28
network generation model
– generalization 32
– unification 32
network Hamiltonian 27ff.
network mapping 271ff.
– method 273
network model
– spectral scaling approach 67
– unified 32ff.
network optimization problem (NOP) 175ff.
network organization, universal law 176
network science 265
neural connectivity 252
– prediction 252
neural network 251
neuron, inhibitory 254
neuronal network 245
– property 246
neuronal system 264
neutral network 110f.
Nikiforov's theorem 156
no-idle requirement 301
no-wait requirement 301
node 24ff., 39, 58, 247, 259, 379
– flow storage 384
– merging 116
– random 447
– removal 257
– splitting 116
node centrality 47ff., 58
node degree 48
node-repulsion graph 80
nonextensivity 31
normal distribution 438ff.
NP-complete problem 319
NP-complete subproblem 304
null model 408
NUTS (Nomenclature of Territorial Units for Statistics) 403

o
object classification, graph-based 115
offspring distribution 429
OPEN 119ff.
open edge 353
open reading frame (ORF) 272
open shop problem 301
OpenBabel 232
OpenEyes 232
optical character recognition 114

optimal dynamic flow 377ff.
optimal dynamic multicommodity flow problem 392
optimal dynamic single-commodity flow problem 378ff.
orbit 2ff.
– approximating 11
– graph 12
– size 9
– vertex 12
order relation, path 179
organization projection property 406
orientation 198

p
parallel machine (PM) 300
– problem 300
parameter 414
partial cube 325, 335
partial Hamming graph 325
particle 427
partition function 36, 60
– subgraph centrality 60
partitioning 184
path 175ff.
– disjoint 102f.
– edit 117
– geodesic 179
pathway
– identifying conserved pathways 285
– mapping 285
– visible and hidden holes 286
pattern 271ff.
– handling cycle 280
pattern graph ordering 279
pattern vertex deletion 281ff.
p-batch machine 301f.
PDB, *see* Protein Data Base
PDB ligands 232
periphery 181, 199
Perron–Frobenius eigenvalue 57
phase, network 28
phylogenetic validation 285
phylogenetics 346
PISA (Protein Interfaces, Surfaces, and Assemblies) 231
planar graph 445
planar subgraph 237
Plant Location Problem (PL problem, PLP) 298f.
(R,q)-polycycle 351ff.
– boundary 353
– elementary 351ff.
– elliptic 351
– fixed-point-free 352
– hyperbolic 351ff.
– kernel-elementary 355
– nonextensible 354
– parabolic 351ff.
– simply connected 352
– with holes 352
$(R,q)_{simp}$-polycycle 352ff.
– kernelable 356
({2,3},4)-polycycle
– elementary 359
({2,3},5)-polycycle
– elementary 359
– sporadic elementary 371ff.
$(\{2,3\},5)_{simp}$ -polycycle 361f.
({2,3,4,5},3)-polycycle
– elementary 356
– sporadic elementary 364ff.
polymorphic categorization 196
polynomial 6ff.
– solvability 318
– time 306ff.
polynomial-time solvable subproblem 305
posteromedial suprasylvian arean (PMSL) 255
prescription 296
principal component analysis (PCA) 131ff.
probability 33, 60, 88, 100, 414
probability distribution 20f.
probability space 88
problem formulation 277
product graph 342
product median graph 342
profile 431
project projection property 407
projection length distribution 251
proper faces 351
Protein Data Base (PDB) 222ff.
protein quaternary structure (PQS) 229
protein secondary structure network 68
protein structure, comparison and 3D alignment 225
protein-protein interaction network 56
prototype selection 131
provider 298
PS-completeness 318
pseudo-median 334
pseudograph

– edge 181
– rooted 181
PubChem 232

q
quality dimension 204
quasi-clique cluster 78
quasi-bipartite cluster 78
quasi-median graph 334
quasi-semimedian graph 335

r
R&D (research and development) 402
random graph 28, 425
– model of Erdős and Rényi 438
random mapping 436
random network 23f.
– degree distribution 39
– ensemble 39
– superstatistics 39
random tree 432
real-world network 24, 65, 401
relation 327
relation database management system (RDBMS) 234
relative 302
removal
– edge 256
– node 257
Riemann zeta function 430
RNA string 85, 106
RNA structure 85, 110
– induced subcube 85
root
– cube polynomial 340
– rational 340
– real 341
RQA (recurrence quantification system) 261
runtime analysis 279

s
Saccharomyces cerevisiae 282ff.
3-SAT problem 305
scale-free feature 248
scale-freeness 65
scanning, vertex 314
scheduling 301
– feasible 302
– no-idle 300
– unit job 301
scheduling problem 295ff.
secondary structure element (SSE) 225ff.
semantic space 204
semicube 326
semiotic network 175
semiotic system 211f.
– conceptual graph 212
semiregular bipartite graph 149
sequence, binary 346
set of smallest cycles at edges (SSCE) 235
Shannon equation 47
Shannon expression 60
short cut 193
short cut edge 196
shortest path generalized tree (SPGT) 193
shortest path tree (SPT) 190ff.
single source problem 190
skeleton 181
small-world connectivity descriptor 49
small-world connectivity index B2 49
small-world feature 247
small-world network 23f., 258ff.
small-worldness 65
smallest set of smallest rings (SSSR) 235
SMARTS 225
SMILES 224
Soares *et al.* limit 35
social network 16, 68, 401ff.
social tagging 215
solution, connected 299
spanning peripheral edge algorithm 186
spanning peripheral arcs algorithm 201
sparse graph 190
spatial growth 258
spatial layout 250
spatial range 259
spectral scaling method 63
spectral graph theory 56
spectral measure 58
spectrum 145
split 326, 345
– compatible 330
– incompatible 330
split system 326, 345
– full circular 346
sporadic elementary ({2,3},5)-polycycle 371ff.
sporadic elementary ({2,3,4,5},3)-polycycle 364ff.

spreading, topological inhibition 262
SQL (Structured Query Language) 232ff.
star 55
state 31
static network 389ff.
statistical mechanics 23ff.
strength 60
– interaction 60
stress centrality 19
string edit distance 133
strong deletion 281
structural brain network 249
structural complexity 19
structural damage 255
structural dynamics 255
structuralistic interpretation 213
structured value 205
subcomponent (sc) 95
subcube 85
subgraph
– convex 325
– isometric 325
– isomorphism solution 232
– planar 237
– random induced 85ff.
subgraph centrality 59
– partition function 60
submatrix, visual cortex 253
substitution 116
Sudarshan–Glauber representation 40
support vector machine (SVM) 136
symmetry 1, 206
– group 354
syncytium 260
sysres EUPRO database 402

t
technological system 68
temperature 29
– superposition 29
text 271
text graph 274
– preprocessing 278
text networking 214
thematic centralization 216
thematic condensation 216
thematic progression 216
thematic shortcut 216
theory of graph energy 146
thermodynamics of network 25
Thermus thermophilus 282ff.
time window 259
time-expanded network 381ff., 394ff.
total cost
– dynamic flow 380
– dynamic multicommodity flow 394
total wiring length 251
transit function 390
transit time function 378, 393
transition 36
tree 55, 161ff., 426ff.
– (p,q)-bipartition 166
– equienergetic 161
– height 430
– induced 95
– initial 311
– maximum energy 164
– minimum energy 164f.
– planted plane 426ff.
– profile 431
– n-regulated 100
– second minimum energy 164
– simply generated 427f.
– n-vertex 164
tree-like equality 330
triangle inequality 206
triangulated graph 336
Tsallis entropy 26ff.

u
unified model, limit 35
unipartite model 213
unit job 301
– scheduling 301
utility 36

v
variable
– construction 415
– dependent 415
– FP experience of organization 415
– geographical effect 415
– relational effect 416
ventricular fibrillation (VF) 261
vertex 12ff., 63, 275, 361
– betweenness centrality 15
– bipartition 149
– boundary 87ff.
– degree 13, 47ff., 155
– degree distribution 48
– dependency matrix 17
– disjoint 97ff.
– disjoint path 102ff.
– distance 51
– distance distribution 48f.

– feedback set 278
– geodesically equidistant 207
– independent 19
– interior 351
– intermediate 300
– orbit 12
– point-deleted neighborhood 14
– rooted 448
– scanning 314
– top-level 210
vertex-to-vertex mapping 290

w
walk 69
web page database 127
width 434
Wright's method 438

z
Zachary karate club 74
zero
– primed 122
– starred 122